Infectious Diseases

This volume collects selected topical entries from the *Encyclopedia of Sustainability Science and Technology* (ESST). ESST addresses the grand challenges for science and engineering today. It provides unprecedented, peer-reviewed coverage of sustainability science and technology with contributions from nearly 1,000 of the world's leading scientists and engineers, who write on more than 600 separate topics in 38 sections. ESST establishes a foundation for the research, engineering, and economics supporting the many sustainability and policy evaluations being performed in institutions worldwide.

Phyllis Kanki • Darrell Jay Grimes
Editors

Infectious Diseases

Selected Entries from the Encyclopedia of Sustainability Science and Technology

Springer

Editors
Phyllis Kanki
Department of Immunology
and Infectious Disease
Harvard School of Public Health
651 Huntington Avenue
Boston, MA, USA

Darrell Jay Grimes
Department of Coastal Sciences
The University of Southern Mississippi
703 East Beach Drive
Ocean Springs, MS, USA

This book consists of selections from the Encyclopedia of Sustainability Science and Technology edited by Robert A. Meyers, originally published by Springer Science+Business Media New York in 2012.

ISBN 978-1-4899-9494-3 ISBN 978-1-4614-5719-0 (eBook)
DOI 10.1007/978-1-4614-5719-0
Springer New York Heidelberg Dordrecht London

Printed on acid-free paper

Springer is part of Springer Science+Business Media (www.springer.com)

Contents

Chapter 1
Infectious Diseases, Introduction

Phyllis J. Kanki

Infectious diseases of humans and animals are illnesses resulting from an infection caused by the presence or growth of a biological organism, often termed a pathogen, for its disease-causing behavior. The term derives from the transmissibility of the pathogen to others and when this results in large numbers of infections in a region can be responsible for epidemics. Pathogens responsible for infectious diseases can be viruses, bacteria, protozoa, fungi, multicellular parasites, and prions. While antibiotics and vaccines have made major progress in the treatment and prevention of major infectious diseases, largely in the developed world, the developing world still bears a significant burden of disease due to infectious disease pathogens such as malaria, tuberculosis, and the Human Immunodeficiency Virus (HIV). Changes in the environment, zoonotic pathogens and their interaction with human populations, and medical practice including treatment and vaccines are just some examples of determinants that can modulate the impact of infectious diseases, in terms of spread, ability to cause disease, or even response to prevention or treatment measures. The ever-changing dynamic nature of infectious diseases is not only due to some pathogen's intrinsic propensity for diversity and fitness but also complex lifecycles involving intermediate nonhuman hosts. Therefore, our ability to control or eradicate various infectious diseases must entail new technologies and analytic methods.

There is significant disparity in the burden of infectious diseases globally. According to the 2008 Global Health Observatory report, infectious diseases only account for one of the top ten causes of death in high-income countries of the world, whereas in low- and middle-income countries, four of the ten leading causes of

---

This chapter, which has been modified slightly for the purposes of this volume, was originally published as part of the Encyclopedia of Sustainability Science and Technology edited by Robert A. Meyers. DOI:10.1007/978-1-4419-0851-3

P.J. Kanki (✉)
Department of Immunology and Infectious Diseases, Harvard School of Public Health, 651 Huntington Avenue, 02115 Boston, MA, USA
e-mail: pkanki@hsph.harvard.edu

P. Kanki and D.J. Grimes (eds.), *Infectious Diseases: Selected Entries from the Encyclopedia of Sustainability Science and Technology*,
DOI 10.1007/978-1-4614-5719-0_1,

death are infectious diseases [1]. However, the mobility of populations globally has resulted in infectious disease outbreaks such as the H1N1 influenza outbreak in Mexico in March 2009 that led to 28,000 confirmed cases in the United States just 3 months later. The WHO raised the pandemic alert level to phase 6, the highest level indicating a global pandemic, because of widespread infection beyond North America to Australia, the United Kingdom, Argentina, Chile, Spain, and Japan. Six months after the initial outbreak in Mexico, H1N1 infection had been confirmed in over 200,000 people from more than 100 countries and several thousand deaths [2]. While influenza virus infections are found in both high- and low- and middle-income countries, the virus responsible for this pandemic appeared to be a novel virus with characteristics of North American and Asian swine influenza viruses, as well as human and avian influenza viruses. Thus, the viruses' propensity for variation through genetic reassortment, various animal reservoirs and their contact with human populations, and the mobility of populations led to an epidemic of global proportions in a short time period.

In the past decade, international donor agencies have supported large-scale programs to address the gap in prevention and treatment of HIV/AIDS, malaria and tuberculosis. The burden of these three infectious diseases is disproportionately high in Africa, where health systems are weak and heavily dependent on foreign aid. The President's Emergency Plan For AIDS Relief initiated in 2005 is the single largest funded program for a disease in the history of US government support, active in 30 countries primarily in Africa and responsible for the initiation of antiretroviral therapy to 3.2 million adults and children with AIDS. A summary of the "HIV/AIDS Global Epidemic" describes the many challenges posed by the HIV virus, first described in the early 1980s. The HIV/AIDS epidemic most severe in Africa has also led to a concurrent increase in tuberculosis, where the presence of either infection increases the risk of coinfection, and as a result in the past decade, TB incidence has tripled. HIV and TB coinfected patients are more difficult to treat and are responsible for the highest mortality rates in both untreated and treated populations. The complex "Tuberculosis, Epidemiology of" described by Mario Raviglione and colleagues illustrates both the severity of the public health problem and the efforts by the WHO Stop TB alliance in its control. Development of improved, cost-effective, and point-of-care diagnostics is an emphasis for all three of these pathogens.

The development of drug resistance is another feature common to many infectious disease pathogens. The widespread use of chloraquine to treat malaria in the 1940s and 1950s, led to the detection of chloraquine resistant malaria first in South America and Asia and later in Africa by the late 1970s. It became widespread across Africa within a decade. Continued surveillance for drug resistance is critical to adjust treatment policies and the need for more effective drugs is ever present. In 2006 in Tugela Ferry, South Africa, the interaction between tuberculosis and HIV resulted in the recognition of an "extensively drug resistant" tuberculosis strain (XDR), where the bacteria was not only resistance to the common first line drugs, rifampicin and isoniazid but also to drugs in the quinolone family and at least one of the second line drugs [3]. The XDR tuberculosis epidemic in Tugela Ferry was unusually severe with rapid ($\sim$2 weeks) mortality, demonstrating the grave consequences of pathogens that readily evolve under drug pressure. As a result of these biologic propensities, the need for new drugs that target resistant strains is an ongoing process. The cost of second and third line drug therapy is prohibitive in most low-income countries and the need for more efficacious and cost-effective drugs is an important priority. Unfortunately, despite the importance of these pathogens like malaria and TB primarily in low-income countries, major biotechnology firms do not prioritize these diseases agents for diagnostic, vaccine, or drug development. The example of malaria and the structural barriers to solutions for low-income (tropical) settings is well described by J. Kevin Baird in "Tropical Health and Sustainability."

It is widely believed that prevention measures including vaccines are the most effective means of combating infectious diseases whenever possible and this becomes of paramount importance in infectious diseases with high burden and mortality. In the case of malaria, the ubiquity of the mosquito vector, difficulty in its control, and prevalent drug resistance all lend support for the search for an effective malaria vaccine. As described by Christopher Plowe in "Malaria Vaccines," a study conducted in a single African village, documented more than 200 variants of blood stage malaria antigens. Thus evidence of the difficulty in developing vaccines that must elicit cross-protective immunity to an ever-expanding set of antigens, such as the multiple parasitic stages of malaria. Despite these many challenges, Christopher Plowe describes progress toward a malaria vaccine that would reduce parasite burden, rather than sterilizing protection, such a vaccine would be an important milestone to be reached in the short-term future of malaria control.

While effective vaccines against poliomyelitis have been available since the 1950s, the global eradication campaign is still in effect, with >99% reduction in the number of cases since 1988 and the inception of the Global Poliomyelitis Eradication Initiative by the World Health Assembly. Indigenous poliovirus remains in only four countries of the world, including Afghanistan, Pakistan, Nigeria, and India. "Polio and Its Epidemiology" by Lester Shulman describes the complexity of a disease system with both natural and vaccine strains of the poliovirus, and the many challenges to its future eradication. The use of the live oral polio vaccine has generated vaccine-derived poliovirus, which contributes to the complex molecular epidemiology of polioviruses in countries with residual infection. The cost and implementation considerations for polio's ultimate eradication are therefore far from simple. It is possible that alternative inactivated vaccines may need to be developed if the ultimate phase out of the current oral polio vaccine is to be considered.

Worldwide, one billion people are infected with pathogens termed neglected tropical diseases, largely in low-income countries. Many infectious diseases in this category are considered waterborne. A comprehensive review of major waterborne diseases is covered in "Waterborne Infectious Diseases, Approaches to Control" by Fenwick and colleagues. Where the water serves as the habitat for the intermediate animal host or vector and the proximity of human populations facilitates the lifecycle. These include diseases such as schistosomiasis, a protozoa transmitted by snails and guinea worm, transmitted by contaminated water, onchocerciasis or river blindness transmitted by flies, as examples. Protozoal and parasitic infections often have complex lifecycles involving multiple hosts, creating challenges to prevention and treatment. Since 1986, the Carter Foundation has devoted its efforts to neglected tropical diseases such as guinea worm in Africa. More than 3.5 million people were affected by this parasitic roundworm untreatable infection caused by *Dracunculus medenisis* in the 1980s and today, the eradication of this disease through prevention is imminent, despite its neglect in the global health agenda.

Zoonotic diseases are infectious diseases transmitted from animals to humans, and constitute more than half of infectious diseases to humans [4]. There are examples of viruses, bacteria, protozoa, parasites, and prions (transmissible proteins) that are considered zoonotic diseases, where their biology and epidemiology are influenced by the animal host, its behavior, and ecology. Examples such as anthrax (*Bacillus anthrasis*), bovine tuberculosis (*Mycobacterium bovis*), brucellosis (*Brucella* sp), cysticercosis (*Taenia solium*, the pork tapeworm), echinococcosis (*Echinococcus* sp), and rabies virus are endemic in many developing countries of Africa, Asia, and South and Central America. Many of which have poor human and veterinary infrastructure to control these important pathogens. Interdisciplinary research is needed to develop novel and more effective control measures. The divided responsibilities between veterinary and medical governing bodies and resources needs to be further integrated as envisioned by the "One Health" initiatives that study the risks of biological pollution on wildlife and humans.

Climate change has long been considered an important determinant of many infectious diseases but the field has been recently expanding in its scope. Pathogens requiring an intermediate host or insect vector may be particularly sensitive to climate change. Warmer temperatures will be predicted to provide an expanded environment for vectors such as mosquitoes, potentially changing the distribution of vector borne human disease. Climate change has also been associated with the frequency or magnitude of outbreaks of food poisoning due to salmonellosis in meat or Vibrio infection in shellfish. This field is expanding to consider infectious diseases that are nonvector borne with consideration of climate's impact on seasonality, pathogen replication, dispersal, and survival. However, as described in "Infectious Diseases, Climate Change Effects on" by Matthew Baylis and Claire Risley, the methodology for predicting climate change's impact on disease is yet to be fully developed and more research is needed to collect data on pathogens that might be influenced.

Disease control in humanitarian emergencies should rely on joint situation analysis and technical support involving experts from related specialties and include the development of standards, guidelines, and tools adapted for field use. Communicable disease epidemiological profiles and risk assessments specific to countries or crisis situations prioritize interventions and provide policy guidance to national authorities and humanitarian partner agencies for the control of communicable diseases in specific settings [5]. As an example, in an 8-week period in 2011, a cholera outbreak was reported in the Democratic Republic of Congo (DRC) and Republic of Congo, a poliomyelitis outbreak in Pakistan, and cases of avian influenza in humans in Egypt. Thus highlighting the ever-changing threat of infectious disease infections on a global and temporal scale.

The dynamic nature of various infectious disease agents is thus evident from a variety of examples, and the harnessing of new technologies for the rapid diagnosis and response to infectious disease agents is described in "Infectious Diseases, Vibrational Spectroscopic Approaches to Rapid Diagnostics" by Jeremy Driskell and Ralph Tripp. These new high-resolution approaches are being developed and evaluated for both bacterial and viral pathogens. Their further instrumentation and commercialization envisions point-of-care, mobile, and cost-effective spectroscopic based diagnostic methods, which has great potential for the sustainability of infectious disease agent control in our ever-changing environment.

The development of new treatments for current, emerging, and drug resistant infectious disease pathogens is also a priority. In "Antibiotics for Emerging Pathogens," Vinayak Agarwal and Satish Nair describe improvements and innovations to the approach of identification of antibiotics through metabolic connections between the host and microbe, as well as synthesis and mining of new potential antibiotic candidates. Added to these more conventional approaches is the use of genomics and bioinformatics to identify antibiotic gene clusters and microbial ecological evaluations to better understand the interactions of natural antibiotic with their microbial targets. Future emphasis on narrow spectrum antibiotics coupled with more discriminating diagnostic methods may reduce the emergence of drug resistance already associated with use of broad-spectrum agents.

"Infectious Disease Modeling," as described by Angela McLean, has become an important methodology to characterize disease spread, both in populations and within a single host. While within-host modeling, often considers the spread of infection within an individual and its interface with the host's immune responses, newer models employ multiple levels simultaneously; such as within-host dynamics and between host transmissions. Modeling has become an even more important tool in characterizing infectious diseases particularly with the challenges of growing population and densities. These methods can organize available data and identify critical missing data. Perhaps most important is the use of modeling techniques to compare or project the impact of various intervention strategies.

To complete the coverage of this volume, we are pleased to have contributions on Human Bacterial Diseases from Ocean; Marine and Freshwater Fecal Indicators and Source Identification; Waterborne Diseases of the Ocean, Enteric Viruses; and Waterborne Parasitic Diseases in Ocean.

Globally, infectious diseases account for more than 17 million deaths each year. While modern medicine and technology have diminished the threat of many of these pathogens in high-income countries, the ever present threats of reemerging infections, population mobility, and pathogen genetic variability are but some of the reasons for the dynamic threat of this broad category of risks to human health. The majority of infectious disease burden remains in the tropics, in low- and middle-income countries with scare resources, infrastructure, and health systems to mount or sustain control efforts in the absence of outside support. It is therefore critical that efforts from the scientific research community and international donor agencies continue to increase their efforts with integrated goals of vigilant surveillance, improved and cost-effective diagnostics, and treatment with a goal of sustainable control.

Bibliography

1. WHO. http://www.who.int/gho/mortality_burden_disease/causes_death_2008/en/index.html
2. WHO. Influenza A (H1N1): special insights. http://www.who.int/en/
3. Gandhi NR, Moll A, Sturm AW, Pawinski R, Govender T, Lalloo U, Zeller K, Andrews J et al (2006) Extensively drug-resistant tuberculosis as a cause of death in patients co-infected with tuberculosis and HIV in a rural area of South Africa. Lancet 368:1575–1580
4. Taylor LH, Latham SM, Woolhouse MEJ (2001) Risk factors for human disease emergence. Philos Trans R Soc Lond B 356:983–989
5. WHO. Global alert and response. http://www.who.int/csr/en/

Chapter 2
Antibiotics for Emerging Pathogens

Vinayak Agarwal and Satish K. Nair

Glossary

Microorganism	A microscopic or submicroscopic organism, too small to be seen by unaided human eye, comprising of bacteria, virus, yeast, protozoa and fungi.
Pathogen	A disease-causing microorganism which may or may not be infectious.
Antibiotic	A chemical substance, usually organic in nature, capable of destroying or inhibiting the growth of pathogenic microorganisms.
Antibiotic resistance	The developed or acquired ability of antibiotic susceptible pathogenic microorganisms to grow and survive despite the inhibitory action of the antibiotic molecules.
Plasmid	Small, linear or circular genetic elements which can replicate independently of the chromosomal DNA inside a cell.
Tuberculosis	A highly contagious bacterial disease of humans and animals caused by various strains of *Mycobacterium*, which normally affects the lungs but can also spread to other organs of the body.

This chapter, which has been modified slightly for the purposes of this volume, was originally published as part of the Encyclopedia of Sustainability Science and Technology edited by Robert A. Meyers. DOI:10.1007/978-1-4419-0851-3

V. Agarwal
Center for Biophysics and Computational Biology, University of Illinois at Urbana-Champaign, 600 S. Mathews Ave, Urbana, IL 61801, USA
e-mail: vagarwa2@uiuc.edu

S.K. Nair (✉)
Department of Biochemistry and Center for Biophysics and Computational Biology, University of Illinois at Urbana-Champaign, 600 S. Mathews Ave, Urbana, IL 61801, USA
e-mail: snair@uiuc.edu

P. Kanki and D.J. Grimes (eds.), *Infectious Diseases: Selected Entries from the Encyclopedia of Sustainability Science and Technology*,
DOI 10.1007/978-1-4614-5719-0_2,

Malaria	A highly contagious protozoal disease caused by various strains of mosquito-borne *Plasmodium*.
Peptide	Short chains of amino acids connected by peptide bonds.
Polyketide	Compounds characterized by more than two carbonyl groups connected by single intervening carbon atoms.

Definition of the Subject

Antibiotics are organic small molecules, many of which are from natural sources, which are used to treat human infections caused by pathogenic microorganisms. While most validated antibiotics are initially very useful clinically, the pathogenic microorganisms that these compounds target are able to evade the action of the antibiotic by development of resistance mechanisms, which eventually render these antibiotics ineffective. Moreover, these resistance mechanisms can be passed on among different types of bacteria in a very simplistic manner that further compromises the usefulness of antibiotics. As a consequence, many diseases that were thought to have been eradicated by antibiotics (such as tuberculosis) have reemerged within these antibiotic-resistant strains. Hence, there is a constant need for the development of new and better antibiotic molecules that can be used to target these drug-resistant microbial populations.

Traditionally, the discovery and development of new antibiotics and other drug molecules has relied on two parallel and complimentary approaches: the discovery of new molecules of clinical relevance from natural sources, and the use of synthetic chemistry methods to derive compounds, based on the scaffold of naturally occurring molecules, with enhanced favorable antimicrobial properties. More recent development includes the identification of molecules without any natural precedents, based on the use of high-throughput screening methods of potential antibiotic targets against a large number of synthetic compound libraries to identify new classes of antibiotic molecules. Continued developments in all of these approaches, by an amalgam of both academic and industrial efforts, are essential for the development of compounds aimed at treating the evolution of drug-resistant pathogenic microorganisms.

Introduction

The interaction of prokaryotic microorganisms with human beings is widespread and significant, and the number of bacterial cells within the human body far exceeds the number of the human eukaryotic cells. This interaction between humans and bacteria is designed to be mutually beneficial to both [1, 2]. However, a small number of these bacterial microorganisms are known to cause disease in human beings and are hence called pathogens. The mechanisms by which these pathogenic organisms cause disease are extremely diverse and affect nearly all types of tissues and organs. The advancement of medical science has resulted in the development of two frontline defense mechanisms to fight infections and diseases caused by microbes. The first line of defense is the vaccines that specifically elicit the human host's adaptive immune response to recognize and eliminate pathogenic and harmful microbes, while selectively preserving the beneficial microbe population. Vaccines are generally biological macromolecules, derived from the pathogenic microbes themselves, which provide the adaptive immune response with the information necessary to elicit an antibody or cellular immune response against future invasions by the pathogenic microbes. Vaccines have been developed against several pathogens and have led to very significant reductions of deadly diseases such as tuberculosis and tetanus. However, vaccinations are not currently available against all pathogens, and the natural evolution of the pathogens themselves may render these vaccinations ineffective against future infections, as exemplified by the inherent difficulties in vaccine development against viral infections such as influenza and HIV/AIDS.

The second frontline defense against pathogenic infections is the discovery and development of small molecule antibiotics that are designed to target and kill the harmful microorganisms with minimal side effects against the human host. Antibiotics are naturally occurring molecules produced by a number of different bacterial species to give them a selective competitive advantage for nutrients or other growth necessities over other microorganisms with which they share their surroundings. Serendipitous discovery and exploitation of these molecules dates back to ancient history [3–5] with numerous early reports, such as the use of the bark from cinchona tree, which contains the antibiotic quinine, used to treat malaria. Antibiotics generally have well-defined mechanisms of action against microbes and target specific biological processes to either retard growth (bacteriostatic antibiotics) or kill the microbe (bacteriocidal antibiotics).

The first report of an isolated antibiotic molecule used to effectively treat infections was the development and use of penicillin from the fungus *Penicillium notatum* by Alexander Fleming in 1928. Penicillin was used to treat diseases such as syphilis and other infections caused by staphylococci and streptococci. The term "antibiotic" was coined by the microbiologist Selman Waksman in 1942 [6] and is derived from the Greek *antibiosis* ("against life"). Hailed as a wonder drug, the discovery of penicillin marked the beginning of the antibiotic age, during which time antibiotics were developed to treat almost all major human infectious diseases. During this era, thousands of antibiotic molecules were discovered and these compounds were applied to fight bacterial, fungal, protozoan, and yeast infections. Concurrently, synthetic organic chemistry was also applied for the production of derivatives of naturally occurring antibiotics, and these synthetic compounds exhibited more desirable properties, such as decreased incidences of allergic reactions in humans. The discovery and development of these molecules marked a decreased incidence of outbreaks of the most common infectious diseases and contributed to an increase in life expectancy around the world.

However, repeated challenge of the pathogenic microorganisms with these antibiotics, in combination with the natural evolution of their genetic information, has resulted in the development of antibiotic resistance in these pathogens [7–9]. This process has led to the development of microbial resistance against nearly all available antibiotic molecules and has precipitated to an immense medical and financial challenge. The arsenal available to fight infections is becoming increasingly limited, and the rate of development of new molecules has not kept pace with the emergence of antibiotic resistance.

Factors Contributing to Increased Antibiotic Resistance

1. Widespread and indiscriminate overuse of antibiotics in the clinical environment, which exposes pathogenic microorganisms to an ever increasing concentration of different antibiotic molecules, which in turn increases the selection pressure to develop resistance against these antibiotic molecules [10, 11].
2. Use of antibiotics in agriculture and livestock rearing also leads to increased exposure of pathogenic bacteria to antibiotic molecules. Many of the antibiotics in use in agriculture and livestock feeds are also used for treating humans, and hence, development of resistant bacterial strains can be transferred from one setting to the other very easily [12–14].
3. Genetic recombination and transfer of genes from an antibiotic-resistant microbe to a non-resistant microbe, which leads to spread of resistance among microbial species. Antibiotic resistance genes are usually borne on highly mobile genetic vehicles called plasmids, which can easily be transmitted among microbes [15]. Microorganisms are able to sequester multiple plasmids, resulting in the evolution of plasmids that contain more than one resistance determinant. For example, the *Shigella* epidemic that caused nearly a quarter million deaths in Guatemala in 1968 was caused by a pathogen that contained a plasmid with resistance against four of the most commonly used antibiotics [16–18].

Mechanisms by Which Microorganisms Develop Resistance to Antibiotics

1. Efflux pumps in the microbial cell walls and membranes which actively transport the antibiotic molecule out of the cell and hence decrease its concentration within the cell. Consequently, increasing doses of antibiotics are required to target such microbes [19–21]. This mechanism is typified by the resistance to the broad-spectrum antibiotic tetracycline, which is mediated by the genetically mobile *tet* genes that encode for active efflux pumps for tetracycline and its derivatives [22–24].

2. Chemical modification of the antibiotic molecule by specific enzymes within the target pathogenic microbe, which render the antibiotic ineffective [25]. This mechanism is of particular concern against β-lactam antibiotics, such as penicillin and its derivatives [26, 27].
3. Modulation of the antibiotic target within the target cell so that the antibiotic is no longer able to bind and engage the target. Random mutations within the antibiotic target genes may lead to the emergence and subsequent selection of microbe species with a desensitized antimicrobial target. This mechanism of resistance is operative for development of resistance against the macrolide antibiotic erythromycin, and resistance results from the modification of its target, the bacterial ribosome, such that the antibiotic is no longer able to find the target [28, 29]. Another example is resistance to the "drug of last resort" vancomycin, which results as a change in its target of the D-Ala- D-Ala dipeptide moiety of the bacterial cell wall to D-Ala- D-Lac or to D-Ala-D-Ser [30–32].
4. Alteration of the metabolic pathways that are targeted by the antibiotic molecule so that inhibition of these metabolic pathways is not inhibitory or lethal to the microbe. A primary example of this class of resistance is the alterations in the sterol biosynthesis pathway in fungi which confer resistance to azole antibiotics [33, 34].

In light of the above factors, there is an ever-growing need for academic laboratories as well as for the pharmaceutical industries to search for antibiotics that have new and as yet unexplored mode of actions. New chemical classes of drug molecules are needed to mitigate the effects of resistance against the current drug chemical entities.

Emerging Targets of Antibiotic Molecules

Several new classes of antimicrobial targets have been identified, and drug development based on these findings, though still in its infancy, have taken major leaps forward.

Microbial Fatty Acid Biosynthesis Inhibition

The microbial fatty acid biosynthesis pathway presents numerous targets for drug development, particularly against the causative agent of tuberculosis – *Mycobacterium tuberculosis*. The metabolic pathways and chemical mechanisms for the synthesis of fatty acids are shared between a majority of prokaryotes and eukaryotes, including humans. However, as a result of differences in protein sequence and different arrangement of the active sites, compounds that target the bacterial fatty acid synthesis pathway are not cross-reactive or toxic against humans.

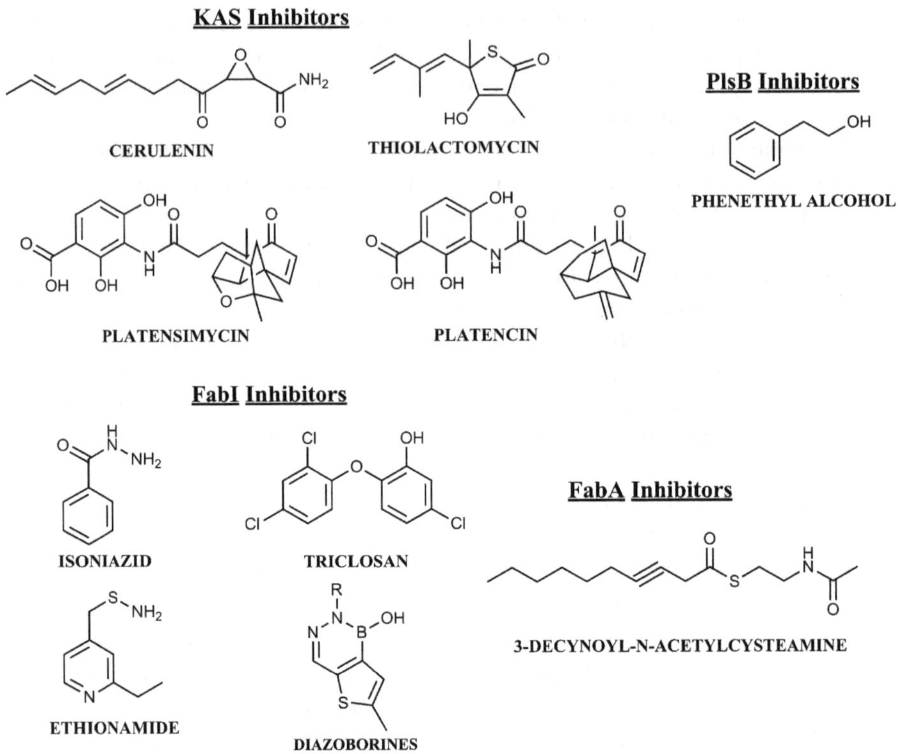

Fig. 2.1 Chemical structures of representative microbial fatty acid inhibitor molecules

The most well-characterized class of fatty acid biosynthesis inhibitor molecules is the antituberculosis molecule isoniazid, which inhibits the biosynthesis of mycolic acid, an essential component of the mycobacterial cell wall. Triclosan, an extensively used molecule in consumer products, such as toothpastes and mouthwashes, is also a fatty acid biosynthesis inhibitor.

An example of an emerging molecule in this class of antimicrobials is platensimycin (Fig. 2.1), produced by the organism *Streptomyces platensis* [35], which inhibits the growth of several pathogenic Gram-positive bacteria such as *Staphylococcus aureus, Enterococcus faecalis, and Staphylococcus pneumonia* [36]. Platensimycin inhibits the 3-ketoacyl-ACP synthase (KAS) enzyme FabF. Continued high-throughput screening and host genetic engineering efforts led to the identification of a platensimycin analog, platencin [37–39], which, in addition to FabF, inhibits another KAS enzyme called FabH and displays enhanced pharmacodynamic properties as compared to platensimycin. Chemical structures of microbial fatty acid biosynthesis inhibitor molecules discussed in this section are illustrated in Fig. 2.1.

Other KAS inhibitor molecules include cerulenin and thiolactomycin [40, 41]. Cerulenin is a broad-spectrum antifungal antibiotic produced by fungus

Caephalosporium caerulens and preferentially inhibits the KAS enzymes FabB and FabF. Cerulenin forms a stable covalent adduct with the active site cysteine residue of these enzymes, and thus inhibits malonyl-group incorporation in the growing fatty acid chains [40, 42]. Thiolactomycin and related compounds have been isolated from various *Streptomycetes* and show potent inhibitory activity against fatty acid biosynthesis and mycolic acid biosynthesis in mycobacteria as well as against protozoan parasites such as malaria-causing *Plasmodium falciparum*. Thiolactomycin mimics the binding of malonyl coenzyme A molecule in the active site of KAS enzymes FabB, FabF, and FabH [41, 43]. Attempts for the preparation of semisynthetic derivatives of thiolactomycin are currently underway, and some of these derivatives are expected to improve upon its antimicrobial properties of thiolactomycin [44].

Molecules such as triclosan, isoniazid, diazaborines, and ethionamide target the enoyl-ACP reductase enzyme FabI of the fatty acid biosynthesis pathway. Each of these molecules binds in the active site of the enzyme and forms a tight complex with the nicotinamide adenine dinucleotide (NAD+) cofactor required for catalysis by FabI. However, pseudomonads and *S. pneumoniae* contain an additional enoyl-ACP reductase enzyme, FabK, which is not susceptible to inhibition by these compounds, which allows these pathogens to overcome the antibacterial activities of these compounds.

The fatty acid biosynthesis pathways also present several other enzymes that can be targeted for drug development. The isomerase activity of the enzyme FabA can be inhibited by the drug candidate compound 3-decynoyl-*N*-acetylcysteamine. Metabolic interaction of the acetyl coenzyme A carboxylase enzyme with the fatty acid biosynthetic pathway also poses a potential target for inhibitor screening. Finally, compounds such as phenethyl alcohol, which inhibit the enzyme PlsB, responsible for the link between fatty acid biosynthesis and phospholipid biosynthesis, are also antimicrobial agents.

Isoprenoid Biosynthesis Inhibition

Isoprenoids are naturally occurring diverse organic compounds derived from the condensation of two or more isoprene monomeric units. They represent the largest class of small molecules on Earth. The isoprenoid pathways between prokaryotes and eukaryotes are very divergent: Humans synthesize isoprenoids by the mevalonate pathway, while microbes synthesize isoprenoids via the nonmevalonate pathways. Both pathways present numerous potential drug targets. One of the most widely used cholesterol-lowering drug, Lipitor, targets the mevalonate pathway in humans which is responsible for cholesterol biosynthesis. Antimicrobials such as terbinafine target the nonmevalonate pathways in fungi and yeast.

Several promising lead compounds that target the nonmevalonate pathway in the malaria parasite *Plasmodium falciparum* have been characterized. Principal among these are the phosphonate molecules fosmidomycin, first isolated from *Streptomyces lavendulae*, and its derivative FR900098, isolated from *Streptomyces rubellomurinus* [45, 46]. Fosmidomycin inhibits the enzyme 1-deoxy-D-xylulose 5-phosphate reductoisomerase (DXR) of the malaria parasite and has progressed to the stage of human phase II clinical trials for use in combination therapy with lincosamide antibiotic clindamycin [47–49]. However, the bioavailability of this molecule is low in bacterial models, with tuberculosis-causing bacterial pathogen essentially resistant to this molecule due to lack of uptake [50]. FR900098 is a structural analog of fosmidomycin and has been shown to demonstrate better efficacy in vivo than the parent compound. Synthetic and biosynthetic routes for the production and manipulation of this compound are currently being pursued [51, 52], as is the synthesis of several other structural analogs of fosmidomycin [53]. Fosmidomycin and FR900098 belong to the phosphonate class of antimicrobials, which are characterized by the presence of a stable carbon–phosphorous covalent linkage [54].

In addition to the DXR enzyme, a second potential drug target is the isoprenoid biosynthesis enzyme IspH, which is present in the vast majority of pathogenic bacteria and the malaria parasite but not in humans [55]. IspH is an attractive target for the development of inhibitors which could be potential drug and herbicide candidates. A series of diphosphate and bisphosphonate compounds have been synthesized, some of which inhibited the IspH enzyme at nanomolar concentrations [56, 57].

Other isoprenoid biosynthetic drug targets include the enzyme dehydrosqualene synthase CrtM, from the human pathogen *Staphylococcus aureus*, which can be inhibited by phosphonosulfonates and phosphonoacetamide compounds [58]. These compounds lead to inhibition of production of the *S. aureus* virulence factor staphyloxanthin, which makes these otherwise recalcitrant pathogenic cells susceptible to reactive oxygen species and subsequent clearance by the human immune system [59, 60]. Finally, the antiarrythmia drug amiodarone, which blocks various metal ion channels in the cell membrane [61], in conjunction with itraconazole, a fungal cytochrome P450 inhibitor [62], is also being reevaluated for the treatment of Chagas disease and cutaneous leismaniasis [63].

Cell Wall Biosynthesis Inhibition

Inhibition of cell wall biosynthesis has been one of the earliest validated targets for antibiotic development and still remains one of the most fruitful areas for the development of new drugs. However, drug development against cell wall biosynthesis is particularly challenging due to the presence of an additional lipopolysaccharide layer in Gram-negative bacteria and the fungal chitin cell wall in fungi, both of which are extremely rigid and impermeable to most small molecule drugs. The following section focuses on some new nucleosidic antibiotic molecules

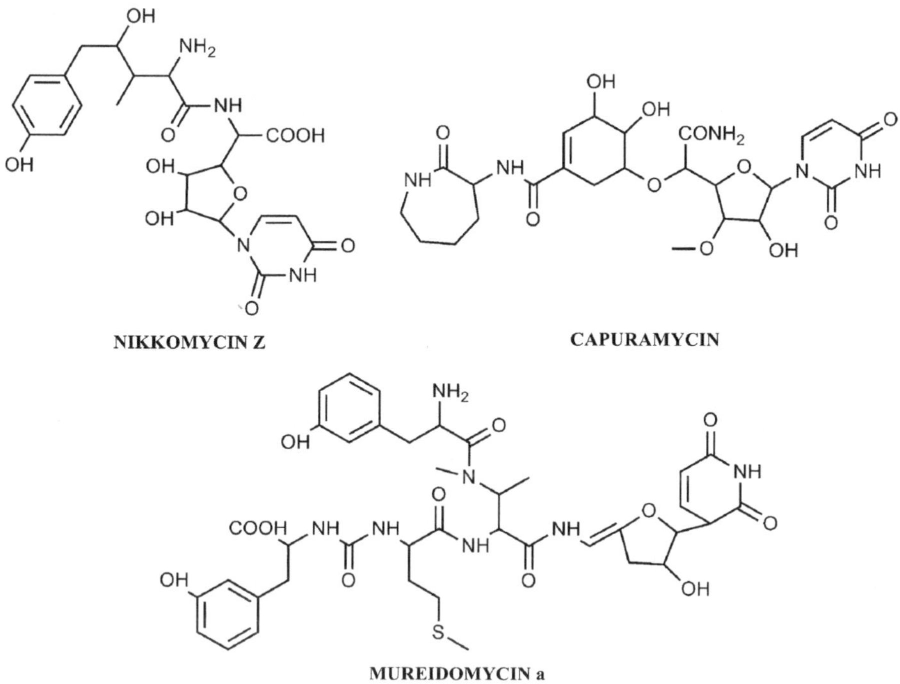

Fig. 2.2 Chemical structures of representative microbial cell wall biosynthesis inhibitor molecules

that inhibit microbial cell wall biosynthesis and hold promise for future development. Chemical structures of microbial cell wall biosynthesis inhibitor molecules discussed in this section are illustrated in Fig. 2.2.

Nucleoside antifungal polyoxin molecules, such as nikkomycins, were isolated from several strains of Streptomyces [64] and have been shown to inhibit the biosynthesis of the fungal chitin cell wall. Nikkomycins display a wide spectrum of bioactivity while being nontoxic to animals and plants and have traditionally been used for controlling plant fungal diseases in agriculture. The parent nikkomycin compounds show relatively weak activity against opportunistic human pathogenic fungi, and semisynthetic derivatives are now being prepared which may display better pharmacological properties [65].

Capuramycin was first isolated from *Streptomyces griseus* in the 1980s and shows potent antibacterial properties against the human pathogens *Streptococcus pneumoniae and Mycobacterium smegmatis* [66]. Structure activity relationships have been explored by the synthesis of capuramycin derivatives [67, 68]. In in vivo studies against *Mycobacterium tuberculosis*, capuramycin demonstrated unique phenotypic changes in the pathogen which are not shown by any other currently available antibiotic and which may signify a mode of action distinct from those of known antibiotics [69].

Structurally related uridylpeptide molecules such as mureidomycins, pacidamycins, napsamycins, and sansanmycins have demonstrated activity against infections of the opportunistic human pathogen *Pseudomonas aeruginosa* in rodent models. Such infections are difficult to treat in immune-compromised patients as this pathogen can form a protective biofilm layer resulting in chronic infections [70–72]. These molecules do not show any cross-reactivity with mammalian cells.

Emerging Chemical Classes of Antibiotic Molecules

Peptidic Antibiotics

Peptidic antimicrobials are typically synthesized by the ribosome and consist of short polymers of the 20 naturally occurring amino acids linked together by amide bonds. A second class of peptide antimicrobials are generated by a molecular assembly line, called nonribosomal peptide synthases (NRPS), which can incorporate nonnatural amino acids. The use of peptides to target pathogenic microbes is exemplified by the production of short cationic peptides by the human and plant immune systems, which either recruit the immune system for pathogen clearance or disrupt the outer cell and organelle cell membranes of microbes. A common conserved feature among all peptidic antimicrobials is posttranslational modifications conferred upon the peptide backbone, which are essential for their biological activities [73–75].

A well-known example of one class of ribosomally produced antimicrobial peptides is lantibiotics. Although these molecules are synthesized on the ribosome, they undergo posttranslational processing, resulting in the formation of several ring-like structures that add rigidity [76]. These compounds can be further modified on the amino and carboxy termini of these peptides as well, as shown in the structures of the lantibiotics epilancin [77] and microbisporicin [78]. Lantibiotics demonstrated multiple modes of action and exert their biological activities by binding to the lipid II cell wall moiety in bacteria [79], which leads to both inhibition of peptidoglycan biosynthesis as well as cell membrane permeabilization.

The most extensively used and well-characterized lantibiotic to date is Nisin (Fig. 2.3), initially identified as a by-product of milk fermentation, which contains seven lantibiotic rings, and has been used in the livestock industry for the treatment of bovine mastitis [80, 81]. Nisin, produced by the bacterium *Lactococcus lactis*, shows no adverse effects in humans and is being evaluated in human trials for the treatment of staphylococcal mastitis in lactating women [82].

Among the NRPS peptides, the most promising drug candidates are the lipopeptide antibiotic daptomycin and glycopeptides vancomycin and teicoplanin. Lipopeptide antibiotic such as daptomycin (Fig. 2.4) contains a long lipid chain attached to the peptide, which helps in targeting the bacterial cell wall. Daptomycin is produced by *Streptomyces roseosporus*, and genetic manipulations of the producer organism have led to the production of a combinatorial library of daptomycin derivatives [83–86]. Daptomycin shows potent antimicrobial activity against

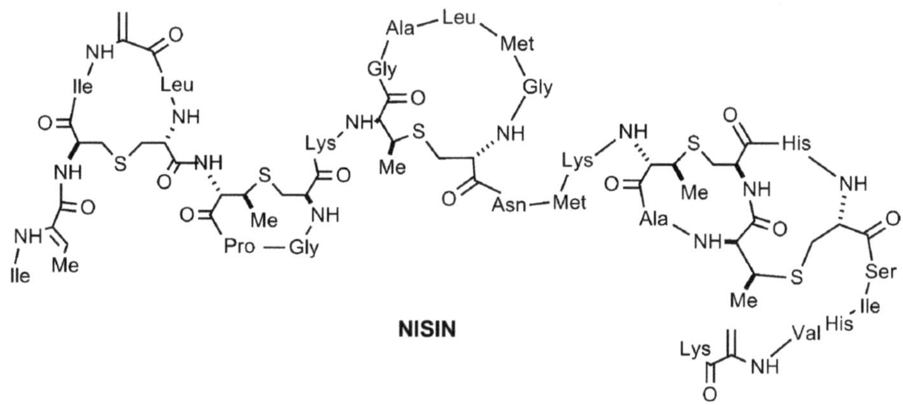

Fig. 2.3 Chemical structures of lantibiotic nisin. Some amino acids are abbreviated by their three-letter codes

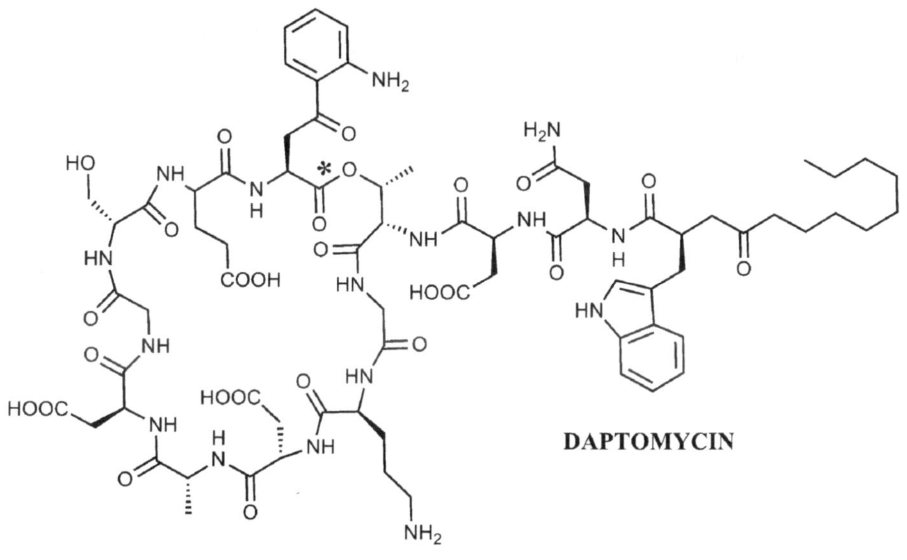

Fig. 2.4 Chemical structures of lipopeptide antibiotic daptomycin

methicillin-resistant *Staphylococcus aureus* (MRSA) [87], infections by which are a rising global concern, and is already registered for the treatment of skin and soft tissue infections [88], bacteremia, and bacterial endocarditis [89].

The glycopeptide antibiotics are produced as cyclic aglycones (without sugars) and are then decorated by a number of different sugar molecules that are covalently attached by dedicated glycosyltransferase enzymes [90–92]. Among these, vancomycin (Fig. 2.5) [93], which prevents cell wall biosynthesis in Gram-positive bacteria, is clinically approved as the "drug of last resort" against MRSA infections [94].

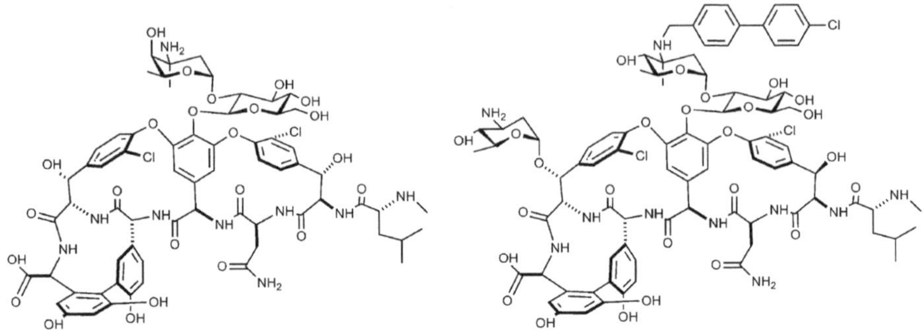

VANCOMYCIN ORITAVANCIN

Fig. 2.5 Chemical structures of representative glycopeptides and lipoglycopeptide antibiotics vancomycin and oritavancin

Oritavancin (Fig. 2.5), a semisynthetic derivative of vancomycin, is potent against vancomycin-resistant enterococci (VRE) and is currently under human trials for the treatment of soft tissue infections by Gram-positive bacteria [95]. Also of note are the cyclic lactone linkage–containing depsipeptides katanosins/plusbacins [96] and glycolipodepsipeptide ramoplanin [97], which are also active against MRSA.

Polyketide Antibiotics

Polyketide antibiotics are similar to nonribosomal peptide antibiotics as they are synthesized by multimodular polyketide synthases. These enzymes are analogous to the nonribosomal peptide synthases but typically catalyze the condensation of malonyl coenzyme A–derived monomeric units into cyclic lactone rings, onto which further modifications such as hydroxylations, glycosylations, and methylations are added. The most well-known polyketide antibiotic is erythromycin (Fig. 2.6), produced by *Streptomyces erythreus*, which has been in clinical use for more than 50 years. Several derivatives of erythromycin, such as clarithromycin and azithromycin (Fig. 2.6), have been developed for clinical use [98]. The mode of action of these compounds is inhibition of protein synthesis through binding to the various subunits of the bacterial ribosome [99].

Advancements in genetic technologies have opened up avenues for the manipulation and reprogramming of polyketide synthases for the production of new and novel molecules, which may yield drug candidates with enhanced pharmacological properties [100–102]. Of particular note are the ansamycin group of macrolide molecules, which differ from the molecules previously discussed in that the monomeric unit is 3-amino-5-hydroxybenzoic acid rather than a malonyl coenzyme A derivative. Various ansamycin molecules, such as rifamycin, geldanamycin, herbimycin, and ansamitocins, have been isolated, and their

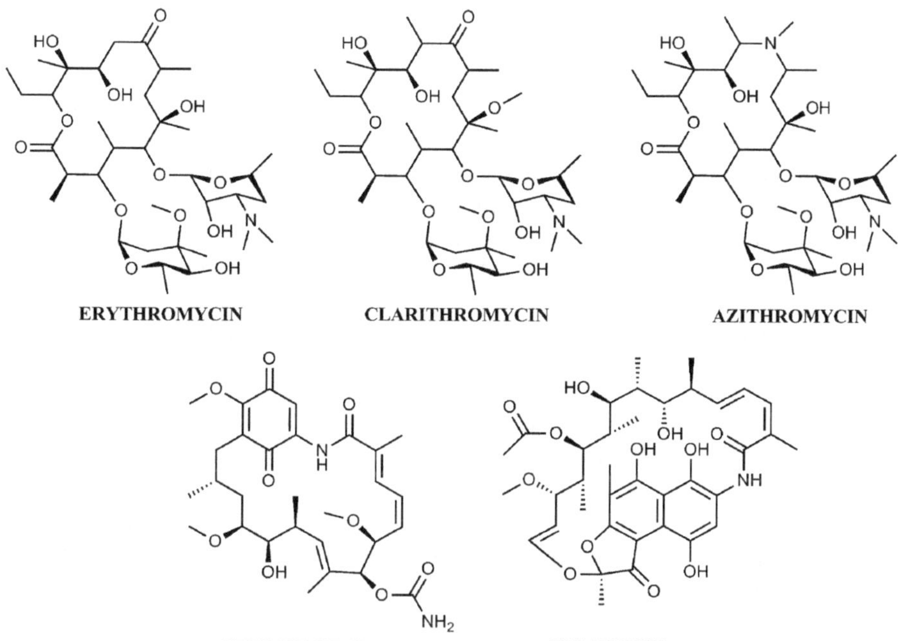

Fig. 2.6 Chemical structures of representative polyketide antibiotics

biosynthetic routes have been characterized. Of these, rifamycin, which inhibits bacterial RNA polymerase [103], is widely used in the treatment of tuberculosis. Geldanamycin, biosynthesized by *Streptomyces hygroscopicus*, and its analogs [104] inhibit heat shock protein 90 (Hsp90) [105] and are being evaluated in human clinical trials as antitumor agents [106].

Trojan Horse Antibiotics

A limiting factor for many antimicrobials and antibiotics is the lack of permeability across the microbial cell membranes. The outer membrane of the Gram-negative bacteria presents a formidable obstacle for drug entry [107]. Such Gram-negative bacteria employ a variety of active importer channels and pores to internalize requisite nutrient material. Nature has ingeniously designed several antimicrobial compounds that utilize a "Trojan horse" strategy to overcome permeability barriers; specifically, these drug molecules are covalently attached to a carrier that facilitates recognition and import through the active transporters. Once inside the target cell, the carrier is removed to release the active drug, which then exerts its antimicrobial activities. This strategy for drug delivery has several advantages. First, active transport of the molecule inside susceptible cells alleviates the need for very high doses for the drug. Second, targeting of the molecule is highly specific only for those microbial cells that have the required transporter to internalize the molecules. Consequently, these "Trojan horse" are not taken up in humans and can be effective microbials even at very low doses.

Notable among this class of compounds are the sideromycins [108, 109], which are characterized by the attachment of the antibiotic molecule to ferric ion–binding chemical components called siderophores. The siderophore–iron complexes are specifically recognized and internalized by dedicated transport machinery in bacteria. Sideromycins consists of structurally diverse molecules including danomycins, salmycins, albomycins, ferrimycins, and microcins. Bioactivities for all these molecules have been established against several human pathogens [110]. Interestingly, as rapidly dividing cancerous cells also display siderophore uptake receptors on their cell surfaces, these molecules are also being evaluated as potential anticancer compounds [109]. Several synthetic conjugates of antibiotic molecules with siderophores have also been generated for evaluation of antimicrobial properties [111].

A different class of "Trojan horse" molecules consists of drugs that are attached to peptide carrier molecules. An example of one such drug is the compound (Z)-l-2-amino-5-phosphono-3-pentenoic acid (APPA), which is a potent inhibitor of the metabolic enzyme threonine synthase, and the "Trojan horse" version consists of APPA attached to a dipeptide. Depending on the amino acids in the dipeptide, the resultant conjugate can either be an antifungal (rhizocticins) or antibacterial (plumbemycin) [112]. Thus, a variety of activities can be encoded in the identical molecule by simply altering the nature of the conjugated peptide. A similar strategy for peptide attachment is utilized in the protein synthesis inhibitor microcin C7 [113].

Future Directions

Maintenance of global health in light of the reemergence of life-threatening infectious diseases requires a continuous influx of new and potent antibiotic molecules that are efficacious and safe for human and animal clinical use. Pathogenic organisms have displayed a remarkable resilience in both developing and propagating mechanisms for overcoming the activities of these antibiotic molecules. The development of new antibiotic molecules relies on two principal paradigms: the characterization of new microbial targets against which new drug molecules can be developed, and the identification of new chemical entities which can be used as antibiotics. Both routes have resulted in the production of very successful drug molecules, and future prospects for antibiotic development are reliant on further progress in these directions. Recent biotechnological advances in the area of microbial genome sequencing have led to a rapid increase in exploration of the microbial metabolome for both novel pathways to target in pathogens and new candidates for use as antibiotics. Semisynthetic chemical derivatization of naturally occurring antibiotics has further expanded upon the repertoire of candidate drugs, and combinatorial synthetic approaches have generated vast chemical libraries which may be screened for compounds with desired biological properties.

Future antibiotic development will require an amalgamation of various academic and industrial approaches such as (1) systems biology to explore metabolic connections and effects of antibiotic molecules on the human host and targeted pathogenic microbes, (2) synthetic biochemistry to generate and diversify arrays of compounds, (3) genomics and bioinformatics to identify and analyze antibiotic biosynthetic gene clusters, and (4) microbial ecological evaluations to better understand the role of naturally occurring antibiotic scaffolds in their native microbial niches. Another promising developing area is combinatorial therapy, in which multiple antibiotic molecules are administered together in order to overcome microbial resistance mechanisms and achieve better efficacy and pathogen clearance. This approach has proven to be very successful in antiviral and oncological applications, and combination regimens have been used for the treatment of tuberculosis for several decades now. Another area of rapid progress in the future promises to be the development of narrow-spectrum antibiotics and better diagnostic tools which would aid in curtailing the indiscriminate use of broad-spectrum antibiotics, and thus reduce the incidence of microbial resistance against antibiotics.

Acknowledgments Research in the Nair lab is supported by the NIGMS. We thank Neha Garg, Yue Hao, and Zhi Li for stimulating discussions.

Bibliography

1. Sears CL (2005) A dynamic partnership: celebrating our gut flora. Anaerobe 11(5):247–251
2. Backhed F et al (2005) Host-bacterial mutualism in the human intestine. Science 307(5717):1915–1920
3. Lindblad WJ (2008) Considerations for determining if a natural product is an effective wound-healing agent. Int J Low Extrem Wounds 7(2):75–81
4. Forrest RD (1982) Early history of wound treatment. J R Soc Med 75(3):198–205
5. Forrest RD (1982) Development of wound therapy from the dark ages to the present. J R Soc Med 75(4):268–273
6. Waksman SA (1947) What is an antibiotic or an antibiotic substance? Mycologia 39(5):565–569
7. Davies J, Davies D (2010) Origins and evolution of antibiotic resistance. Microbiol Mol Biol Rev 74(3):417–433
8. Allen HK et al (2010) Call of the wild: antibiotic resistance genes in natural environments. Nat Rev Microbiol 8(4):251–259
9. Moellering RC Jr (1998) Antibiotic resistance: lessons for the future. Clin Infect Dis 27(Suppl 1):S135–S140 discussion S141–2
10. Goossens H et al (2005) Outpatient antibiotic use in Europe and association with resistance: a cross-national database study. Lancet 365(9459):579–587

11. Tacconelli E et al (2008) Does antibiotic exposure increase the risk of methicillin-resistant *Staphylococcus aureus* (MRSA) isolation? A systematic review and meta-analysis. J Antimicrob Chemother 61(1):26–38
12. Mathew AG, Cissell R, Liamthong S (2007) Antibiotic resistance in bacteria associated with food animals: a United States perspective of livestock production. Foodborne Pathog Dis 4(2):115–133
13. Angulo FJ, Nargund VN, Chiller TC (2004) Evidence of an association between use of anti-microbial agents in food animals and anti-microbial resistance among bacteria isolated from humans and the human health consequences of such resistance. J Vet Med B Infect Dis Vet Public Health 51(8–9):374–379
14. Singer RS et al (2003) Antibiotic resistance–the interplay between antibiotic use in animals and human beings. Lancet Infect Dis 3(1):47–51
15. Koch AL (1981) Evolution of antibiotic resistance gene function. Microbiol Rev 45(2):355–378
16. Mendoza MC et al (1988) Plasmid typing of *Shigella sonnei* epidemic strains and molecular relationship of their R-plasmids. Eur J Epidemiol 4(2):158–163
17. Yagupsky P et al (1991) Use of multiple markers for investigation of an epidemic of *Shigella sonnei* infections in Monroe County, New York. J Clin Microbiol 29(12):2850–2855
18. Kariuki S et al (1996) Molecular typing of multi-drug resistant *Shigella dysenteriae* type 1 by plasmid analysis and pulsed-field gel electrophoresis. Trans R Soc Trop Med Hyg 90(6):712–714
19. Li XZ, Nikaido H (2009) Efflux-mediated drug resistance in bacteria: an update. Drugs 69(12):1555–1623
20. Lin L, Ling BD, Li XZ (2009) Distribution of the multidrug efflux pump genes, adeABC, adeDE and adeIJK, and class 1 integron genes in multiple-antimicrobial-resistant clinical isolates of *Acinetobacter baumannii-Acinetobacter calcoaceticus* complex. Int J Antimicrob Agents 33(1):27–32
21. Van Bambeke F, Balzi E, Tulkens PM (2000) Antibiotic efflux pumps. Biochem Pharmacol 60(4):457–470
22. Chopra I, Roberts M (2001) Tetracycline antibiotics: mode of action, applications, molecular biology, and epidemiology of bacterial resistance. Microbiol Mol Biol Rev 65(2):232–260, second page, table of contents
23. Rodrigues L et al (2009) The role of efflux pumps in macrolide resistance in *Mycobacterium avium* complex. Int J Antimicrob Agents 34(6):529–533
24. Pages JM, Amaral L (2009) Mechanisms of drug efflux and strategies to combat them: challenging the efflux pump of gram-negative bacteria. Biochim Biophys Acta 1794(5): 826–833
25. Wright GD (2005) Bacterial resistance to antibiotics: enzymatic degradation and modification. Adv Drug Deliv Rev 57(10):1451–1470
26. Bradford PA (2001) Extended-spectrum beta-lactamases in the 21st century: characterization, epidemiology, and detection of this important resistance threat. Clin Microbiol Rev 14(4):933–951
27. Jacoby GA, Sutton L (1985) Beta-Lactamases and beta-lactam resistance in *Escherichia coli*. Antimicrob Agents Chemother 28(5):703–705
28. Lovmar M et al (2009) Erythromycin resistance by L4/L22 mutations and resistance masking by drug efflux pump deficiency. EMBO J 28(6):736–744
29. Walsh C (2000) Molecular mechanisms that confer antibacterial drug resistance. Nature 406(6797):775–781
30. Bugg TD et al (1991) Molecular basis for vancomycin resistance in *Enterococcus faecium* BM4147: biosynthesis of a depsipeptide peptidoglycan precursor by vancomycin resistance proteins VanH and VanA. Biochemistry 30(43):10408–10415
31. Bugg TD et al (1991) Identification of vancomycin resistance protein VanA as a D-alanine: D-alanine ligase of altered substrate specificity. Biochemistry 30(8):2017–2021

32. Walsh CT et al (1996) Bacterial resistance to vancomycin: five genes and one missing hydrogen bond tell the story. Chem Biol 3(1):21–28
33. Joseph-Horne T, Hollomon DW (1997) Molecular mechanisms of azole resistance in fungi. FEMS Microbiol Lett 149(2):141–149
34. Lupetti A et al (2002) Molecular basis of resistance to azole antifungals. Trends Mol Med 8(2):76–81
35. Herath KB, Attygalle AB, Singh SB (2007) Biosynthetic studies of platensimycin. J Am Chem Soc 129(50):15422–15423
36. Wang J et al (2006) Platensimycin is a selective FabF inhibitor with potent antibiotic properties. Nature 441(7091):358–361
37. Smanski MJ et al (2009) Engineered *Streptomyces platensis* strains that overproduce antibiotics platensimycin and platencin. Antimicrob Agents Chemother 53(4):1299–1304
38. Jayasuriya H et al (2007) Isolation and structure of platencin: a FabH and FabF dual inhibitor with potent broad-spectrum antibiotic activity. Angew Chem Int Ed Engl 46(25):4684–4688
39. Wang J et al (2007) Discovery of platencin, a dual FabF and FabH inhibitor with in vivo antibiotic properties. Proc Natl Acad Sci USA 104(18):7612–7616
40. Siggaard-Andersen M et al (1994) The fabJ-encoded beta-ketoacyl-[acyl carrier protein] synthase IV from *Escherichia coli* is sensitive to cerulenin and specific for short-chain substrates. Proc Natl Acad Sci USA 91(23):11027–11031
41. Machutta CA et al (2010) Slow onset inhibition of bacterial beta-ketoacyl-acyl carrier protein synthases by thiolactomycin. J Biol Chem 285(9):6161–6169
42. Johansson P et al (2008) Inhibition of the fungal fatty acid synthase type I multienzyme complex. Proc Natl Acad Sci USA 105(35):12803–12808
43. Nishida I, Kawaguchi A, Yamada M (1986) Effect of thiolactomycin on the individual enzymes of the fatty acid synthase system in *Escherichia coli*. J Biochem 99(5):1447–1454
44. McFadden JM et al (2005) Application of a flexible synthesis of (5R)-thiolactomycin to develop new inhibitors of type I fatty acid synthase. J Med Chem 48(4):946–961
45. Okuhara M et al (1980) Studies on new phosphonic acid antibiotics. III. Isolation and characterization of FR-31564, FR-32863 and FR-33289. J Antibiot (Tokyo) 33(1):24–28
46. Okuhara M et al (1980) Studies on new phosphonic acid antibiotics. I. FR-900098, isolation and characterization. J Antibiot (Tokyo) 33(1):13–17
47. Lell B et al (2003) Fosmidomycin, a novel chemotherapeutic agent for malaria. Antimicrob Agents Chemother 47(2):735–738
48. Missinou MA et al (2002) Fosmidomycin for malaria. Lancet 360(9349):1941–1942
49. Wiesner J, Jomaa H (2007) Isoprenoid biosynthesis of the apicoplast as drug target. Curr Drug Targets 8(1):3–13
50. Brown AC, Parish T (2008) Dxr is essential in *Mycobacterium tuberculosis* and fosmidomycin resistance is due to a lack of uptake. BMC Microbiol 8:78
51. Fokin AA et al (2007) Synthesis of the antimalarial drug FR900098 utilizing the nitroso-ene reaction. Org Lett 9(21):4379–4382
52. Johannes TW et al (2010) Deciphering the late biosynthetic steps of antimalarial compound FR-900098. Chem Biol 17(1):57–64
53. Kurz T et al (2007) Conformationally restrained aromatic analogues of fosmidomycin and FR900098. Arch Pharm (Weinheim) 340(7):339–344
54. Metcalf WW, van der Donk WA (2009) Biosynthesis of phosphonic and phosphinic acid natural products. Annu Rev Biochem 78:65–94
55. Rohrich RC et al (2005) Reconstitution of an apicoplast-localised electron transfer pathway involved in the isoprenoid biosynthesis of *Plasmodium falciparum*. FEBS Lett 579(28):6433–6438
56. Wang K et al (2010) Inhibition of the Fe(4)S(4)-cluster-containing protein IspH (LytB): electron paramagnetic resonance, metallacycles, and mechanisms. J Am Chem Soc 132(19):6719–6727

57. Wang W et al (2010) Bioorganometallic mechanism of action, and inhibition, of IspH. Proc Natl Acad Sci USA 107(10):4522–4527
58. Liu CI et al (2008) A cholesterol biosynthesis inhibitor blocks *Staphylococcus aureus* virulence. Science 319(5868):1391–1394
59. Song Y et al (2009) Inhibition of staphyloxanthin virulence factor biosynthesis in *Staphylococcus aureus*: in vitro, in vivo, and crystallographic results. J Med Chem 52(13):3869–3880
60. Song Y et al (2009) Phosphonosulfonates are potent, selective inhibitors of dehydrosqualene synthase and staphyloxanthin biosynthesis in *Staphylococcus aureus*. J Med Chem 52(4):976–988
61. Kodama I, Kamiya K, Toyama J (1999) Amiodarone: ionic and cellular mechanisms of action of the most promising class III agent. Am J Cardiol 84(9A):20R–28R
62. Isoherranen N et al (2004) Role of itraconazole metabolites in CYP3A4 inhibition. Drug Metab Dispos 32(10):1121–1131
63. Paniz-Mondolfi AE et al (2009) Amiodarone and itraconazole: a rational therapeutic approach for the treatment of chronic Chagas' disease. Chemotherapy 55(4):228–233
64. Chen W, Zeng H, Tan H (2000) Cloning, sequencing, and function of sanF: A gene involved in nikkomycin biosynthesis of *Streptomyces ansochromogenes*. Curr Microbiol 41(5): 312–316
65. Stauffer CS et al (2007) Total synthesis and antifungal activity of a carbohydrate ring-expanded pyranosyl nucleoside analogue of nikkomycin B. J Org Chem 72(26):9991–9997
66. Yamaguchi H et al (1986) Capuramycin, a new nucleoside antibiotic. Taxonomy, fermentation, isolation and characterization. J Antibiot (Tokyo) 39(8):1047–1053
67. Hotoda H et al (2003) Synthesis and antimycobacterial activity of capuramycin analogues. Part 2: acylated derivatives of capuramycin-related compounds. Bioorg Med Chem Lett 13(17):2833–2836
68. Hotoda H et al (2003) Synthesis and antimycobacterial activity of capuramycin analogues. Part 1: substitution of the azepan-2-one moiety of capuramycin. Bioorg Med Chem Lett 13(17):2829–2832
69. Reddy VM, Einck L, Nacy CA (2008) In vitro antimycobacterial activities of capuramycin analogues. Antimicrob Agents Chemother 52(2):719–721
70. Isono F et al (1989) Mureidomycins A-D, novel peptidylnucleoside antibiotics with spheroplast forming activity. III. Biological properties. J Antibiot (Tokyo) 42(5):674–679
71. Isono F et al (1989) Mureidomycins A-D, novel peptidylnucleoside antibiotics with spheroplast forming activity. II. Structural elucidation. J Antibiot (Tokyo) 42(5):667–673
72. Inukai M et al (1989) Mureidomycins A-D, novel peptidylnucleoside antibiotics with spheroplast forming activity I Taxonomy, fermentation, isolation and physico-chemical properties. J Antibiot (Tokyo) 42(5):662–666
73. Hancock RE, Sahl HG (2006) Antimicrobial and host-defense peptides as new anti-infective therapeutic strategies. Nat Biotechnol 24(12):1551–1557
74. Sahl HG (2006) Optimizing antimicrobial host defense peptides. Chem Biol 13(10):1015–1017
75. Peschel A, Sahl HG (2006) The co-evolution of host cationic antimicrobial peptides and microbial resistance. Nat Rev Microbiol 4(7):529–536
76. Willey JM, van der Donk WA (2007) Lantibiotics: peptides of diverse structure and function. Annu Rev Microbiol 61:477–501
77. Ekkelenkamp MB et al (2005) Isolation and structural characterization of epilancin 15X, a novel lantibiotic from a clinical strain of *Staphylococcus epidermidis*. FEBS Lett 579(9):1917–1922
78. Castiglione F et al (2008) Determining the structure and mode of action of microbisporicin, a potent lantibiotic active against multiresistant pathogens. Chem Biol 15(1):22–31
79. Bauer R, Dicks LM (2005) Mode of action of lipid II-targeting lantibiotics. Int J Food Microbiol 101(2):201–216

80. Cao LT et al (2007) Efficacy of nisin in treatment of clinical mastitis in lactating dairy cows. J Dairy Sci 90(8):3980–3985
81. Wu J, Hu S, Cao L (2007) Therapeutic effect of nisin Z on subclinical mastitis in lactating cows. Antimicrob Agents Chemother 51(9):3131–3135
82. Fernandez L et al (2008) The bacteriocin nisin, an effective agent for the treatment of staphylococcal mastitis during lactation. J Hum Lact 24(3):311–316
83. Nguyen KT et al (2006) Combinatorial biosynthesis of novel antibiotics related to daptomycin. Proc Natl Acad Sci USA 103(46):17462–17467
84. Penn J et al (2006) Heterologous production of daptomycin in *Streptomyces lividans*. J Ind Microbiol Biotechnol 33(2):121–128
85. Baltz RH, Miao V, Wrigley SK (2005) Natural products to drugs: daptomycin and related lipopeptide antibiotics. Nat Prod Rep 22(6):717–741
86. Miao V et al (2005) Daptomycin biosynthesis in *Streptomyces roseosporus*: cloning and analysis of the gene cluster and revision of peptide stereochemistry. Microbiology 151(Pt 5): 1507–1523
87. Tally FP, DeBruin MF (2000) Development of daptomycin for Gram-positive infections. J Antimicrob Chemother 46(4):523–526
88. Henken S et al (2010) Efficacy profiles of daptomycin for treatment of invasive and noninvasive pulmonary infections with *Streptococcus pneumoniae*. Antimicrob Agents Chemother 54(2):707–717
89. Fowler VG Jr et al (2006) Daptomycin versus standard therapy for bacteremia and endocarditis caused by *Staphylococcus aureus*. N Engl J Med 355(7):653–665
90. Walsh C, Freel Meyers CL, Losey HC (2003) Antibiotic glycosyltransferases: antibiotic maturation and prospects for reproGramming. J Med Chem 46(16):3425–3436
91. Sussmuth RD, Wohlleben W (2004) The biosynthesis of glycopeptide antibiotics–a model for complex, non-ribosomally synthesized, peptidic secondary metabolites. Appl Microbiol Biotechnol 63(4):344–350
92. Donadio S et al (2005) Comparative analysis and insights into the evolution of gene clusters for glycopeptide antibiotic biosynthesis. Mol Genet Genomics 274(1):40–50
93. Levine DP (2006) Vancomycin: a history. Clin Infect Dis 42(Suppl 1):S5–S12
94. Gonzalez C et al (1999) Bacteremic pneumonia due to *Staphylococcus aureus:* A comparison of disease caused by methicillin-resistant and methicillin-susceptible organisms. Clin Infect Dis 29(5):1171–1177
95. Scheinfeld N (2007) A comparison of available and investigational antibiotics for complicated skin infections and treatment-resistant *Staphylococcus aureus* and enterococcus. J Drugs Dermatol 6(1):97–103
96. Maki H, Miura K, Yamano Y (2001) Katanosin B and plusbacin A(3), inhibitors of peptidoglycan synthesis in methicillin-resistant *Staphylococcus aureus*. Antimicrob Agents Chemother 45(6):1823–1827
97. Fulco P, Wenzel RP (2006) Ramoplanin: a topical lipoglycodepsipeptide antibacterial agent. Expert Rev Anti Infect Ther 4(6):939–945
98. Weissman KJ, Leadlay PF (2005) Combinatorial biosynthesis of reduced polyketides. Nat Rev Microbiol 3(12):925–936
99. Menninger JR (1995) Mechanism of inhibition of protein synthesis by macrolide and lincosamide antibiotics. J Basic Clin Physiol Pharmacol 6(3–4):229–250
100. McDaniel R et al (1995) Rational design of aromatic polyketide natural products by recombinant assembly of enzymatic subunits. Nature 375(6532):549–554
101. Katz L (1997) Manipulation of modular polyketide synthases. Chem Rev 97(7):2557–2576
102. Ruan X et al (1997) Acyltransferase domain substitutions in erythromycin polyketide synthase yield novel erythromycin derivatives. J Bacteriol 179(20):6416–6425
103. Floss HG, Yu TW (2005) Rifamycin-mode of action, resistance, and biosynthesis. Chem Rev 105(2):621–632

104. Kim W et al (2009) Rational biosynthetic engineering for optimization of geldanamycin analogues. Chembiochem 10(7):1243–1251
105. Bedin M et al (2004) Geldanamycin, an inhibitor of the chaperone activity of HSP90, induces MAPK-independent cell cycle arrest. Int J Cancer 109(5):643–652
106. Neckers L, Schulte TW, Mimnaugh E (1999) Geldanamycin as a potential anti-cancer agent: its molecular target and biochemical activity. Invest New Drugs 17(4):361–373
107. Delcour AH (2009) Outer membrane permeability and antibiotic resistance. Biochim Biophys Acta 1794(5):808–816
108. Mollmann U et al (2009) Siderophores as drug delivery agents: application of the "Trojan Horse" strategy. Biometals 22(4):615–624
109. Braun V et al (2009) Sideromycins: tools and antibiotics. Biometals 22(1):3–13
110. Ballouche M, Cornelis P, Baysse C (2009) Iron metabolism: a promising target for antibacterial strategies. Recent Pat Antiinfect Drug Discov 4(3):190–205
111. Wencewicz TA et al (2009) Is drug release necessary for antimicrobial activity of siderophore-drug conjugates? Syntheses and biological studies of the naturally occurring salmycin "Trojan Horse" antibiotics and synthetic desferridanoxamine-antibiotic conjugates. Biometals 22(4):633–648
112. Borisova SA et al (2010) Biosynthesis of rhizocticins, antifungal phosphonate oligopeptides produced by *Bacillus subtilis* ATCC6633. Chem Biol 17(1):28–37
113. Vondenhoff GH et al (2011) Characterization of peptide chain length and constituency requirements for YejABEF-mediated uptake of Microcin C analogues. J Bacteriol 193(14): 3618–3623

Chapter 3
HIV/AIDS Global Epidemic

Phyllis J. Kanki

Glossary

Acquired immunodeficiency syndrome (AIDS)	A clinical syndrome caused by the human immunodeficiency virus (HIV). Its pathogenesis is related to a qualitative and quantitative impairment of the immune system, particularly a reduction of the CD4+ helper T lymphocyte cell count (surrogate marker of the disease). After an average of 10 years, if untreated, HIV + individuals can develop opportunistic diseases (i.e., infections and cancers rarely detected in people with normal immune systems). The natural history of the disease can be dramatically modified with administration of combination therapy composed of antiretroviral (ARV) drugs.
CCR-5	A cell membrane protein expressed on several cell types including peripheral blood-derived dendritic cells, CD34+ hematopoietic progenitor cells, and certain activated/memory Th1 lymphocytes. This receptor is well defined as a major coreceptor in conjunction with CD4+, implicated in susceptibility to HIV-1 infection.

This chapter, which has been modified slightly for the purposes of this volume, was originally published as part of the Encyclopedia of Sustainability Science and Technology edited by Robert A. Meyers. DOI:10.1007/978-1-4419-0851-3

P.J. Kanki (✉)
Department of Immunology and Infectious Diseases, Harvard School of Public Health,
651 Huntington Avenue, Boston, MA 02115, USA
e-mail: pkanki@hsph.harvard.edu

P. Kanki and D.J. Grimes (eds.), *Infectious Diseases: Selected Entries from the Encyclopedia of Sustainability Science and Technology*,
DOI 10.1007/978-1-4614-5719-0_3, © Springer Science+Business Media New York 2013

CD4+	A large glycoprotein that is found on the surface of helper T lymphocyte cells, regulatory T cells, monocytes, and dendritic cells. Its natural function is as a coreceptor that assists the T cell receptor (TCR) to activate its T cell following an interaction with an antigen-presenting cell. CD4+ is a primary receptor used by HIV-1 to gain entry into host T cells.
CD4+ T cell	An immune cell, lymphocyte (white blood cell) characterized by the CD4+ antigen (protein) on its surface. This is a T lymphocyte considered to have a “helper” function to enhance the cellular immune response. The CD4+ is the primary receptor for the HIV virus, and upon infection, the virus can destroy the CD4+ cell. In HIV-infected people, the drop in CD4+ T lymphocyte cells is a major determinant of the progression of HIV infection to AIDS.
Coreceptor (CCR-5 or CXCR-4)	Protein molecules on the surface of lymphocytes or monocytes that bind to the gp120 protein of HIV and facilitate, with CD4, binding, fusion, and entry of the virus into the susceptible cell.
CXCR-4	An alpha-chemokine receptor specific for stromal-derived factor-1 (SDF-1 also called CXCL12), a molecule endowed with potent chemotactic activity for lymphocytes. This coreceptor is one of several chemokine receptors that HIV isolates can use to specifically infect CD4+ T cells.
DNA (deoxyribonucleic acid)	A nucleic acid that contains the molecular basis of heredity for all known living organisms and some viruses and is found in the nuclei and mitochondria of eukaryotes. Chemically, DNA consists of two polymer strands of units called nucleotides made up of one of four possible bases plus sugar and phosphate groups. The polymers are joined at the bases by hydrogen bonds to form a double helix structure.
Fusion of virus and cell membranes	A merging of cell and virus membranes that permits HIV proteins and nucleic acids to enter the host cell.
Fusion/entry inhibitors	A class of ART drugs that interferes with the virus’ ability to fuse with the target cell’s outer membrane, thereby blocking entry of the HIV into the host cell.
gp120	The major HIV envelope glycoprotein having a molecular weight of 120 that protrudes from the

	outer surface of the virion. This glycoprotein binds to a CD4+ receptor on a T cell to facilitate entry of the virus into the cell.
Human immunodeficiency virus (HIV)	The virus that causes acquired immunodeficiency syndrome (AIDS). It is a lentivirus belonging to *Retroviridae* family and was discovered in 1983 by Robert Gallo and Luc Montagnier. HIV infects and destroys helper T cells of the immune system causing a marked reduction in their numbers. Loss of CD4 cells leads to generalized failure of the immune system and susceptibility to life-threatening opportunistic infections. It is transmitted mainly through sexual intercourse, exchange of contaminated syringes among intravenous drug users, and contaminated blood transfusion. HIV-1 is the HIV type most frequently detected HIV worldwide and responsible for the global pandemic.
HIV-1 subtypes or clades	Genetically related HIV strains that are essentially phylogenetically equidistant, generating a starlike phylogeny. Subtypes A, B, C, D, F, G, H, J, and K are currently known subtypes A, B, C, and D are highly prevalent others have low prevalence and limited geographic distributions.
HIV-2	The second HIV virus discovered in West Africa in 1984, the virus is more closely related to the simian immunodeficiency virus of primates. Although HIV-2 can cause AIDS, it has a distinct epidemiology, lower rate of transmission, and slower progression to disease.
Incidence	Rate describing the number of new cases of disease occurring within a given time period, expressed as new cases per person-time.
Integrase	An enzyme found in retroviruses including HIV that permits the reverse transcribed viral DNA to be integrated into the infected cell's DNA. Integrase is an enzyme encoded by the polymerase gene of HIV.
Integrase inhibitors	A class of ART drugs that blocks the viral integrase the enzyme HIV uses to integrate its genetic material into its target host cell DNA.
Nucleus	A membrane-enclosed central compartment of a cell that functions to contain the genomic DNA and to regulate gene expression.

Prevalence — Number of cases of disease in a defined population at a specific point in time it is often expressed as a percentage.

Protease — An enzyme that hydrolyzes or cleaves the polyproteins into proteins and is important in the final steps of HIV maturation. In HIV, the protease enzyme is encoded by the *polymerase* gene.

Protease inhibitors — A class of ART drugs that interferes with the viral protease enzyme of HIV by inhibiting the viral polyproteins from being cleaved, which would allow the individual viral proteins to produce infectious viral particles.

Reverse transcriptase — An enzyme found in HIV that creates double-stranded DNA using viral RNA as a template and host tRNA as primers. The reverse transcriptase enzyme is encoded by the *polymerase* gene of HIV.

Reverse transcriptase (RT) inhibitors — A class of ART drugs that interfere with the reverse transcription step during the HIV life cycle. During this step, the HIV enzyme RT converts HIV RNA to HIV DNA. There are two main classes of RT inhibitors that are used as ART drugs.

Nucleoside/nucleotide RT inhibitors (NRTI) are faulty DNA building blocks. When these faulty pieces are incorporated into the HIV DNA (during the process when HIV RNA is converted to HIV DNA), the DNA chain cannot be completed, thereby blocking HIV from replicating in a cell.

Nonnucleoside RT inhibitors (NNRTI) bind to RT, interfering with its ability to convert the HIV RNA into HIV DNA.

RNA (ribonucleic acid) — A universal form of genetic material typically transcribed from DNA, it differs from DNA in that it contains ribose and uracil as structural components. In retroviruses like HIV, RNA is their primary genetic material and is found in a mature virus particle.

T-lymphotropic — A characteristic of a virus that infects and replicates in T lymphocytes, a type of immune cell. This was the descriptor of the human T cell leukemia virus (HTLV), a human retrovirus that causes T cell leukemia and lymphoma and is T-lymphotropic like HIV. HIV was originally called human T-lymphotropic virus type III (HTLV-III) by Gallo and colleagues.

Tuberculosis	The infectious disease caused by *Mycobacterium tuberculosis*. It usually involves the lungs (pulmonary tuberculosis) but can also affect other organs (i.e., kidneys, central nervous system, lymph nodes, bones, etc.; extrapulmonary tuberculosis). Pulmonary tuberculosis, which is the most frequent clinical form, can be classified as smear positive or smear negative according to the result of the sputum bacteriological examination. The former is a major public health problem being highly contagious. Only a few individuals develop tuberculosis after a mycobacterial infection, and most of them soon after infection: it is estimated that the lifetime risk is 5–10% in HIV negatives and 5–15% yearly in HIV positives.
Virion	A single and complete extracellular infective form of a virus that consists of an RNA or DNA core and in the case of HIV with a glycoprotein coat or "envelope."

Definition of the Problem

The HIV/AIDS epidemic is now in its third decade since the discovery of the virus responsible for the disease in 1981. While the first cases of AIDS were first recognized in young men who have sex with men in the United States and Europe in the 1980s, it soon became clear that the virus could be spread through contaminated blood products and heterosexual sex. At the time of its discovery, acquired immunodeficiency syndrome (AIDS) was a new disease with high mortality, and the discovery of a new human virus as its cause in 1983 created new challenges for prevention, treatment, and vaccine efforts, many of which remain unmet today. Human immunodeficiency virus type 1 (HIV-1), as the causative agent of AIDS, has been the subject of intense research over the past three decades, in an effort to understand the biological properties of this new virus, its relatedness to other known retroviruses, characterize its epidemiology, and discover drugs and vaccines to control the epidemic.

In the early 1980s, it was recognized that HIV could be spread through blood, blood products, and sexual and perinatal transmission routes. High rates of HIV infection and its accompanying disease began to be recognized throughout the world in the late 1980s as the HIV/AIDS global pandemic became a frightening reality to the international community. The disproportionate burden of infection and disease first recognized in sub-Saharan Africa and then Asia with growing rates of disease and mortality led to estimates of global infections growing in the tens of millions with no magic bullet to end the spread. While global estimates are necessarily fraught with numerous assumptions and subjected to poor reporting, UNAIDS revised down their global estimates in 2007 [1]. Nonetheless, at the end of 2009, an estimated 33.3 million people worldwide were living with HIV; 2.5 million of these were children, with almost two thirds of them in low- and middle-income countries [2]. While stunning in magnitude, these estimates now reflect a continued slowing of the growth trajectory of the pandemic, with lower rates of new infections and mortality, the latter being largely due to the continued scale-up of provision of antiretroviral therapy (ART) that began in 2004.

We are therefore at an important point in the time for the global HIV pandemic; the peak of infections in most countries occurred about a decade ago, and many high-burden countries are seeing a plateau or decrease in HIV prevalence rates. Large-scale treatment programs for low-income countries have succeeded in providing complex and relatively expensive ART drugs to almost one third of patients in need of these life-saving therapies; 5 million of the 15 million in need. Recent strides in identifying new and efficacious methods for prevention have been encouraging as the development of an effective vaccine is still awaited.

Introduction

AIDS was first recognized as a new and distinct clinical entity in 1981 [3–5]. The first cases were recognized because of an unusual clustering of diseases such as Kaposi's sarcoma and *Pneumocystis carinii* pneumonia in young homosexual men. Although such syndromes were occasionally observed in distinct subgroups of the population – such as older men of Mediterranean origin in the case of Kaposi's sarcoma or severely immunosuppressed cancer patients in the case of *Pneumocystis carinii* pneumonia – the occurrence of these diseases in previously healthy young people was unprecedented. Since most of the first cases of this newly defined clinical syndrome involved homosexual men, lifestyle practices were first implicated as the cause of the disease.

AIDS cases were soon reported in other populations as well, including intravenous (IV) drug users [6] and hemophiliacs [7–9]. Hemophiliacs used clotting factor preparations which were prepared from the pooled blood of a huge number of donors, and IV drug users often used needles contaminated with small amounts of blood from previous users, thereby increasing their exposure to foreign tissue antigens. Asymptomatic hemophiliacs and intravenous drug users were often found to have abnormally low CD4 helper lymphocytes and higher than normal T suppressor cells, similar to the gay men with AIDS. The increase in T suppressor cells was presumably due to frequent antigenic stimulation; the decrease in CD4+ T helper cells was the more direct effect of the yet-to-be-discovered causative agent.

Three new categories of AIDS patients were soon observed: blood transfusion recipients [10, 11], adults from Central Africa [12–14], and infants born to mothers who themselves had AIDS or were IV drug users [15, 16]. The transfusion-associated cases had received blood donated from an AIDS patient at least 3 years before they began showing symptoms [10, 11].

Based on the disparate populations afflicted with this new malady and the emerging epidemiology of the disease, the possible infectious etiology for AIDS was considered [17]. Multiple studies were initiated to determine the possible role of various microorganisms, especially viruses in causing AIDS. These studies measured and compared seroprevalence rates for suspect viruses in AIDS patients and controls. The short list of candidate viruses included cytomegalovirus (CMV), because it was already associated with immunosuppression in kidney transplant patients; Epstein-Barr virus (EBV), presumably because it was a lymphotropic virus; and hepatitis B (HBV), because infection with this virus was known to occur at elevated rates in both homosexual men and recipients of blood or blood products. However, based on the unique clinical syndrome and unusual epidemiology of AIDS, if the etiology was an already known virus, it would presumably have to be a newly mutated or recombinant genetic variant.

Max Essex [18, 19], Bob Gallo [20, 21], and Luc Montagnier, Francoise Barre-Sinoussi, and Jean-Claude Chermain [22] postulated that a variant T-lymphotropic retrovirus (HTLV) might be the etiologic agent of AIDS. Among the most compelling reasons for this hypothesis was that the human T-lymphotropic retrovirus (HTLV), discovered by Gallo and his colleagues [23] in 1980, was the only human virus known to infect T helper lymphocytes at that time. This fit with the new disease where T helper lymphocytes were selectively depleted by the causative agent [24–26]. AIDS patient blood samples were repeatedly cultured in an attempt to find a virus related to HTLV-I or HTLV-II [27]; however, these studies were only partially successful. Although antibodies cross-reactive with HTLV-I and HTLV-related genomic sequences were found in a minority of AIDS patients [18, 21, 22, 28], the reactivity was weak, suggesting either the coinfection of AIDS patients with an HTLV, or that a distant, weakly reactive virus was the causative agent. Proof that the disease was linked to a T-lymphotropic retrovirus was obtained by Gallo and his colleagues [29–31]. Further characterization of the agent – now termed human immunodeficiency virus type 1 (HIV-1) – revealed that it was the same as the isolate detected earlier by Montagnier and his colleagues [22]. Despite controversy over the names and identity of certain isolates, it is now clear that this new and unique human pathogen was not only a distant genetic relative of the known HTLV but also a virus that may have been more recently introduced into the humans from a primate reservoir.

A second HIV was discovered in 1984 based on antibodies from West African commercial sex workers that recognized proteins not only from the simian immunodeficiency virus (SIV) but also from HIV-1 [32, 33]. It is now known that HIV-1 is more closely related to SIVcpz, found mainly in the *Pan troglodytes troglodytes* chimpanzee species [34], while HIV-2 is related to SIVsm found in sooty mangabey monkeys (*Cercocebus atys*) [35]. HIV-2 is the second human immunodeficiency virus and constitutes the closest known human virus related to the prototype AIDS virus, HIV-1. HIV-2 shares many virologic and biologic features with HIV-1; however, its ability to transmit and cause disease is much lower [36–38].

HIV-2 infection is much less prevalent in the world; it is found primarily in West Africa and other parts of the globe with connections to West Africa, such as India, parts of South America, and urban centers with high rates of immigration from West Africa [39, 40].

Basic Virology of HIV

Human immunodeficiency virus types 1 and 2 (HIV-1 and HIV-2) are members of the *Lentivirus* genus of the *Retroviridae* family of RNA viruses and are 50% similar at the genetic level. Retroviruses are enveloped viruses that infect a wide range of vertebrate hosts in a species-specific manner [41]. Steps in the HIV replication cycle (Fig. 3.1) include (1) recognition and binding of the virion to the host cell surface via specific primary and coreceptors and fusion to the host cell membrane; (2) uncoating of the virion and release of the HIV genetic material and other viral proteins including enzymes such as reverse transcriptase, integrase, and protease; (3) HIV RNA is reverse transcribed by the enzyme reverse transcriptase to make a double-stranded DNA copy; (4) HIV viral DNA is transported to the cell nucleus, where it integrates into the host cell's DNA, using the HIV viral integrase; (5) a new copy of the HIV viral RNA is produced which will become the genetic material for new HIV virions; (6) new viral RNA and proteins produced by the host cell move to the cell surface, and new, immature virions are packaged; and (7) the viral envelope proteins are inserted into the host cell membrane, and the virions bud from the cell surface, encapsulated by viral envelope, forming a mature HIV virion [42, 43].

Like other retroviruses, HIV's genetic material becomes permanently integrated in the host cell's DNA, resulting in lifelong infection. The enzyme responsible for viral replication and encoded by the *polymerase* gene is the reverse transcriptase, which as an enzyme is error-prone, which results in considerable genetic variation. HIV enters susceptible cells via a primary receptor, the CD4+ on immune cells such as T lymphocytes. However, secondary coreceptors are also required for HIV's entry into a cell [44–47]. These are normal cellular membrane proteins, part of a seven-transmembrane spanning protein family, called chemokine receptors, and involved in the recruitment of chemokines for normal cellular function. CCR-5 and CXCR-4 are chemokine receptors that also serve as major coreceptors for the HIV virus. In conjunction with the primary receptor, CD4 molecule, they allow the virion to fuse with the susceptible host cell and allow virion entry. The CCR-5 coreceptor along with CD4 allows for HIV to infect CCR-5 bearing cells usually of the macrophage/monocyte lineage. The CXCR-4 coreceptor along with CD4 allows for HIV's fusion and entry into T cells [44, 48, 49].

The HIV genome, or genetic makeup, has been characterized and is 9.8 kilobases in length and contains nine different genes, which encode 15 proteins. There are three major classes of proteins: structural, regulatory, and

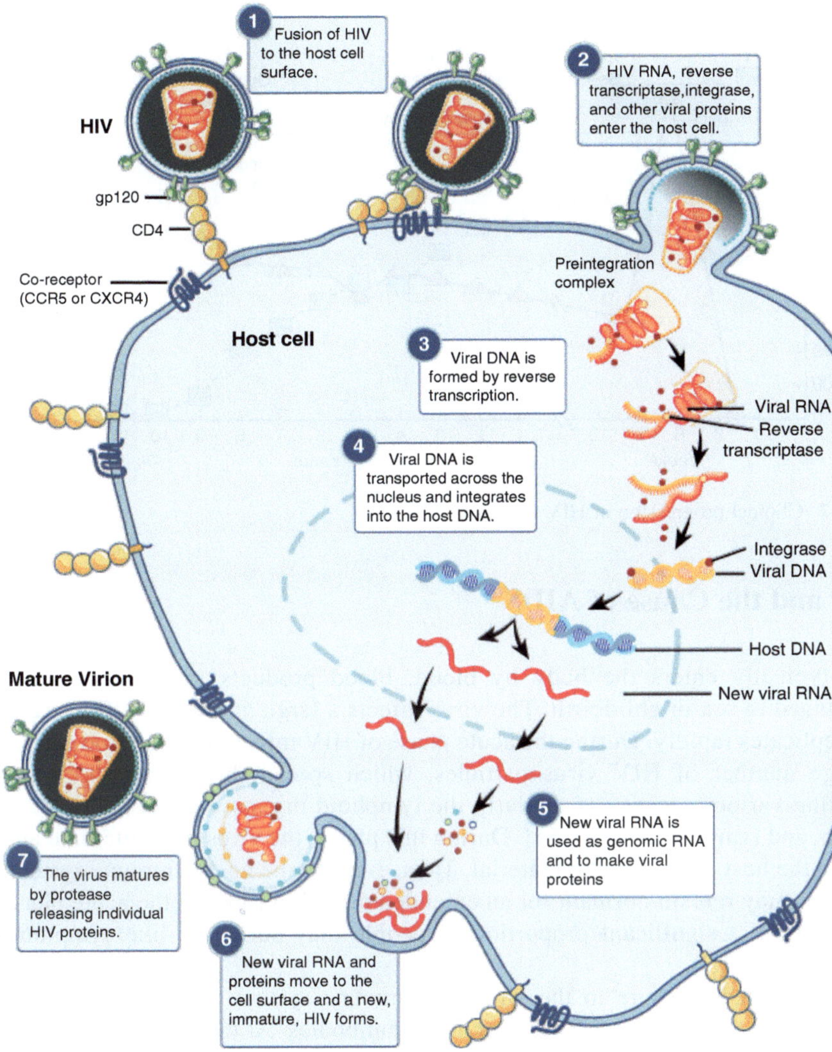

Fig. 3.1 HIV replication cycle [43]

accessory [42]. Three genes, *gag*, *pol*, and *env*, each encode multiple viral structural proteins. The *gag* (group-specific antigen) gene produces a 55-kd precursor protein, which is cleaved during viral maturation into the matrix (MA; p17), capsid (CA; p24), nucleocapsid (NC; p9), p1, p7, and p6 proteins. The *pol* gene (polymerase) encodes the enzymes including protease, reverse transcriptase (RT), and integrase, while the *env* gene (envelope) gives rise to the envelope proteins gp120 and gp41. The regulatory proteins are Tat and Rev, while the accessory proteins include Vpu, Vpr, Vif, and Nef [42].

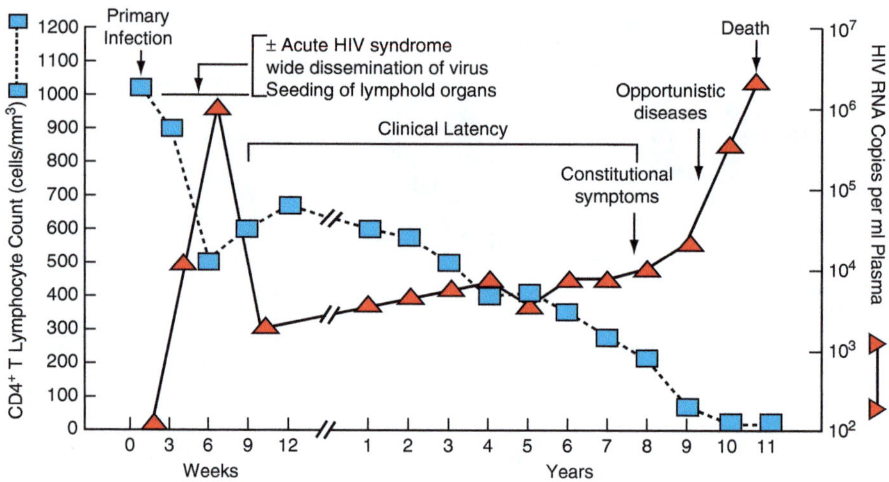

Fig. 3.2 Clinical progression of HIV infection [52]

HIV and the Cause of AIDS

HIV typically enters the body by blood, blood products, or through fluids exchanged in sex or childbirth. The virus infects a large number of CD4+ cells and replicates rapidly. During the acute phase of HIV infection, the blood carries a large number of HIV virus particles, which spread throughout the body, infecting various organs, particularly the lymphoid organs such as the thymus, spleen, and lymph nodes [50, 51]. During this phase, the virus may integrate and hide in the host cell's genetic material. Thus, evading the host's immune system, the virus may remain dormant for an extended period of time. In the acute phase of infection, a significant proportion of people may suffer flu-like symptoms [51, 52] (Fig. 3.2).

Weeks after exposure to the virus, the immune system responds with killer T cells (CD8+ T cells) and B cell–produced antibodies. At the same time, CD4+ T cell counts rebound. The virus levels may decrease at this point in time perhaps in response to the first host immune response. During this latency phase, a person infected with HIV may not experience HIV-related symptoms for several years despite the fact that the HIV continues to replicate [53]. It is this long course of dormancy that has given the lentiviruses ("lenti" – meaning slow) their name.

The immune system eventually deteriorates to the point that the human body is unable to fight off other infections; the timing and reasons for this are still subject to research. The level of HIV virus in the blood may dramatically increase, while the number of CD4+ T cells drops to dangerously low levels. An HIV-infected person is diagnosed with AIDS when he or she has one or more opportunistic infections, such as tuberculosis, and CD4+ T cells drop below 200 cells per cubic millimeter of blood, where the normal would be ~1,000 CD4+ T cells.

Epidemiology of the Global HIV Pandemic

HIV and AIDS is a global pandemic. Although the means by which the virus can spread between people remains unchanged, the populations that are highly infected in different parts of the world are different. Transmission of the virus by blood and blood products for the treatment of various diseases such as hemophilia was recognized in the early 1980s. Ensuring that blood banks and blood products were free of HIV was accomplished relatively early in the pandemic in high-income countries, but is yet to be perfected in many low-income countries where screening of HIV and the high burden of disease pose distinct challenges to maintaining safe blood banks and blood products.

Spread of HIV infection through intravenous drug use was also recognized in high-income countries early in the epidemic. These often disenfranchised and stigmatized populations were difficult to identify to provide necessary diagnosis, care, and treatment. This continues to be a challenge for prevention efforts throughout the world, most significant in parts of Asia, where the spread of HIV through this mode of transmission is both prevalent and difficult to control.

Sexual transmission of HIV in most of the high-income countries remains a threat to men who have sex with men, where the virus and its disease were first recognized. In contrast, in most middle- and low-income countries, the risk of HIV transmission is predominantly through heterosexual sex. This means that the risk to women and their offspring is more significant in these geographic locations and represents a distinction in the epidemiology of HIV/AIDS in these settings. As a consequence, the risk of mother-to-child transmission in middle- and low- income countries was most severe in nations already burdened with high infant mortality. In 2009, 370,000 (230,000–510,000) children were infected with HIV through mother-to-child transmission [2]. This represents a decrease of 24% from 5 years earlier due to better methods of preventing transmission through identification of infected pregnant women and provision of more effective prophylaxis. Inadequate access to antenatal and postnatal services remains a barrier in providing these methods to all those in need [54].

UNAIDS has just completed their 2010 report which reports on HIV and AIDS statistics from around the globe ending in 2009 [2] (Table 3.1). Extensive analysis of these current rates and the trends from previous years allows for some cautious optimism on the status of the pandemic. This report continues to support the notion that the peak of new infections occurred in 1999 and rates of new infections continue to decline or plateau [55, 56]. In 2009, an estimated 2.6 million (2.3–2.8 million) people were newly infected with HIV. In 33 countries, most in sub-Saharan Africa, HIV incidence showed significant decreases compared to 2001 [2].

Globally, an estimated 33 million people are infected with HIV. In earlier years treatment was only provided in high-income countries; more recently local governments and international programs supporting antiretroviral therapy in low-income countries have had an impact on the annual death toll. A cumulative total of

Table 3.1 UNAIDS regional estimates for adults and children living with HIV (Adapted from [2])

	People living with HIV	
Geographic region	2009	2001
Sub-Saharan Africa	22,500,000	20,300,000
Middle East and North Africa	460,000	180,000
South and Southeast Asia	4,100,000	3,800,000
East Asia	770,000	350,000
Oceania	57,000	29,000
Central and South America	1,400,000	1,100,000
Caribbean	240,000	240,000
Eastern Europe and Central Asia	1,400,000	760,000
Western and Central Europe	820,000	630,000
North America	1,500,000	1,200,000

24 million people have died from AIDS between 1980 and 2007, and Bongaarts and colleagues project that this will reach 75 million by 2030 [56].

Regional HIV statistics for both 2009 and 2001 are provided in Table 3.1. These demonstrate the heterogeneity of the global pandemic. It is beyond the scope of this chapter to discuss each of the region's HIV epidemiology in depth. Summarized findings are therefore provided with more emphasis on sub-Saharan Africa where the highest burden of infection and disease is found.

Sub-Saharan Africa

Over 22 million people in sub-Saharan Africa are infected with HIV, although there is considerable variability among nations in rates of infection and disease [2] (Fig. 3.3). The countries of Southern Africa (Angola, Botswana, Lesotho, Malawi, Mozambique, Namibia, South Africa, Swaziland, Zambia, and Zimbabwe) have over 11 million people currently infected. Thus, this region of the continent alone represents over one third of the global pandemic in terms of adult HIV infections, new HIV infections, and deaths due to HIV. Swaziland, in 2009, reported a prevalence of 25.9%, representing the highest adult HIV prevalence rate for a given country [57]. In South Africa, where over five million people are living with HIV in 2009, AIDS is the major cause of maternal mortality and attributed to cause over one third of deaths in children under the age of 5 years old.

Other large countries, such as Nigeria with a population estimated at 150 million and Ethiopia with a population 82 million, have relatively low HIV prevalence rates below 4%, but still represent a significant portion of the global burden of disease because of their large populations. By contrast, many of the countries of Southern Africa are small, but with high prevalence rates. Thus, the distinct impact of the epidemics in such diverse settings can be appreciated.

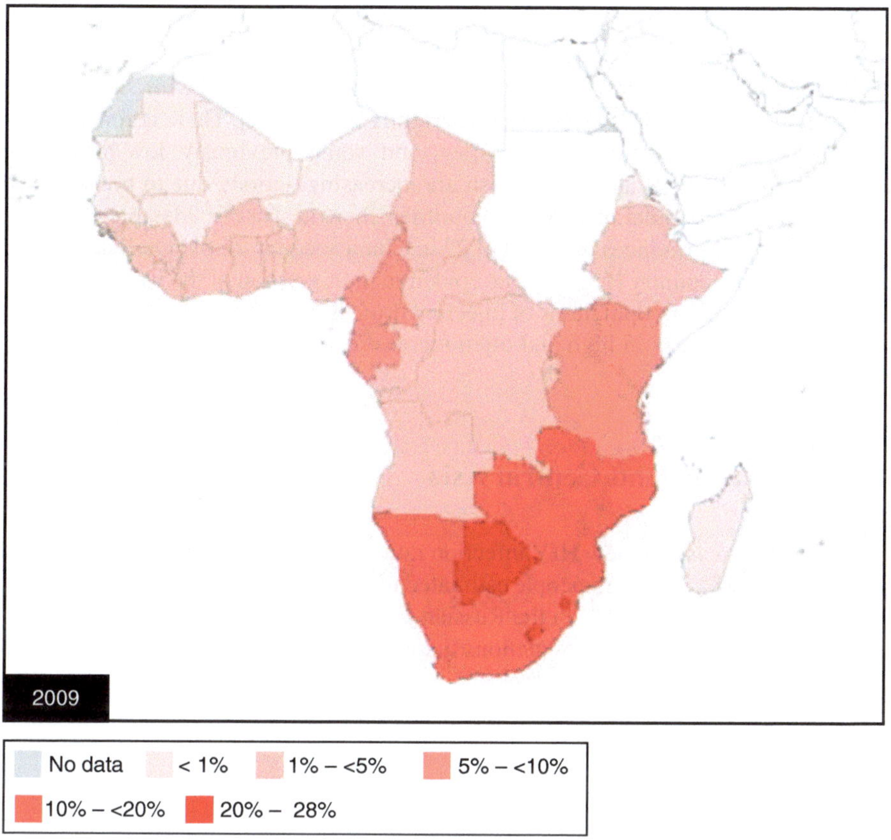

Fig. 3.3 HIV prevalence rates among adults aged 15–49 years old in sub-Saharan Africa, 2009 (Adapted from [2])

As of 2009, the rates of HIV infection in the countries of East Africa have been declining. Uganda's HIV prevalence has been stable in the 6–7% range since 2001. Kenya and Tanzania have shown declining prevalence rates, and Rwanda has been stable at 3% since 2005 [2].

In West and Central Africa, countries have maintained some of the lowest HIV infection rates of the continent. In 2009, 12 countries had rates below 2%; these included: Benin, Burkina Faso, Democratic Republic of the Congo, Gambia, Ghana, Guinea, Liberia, Mali, Mauritania, Niger, Senegal, and Sierra Leone. Nigeria's HIV infection rate has been declining for the past 6 years with the 2009 estimate at 3.6% (3.3–4.0%). The prevalence of HIV is highest in Cameroon at 5.3% (4.9–5.8%); Central African Republic, 4.7% (4.2–5.2%); Côte d'Ivoire, 3.4% (3.1–3.9%); and Gabon, 5.2% (4.2–6.2%) [2].

Asia

HIV infection rates in 2009 for the region are similar to 2004, with an estimated 4.9 million (4.5–5.5 million) people are currently infected [2]. There is considerable heterogeneity of infection by countries, and some previously low-prevalence countries like Pakistan and Bangladesh are increasing, largely due to transmission in drug-injecting populations. Within countries, there is also considerable variation; for instance, five provinces of a total of 22 in China account for over half of the HIV infections of the country [58, 59]. The epidemic in this region of the world is largely due to significant subpopulations of injecting drug users and men who have sex with men, where rates remain high and often increasing.

Eastern Europe and Central Asia

The largest increases in new HIV infection rates are seen in this region of the world with a burden of 1.4 million people estimated to be living with the virus in 2009 [2]. HIV prevalence exceeds 1% in the Russian Federation and Ukraine. The epidemic is concentrated in high-risk populations that use injection drugs and are involved in sex work and their sexual partners [60].

Caribbean

This region of the world has a relatively low burden of people living with HIV (240,000), yet because of their small population size, their HIV prevalence of ~1% (0.9–1.1%) is still considered high for the region [2]. There is considerable variability between countries and within countries. Haiti has highly variable rates, including 12% in pregnant women from some urban settings [61, 62], whereas Cuba's prevalence is exceptionally low, at 0.1% (0.08–0.13%) [2]. Unprotected paid sex is considered the major mode of HIV transmission, and infection rates in women are over half of those infected with the virus.

Central and South America

The HIV epidemic appears stable in 2009 compared to previous years, and the number of children infected is declining with ~4,000 children newly infected in 2009 [2]. There are 1.4 million people living with the infection in this region, with concentrated infection rates in men who have sex with men and sex workers.

North America and Western and Central Europe

The burden of HIV infection is 2.3 million (2.0–2.7 million) in these higher-income countries, representing a 30% increase from 2001 [2]. It is believed that unprotected sex among men having sex with men is responsible for these disturbing increasing trends in Canada, the USA, and parts of Europe [63]. General decreases are seen in injecting drug user subpopulations overall, but certain countries continue to see this as a driver of the epidemic [64–66]. Immigrant populations and individuals acquiring HIV from more endemic regions are also contributing to the burden of infection particularly in urban centers [67].

Middle East and North Africa

This region of the world has historically had low rates of HIV infection, which may be due to inadequate reporting. In 2009, an estimated 460,000 people are living with HIV, which is a substantial increase from the 2001 reported statistics [2]. Transmission of the virus from contaminated drug-injection equipment is considered the major mode of transmission in the countries of this region.

Oceania

The burden of HIV infection in this region is small at 57,000 (50,000–64,000) but has increased from the 2001 statistics [2]. Papua New Guinea has the largest infection rates in this region, with a national adult HIV prevalence of 0.9% (0.8–1.0%). Unprotected heterosexual sex is more predominant in Papua New Guinea, whereas unprotected sex in men who have sex with men is more common in New Zealand and Australia [68].

Prevention

The optimal biomedical means of preventing infection with an infectious disease pathogen would be the use of an effective vaccine. As efforts to develop such a vaccine are still underway, public health officials around the world have resorted to other methods of preventing HIV transmission [69]. These revolve around the various types of HIV transmission such as men who have sex with men, sex workers, intravenous drug users, and heterosexual adults in places such as Africa. Education on risks of transmission, methods of prevention, and behavior change are at the core of all prevention programs spanning on populations at risk.

For many of these subpopulations, stigma, lack of access to services, or even legislation can diminish the potential of prevention programs and messages for behavior change.

Access to HIV testing and counseling is a foundation to all prevention methods, which provides access to comprehensive HIV care. Education and behavior change are critical components of prevention methods, and provision is often most effective when provided by community groups. In the early 1980s in the USA, pioneer groups such as the San Francisco AIDS Foundation, Gay Men's Crisis in New York, and AIDS Project Los Angeles demonstrated the role of education in promoting safer sex practices among gay and bisexual men [70]. Recent increases in new infections in young gay men in North America and Europe are of concern and suggest the need to enhance prevention efforts [2, 63].

Globally, HIV acquired through injecting drug use is estimated to represent 20% of the people living with HIV in 2009 [2]. Efforts to make injecting drug use safer through provision of clean needles have proven efficacy in lowering transmission through this route. UNAIDS reports in 2009 that in Eastern Europe and Central Asia where injecting drug use is a major driver of the HIV epidemic, five of nine countries report that more than 80% of people injecting drugs used sterile injecting equipment at their last injection. Eight of twelve countries in South and Southeast Asia reported the same [2]. However, the various governmental responses across the globe have often prevented the widespread adoption and support of such programs [65]. While provision of clean needles is inexpensive, it is believed that many programs provide inadequate numbers of clean needles to insure optimal prevention [66]. In addition, the substance abuse treatment programs, such as methadone programs, may help curtail the intravenous drug–using behavior and therefore eliminate the risk of the transmission [64].

Prevention methods directed at commercial sex work are needed worldwide and again represent a targeted high-risk population that is disenfranchised and difficult to reach through conventional health system structures. The prevention methods focus on HIV counseling and testing, safe sex through proper condom use, and concomitant diagnosis and treatment of sexually transmitted infections (STIs), which are known to facilitate HIV transmission [71]. A long-standing program for sex workers in Southern India, called Avahan, provided a combined prevention approach of community outreach, empowerment, condom programming, and STI and HIV testing services over an 8-year period with a decrease in HIV prevalence from 20% to 16% in these high-risk women [71, 72].

In many parts of the resource-limited countries, heterosexual transmission is the major risk factor for HIV acquisition. Testing and access to HIV care is again critical to these populations along with appropriate counseling on behavior change, proper condom use, and STI diagnosis and treatment. Proven prevention measures for this group would include abstinence, mutual monogamy, reduced number of sexual partners, and consistent condom use. For young people, abstinence and older age of sexual debut are considered part of prevention messages [69].

Since 2000, a number of epidemiologic surveys and studies considered the possible benefit of male circumcision (surgical removal of the foreskin) in decreasing HIV transmission [73]. The penile foreskin is thought to be sensitive to epithelial tears, which carry a high density of CD4-bearing cells as targets for virus infection and subsequent replication [74]. In addition, uncircumcised men are at higher risk for STIs, which would facilitate HIV transmission [75–77]. In 2005–2006, randomized clinical trials conducted in South Africa, Kenya, and Uganda demonstrated a 55–76% lower HIV incidence compared to men still on the wait list of circumcision [78–80]. It is still unknown whether male circumcision will decrease male-to-female transmission of the virus, but these studies are underway [80]. There are other health benefits to male circumcision, but the surgical procedure is not without some risk, although serious complications are rare [76]. Nonetheless, for populations where men or boys are not circumcised, such as much of Southern and Eastern Africa, this becomes an important prevention tool [81].

Mother-to-child transmission of HIV is in large part responsible for the 2.5 million children living with HIV worldwide. Prevention of mother-to-child transmission (PMTCT) consists of identification of HIV infection in pregnant women and provision of a short course of antiretroviral drugs in a prophylactic manner (rather than treatment of the mother for her own disease to lower the viral burden in the mother and to decrease the risk of transmission to the baby). In resource-rich countries, most HIV-infected pregnant women would receive full ART (three drugs) as a means of preventing mother-to-child transmission; however, in most resource-limited settings, mono- or bi-therapy ART has been used due to cost and international recommendations at the time [82, 83]. In 2010, WHO revised their PMTCT guidelines based on clinical trial data from Africa, indicating that full ART could virtually eliminate mother-to-child transmission [84]. In settings where breast-feeding is judged to be the safest infant feeding option, ART prophylaxis continued to the mother and baby during this period could severely limit this mode of transmission [82–84]. While access to any PMTCT services remains a problem in a number of resource-limited countries and settings, only 54% access for sub-Saharan Africa, the potential for virtual elimination of mother-to-child transmission as these new guidelines are implemented becomes a real possibility [2, 82].

Postexposure prophylaxis (PEP) was the terminology frequently used to describe the use of short-course ART drugs given after an occupational setting exposure such as needle sticks with potentially HIV-infected blood. Similarly, PEP has been used in rape cases in settings where HIV infection is high and the potential for sexual transmission can be potentially decreased [85, 86]. Preexposure prophylaxis (PrEP) is the use of an ART drug administered to an HIV-negative individual prior to or at the time of the potential transmission event in an effort to decrease the viral inoculum and limit the potential for infection.

A recent study reported from South Africa this past year described the use of tenofovir-based gel as a microbicide to be used once 12 h before sex and again 12 h after sex as both safe and effective in reducing HIV transmission and STIs [87–89]. PrEP has also been successful in decreasing transmission in men who have sex with men and in heterosexual discordant couple studies [90, 91].

This paves the way for additional biomedical means of preventing infection, which can be coupled with other prevention methods depending on the setting and populations at risks [92, 93]. Further trials are underway to test this new and promising strategy to prevent HIV transmission.

Genetic Diversity of HIV

Early in the history of HIV research, it was recognized that each viral isolate varied from another in its nucleotide sequence, and multiple mechanisms for this genetic variation were considered [42, 94]. Part of the HIV replication cycle requires the reverse transcription of viral RNA by the virus' reverse transcriptase. This enzyme is error-prone, thereby generating approximately 1 error per 10^4 nucleotides, or viral genome. Estimates of HIV replication in people have suggested 10^{10}–10^{12} new viral particles produced per day [95, 96]. Thus, the potential for viral variation is significant even if the immune system, tissue compartmentalization, or antiretroviral therapy does not select for particular viral variants. Genetic variability between viral isolates from different patients [97, 98], frequently termed interisolate or interpatient variability, can vary by up to 5%. And in turn, this was also distinguishable from the genetic variation that was seen at the level of an individual patient, often termed intrapatient variability and variability up to 1% [99]. At the level of the individual patient, a swarm or quasispecies of highly related but distinguishable viral variants has been demonstrated throughout the course of HIV infection [100–102]. Thus, the genetic variation of HIV is hierarchical as depicted in the simplified schematic of subtype variation, interpatient variation, and intrapatient variation (Fig. 3.4).

Phylogenetic analyses of viruses from various geographic regions were used to identify three distinct groups M, N, and O within HIV-1 [34, 103, 104]. Most HIV-1 sequences belong to group M (major). A divergent subset of viruses identified in Cameroon in 1994, which did not cluster with group M viruses, were classified as group O (outlier) [105], and in 1998, another set of viruses which did not cluster with group M or O viruses were termed group N viruses [106]. A recent virus characterized from a gorilla has been designated group P [107]. It is thought that all virus groups were introduced by independent SIVcpz or SIVgor transmissions into the human population in the early part of the twentieth century [34, 104, 108].

The fruits of international HIV research have provided information of HIV viruses from different parts of the globe [109] (Fig. 3.5). The HIV-1 group M has been subdivided into genetic subtypes, defined originally on comparison of gag and env sequences of the virus and then later with full-length sequencing of the virus [110, 111]. Within a subtype, the average env genetic variation ranges from 5% to 15%, whereas the average env genetic variation between subtypes ranges from 20% to 30% [112, 113]. Remarkably, all the HIV-1s from the USA and Western Europe have been of a single subtype, B. Most of the diverse subtypes of HIV-1 including

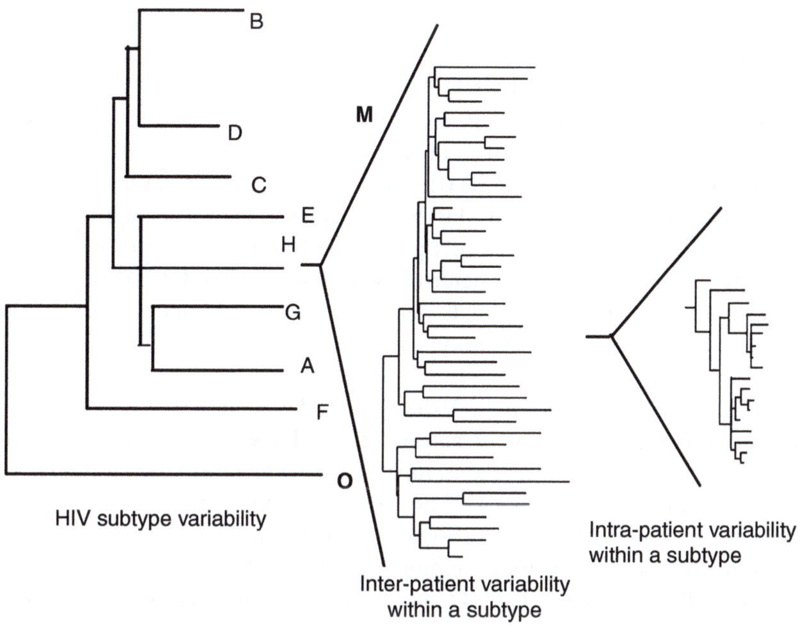

Fig. 3.4 Phylogenetic trees depict differences between HIV subtypes, variation between viruses from different patients, interpatient variability, and variation between viruses within the same patient, intrapatient variability

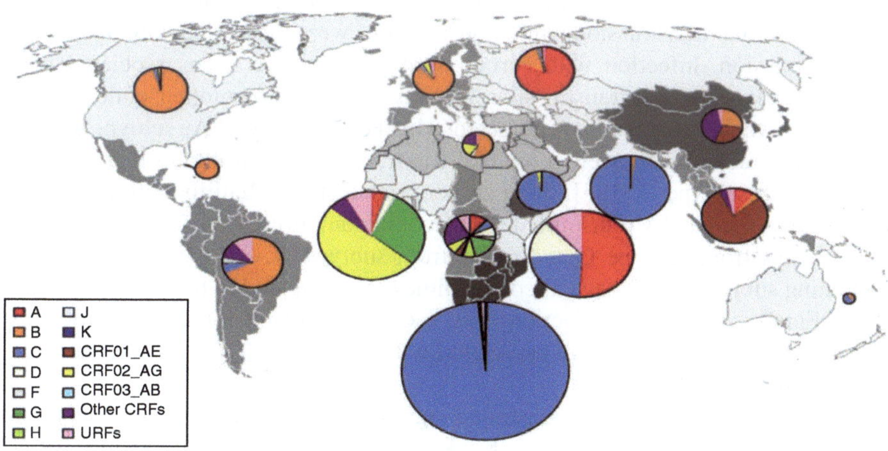

Fig. 3.5 Regional distribution of HIV-1 subtypes and recombinant viruses, 2004–2007. Fifteen regions of the world are depicted with *pie charts* representing the distribution of HIV-1 subtypes and recombinants from 2004 to 2007. The size of the pie charts corresponds to the relative numbers of people living with HIV in the region. *CRF* circulating recombinant form, *URF*, unique recombinant form (Adapted from [109])

circulating recombinant forms (CRFs) and unique recombinant forms (URFS) have been found in sub-Saharan Africa and, to a lesser degree, Asia, now accounting for 20% of the worldwide HIV viruses [109].

In Thailand, HIV-1 subtype B was detected in IV drug users during the mid-1980s. During the late 1980s, subtype E was first detected. By the early to mid-1990s, HIV-1 subtype E (CRF01_AE) had spread very rapidly throughout heterosexuals in Thailand, with the highest rates in the northern regions of the country [114]. Although apparently present earlier in the region, HIV-1 subtype B never spread to cause a major heterosexual epidemic, as did HIV-1 subtype E. Full-length sequencing of subtype E determined that it is actually a circulating recombinant form, consisting of subtypes A and E, now referred to as CRF01_AE [115, 116].

A similar situation occurred in India, with HIV-1 subtypes B and C. While B appeared to be introduced earlier and it expanded among intravenous drug users, this subtype did not appear to spread as rapidly among heterosexuals as did HIV-1 C. Previously associated with the massive heterosexual epidemic in southeastern Africa, subtype C also caused a rapid heterosexual epidemic in Western India, apparently initially spreading from the Bombay region [114, 117]. The results in Africa and Asia suggest that HIV-1 subtypes A, C, D, and CRF01_AE are well adapted for heterosexual transmission compared to subtype B, perhaps due to differential ability to infect dendritic cells in the sexual mucosa [118].

Various studies have shown associations between viral genotype and biology [119]. There is some evidence suggesting a relationship between subtype and modes of transmission. Studies in Cape Town [120], Finland [121], Thailand [122, 123], and Australia [124] found that most subtype B strains were associated with homosexual transmission while non-B strains were associated with heterosexual transmission. Infection with certain subtypes has also been associated with increased risk of vertical transmission. A study conducted on mother-child pairs in Tanzania revealed that mothers infected with HIV-1 subtype A, subtype C, and intersubtype recombinant were more likely to transmit to their infants than mothers infected with subtype D [125]. In an earlier related study, Renjifo et al. [126] also found that in perinatally transmitted C/D recombinant viruses, the V3 regions (*env*) were always from subtype C and never from subtype D, suggesting that viruses containing subtype D-V3 have a reduced fitness as compared to those with subtype C-V3. Finally, a study of injection drug users in Thailand found a significantly higher transmission probability associated with subtype E (CRF_01 AE) as compared to subtype B [127].

Some studies have also demonstrated significant differences between subtypes with regards to disease progression. Kanki et al. [128] found that women infected with a non-A subtype were eight times more likely to develop AIDS than were those infected with subtype A. Similarly, Kaleebu et al. [129, 130] reported that subjects with subtype A had a slower progression to disease than those with subtype D. Clinical and immunological differences have also been found between subtypes. In Kenya, where subtypes A, C, and D were all cocirculating within the same population, Neilson et al. [131] found that high plasma RNA levels and low CD4 counts were significantly associated with subtype C infection. In a prospective study conducted at a methadone treatment clinic in Thailand, people infected with CRF01_AE were found to have higher viral loads in early infection than those infected with subtype B [132]. However, this difference decreased over time such that the viral loads were similar at 12, 18, and 24 months postseroconversion [132]. Similarly, a study in our laboratory indicated that women infected with CRF02_AG had a significantly higher viral load during the early stage of infection than women not infected with CRF02_AG [133]. Infection with multiple subtypes has also been associated with higher viral load and lower CD4+ T cell counts [134].

Complete sequence analysis of viruses from around the world has further characterized an HIV pandemic that is increasing in complexity with a higher prevalence of recombinant viruses, currently at 20% of all HIV viruses [109]. In West Africa, Asia, and South America, the predominant virus is a circulating recombinant form. The predominant virus of West Africa is CRF02_AG, which appears to represent a recombinant event that occurred early in the divergence of subtypes A and G. In contrast, the BC and BF recombinants of China and South America, respectively, are derived from more recent recombinants between contemporary parental virus lineages [135]. An understanding of the genetic diversity of predominant HIV-1 subtypes, sub-subtypes, and circulating recombinant forms in a given population may be important in designing effective HIV vaccines [136, 137]. Although the importance of matching a vaccine candidate to regional circulating strains is yet unclear, incorporation of local strains might maximize the efficacy of a potential vaccine candidate [137].

Treatment

Antiretroviral therapy (ART) drugs that block or reduce HIV virus replication in the infected patient are used for the treatment of AIDS. The first ART drug was put into use in 1986–1987: zidovudine (AZT), a nucleoside reverse transcriptase inhibitor (NRTI) that because of its chemical similarity to the host's nucleoside, inhibited the growing DNA chain and served as a chain terminator such that virus replication could not occur. Other NRTIs act in a similar manner, but another class of ART drugs includes nonnucleoside reverse transcriptase inhibitors (NNRTI), which bind to the virus' reverse transcriptase enzyme in a specific position and manner, blocking the enzyme from completing the transcription. NRTI and NNRTIs are often used in combination as the typical first-line regimens for patients with AIDS [138–140]. Since AZT, there has been tremendous progress in developing improved and more efficacious ART drugs and combination regimens for treating AIDS patients in high-income regions of the world; however, the cost of many of these newer drugs has been prohibitive for most middle- and low-income countries despite their disproportionate share of the AIDS burden.

Currently, there are a total of five classes of FDA-approved ART drugs that act on different parts of the HIV virus life cycle: NRTIs and NNRTIs act on the reverse transcription; other drugs act by inhibiting the viral protease, the viral integrase, or blocking virus entry at the level of the coreceptor. ART is typically prescribed in combination with current recommendations requiring at least two active classes of drugs [138–140]. Side effects and toxicities for each of these drugs may be serious and result in significant complications, requiring the patient to switch drug regimens.

The error-prone reverse transcription of HIV replication and its viral dynamics suggest that every possible single point mutation in the HIV genome occurs between 10^4 and 10^5 times per day in an untreated HIV-1-infected individual [141]. Therefore, the propensity for development of drug resistance mutations is high and the reason that strict adherence to taking the prescribed drugs is necessary to maintain viral suppression. In an effort to improve adherence and decrease pill burden to the patients, fixed-dose combination pills are available for most of the typical first-line ART regimens, allowing all drugs to be included in a single pill.

By the late 1990s, treatment for AIDS was widely available in high-income countries, but the vast majority of patients in middle- and low-income countries could not access these expensive drugs. In 2000, at the United Nations Millennium Summit, Kofi Annan declared a call for action to address the gap in treatment and to promote prevention and research in response to the AIDS epidemic [142]. In 2001, the UN General Assembly adopted a Declaration of Commitment on HIV/AIDS, endorsing equitable access to care and treatment as a fundamental component of a comprehensive and effective global HIV response [143]. UNAIDS had reported at the end of 2002 that developing countries had ~300,000 AIDS patients receiving ART with an unmet need estimated at 5.5 million [144]. The international donor community moved toward addressing the goals of equitable access through initiation of large-scale treatment and care programs. The President's Emergency Plan for AIDS Relief (PEPFAR), a program launched in late 2003, authorized over $15 billion dollars in 15 target countries, the largest commitment to a single disease in the history of US government support [145]. Other programs such as the Global Fund to Fight AIDS, Tuberculosis and Malaria [146], the Bill & Melinda Gates Foundation [147], Clinton Foundation [148], World Bank [149], and Medecins Sans Frontieres [150], to name a few, supported this effort with advocacy to country governments, and the rapid scale-up of providing ART to patients in need moved from rhetoric to action.

UNAIDS reported at the end of 2009 that 5.2 million people in low- and middle-income countries received ART, a significant increase from 2008 and unimaginable in 2003 [2]. However, the pandemic has continued, and in these countries, this accomplishment only represents 36% of the 15 million AIDS patients in these countries in need. Children and patients coinfected with TB remain an ART priority in these settings. Many of the original ART drugs developed in the late 1980s went off patent and have now been manufactured in generic form and meet international standards [151]. The Clinton Foundation in mid-2000 helped negotiate important drug cost reductions so that first-line drug per patient approximated $350 per year, rather than the $1,000 per month of the brand name drugs in 2000 [148]. However, the improvement in ART therapy in high-income countries was made possible through newer drugs and new classes of drugs like integrase inhibitors, or entry blockers, which are still cost-prohibited in most middle- and low-income country ART programs. Thus, although patients are receiving ART, the pharmacy of drugs available for toxicities or second- and third-line therapy remains limited in most low-income countries.

ART is not just the provision of pills; the diagnosis of AIDS requires laboratory tests and the clinical management of an AIDS patient entails regular monitoring of immune function, diagnosis, and treatment of comorbidities and measurement of viral load (virus levels in the bloodstream) when possible [142]. These laboratory tests often require sophisticated lab equipment and costs per test that can be quite high, almost the same cost of annual drug costs. In many low-income settings, these diagnostic or monitoring tests were not even available; in some instances where patients were required to pay for clinical tests, the costs were prohibitive to the typical AIDS patient that already had significant economic burdens. In recent years, a major contribution of large-scale care and treatment support from PEPFAR and others has built the infrastructure and provided the training to perform these critical laboratory tests. Provision of laboratory equipment, insuring adequate clean water and electricity to run the equipment and laboratory test specific requirements for cold chain, and short expiration remain challenges for both start-up and sustainability. In some instances, the costs of the tests have been reduced, but there is a critical need for cheaper, efficient, and point-of-care tests that will increase their availability and use for patient management.

Further, the complexity of clinical management and care of HIV infection requires training, infrastructure building, and overall strengthening of the existing health-care system in most developing country settings. Again, the leadership in individual countries coupled with support from the international donor communities has provided the training necessary for the complex administration of HIV care and treatment. However, human resources in many countries are limiting, and further efforts to encourage task-shifting within the medical team and bolstering of the medical education system are important to achieving the provision of ART to the many patients that are still in need.

Vaccine

Since the discovery of the virus and its initial characterization in the early 1980s, the search for an effective vaccine has been the subject of intense research. HIV infection can occur via free virus and/or virus-infected cells. Once infection has occurred with integration of viral genetic material in the target host cell, elimination of the virus by the host's immune mechanisms is not possible. The approach to vaccine design has often sought to mimic the protective immune responses, often identified in animal model systems or in individuals that have demonstrated protection [152, 153]. To date, the correlates of protective immunity from HIV have yet to be identified. Vaccine scientists therefore began with more conventional approaches of seeking to design vaccine candidates that would elicit strong virus-neutralizing antibodies or robust cell-mediated responses or both [154, 155]. The proof of principle has been shown for both of these approaches in nonhuman primate model [156, 157]. However, the model has not been predictive of HIV vaccine efficacy in humans; differences in routes of exposures, strains, and amount of virus are just some of the complexities of this model. Dozens of vaccine candidates have been developed, tested in nonhuman primate models and even early clinical trials in humans [157, 158].

AIDS vaccine candidates are evaluated in a stepwise manner in a series of clinical trials known as phases I, II, and III. Phase I and II trials generally involve a small number of volunteers and provide researchers with critical information about the safety and immunogenicity of the vaccine. The cost of phase III trials can be in hundreds of millions of dollars, and the enrollment for the trial is large, so the AIDS vaccine community moved to an intermediate "test of concept" trial where lower numbers of trial participants were needed; the statistical power of the trial design would allow evaluation of immunogenicity and low ranges of efficacy. It is not until phase III trials that the efficacy of the vaccine is truly assessed.

Candidate HIV vaccines that were tested in phase I and II trials over the past 20 years included inactivated whole virus, or virus particles, subunit HIV envelope vaccines delivered in novel adjuvant or delivery systems, DNA vaccines and live recombinant vector vaccines, and various combination approaches. The results of phase I and II trials indicated that they were safe but did not generate sufficient immune responses to warrant proceeding to phase III trials [157, 159–163].

By 2008, only four candidate HIV vaccines had proceeded to phase IIb and III efficacy trials. Two of the trials were subunit gp120 (envelope) vaccines tested by VaxGen in the USA, Canada, and the Netherlands with 5,400 volunteers, largely men who have sex with men [164]. A second trial was conducted in 2,545 intravenous drug users in Thailand, using recombinant envelope from subtype B and also CRF01_AE [165, 166]. Although neutralizing antibodies were induced by the vaccine, neither trial showed evidence of protection. Subsequently, two trials assessed vaccine candidates based on the premise that cellular immune responses would be protective; the viral antigens were delivered in an adenovirus 5 vector; in order to enhance the generation of the cellular responses, the candidates were tested in the "Step" and "Phambili" trials conducted by Merck [167–169]. Adenovirus 5 is considered a harmless virus, but it does infect humans, and many populations mount antibodies to this virus, frequently without any signs of disease. Some of the volunteers in the Step and Phambili trials did not have antibodies to adenovirus 5, and in those volunteers, the vaccine showed no effect. However, in volunteers with antibodies to adenovirus 5, the vaccinated group showed higher rates of HIV infection, and thus, the vaccine had increased rather than decreased the chances of HIV infection [167–170]. Both trials were interrupted as a result of these unexpected results.

In 2008, a Summit on HIV Vaccine Research and Development determined that research toward a vaccine needed to place greater emphasis on basic research and slowed the pace of candidates, moving toward expensive clinical trials in people [171, 172].

In late 2009, the results of the US military vaccine group's RV144 trial in Thailand were announced. A phase IIb trial was conducted with a recombinant vector approach using the ALVAC virus (an avian poxvirus) used to express HIV proteins to prime the immune response combined with recombinant gp120 protein; both of these entities alone had been tested in people and shown poor immunogenicity [173]. The RV144 trial conducted in 16,400 men and women in Thailand starting in 2003, nonetheless, showed a modest efficacy of 31% protection after 3.5 years of follow-up; there were 51 infected in the vaccine group compared with 74 in similarly sized placebo group [174]. Further analysis of those protected failed to show significant neutralizing antibodies or even CD8+ T cell responses although more had favorable CD4+ T cell responses. Thus, the trial, while giving some optimism for vaccines, has failed to clearly indicate what specific type of immunity was responsible for the protection seen; further analysis will continue to evaluate the trial in hopes of guiding more improved candidates for the future [175].

While the search for a vaccine continues, the research thus far has given not only the scientific community a deeper understanding of both the HIV virus but also the human immune system and its interaction with this unique virus [157]. This will not only inform vaccine efforts for other infectious diseases but also facilitate the use of novel immunomodulatory entities and delivery systems that have been developed and tested. The development of an effective HIV vaccine will be used in conjunction with the many established methods of prevention that have been described, and new modalities such as male circumcision and ART treatment as prevention, which show great promise for additional preventative benefits [176]. Advances in these areas have now merged the interventions of treatment, vaccine, and prevention, to form a future strategy of combined prevention [69].

Future Directions

Thirty years ago, a mysterious disease affecting men who have sex with men in the USA initiated the world's appreciation of a new human virus with disease-causing potential and global spread that was unprecedented. Since that time, it is estimated that 60 million people have been infected with HIV responsible for over 25 million deaths. The history of HIV's discovery is instructive to our overall outlook on infectious diseases in general. HIV originated in closely related viruses of primates with multiple variants that still resemble close relationship to the viruses of great apes. The virus and its many genetic variants showed distinct differences in its epidemiology based on geographic location, risk groups affected, spread, and disease. In addition, the increasing complexity and higher prevalence of recombinant forms of the HIV epidemic suggests that more vigilant surveillance with improved technologies will be important. The marked dichotomy between the epidemics of high-income versus middle- and low-income countries has challenged public health efforts to address prevention and treatment needs. ART drugs and laboratory tests developed in high-income countries are high in cost and complexity of use; this continues to be a major obstacle to the largest number of people infected and suffering from HIV and AIDS. In the past decade, an international effort to provide equal access to prevention, care, and treatment has demonstrated that the gap in access can be narrowed. Newer and more efficacious regimens to prevent mother-to-child transmission have just been adopted in the 2010 WHO guidelines and hold the potential of eliminating transmission as the guidelines are rolled out [84]. WHO has also revised the 2010 ART guidelines, which will initiate treatment for patients with less severe disease, CD4+ cell counts below 350, rather than 200 cells per cubic milliliter of blood [177]. While the increased costs and actual implementation of these new guidelines will pose an obstacle for many poor countries of the world, the ultimate realization of these guidelines will further decrease the burden of people living with HIV and increase the longevity of those already on treatment.

The historic and unprecedented support and efforts from the international donor community have made major progress in scaling up prevention and treatment of HIV in areas of the world with the most need. They have begun the difficult task of encouraging national leadership, building up broken or nonexistent health systems, in order to address the pandemic, but the burden of disease is still significant, and efforts must be continued with a perspective of long-term sustainability.

Bibliography

1. UNAIDS (2007) AIDS epidemic update. http://data.unaids.org/pub/epislides/2007/2007_epiupdate_en.pdf
2. UNAIDS (2009) AIDS epidemic update. http://data.unaids.org/pub/Report/2009/JCI700/Epi Update 2009 en.pdf
3. Gottlieb MS, Schroff R, Schanker HM, Weisman JD, Fan PT, Wolf RA, Saxon A (1981) Pneumocystis carinii pneumonia and mucosal candidiasis in previously healthy homosexual men: evidence of a new acquired cellular immunodeficiency. N Engl J Med 305:1425–1431
4. Masur H, Michelis MA, Greene JB, Onorato I, Stouwe RA, Holzman RS, Brettman L, Lange M, Murray HW, Cunningham-Rundles S (1981) An outbreak of community-acquired *Pneumocystis carinii* pneumonia: initial manifestation of cellular immune dysfunction. New Eng J Med 305:1431–1438
5. Siegal FP, Lopez C, Hammer GS, Brown AE, Kornfeld SJ, Gold J, Hassett J, Hirschman SZ, Cunningham-Rundles C, Adelsberg BR et al (1981) Severe acquired immunodeficiency in male homosexuals, manifested by chronic perianal ulcerative herpes simplex lesions. New Eng J Med 305:1439–1444

6. Anonymous (1982) Epidemiologic aspects of the current outbreak of Kaposi's sarcoma and opportunistic infections. New Eng J Med 306:248–252
7. Davis KC, Horsburgh CR Jr, Hasiba U, Schocket AL, Kirkpatrick CH (1983) Acquired immunodeficiency syndrome in a patient with hemophilia. Ann Int Med 98:284–286
8. Poon MC, Landay A, Prasthofer EF, Stagno S (1983) Acquired immunodeficiency syndrome with *Pneumocystis carinii* pneumonia and Mycobacterium avium-intracellulare infection in a previously healthy patient with classic hemophilia. Clinical, immunologic, and virologic findings. Ann Int Med 98:287–290
9. Elliott JL, Hoppes WL, Platt MS, Thomas JG, Patel IP, Gansar A (1983) The acquired immunodeficiency syndrome and Mycobacterium avium-intracellulare bacteremia in a patient with hemophilia. Ann Int Med 98:290–293
10. Curran JW, Lawrence DN, Jaffe H, Kaplan JE, Zyla LD, Chamberland M, Weinstein R, Lui KJ, Schonberger LB, Spira TJ et al (1984) Acquired immunodeficiency syndrome (AIDS) associated with transfusions. New Eng J Med 310:69–75
11. Jaffe HW, Francis DP, McLane MF, Cabradilla C, Curran JW, Kilbourne BW, Lawrence DN, Haverkos HW, Spira TJ, Dodd RY, Gold J, Armstrong D, Ley A, Groopman J, Mullins JI, Lee TH, Essex M (1984) Transfusion-associated AIDS: serologic evidence of human T-cell leukemia virus infection of donors. Science 223:1309–1312
12. Piot P, Quinn TC, Taelman H, Feinsod FM, Minlangu KB, Wobin O, Mbendi N, Mazebo P, Ndangi K, Stevens W (1984) Acquired immunodeficiency syndrome in a heterosexual population in Zaire. Lancet 2:65–69
13. Van de Perre P, Rouvroy D, Lepage P, Bogaerts J, Kestelyn P, Kayihigi J, Hekker AC, Butzler JP, Clumeck N (1984) Acquired immunodeficiency syndrome in Rwanda. Lancet 2:62–65
14. Clumeck N, Mascart-Lemone F, de Maubeuge J, Brenez D, Marcelis L (1983) Acquired immune deficiency syndrome in Black Africans (letter). Lancet 1:642
15. Oleske J, Minnefor A, Cooper R et al (1983) Immune deficiency syndrome in children. J Am Med Assoc 249:2345–2349
16. Scott GB, Buck BE, Leterman JG et al (1984) Acquired immunodeficiency syndrome in infants. New Eng J Med 310:76–81
17. Francis DP, Curran JW, Essex M (1983) Epidemic acquired immune deficiency syndrome (AIDS): epidemiologic evidence for a transmitted agent. J Natl Cancer Inst 71:1–6
18. Essex M, McLane MF, Lee TH, Falk L, Howe CWS, Mullins J, Cabradilla C, Francis DP (1983) Antibodies to cell membrane antigens associated with human T-cell leukemia virus in patients with AIDS. Science 220:859–862
19. Essex M, McLane MF, Lee TH, Tachibana N, Mullins JI, Kreiss J, Kasper CK, Poon M-C, Landay A, Stein SF, Francis DP, Cabradilla C, Lawrence DN, Evatt BL (1983) Antibodies to human T-cell leukemia virus membrane antigens (HTLV-MA) in hemophiliacs. Science 221:1061–1064
20. Gelmann EP, Franchini G, Manzari V, Wong-Staal F, Gallo RC (1984) Molecular cloning of a unique human T-cell leukemia virus (HTLV-IIMo). Proc Natl Acad Sci USA 81:993–997
21. Gallo RC, Sarin PS, Gelmann EP, Robert-Guroff M, Richardson E, Kalyanaraman VS, Mann D, Sidhu GD, Stahl RE, Zolla-Pazner S, Leibowitch J, Popovic M (1983) Isolation of human T-cell leukemia virus in acquired immune deficiency syndrome (AIDS). Science 220:865–867
22. Barre-Sinoussi F, Chermann J-C, Rey F, Nugeyre MT, Chamaret S, Gruest J, Dauguet C, Axler-Blin C, Vezinet-Brun F, Rouzioux C, Rozenbaum W, Montagnier L (1983) Isolation of T-lymphotropic retrovirus from a patient at risk for acquired immune deficiency syndrome (AIDS). Science 220:868–870
23. Poiesz BJ, Ruscetti FW, Reitz MS, Kalyanaraman VS, Gallo RC (1981) Isolation of a new type C retrovirus (HTLV) in primary uncultured cells of a patient with Sezary T-cell leukaemia. Nature 294:268–271

24. Ammann AJ, Abrams D, Conant M, Chudwin D, Cowan M, Volberding P, Lewis B, Casavant C (1983) Acquired immune dysfunction in homosexual men: immunologic profiles. Clin Immunol Immunopathol 27:315–325
25. Fahey JL, Prince H, Weaver M, Groopman J, Visscher B, Schwartz K, Detels R (1984) Quantitative changes in T helper or T suppressor/cytotoxic lymphocyte subsets that distinguish acquired immune deficiency syndrome from other immune subset disorders. Amer J Med 6:95–100
26. Lane HC, Masur H, Gelmann EP, Longo DL, Steis RG, Chused T, Whalen G, Edgar LC, Fauci AS (1985) Correlation between immunologic function and clinical subpopulations of patients with the acquired immune deficiency syndrome. Am J Med 78:417–422
27. Kalyanaraman VS, Sarngadharan MG, Robert-Guroff M, Miyoshi I, Golde D, Gallo RC (1982) A new subtype of human T-cell leukemia virus (HTLV-II) associated with a T-cell variant of hairy cell leukemia. Science 218:571–573
28. Gelmann EP, Popovic M, Blayney D, Masur H, Sidhu G, Stahl RE, Gallo RC (1983) Proviral DNA of a retrovirus, human T-cell leukemia virus, in two patients with AIDS. Science 220:862–865
29. Popovic M, Sarin PS, Robert GM, Kalyanaraman VS, Mann D, Minowada J, Gallo RC (1983) Isolation and transmission of human retrovirus (human T-cell leukemia virus). Science 219:856–859
30. Schupbach J, Popovic M, Gilden RV, Gonda MA, Sarngadharan MG, Gallo RC (1984) Serological analysis of a subgroup of human T-lymphotropic retroviruses (HTLV-III) associated with AIDS. Science 224:503–505
31. Sarngadharan MG, Popovic M, Bruch L, Schupbach J, Gallo RC (1984) Antibodies reactive with human T-lymphotropic retroviruses (HTLV-III) in the serum of patients with AIDS. Science 224:506–508
32. Barin F, M'Boup S, Denis F, Kanki P, Allan JS, Lee TH, Essex M (1985) Serological evidence for virus related to simian T-lymphotropic retrovirus III in residents of west Africa. Lancet 2:1387–1389
33. Kanki P, Barin F, Mboup S, Allan JS, Romet-Lemonne JL, Marlink R, McLane MF, Lee TH, Arbeille B, Denis F, Essex M (1986) New human T-Lymphotropic retrovirus related to simian T-lymphotropic virus type III (STLV-III_{AGM}). Science 232:238–243
34. Gao F, Bailes E, Robertson DL, Chen Y, Rodenburg CM, Michael SF, Cummins LB, Arthur LO, Peeters M, Shaw GM, Sharp PM, Hahn BH (1999) Origin of HIV-1 in the chimpanzee Pan troglodytes troglodytes (see comments). Nature 397:436–441
35. Hirsch VM, Olmsted RA, Murphey Corb M, Purcell RH, Johnson PR (1989) An African primate lentivirus (SIVsm) closely related to HIV-2. Nature 339:389–392
36. Marlink R, Kanki P, Thior I, Travers K, Eisen G, Siby T, Traore I, Hsieh SS, Dia MC, Gueye EH, Hellinger J, Gueye NA, Sankalé JL, Ndoye I, Mboup S, Essex M (1994) Reduced rate of disease development with HIV-2 compared to HIV-1. Science 265:1587–1590
37. Kanki PJ, De Cock KM (1994) Epidemiology and natural history of HIV-2. AIDS 8:S1–S9
38. Kanki P, Peeters M, Gueye-NDiaye A (1997) Virology of HIV-1 and HIV-2. AIDS 2(Suppl B):S33–S42
39. Kanki P (1989) HIV-2 infection in West Africa. In: Volberding P, Jacobson M (eds) 1988 AIDS clinical reviews. Marcel Dekker, New York, pp 95–108
40. Kanki P, Meloni S (2009) Biology and variation in HIV-2 and HIV-1. In: Marlink RG, Teitelman S (eds) From the ground up: building comprehensive HIV/AIDS care programs in resource-limited settings. Elizabeth Glaser Pediatric AIDS Foundation, Washington, DC, pp 1–24
41. Katzourakis A, Tristem M, Pybus OG, Gifford RJ (2007) Discovery and analysis of the first endogenous lentivirus. Proc Natl Acad Sci USA 104:6261–6265
42. Wong-Staal F (1990) Human immunodeficiency viruses and their replication. In: Fields DM et al (eds) Virology. Raven, New York, pp 1529–1540

43. NIAID (2011) HIV Replication cycle. http://www.niaid.nih.gov/topics/HIVAIDS/Understanding/Biology/Pages/hivReplicationCycle.aspx
44. Jiang S (1997) HIV-1–co-receptors binding (letter; comment) (published erratum appears in Nat Med 1997 Aug; 3(8):817). Nat Med 3:367–368
45. Feng Y, Broder CC, Kennedy PE, Berger EA (1996) HIV-1 entry cofactor: functional cDNA cloning of a seven-transmembrane G protein-coupled receptor. Science 272: 872–877
46. Deng H, Liu R, Ellmeier W, Choe S, Unutmaz D, Burkhart M, Di Marzio P, Marmon S, Sutton RE, Hill CM, Davis CB, Peiper SC, Schall TJ, Littman DR, Landau NR (1996) Identification of a major co-receptor for primary isolates of HIV-1. Nature 381:661–666
47. Cocchi F, DeVico AL, Garzino-Demo A, Cara A, Gallo RC, Lusso P (1996) The V3 domain of the HIV-1 gp120 envelope glycoprotein is critical for chemokine-mediated blockade of infection. Nat Med 2:1244–1247
48. Dittmar M, McNight A, Simmons G, Clapham P, Weiss R (1997) HIV-1 tropism and co-receptor use. Nature 385:495–496
49. Moore PS, Boshoff C, Weiss RA, Chang Y (1996) Molecular mimicry of human cytokine and cytokine response pathway genes by KSHV. Science 274:1739–1744
50. Cooper DA, Imrie AA, Penny R (1987) Antibody response to human immunodeficiency virus after primary infection. J Infect Dis 155:1113–1118
51. Loes S, de Saussure P, Saurat JH, Stalder H, Hirschel B, Perrin LH (1993) Symptomatic primary infection due to human immunodeficiency virus type 1: review of 31 Cases. Clin Infect Dis 17:59–65
52. Fauci AS, Pantaleo G, Stanley S, Weissman D (1996) Immunopathogenic mechanisms of HIV infection. Ann Intern Med 124:654–663
53. Pantaleo G, Graziosi C, Demarest JF, Butini L, Montroni M, Fox CH, Orenstein JM, Kotler DP, Fauci AS (1993) HIV infection is active and progressive in lymphoid tissue during the clinically latent stage of disease. Nature 362:355–358
54. Cavarelli M, Scarlatti G (2011) Human immunodeficiency virus type 1 mother-to-child transmission and prevention: successes and controversies. J Intern Med 270(6):561–579
55. Bongaarts J, Sinding S (2011) Population policy in transition in the developing world. Science 333:574–576
56. Bongaarts J, Pelletier F, Gerland P (2010) How many more AIDS deaths? Lancet 375:103–104
57. Root R (2010) Situating experiences of HIV-related stigma in Swaziland. Glob Public Health 5:523–538
58. Wu Z, Sullivan SG, Wang Y, Rotheram-Borus MJ, Detels R (2007) Evolution of China's response to HIV/AIDS. Lancet 369:679–690
59. Wang L (2007) Overview of the HIV/AIDS epidemic, scientific research and government responses in China. AIDS 21(Suppl 8):S3–7
60. UNAIDS (2010) Global report Europe and Central Asia. http://www.unaids.org/en/regionscountries/regions/easterneuro peandcentralasia/
61. Guillard EM, Eustache L (2007) Estimation de la séroprevalence du VIH en Haiti en 2007 selon le milieu de residence urbain et rural pour chacun des 10 départments. United States Agency for International Development, Washington, DC
62. Anonymous (2008) US Virgin Islands and Caribbean HIV epidemic need more attention, researchers say. HIV infection rate is high among sex workers. AIDS Alert 23:42–44
63. Bezemer D, de Wolf F, Boerlijst MC, van Sighem A, Hollingsworth TD, Prins M, Geskus RB, Gras L, Coutinho RA, Fraser C (2008) A resurgent HIV-1 epidemic among men who have sex with men in the era of potent antiretroviral therapy. AIDS 22:1071–1077
64. Mathers BM, Degenhardt L, Ali H, Wiessing L, Hickman M, Mattick RP, Myers B, Ambekar A, Strathdee SA (2010) HIV prevention, treatment, and care services for people who inject drugs: a systematic review of global, regional, and national coverage. Lancet 375:1014–1028

65. Strathdee SA, Hallett TB, Bobrova N, Rhodes T, Booth R, Abdool R, Hankins CA (2010) HIV and risk environment for injecting drug users: the past, present, and future. Lancet 376:268–284
66. Stockman JK, Strathdee SA (2010) HIV among people who use drugs: a global perspective of populations at risk. J Acquir Immune Def Syn 55(Suppl 1):S17–22
67. de Felipe B, Perez-Romero P, Abad-Fernandez M, Fernandez-Cuenca F, Martinez-Fernandez FJ, Trastoy M, Mata Rdel C, Lopez-Cortes LF, Leal M, Viciana P, Vallejo A (2011) Prevalence and resistance mutations of non-B HIV-1 subtypes among immigrants in Southern Spain along the decade 2000–2010. Virol J 8:416
68. Ryan CE, Gare J, Crowe SM, Wilson K, Reeder JC, Oelrichs RB (2007) The heterosexual HIV type 1 epidemic in Papua New Guinea is dominated by subtype C. AIDS Res Hum Retroviruses 23:941–944
69. Centers for Disease Control and Prevention (2011) High-impact HIV prevention – CDC's approach to reducing HIV infections in the United States. CDC, Atlanta
70. Katoff L, Dunne R (1988) Supporting people with AIDS: the Gay Men's health crisis model. J Palliat Care 4:88–95
71. Laga M, Galavotti C, Sundararaman S, Moodie R (2010) The importance of sex-worker interventions: the case of Avahan in India. Sex Transm Infect 86(Suppl 1):i6–7
72. Verma R, Shekhar A, Khobragade S, Adhikary R, George B, Ramesh BM, Ranebennur V, Mondal S, Patra RK, Srinivasan S, Vijayaraman A, Paul SR, Bohidar N (2010) Scale-up and coverage of Avahan: a large-scale HIV-prevention programme among female sex workers and men who have sex with men in four Indian states. Sex Transm Infect 86(Suppl 1):i76–i82
73. Tobian AA, Gray RH (2011) The medical benefits of male circumcision. J Am Med Assoc 306:1479–1480
74. Kreiss JK, Hopkins SG (1993) The association between circumcision status and human immunodeficiency virus infection among homosexual men. J Infect Dis 168:1404–1408
75. Weiss HA, Quigley MA, Hayes RJ (2000) Male circumcision and risk of HIV infection in sub-Saharan Africa: a systematic review and meta-analysis. AIDS 14:2361–2370
76. Siegfried N, Muller M, Deeks JJ, Volmink J (2009) Male circumcision for prevention of heterosexual acquisition of HIV in men. Cochrane Database Syst Rev: CD003362
77. Weiss HA, Hankins CA, Dickson K (2009) Male circumcision and risk of HIV infection in women: a systematic review and meta-analysis. Lancet Infect Dis 9:669–677
78. Auvert B, Taljaard D, Lagarde E, Sobngwi-Tambekou J, Sitta R, Puren A (2005) Randomized, controlled intervention trial of male circumcision for reduction of HIV infection risk: the ANRS 1265 Trial. PLoS Med 2:e298
79. Gray RH, Kigozi G, Serwadda D, Makumbi F, Watya S, Nalugoda F, Kiwanuka N, Moulton LH, Chaudhary MA, Chen MZ, Sewankambo NK, Wabwire-Mangen F, Bacon MC, Williams CF, Opendi P, Reynolds SJ, Laeyendecker O, Quinn TC, Wawer MJ (2007) Male circumcision for HIV prevention in men in Rakai, Uganda: a randomised trial. Lancet 369:657–666
80. Wawer MJ, Makumbi F, Kigozi G, Serwadda D, Watya S, Nalugoda F, Buwembo D, Ssempijja V, Kiwanuka N, Moulton LH, Sewankambo NK, Reynolds SJ, Quinn TC, Opendi P, Iga B, Ridzon R, Laeyendecker O, Gray RH (2009) Circumcision in HIV-infected men and its effect on HIV transmission to female partners in Rakai, Uganda: a randomised controlled trial. Lancet 374:229–237
81. WHO/UNAIDS (2007) Announce recommendations about male circumcision as HIV prevention. Strategy should be employed with care. AIDS Alert 22:66–67
82. Shapiro RL, Hughes MD, Ogwu A, Kitch D, Lockman S, Moffat C, Makhema J, Moyo S, Thior I, McIntosh K, van Widenfelt E, Leidner J, Powis K, Asmelash A, Tumbare E, Zwerski S, Sharma U, Handelsman E, Mburu K, Jayeoba O, Moko E, Souda S, Lubega E, Akhtar M, Wester C, Tuomola R, Snowden W, Martinez-Tristani M, Mazhani L, Essex M (2010) Antiretroviral regimens in pregnancy and breast-feeding in Botswana. N Engl J Med 362:2282–2294
83. Chasela CS, Hudgens MG, Jamieson DJ, Kayira D, Hosseinipour MC, Kourtis AP, Martinson F, Tegha G, Knight RJ, Ahmed YI, Kamwendo DD, Hoffman IF, Ellington SR, Kacheche Z,

Soko A, Wiener JB, Fiscus SA, Kazembe P, Mofolo IA, Chigwenembe M, Sichali DS, van der Horst CM (2010) Maternal or infant antiretroviral drugs to reduce HIV-1 transmission. N Engl J Med 362:2271–2281

84. World Health Organization (2010) Antiretroviral drugs for treating pregnant woman and preventing HIV infections in infants, Recommendations for a public health approach. 2010 revision.
85. Youle M, Wainberg MA (2003) Pre-exposure chemoprophylaxis (PREP) as an HIV prevention strategy. J Int Assoc Physicians AIDS Care (Chic) 2:102–105
86. Young TN, Arens FJ, Kennedy GE, Laurie JW, Rutherford G (2007) Antiretroviral post-exposure prophylaxis (PEP) for occupational HIV exposure. Cochrane Database Syst Rev: CD002835
87. Karim QA, Kharsany AB, Frohlich JA, Baxter C, Yende N, Mansoor LE, Mlisana KP, Maarschalk S, Arulappan N, Grobler A, Sibeko S, Omar Z, Gengiah TN, Mlotshwa M, Samsunder N, Karim SS (2011) Recruitment of high risk women for HIV prevention trials: baseline HIV prevalence and sexual behavior in the CAPRISA 004 tenofovir gel trial. Trials 12:67
88. Karim QA, Kharsany AB, Naidoo K, Yende N, Gengiah T, Omar Z, Arulappan N, Mlisana KP, Luthuli LR, Karim SS (2011) Co-enrollment in multiple HIV prevention trials – experiences from the CAPRISA 004 Tenofovir gel trial. Contemp Clin Trials 32:333–338
89. Abdool Karim Q, Abdool Karim SS, Frohlich JA, Grobler AC, Baxter C, Mansoor LE, Kharsany AB, Sibeko S, Mlisana KP, Omar Z, Gengiah TN, Maarschalk S, Arulappan N, Mlotshwa M, Morris L, Taylor D (2010) Effectiveness and safety of tenofovir gel, an antiretroviral microbicide, for the prevention of HIV infection in women. Science 329:1168–1174
90. Grant RM, Lama JR, Anderson PL, McMahan V, Liu AY, Vargas L, Goicochea P, Casapia M, Guanira-Carranza JV, Ramirez-Cardich ME, Montoya-Herrera O, Fernandez T, Veloso VG, Buchbinder SP, Chariyalertsak S, Schechter M, Bekker LG, Mayer KH, Kallas EG, Amico KR, Mulligan K, Bushman LR, Hance RJ, Ganoza C, Defechereux P, Postle B, Wang F, McConnell JJ, Zheng JH, Lee J, Rooney JF, Jaffe HS, Martinez AI, Burns DN, Glidden DV (2010) Preexposure chemoprophylaxis for HIV prevention in men who have sex with men. N Engl J Med 363:2587–2599
91. Cohen MS, Chen YQ, McCauley M, Gamble T, Hosseinipour MC, Kumarasamy N, Hakim JG, Kumwenda J, Grinsztejn B, Pilotto JH, Godbole SV, Mehendale S, Chariyalertsak S, Santos BR, Mayer KH, Hoffman IF, Eshleman SH, Piwowar-Manning E, Wang L, Makhema J, Mills LA, de Bruyn G, Sanne I, Eron J, Gallant J, Havlir D, Swindells S, Ribaudo H, Elharrar V, Burns D, Taha TE, Nielsen-Saines K, Celentano D, Essex M, Fleming TR (2011) Prevention of HIV-1 infection with early antiretroviral therapy. N Engl J Med 365:493–505
92. Vissers DC, Voeten HA, Nagelkerke NJ, Habbema JD, de Vlas SJ (2008) The impact of pre-exposure prophylaxis (PrEP) on HIV epidemics in Africa and India: a simulation study. PLoS One 3:e2077
93. Veronese F, Anton P, Fletcher CV, DeGruttola V, McGowan I, Becker S, Zwerski S, Burns D (2011) Implications of HIV PrEP trials results. AIDS Res Hum Retroviruses 27:81–90
94. Alizon M, Wain-Hobson S, Montagnier L, Sonigo P (1986) Genetic variability of the AIDS virus: nucleotide sequence analysis of two isolates from African patients. Cell 46: 63–74
95. Wei X, Ghosh SK, Taylor ME, Johnson VA, Emini EA, Deutsch P, Lifson JD, Bonhoeffer S, Nowak MA, Hahn BH, Saag MS, Shaw GM (1995) Viral dynamics in human immunodeficiency virus type 1 infection. Nature 373:117–122
96. Ho DD, Nuemann AU, Perelson AS, Chen W, Leonard JM, Markovitz M (1995) Rapid turnover of plasma virions and CD4 lymphocytes in HIV-1 infection. Nature 373:123–126
97. Starcich BR, Hahn BH, Shaw GM, McNeely PD, Modrow S, Wolf H, Parks ES, Parks WP, Josephs SF, Gallo RC, Wong-Staal F (1986) Identification and characterization of conserved

and variable regions in the envelope gene of HTLV-III/LAV, the retrovirus of AIDS. Cell 45:637–648
98. Wong-Staal F, Gallo RC (1985) Human T-lymphotropic retroviruses. Nature 317:395–403
99. Myers G, Pavlakis GN (1992) Evolutionary potential of complex retroviruses. In: Levy JA (ed) The retroviridae. Plenum, New York, pp 1–37
100. Goodenow M, Huet TH, Saurin W, Kowk S, Sninsky J, Wain-Hobson S (1989) HIV-1 isolates are rapidly evolving quasispecies: evidence for viral mixtures and preferred nucleotide substitutions. J Acquir Immune Def Syn 2:344–352
101. Wolfs TFW, deJong JJ, van den Berg H, Tijnagel JMGH, Krone WJA, Goudsmit J (1990) Evolution of sequences encoding the principal neutralization epitope of human immunodeficiency virus type 1 is host dependent, rapid, and continuous. Proc Natl Acad Sci USA 87:9938–9942
102. Balfe P, Simmonds P, Ludlam CA, Bishop JO, Brown AJ (1990) Concurrent evolution of human immunodeficiency virus type 1 in patients infected from the same source: rate of sequence change and low frequency of inactivating mutations. J Virol 64:6221–6233
103. Essex M, Kanki P (1997) Human immunodeficiency virus type 2 (HIV-2). In: Merigan T, Bartlett J, Bologenesi D (eds) Textbook of AIDS medicine, 2nd edn. Williams & Wilkins, Baltimore, pp 873–886
104. Sharp PM, Bailes E, Chaudhuri RR, Rodenburg CM, Santiago MO, Hahn BH (2001) The origins of acquired immune deficiency syndrome viruses: where and when? Philos Trans R Soc Lond B Biol Sci 356:867–876
105. Gurtler LG, Hauser PH, Eberle J, von Brunn A, Knapp S, Zekeng L, Tsague JM, Kaptue L (1994) A new subtype of human immunodeficiency virus type 1 (MVP-5180) from Cameroon. J Virol 68:1581–1585
106. Simon F, Mauclére P, Roques P, Loussert-Ajaka I, Muller-Trutwin MC, Saragosti S, Georges-Courbot MC, Barre-Sinoussi F, Brun-Vezinet F (1998) Identification of a new human immunodeficiency virus type 1 distinct from group M and group O. Nat Med 4:1032–1037
107. Plantier JC, Leoz M, Dickerson JE, De Oliveira F, Cordonnier F, Lemee V, Damond F, Robertson DL, Simon F (2009) A new human immunodeficiency virus derived from gorillas. Nat Med 15:871–872
108. Keele BF, Van Heuverswyn F, Li Y, Bailes E, Takehisa J, Santiago ML, Bibollet-Ruche F, Chen Y, Wain LV, Liegeois F, Loul S, Ngole EM, Bienvenue Y, Delaporte E, Brookfield JF, Sharp PM, Shaw GM, Peeters M, Hahn BH (2006) Chimpanzee reservoirs of pandemic and nonpandemic HIV-1. Science 313:523–526
109. Hemelaar J, Gouws E, Ghys PD, Osmanov S (2011) Global trends in molecular epidemiology of HIV-1 during 2000–2007. AIDS 25:679–689
110. Louwagie J, McCutchan F, Mascola J (1993) Genetic subtypes of HIV-1. AIDS Res Hum Retroviruses 9(Suppl 1):147s–150s
111. Louwagie J, McCutchan F, Peeters M (1993) Phylogenetic analysis of gag genes from 70 international HIV-1 isolates provides evidence for multiple genotypes. AIDS 7:769–780
112. McCutchan F, Salimen MO, Carr JK, Burke DS (1996) HIV-1 genetic diversity. AIDS 10: S13–S20
113. Burke D, McCutchan F (1997) Global distribution of human immunodeficiency virus-1 clades. In: Vincent T, DeVita J, Hellman S, Rosenberg S (eds) AIDS: biology, diagnosis, treatment and preventions, 4th edn. Lippincott-Raven, Philadelphia, pp 119–126
114. Weniger BG, Takebe Y, Ou C-Y, Yamazaki S (1994) The molecular epidemiology of HIV in Asia. AIDS 8:13s–28s
115. Korber B, Brander C, Moore J, D'Souza P, Walker B, Koup R, Moore J, Haynes B, Myers G (eds) (1996) HIV molecular immunology database 1996. Theoretical Biology and Biophysics Group, Los Alamos National Laboratory, Los Alamos
116. Robertson D, Gao F, Hahn B, Sharp PM (1997) Intersubtype recombinant HIV-1 sequences. In: Korber B, Hahn B, Foley B, Mellors JW, Leitner T, Myers G, McCutchan F, Kuiken

C (eds) Human retroviruses and AIDS 1997. Theoretical Biology and Biophysics Group, Los Alamos National Laboratory, Los Alamos, pp III-25-III30

117. Jain MK, John TJ, Keusch GT (1994) Epidemiology of HIV and AIDS in India. AIDS 8: S61–75
118. Soto-Ramirez LE, Renjifo B, McLane MF, Marlink R, O'Hara C, Sutthent R, Wasi C, Vithayasai P, Vithayasai V, Apichartpiyakul C, Auewarakul P, Pena Cruz V, Chui DS, Osathanondh R, Mayer K, Lee TH, Essex M (1996) HIV-1 Langerhans' cell tropism associated with heterosexual transmission of HIV. Science 271:1291–1293
119. Taylor BS, Sobieszczyk ME, McCutchan FE, Hammer SM (2008) The challenge of HIV-1 subtype diversity. N Engl J Med 358:1590–1602
120. Vanharmelen J, Wood R, Lambrick M, Rybicki EP, Williamson AL, Williamson C (1997) An association between HIV-1 subtypes and mode of transmission in Capetown, South Africa. AIDS 11:81–87
121. Liitsola K, Holmstrom P, Laukkanen T, Brummer-Korvenkontio H, Leinikki P, Salminen MO (2000) Analysis of HIV-1 genetic subtypes in Finland reveals good correlation between molecular and epidemiological data. Scand J Infect Dis 32:475–480
122. Kalish ML, Korber BT, Pillai S, Robbins KE, Leo YS, Saekhou A, Verghese I, Gerrish P, Goh CL, Lupo D (2002) The sequential introduction of HIV-1 subtype B and CRF01 in Singapore by sexual transmission: accelerated V3 region evolution in a subpopulation of Asian CRF01 viruses. Virology 304:311–329
123. Ou CY, Takebe Y, Weniger BG, Luo CC, Kalish ML, Auwanit W, Yamazaki S, Gayle HD, Young NL, Schochetman G (1993) Independent introduction of two major HIV-1 genotypes into distinct high-risk populations in Thailand. Lancet 341:1171–1174
124. Herring BL, Ge YC, Wang B, Ratnamohan M, Zheng F, Cunningham AL, Saksena NK, Dwyer DE (2003) Segregation of human immunodeficiency virus type 1 subtypes by risk factor in Australia. J Clin Microbiol 41:4600–4604
125. Renjifo B, Fawzi W, Mwakagile D, Hunter D, Msamanga G, Spiegelman D, Garland M, Kagoma C, Kim A, Chaplin B, Hertzmark E, Essex M (2001) Differences in perinatal transmission among human immunodeficiency virus type 1 genotypes. J Hum Virol 4:16–25
126. Renjifo B, Gilbert P, Chaplin B, Vannberg F, Mwakagile D, Msamanga G, Hunter D, Fawzi W, Essex M (1999) Emerging recombinant human immunodeficiency viruses: uneven representation of the envelope V3 region. AIDS 13:1613–1621
127. Hudgens MG, Longini IM Jr, Vanichseni S, Hu DJ, Kitayaporn D, Mock PA, Halloran ME, Satten GA, Choopanya K, Mastro TD (2002) Subtype-specific transmission probabilities for human immunodeficiency virus type 1 among injecting drug users in Bangkok, Thailand. Am J Epidemiol 155:159–168
128. Kanki PJ, Hamel DJ, Sankalé JL, Hsieh CC, Thior I, Barin F, Woodcock SA, Guèye-NDiaye A, Zhang E, Montano M, NDoye I, Essex ME, MBoup S (1999) Human immunodeficiency virus type 1 subtypes differ in disease progression. J Infect Dis 179:68–73
129. Kaleebu P, French N, Mahe C, Yirrell D, Watera C, Lyagoba F, Nakiyingi J, Rutebemberwa A, Morgan D, Weber J, Gilks C, Whitworth J (2002) Effect of human immunodeficiency virus (HIV) type 1 envelope subtypes A and D on disease progression in a large cohort of HIV-1-positive persons in Uganda. J Infect Dis 185:1244–1250
130. Kaleebu P, Ross A, Morgan D, Yirrell D, Oram J, Rutebemberwa A, Lyagoba F, Hamilton L, Biryahwaho B, Whitworth J (2001) Relationship between HIV-1 env subtypes A and D and disease progression in a rural Ugandan cohort. AIDS 15:293–299
131. Neilson JR, John GC, Carr JK, Lewis P, Kreiss JK, Jackson S, Nduati RW, Mbori-Ngacha D, Panteleeff DD, Bodrug S, Giachetti C, Bott MA, Richardson BA, Bwayo J, Ndinya-Achola J, Overbaugh J (1999) Subtypes of human immunodeficiency virus type 1and disease stage among women in Nairobi, Kenya. J Virol 73:4393–4403
132. Hu DJ, Vanichseni S, Mastro TD, Raktham S, Young NL, Mock PA, Subbarao S, Parekh BS, Srisuwanvilai L, Sutthent R, Wasi C, Heneine W, Choopanya K (2001) Viral load differences in early infection with two HIV-1 subtypes. AIDS 15:683–691

133. Sarr AD, Eisen G, Gueye-Ndiaye A, Mullins C, Traore I, Dia MC, Sankale JL, Faye D, Mboup S, Kanki P (2005) Viral dynamics of primary HIV-1 infection in Senegal, West Africa. J Infect Dis 191:1460–1467
134. Sagar M, Lavreys L, Baeten JM, Richardson BA, Mandaliya K, Chohan BH, Kreiss JK, Overbaugh J (2003) Infection with multiple human immunodeficiency virus type 1 variants is associated with faster disease progression. J Virol 77:12921–12926
135. Zhang M, Foley B, Schultz AK, Macke JP, Bulla I, Stanke M, Morgenstern B, Korber B, Leitner T (2010) The role of recombination in the emergence of a complex and dynamic HIV epidemic. Retrovirology 7:25–40
136. Ellenberger DL, Li B, Lupo LD, Owen SM, Nkengasong J, Kadio-Morokro MS, Smith J, Robinson H, Ackers M, Greenberg A, Folks T, Butera S (2002) Generation of a consensus sequence from prevalent and incident HIV-1 infections in West Africa to guide AIDS vaccine development. Virology 302:156–163
137. Essex M (2009) The impact of HIV variation on prevention and treatment. In: Kanki PM, Marlink RG (eds) A line drawn in the sand. Harvard University Press, Cambridge, MA, pp 231–242
138. Siegfried N, Uthman OA, Rutherford GW (2010) Optimal time for initiation of antiretroviral therapy in asymptomatic, HIV-infected, treatment-naive adults. Cochrane Database Syst Rev:CD008272
139. Spaulding A, Rutherford GW, Siegfried N (2010) Tenofovir or zidovudine in three-drug combination therapy with one nucleoside reverse transcriptase inhibitor and one non-nucleoside reverse transcriptase inhibitor for initial treatment of HIV infection in antiretroviral-naive individuals. Cochrane Database Syst Rev:CD008740
140. Spaulding A, Rutherford GW, Siegfried N (2010) Stavudine or zidovudine in three-drug combination therapy for initial treatment of HIV infection in antiretroviral-naive individuals. Cochrane Database Syst Rev:CD008651
141. Coffin JM (1996) HIV viral dynamics. AIDS 10:S75–84
142. Kanki P, Marlink RG (2009) A line drawn in the sand: responses to the AIDS treatment crisis in Africa. Harvard University Press, Cambridge, MA
143. United Nations (2001) Secretary-general urges united states business leaders to take concerted action against "Unparalleled Nightmare" of AIDS, U.N. Document
144. UNAIDS (2003) AIDS epidemic. data.unaids.org/publications/inc-pub06/jc943.epiupdate2003_en_pdf
145. PEPFAR (2011) United States President's emergency plan for AIDS relief. http://www.pepfar.gov/about/index.html
146. GFATM (2002) Global fund to fight AIDS, Tuberculosis and Malaria. http://www.theglobalfund.org/en/
147. BMGF (2011) Bill & Melinda Gates Foundation Global Health http://www.gatesfoundation.org/global-health/Pages/overview.aspx
148. CHAI (2011) Clinton Health Access Initiative. http://www.clintonfoundation.org/what-we-do/clinton-health-access-initiative
149. MAP (2011) World Bank Multi-Country HIV/AIDS Program (MAP). http://web.worldbank.org/WBSITE/EXTERNAL/COUNTRIES/AFRICAEXT/EXTAFRHEANUTPOP/EXTAFRREGTOPHIVAIDS/0,,contentMDK:20415735~menuPK:1001234~pagePK:34004173~piPK:34003707~theSitePK:717148,00.html
150. MSF (2011) Medecins sans Frontieres. http://www.msf.org/msf/about-msf/about-msf_home.cfm
151. Holmes C, Coggin W, Jamieson D, Mihm H, Savio P, Hope M, Ryan C, Moloney-Kitts M, Dybul M (2009) Measuring progress in reducing the costs of ARV drugs purchased by the president's emergency plan for AIDS relief, 2005–2007. In: Conference on retroviruses and opportunistic infections. Montreal, Canada

152. Shearer G, Clerici M (2010) Historical perspective on HIV-exposed seronegative individuals: has nature done the experiment for us? J Infect Dis 202(Suppl 3): S329–332
153. O'Brien SJ, Nelson GW (2004) Human genes that limit AIDS. Nat Genet 36:565–574
154. Wu X, Yang ZY, Li Y, Hogerkorp CM, Schief WR, Seaman MS, Zhou T, Schmidt SD, Wu L, Xu L, Longo NS, McKee K, O'Dell S, Louder MK, Wycuff DL, Feng Y, Nason M, Doria-Rose N, Connors M, Kwong PD, Roederer M, Wyatt RT, Nabel GJ, Mascola JR (2010) Rational design of envelope identifies broadly neutralizing human monoclonal antibodies to HIV-1. Science 329:856–861
155. Koup RA, Graham BS, Douek DC (2011) The quest for a T cell-based immune correlate of protection against HIV: a story of trials and errors. Nat Rev Immunol 11:65–70
156. Bradac J, Dieffenbach CW (2009) HIV vaccine development: lessons from the past, informing the future. IDrugs 12:435–439
157. Girard MP, Osmanov S, Assossou OM, Kieny MP (2011) Human immunodeficiency virus (HIV) immunopathogenesis and vaccine development: a review. Vaccine 29:6191–6218
158. Fauci AS, Johnston MI, Dieffenbach CW, Burton DR, Hammer SM, Hoxie JA, Martin M, Overbaugh J, Watkins DI, Mahmoud A, Greene WC (2008) HIV vaccine research: the way forward. Science 321:530–532
159. Corey L, McElrath MJ (2010) HIV vaccines: mosaic approach to virus diversity. Nat Med 16:268–270
160. Caputo A, Gavioli R, Bellino S, Longo O, Tripiciano A, Francavilla V, Sgadari C, Paniccia G, Titti F, Cafaro A, Ferrantelli F, Monini P, Ensoli F, Ensoli B (2009) HIV-1 Tat-based vaccines: an overview and perspectives in the field of HIV/AIDS vaccine development. Int Rev Immunol 28:285–334
161. MacGregor RR, Boyer JD, Ugen KE, Lacy KE, Gluckman SJ, Bagarazzi ML, Chattergoon MA, Baine Y, Higgins TJ, Ciccarelli RB, Coney LR, Ginsberg RS, Weiner DB (1998) First human trial of a DNA-based vaccine for treatment of human immunodeficiency virus type 1 infection: safety and host response. J Infect Dis 178:92–100
162. Pantaleo G, Esteban M, Jacobs B, Tartaglia J (2010) Poxvirus vector-based HIV vaccines. Curr Opin HIV AIDS 5:391–396
163. Graham BS, Koup RA, Roederer M, Bailer RT, Enama ME, Moodie Z, Martin JE, McCluskey MM, Chakrabarti BK, Lamoreaux L, Andrews CA, Gomez PL, Mascola JR, Nabel GJ (2006) Phase 1 safety and immunogenicity evaluation of a multiclade HIV-1 DNA candidate vaccine. J Infect Dis 194:1650–1660
164. Flynn NM, Forthal DN, Harro CD, Judson FN, Mayer KH, Para MF (2005) Placebo-controlled phase 3 trial of a recombinant glycoprotein 120 vaccine to prevent HIV-1 infection. J Infect Dis 191:654–665
165. Gilbert PB, Peterson ML, Follmann D, Hudgens MG, Francis DP, Gurwith M, Heyward WL, Jobes DV, Popovic V, Self SG, Sinangil F, Burke D, Berman PW (2005) Correlation between immunologic responses to a recombinant glycoprotein 120 vaccine and incidence of HIV-1 infection in a phase 3 HIV-1 preventive vaccine trial. J Infect Dis 191:666–677
166. Gilbert PB, Ackers ML, Berman PW, Francis DP, Popovic V, Hu DJ, Heyward WL, Sinangil F, Shepherd BE, Gurwith M (2005) HIV-1 virologic and immunologic progression and initiation of antiretroviral therapy among HIV-1-infected subjects in a trial of the efficacy of recombinant glycoprotein 120 vaccine. J Infect Dis 192:974–983
167. Buchbinder SP, Mehrotra DV, Duerr A, Fitzgerald DW, Mogg R, Li D, Gilbert PB, Lama JR, Marmor M, Del Rio C, McElrath MJ, Casimiro DR, Gottesdiener KM, Chodakewitz JA, Corey L, Robertson MN (2008) Efficacy assessment of a cell-mediated immunity HIV-1 vaccine (the Step Study): a double-blind, randomised, placebo-controlled, test-of-concept trial. Lancet 372:1881–1893
168. McElrath MJ, De Rosa SC, Moodie Z, Dubey S, Kierstead L, Janes H, Defawe OD, Carter DK, Hural J, Akondy R, Buchbinder SP, Robertson MN, Mehrotra DV, Self SG, Corey L,

Shiver JW, Casimiro DR (2008) HIV-1 vaccine-induced immunity in the test-of-concept step study: a case-cohort analysis. Lancet 372:1894–1905
169. Gray G, Buchbinder S, Duerr A (2010) Overview of STEP and Phambili trial results: two phase IIb test-of-concept studies investigating the efficacy of MRK adenovirus type 5 gag/pol/nef subtype B HIV vaccine. Curr Opin HIV AIDS 5:357–361
170. Nicholson O, Dicandilo F, Kublin J, Sun X, Quirk E, Miller M, Gray G, Pape J, Robertson MN, Mehrotra DV, Self S, Turner K, Sanchez J, Pitisuttithum P, Duerr A, Dubey S, Kierstead L, Casimiro D, Hammer SM (2010) Safety and immunogenicity of the MRKAd5 gag HIV type 1 vaccine in a worldwide phase 1 study of healthy adults. AIDS Res Hum Retroviruses 27(5):557–567
171. Voronin Y, Manrique A, Bernstein A (2010) The future of HIV vaccine research and the role of the Global HIV Vaccine Enterprise. Curr Opin HIV AIDS 5:414–420
172. Voronin Y, Phogat S (2010) HIV/AIDS: vaccines and alternate strategies for treatment and prevention. Ann NY Acad Sci 1205(Suppl 1):E1–9
173. Nitayaphan S, Pitisuttithum P, Karnasuta C, Eamsila C, de Souza M, Morgan P, Polonis V, Benenson M, VanCott T, Ratto-Kim S, Kim J, Thapinta D, Garner R, Bussaratid V, Singharaj P, El-Habib R, Gurunathan S, Heyward W, Birx D, McNeil J, Brown AE (2004) Safety and immunogenicity of an HIV subtype B and E prime-boost vaccine combination in HIV-negative Thai adults. J Infect Dis 190:702–706
174. Rerks-Ngarm S, Pitisuttithum P, Nitayaphan S, Kaewkungwal J, Chiu J, Paris R, Premsri N, Namwat C, de Souza M, Adams E, Benenson M, Gurunathan S, Tartaglia J, McNeil JG, Francis DP, Stablein D, Birx DL, Chunsuttiwat S, Khamboonruang C, Thongcharoen P, Robb ML, Michael NL, Kunasol P, Kim JH (2009) Vaccination with ALVAC and AIDSVAX to prevent HIV-1 infection in Thailand. N Engl J Med 361:2209–2220
175. Kresge KJ (2009) Raft of results energizes researchers. IAVI Rep 13:4–5, 7–13, 17
176. Dieffenbach CW, Fauci AS (2011) Thirty years of HIV and AIDS: future challenges and opportunities. Ann Intern Med 154:766–771
177. World Health Organization (2010) Antiretroviral therapy for HIV infection in adults and adolescents: recommendations for a public health approach. 2010 Revision, Geneva

Chapter 4
Human Bacterial Diseases from Ocean

Darrell Jay Grimes, Lisa W. Plano, and Okechukwu Ekenna

Glossary

Allochthonous	Exogenous, alien or nonindigenous; arising from another source or medium.
Ambient	Being of the surrounding area or environment.
Archaea	One of three domains on Earth, including the *Bacteria* and *Eukarya*. *Archaea* are prokaryotes that do not have peptidoglycan cell walls; they lack membrane-bound organelles (e.g., nucleus, mitochondria, endoplasmic reticulum, chloroplasts), possess 70 S ribosomes and have ether-linked lipids in their membranes.
Autochthonous	Indigenous, native, arising from within.

This chapter, which has been modified slightly for the purposes of this volume, was originally published as part of the Encyclopedia of Sustainability Science and Technology edited by Robert A. Meyers. DOI:10.1007/978-1-4419-0851-3

D.J. Grimes (✉)
Department of Coastal Sciences, The University of Southern Mississippi,
703 East Beach Drive, Ocean Springs, MS, USA
e-mail: jay.grimes@usm.edu

L.W. Plano
Departments of Pediatrics & Department of Microbiology and Immunology,
Miller School of Medicine, University of Miami, 1600 N.W. 10th Avenu,
Miami, FL, USA
e-mail: lplano@miami.edu

O. Ekenna
Singing River Health System, Pascagoula, MS, USA

University of South Alabama, Mobile, AL, USA
e-mail: okekennaid@cableone.net

P. Kanki and D.J. Grimes (eds.), *Infectious Diseases: Selected Entries from the Encyclopedia of Sustainability Science and Technology*,
DOI 10.1007/978-1-4614-5719-0_4,

Bacteria	One of three domains on Earth, including the *Archaea* and *Eukarya*. *Bacteria* are prokaryotes that possess peptidoglycan cell walls; they lack membrane-bound organelles (e.g., nucleus, mitochondria, endoplasmic reticulum, chloroplasts), possess 70 S ribosomes and have ester-linked lipids in their membranes.
Biodiversity	The richness or complexity of life forms in an ecosystem, biome or on Earth itself.
Commensal organism	An organism participating in a symbiotic relationship in which one species derives some benefit while the other is unaffected.
Ectotherm	An organism that controls body temperature through external means.
Endotoxin	The lipid component (lipid A) of the outer membrane lipopolysaccharide (LPS) of all gram-negative bacteria. Endotoxin is released into a host or the environment when the cell lyses and its outer membrane breaks up.
Epidemic	The incidence of disease above the normal or endemic incidence.
Eukarya	One of three domains on Earth, including the *Bacteria* and *Archaea*. *Eukarya* possess membrane-bound organelles (e.g., nuclei, mitochondria, chloroplasts), histones associated with their DNA and 80 S ribosomes in their cytoplasm. Plants and animals are eukaryotic.
Exotoxin	Any toxin that is secreted into the cell's immediate environment. Most exotoxins are proteins, and they are made by both gram-negative and gram-positive bacteria.
Facultative organism	An organism that is capable of growth both in the presence and absence of oxygen.
Foodborne disease	A disease that is caused by the ingestion of pathogens conveyed by food.
Food intoxication	Illness caused by the ingestion of food that contains a toxic substance.
Hemolysin	A proteolytic enzyme that lyses red blood cells.
Lysogenic conversion	Insertion of bacterial virus (bacteriophage) DNA into the chromosomal DNA of its bacterial host thereby conferring one or more new traits on the host.
Nosocomial	Infections (and disease) that are acquired in clinical settings (e.g., hospitals, outpatient clinics, emergency rooms, physician offices).
Opportunistic pathogen	Any pathogen that accidently acquires entrance to a host and then only causes disease if one or more risk factors are present in the host.
Pandemic	An epidemic of world-wide proportions.

Pathogenesis	The production or development of a disease, specifically the cellular reactions and other pathologic mechanisms occurring in the progression of the disease.
Pathogenic	The ability of a species to cause disease. However, because pathogenesis is typically caused by one or more than one virulence factors produced by one or more genes, any given pathogenic species will often display different degrees of pathogenesis.
Pathogenicity island	A cluster of virulence genes (and sometimes cryptic genes and other small genetic elements) flanked by direct repeats, insertion sequences or tRNA genes such that the clusters are easily transmitted to other bacteria via a process called *horizontal gene transfer*.
Plasmid	A circular, double-stranded DNA molecule containing specialty genes that, in general, are not essential for survival of the host bacterium or genes that are cryptic (unknown). Plasmids can replicate autonomously or integrate into and replicate with the chromosome. Plasmids are smaller than the chromosome, on average 5% the size of the chromosome.
Point source	A single, identifiable localized source of something.
Quorum sensing	A chemical mechanism used by bacteria to measure their population density. When the chemical signals reach a certain level, special genes are expressed.
Sentinel	An indicator whose presence is directly related to a particular quality in its environment at a given location.
Sequences	The order of nucleotides in a specific length of DNA or RNA.
Virulence	The degree of pathogenicity. Virulence is a compilation of toxins, hemolysins, proteases and lipases that may not be possessed by all strains of a pathogenic species.
Waterborne disease	A disease that is transmitted by water.
Zoonosis	An animal disease transmissible to humans under natural conditions or a human disease transmissible to animals.

Definition of the Subject and Its Importance

Several bacteria that cause human disease can be found in the ocean. The actual incidence of bacterial disease that results from seawater or seafood is not precisely known but is thought to be relatively low in the USA, although some diseases are on the rise. Bacterial disease from the ocean is more prevalent worldwide, especially in developing countries and in countries that derive most of their protein from seafood. Compared to the viruses, bacteria account for a much lower incidence of disease emanating from the ocean, both in the USA and worldwide. However, it is important to understand and mitigate bacterial disease from the ocean, because of such environmental pressures as global warming, antibiotic resistance, pollution, breakdowns in sanitation (e.g., Haiti after the earthquake) and tourism.

Introduction

Myriad bacteria reside in the ocean and most (>99.9%) have not been isolated and are known only by their unique molecular signatures (e.g., 16 S rRNA sequences) [35, 59]. These bacteria belong to the domain *Bacteria* which exhibits extensive *biodiversity*, only exceeded by the biodiversity within the domain *Archaea* and the viruses. The vast majority of bacteria in the ocean do not cause disease and Archaea are not known to cause any human disease [39]; however, a very small number (percentage) of bacterial species found in the ocean cause *pathogenesis* in plants and animals. Some of these pathogens are indigenous to the ocean and are defined as *autochthonous*. Others (*allochthonous* bacteria) are of exogenous or terrestrial origin and are introduced to the ocean via surface runoff, rivers and streams, atmospheric fallout and ocean disposal of wastes (intentional or accidental); the chapter on enteric viruses (See S.NO.12, J. Woods, Waterborne Diseases of the Ocean, Enteric Viruses, this volume) has a nice discussion of allochthonous sources. This chapter will focus on those allochthonous and autochthonous bacteria that cause *waterborne* and *foodborne* disease in humans. Some also cause disease in marine animals (and a few in marine plants) but nonhuman diseases will not be addressed in this chapter.

Autochthonous human pathogens have evolved in the ocean and for largely unknown reasons have the ability to infect and cause disease in humans. Almost all of these diseases result from ingesting seawater, eating seafood, or broken skin contact (swimming, wading, or working in seawater). The allochthonous human pathogens are transients in the ocean with varying abilities to survive in seawater; all of the allochthonous pathogens infect humans via the same routes as autochthonous pathogens, i.e., contaminated seawater, seafood or broken skin contact. There are a few bacteria that may fit both definitions, and there is no current scientific consensus about the place of these bacteria. Examples of these ubiquitous bacteria include the enterococci [159], *Staphylococcus aureus* (see Allochthonous Pathogens in this chapter) and *Pseudomonas aeruginosa* [114].

In the following pages, allochthonous and autochthonous marine bacteria that cause disease(s) in humans will be discussed in terms of their biology, ecology, pathogenesis, and epidemiology. Disease treatment will also be discussed but not in clinical detail.

Autochthonous Pathogens

Autochthonous pathogens generally cause one or more diseases in marine animals and if the disease(s) is transmissible to humans it is classified as a *zoonosis*. It some cases, humans can transmit disease to marine animals.

Vibrio

Vibrios, and specifically *Vibrio cholerae*, were first observed by Pacini in 1854 and later isolated in pure culture from cholera patients by Robert Koch in 1883 [8, 27]. Today, the Vibrios comprise a large genus (>80 species) and belong to the *Gammaproteobacteria* [47]. They are gram-negative rods, often slightly curved (e.g., *V. cholerae*) and most are motile in liquids by means of one or more polar flagella; on solid surfaces they are motile by means of lateral flagella. Unlike most bacteria, Vibrios possess two circular chromosomes (one large and one small) which are of relevance to this chapter because of the distribution of *virulence* genes on the two chromosomes. The large chromosome is usually referred to as Chromosome 1 (Ch1) and it tends to contain housekeeping genes (DNA replication, transcription, translation, flagellar synthesis, metabolic pathways). Chromosome 2 (Ch2) tends to contain accessory genes (pathogenicity, antibiotic resistance, host defense avoidance, survival in adverse environments). Most of the Vibrios can metabolize a large number of organic compounds, including sugars, amino acids, fatty acids, carbohydrates, proteins, lipids, alcohols, and selected aliphatic and aromatic hydrocarbons. Indeed, prior to the development of molecular biology methods, the Vibrios were largely identified and classified by these diverse and extensive metabolic traits [22].

In addition to metabolic diversity, the pathogenic Vibrios also possess an array of *exotoxins*, proteases, transport proteins, attachment mechanisms, and lipases that act as virulence factors. Since Vibrios are gram-negative, they also possess *endotoxin* which differs in toxicity from strain to strain. Indeed, the Vibrios are well equipped to cause disease in their hosts – accidental or otherwise. In general, the Vibrios are opportunistic pathogens – for both humans and marine animals – and they cause systemic infections, skin lesions and gastroenteritis. In fish, the infections often lead to hemorrhagic skin lesions (known as vibriosis) and pathology in the liver, spleen, and kidney. In humans, the diseases arise from contamination of cuts and other skin lesions with seawater (and to a lesser extent marine animals, e.g., stingray barbs and barnacles) and from ingestion of raw, undercooked or cooked but re-contaminated seafood and seawater. Depending on the species, skin infections can remain localized or become systemic and ultimately fatal. Oral ingestion of seafood and seawater leads to various degrees of gastroenteritis and, in some cases (e.g., *V. vulnificus)* life-threatening systemic infections. Although 12 *Vibrio* spp. are considered to be

Table 4.1 Number of *Vibrio* illnesses (excluding toxigenic *V. cholerae*) in the USA, 2008[a]

Species	Patients	Deaths	Isolates	Specimen[b]
V. alginolyticus	99	1	99	Wound and other[c]
V. cholerae (nontoxic)	50	2	50	Stool and blood
V. fluvialis	29	3	29	Stool and wound
V. hollisae[d]	4	1	4	Stool and wound
V. metschnikovii	1	0	1	Blood
V. mimicus	32	0	32	Stool and other
V. parahaemolyticus	270	4	270	Stool and wound
V. vulnificus	85	24	94	Blood and wound
Species not identified	23	0	24	Stool and other
Multiple species	6	0	13	Wound and stool
Total	**599**	**35**	**616**	**Stool most frequent**

[a]Data reported to the Cholera and Other *Vibrio* Illness Surveillance (COVIS) system maintained by the CDC (see http://www.cdc.gov/nationalsurveillance/PDFs/Jackson_Vibrio_CSTE2008_FINAL.pdf)
[b]The predominant specimen types with the most predominant listed first. Additional specimens, including actual numbers, can be seen in the COVIS report[a]
[c]"Other" consists of ear, sputum urine, and other specimens
[d]Reclassified to *Grimontia hollisae* in 2003 [137]

human pathogens, the predominant human pathogens are *V. cholerae, V. parahaemolyticus* and *V. vulnificus*. Table 4.1 presents a recent compilation of *Vibrio* diseases in the USA. In addition to the eight species listed in Table 4.1, the other human pathogens include *V. cincinnatiensis, Photobacterium damselae* (initially named *V. damsela*), *V. furnissii*, and *V. harveyi*; these four Vibrios, along with *G. hollisae* and *V. metschnikovii*, will not be discussed in this chapter, due to their low incidence of disease in humans.

All of the Vibrios require NaCl to grow properly, although this requirement is minimal for some species, e.g., *V. cholerae* and *V. mimicus*. Most if not all Vibrios use a sodium motive force to drive their polar flagella [94, 157] and at least a few species appear to use a sodium motive force to make ATP [37, 138]. Lateral flagella are produced by Vibrios when they grow on solid surfaces; these flagella are driven by a proton motive force [94] and are responsible for swarming. In addition to a requirement for sodium, most Vibrios are mesophiles and, therefore, do best in warmer waters. Recently, it has been suggested that global warming might increase the incidence of Vibrios worldwide [25, 28]. Interestingly, laboratory-confirmed infections with *Vibrio* spp. began to exceed those from *Salmonella*, Shiga toxin-producing *Escherichia coli* O157, *Campylobacter* and *Listeria* in late 2000 and this US trend has increased through 2008 [17]. In 2009, the overall rate of foodborne disease caused by Vibrios was 0.35 per 100,000 population [18]. The relative rates of the other four pathogens have either remained level (*Salmonella*) or shown a decrease (*Shigella* rates decreased 40% and STEC O157 decreased 25%) while Vibrio rates have increased by 47% [17]. Some have suggested that because this trend began just before the El Niño years 2002–2003 and continued through the El Niño years 2006–2007, it was caused by global warming. Intriguing as it may

Table 4.2 Temperature and salinity preferences for the pathogenic Vibrios[a]

	Temperature		Salinity	
Vibrio species	Optimum	Range	Optimum	Range
V. cholerae	25°C [24, 26]	17–40°C [15, 31]	2–14 ppt [26, 31]	<1–60 ppt [15]
V. parahaemolyticus	38°C [27, 33]	8–45°C [16, 33]	17–23 ppt [19, 27, 44]	<1–96 ppt [33]
V. vulnificus	30°C [39]	9–40°C [21, 39, 40, 106]	5–10 ppt [25]	5–35 ppt [25, 40, 106]

[a]Bracketed numbers are supporting citations found in the *Primary Literature*

be, this observation is circumstantial and cannot be confirmed. However, in 1991 there was an outbreak of cholera in Peru, a country that had not seen this disease for over 100 years [136]. The source of the 1991 outbreak was never determined and two more outbreaks occurred in 1993–1995 and 1997–1998. Gil et al. [56] very nicely demonstrated that the Peru outbreaks in summer 1998 correlated (linear regression, $P < 0.001$) with the sea surface temperature peak associated with the strong El Niño that year. Colwell et al. [25] and [28] have made similar observations for the Bay of Bengal. In 2004, an outbreak involving cultured raw oysters and *V. parahaemolyticus* occurred aboard a cruise ship in Prince Edward Sound [96]. Mean water temperatures had been increasing in Prince William Sound (0.21°C per year) since 1997, and in 2004, for the first time, mean daily temperatures in the sound did not drop below 15°C [96]. Johnson et al. [74] reported that when temperatures drop below 15°C *V. parahaemolyticus* and *V. vulnificus* are no longer culturable from water and sediment in Mississippi Sound; similar observations have been reported by others, suggesting that 15°C is a limiting temperature for many of the Vibrios. Clearly, it is beginning to look like increasing ocean temperatures are increasing the incidence of Vibrio disease and as more data become available this Vibrio incidence-climate link may be substantiated. Temperature and sodium preferences for the three principal pathogenic Vibrios are shown in Table 4.2. Finally, in the past decade, scientists are using remotely-sensed satellite data, including temperature and salinity, to predict human health risks from pathogenic vibrios in water and oysters [90, 117].

Vibrio alginolyticus

This *Vibrio* is very common in estuarine and marine waters worldwide; in fact, it is unusual not to isolate *V. alginolyticus* when culturing coastal and estuarine water samples. It is known for its ability to swarm on isolation media [141] and it often overgrows desired isolates. Recently, it was demonstrated that *V. alginolyticus* is capable of quorum sensing and that QS is responsible for biofilm formation (swarming is a prelude to this phenomenon) and development of virulence factors involved with fish disease [64].

The most common disease caused by *V. alginolyticus* in humans is wound infection. Of the 99 infections reported to the CDC in 2008 (Table 4.1), 64 isolates came from wound infections, 5 from blood, 3 from stools and 27 from other specimens. Most infections are mild and self-limiting, although *V. alginolyticus* has been demonstrated in a few cases of severe necrotizing fasciitis that involved patients at risk because of cirrhosis [66]. Cases of gastroenteritis caused by *V. alginolyticus* are rare [140], as indicated in Table 4.1. This Vibrio has also caused eye infections [86].

Vibrio cholerae

During the early cholera pandemics of the 1800s, *V. cholerae* was responsible for millions of deaths and was feared as the most dangerous waterborne human pathogen known [27, 76]. It still causes several thousand deaths annually [153] and a small number of cases occur in the USA each year (see COVIS report cited in Table 4.1). The World Health Organization recently reported that in 2009 the number of cases worldwide (45 countries reporting) was 221,226 with 4,946 deaths (2.24% case fatality rate) [153]. The number of cases in the USA in 2008 was nine and none of these patients died (see COVIS report, Table 4.1). Although molecular detection is very effective and widely used [102] and the genomes of several strains have been fully sequenced, for documentation and epidemiology purposes isolates are still serogrouped by means of their O antigens. The predominant serogroups causing human disease worldwide are O1 and O139; in the USA, the CDC tests for serogroups O1, O75, O139, and O141 (COVIS report cited in Table 4.1).

The classic disease caused by *V. cholerae* is a rapidly developing, profuse, watery diarrhea that is usually accompanied by severe dehydration. The cholera diarrhea is often called a “rice-water” stool because as the disease progresses the frequent stools are little more than water containing flecks of mucous abraded from the intestinal mucosa. When death occurs, it is because of the severe dehydration (water loss) and acidic coma (loss of sodium bicarbonate to the diarrhea). Historically, the death rate often exceeded 50% and death could result in as few as 24–36 h after onset of symptoms. As noted above [153] the death rate for cholera is now below 5% and most cases of cholera today are little more than a transient diarrhea. In addition to classic cholera, *V. cholerae* can also cause self-limiting gastroenteritis and wound infections; it has also been implicated in fish disease (eels and ayu). An interesting and illustrative human case occurred after Hurricanes Katrina and Rita and involved a Louisiana couple who consumed properly cooked (boiled) shrimp that had been placed on the ice used by them to transport the raw shrimp. The husband developed a severe case of cholera that caused renal, pulmonary and cardiac failure. He was given ciprofloxacin and aggressive rehydration therapy and he did not die. Of interest is that the husband had a history of common Vibrio risk factors – history of high blood pressure, alcoholism, diabetes, brain tumor, and renal failure that required frequent dialysis. His wife had mild diarrhea and was treated (ciprofloxacin and rehydration) as an outpatient. *V. cholerae* O1, serotype Inaba, biotype El Tor was isolated from both patients [15].

The main virulence factor associated with cholera is an exotoxin known as the cholera toxin (CTX). CTX is not produced by all strains of *V. cholerae*, and it is a protein that is composed of two subunits – one A subunit and five B subunits. The *ctx* genes that encode for these two subunits, *ctx*A and *ctx*B, are actually genes in a virus (a bacteriophage or phage) that infects *V. cholerae* and establishes a lysogenic relationship with its host [148]. The CTX phage that infects *V. cholerae* is a lambdaphage; and instead of producing more phage when it infects its host, this phage inserts its DNA into the chromosome (Ch1) of the host which then allows the host to produce CTX – a process called *lysogenic conversion.*

V. cholerae produces other virulence factors, including the toxin-coregulated pilus (TCP) produced by the *tcp*A gene, repeat-in-toxins (RTXs), and a heat-stable enterotoxin (NAG-ST) that is related to the heat-stable enterotoxin produced by *E. coli* [135]. The TCP is thought to be necessary for intestinal colonization by *V. cholerae* and may also serve as an attachment site for CTX phage [148]. Type 2 secretion systems are present in *V. cholerae*, and these systems provide a physical conduit for the bacteria to secrete toxins into their host cells (see secretion system discussion in "*Vibrio parahaemolyticus*" section).

In closing, prevention and treatment of cholera deserves mention. Cholera has a very low incidence in developed countries, primarily because of good sanitation, but this is not the case in developing countries [68]. In both developed and developing countries, cholera is usually a self-limiting disease requiring minimal treatment other than rehydration and electrolyte replacement; antibiotics are not usually administered. In developing countries where safe drinking water is not readily available, filtration of drinking water is very effective in preventing cholera. A very simple but highly effective filtration procedure was pioneered in Bangladesh [26]. The procedure involves using folded sari cloth (eight layers of old sari) to filter the drinking water; this procedure removes zooplankton to which the cholera bacilli are attached [26].

Cholera in Haiti

In January, 2010 Haiti was ravaged by a major earthquake that had an epicenter approximately 25 km from the capital city of Port-au-Prince. On October 21, 2010 cholera was confirmed in Haiti by the CDC; the causative agent has been identified as *V. cholerae* O1 serotype Ogawa. As of October 14, 2011, the Hatian Public Health Ministry reported that over 473,649 people have been infected and 6,631 patients have died. The disease peaked in early 2011 but cases and deaths continue. Interestingly, Haiti had not reported a cholera case for over a century and the source of the outbreak was, early on, not known. Clearly, the lack of hygiene, safe drinking water and safe food that followed the earthquake and continues to exist has contributed to the onset and continuation of cholera in Haiti. However, recent molecular evidence reported in the CDC journal "Emerging Infectious Diseases" demonstrated that a UN peacekeeping force from Nepal introduced the cholera strain into the Meille River.

Vibrio fluvialis

The incidence of disease caused by *V. fluvialis* is thought to be very low, both in the USA (Table 4.1) and worldwide. This *Vibrio* primarily causes enteric disease but can cause wound infections and, rarely, other extraintestinal infections.

Vibrio mimicus

This *Vibrio* is similar to *V. cholerae* in many ways and early on was thought to be an atypical *V. cholerae* – hence the species name "mimicus." The primary disease caused by *V. mimicus* is gastroenteritis and some strains carry *ctx* genes as well as other virulence factors found in *V. cholerae* and other Vibrios.

Vibrio parahaemolyticus

First isolated in Japan from a gastroenteritis outbreak traced to the consumption of shirasu, a popular fish product (272 cases and 20 deaths; [54]), *V. parahaemolyticus* is the most common cause of foodborne disease in countries that consume high quantities of seafood [61]; in the USA, it is the most common cause of seafood-borne disease (see Table 4.1). As is the case with all the Vibrios that cause human disease, *V. parahaemolyticus* causes gastroenteritis and wound infections, and it is the most common cause of Vibrio disease in the USA (Table 4.1). *V. parahaemolyticus* can be isolated from most forms of seafood but is most commonly associated with shellfish [12].

The first *V. parahaemolyticus* pandemic began in 1996, and it continues to involve three major serotypes – O3:K6, O4:K68, and O1:K untypable [23]. Recent outbreaks are described in Table 4.3.

All *V. parahaemolyticus* strains possess the *tlh* (thermolabile hemolysin) gene, and this gene is frequently used to rapidly detect and confirm identification of

Table 4.3 Recent outbreaks and cases of *V. parahaemolyticus* gastroenteritis

Date	Location (ref.)	Source	Cases/deaths	Serotype	El Niño
1997	Pacific North West [12]	Oysters	209/1	O1, O4, O5	Yes
1998	Gulf of Mexico, NE, Pacific NW [32]	Oysters	416/?	O3:K6	Yes
1998	Chile [62]	Shellfish	~300/?	O3:K6	Yes
1998	Japan [61]	Seafood	12,318/?	O3:K6	Yes
2004 and 2005	Chile [62]	Seafood	~5,100/?	O3:K6	No
2004	Alaska [96]	Oysters	62/0	O6:K18	No
2006	New York, Oregon, Washington [16]	Oysters	177/0	O4:K12	Yes
2006	Chile [62]	Shellfish	900/?	O3:K59	Yes

the species. However, *tlh* is not unique to *V. parahaemolyticus*; *V. harveyi*, *V. alginolyticus*, and *V. fischeri* have also been shown to contain a *tlh* homologue or a related hemolysin gene [72, 134, 149].

When first discovered, it was shown that *V. parahaemolyticus* had the ability to hemolyze red blood cells in a special culture medium called Watsumaga agar; it was later shown that this hemolysis, called the Kanagawa reaction, was mediated by a hemolysin called thermostable direct hemolysin TDH [67]. Most pathogenic *V. parahaemolyticus* strains possess the *tdh* gene and/or the *trh* (thermostable related hemolysin) gene [73], although some pathogenic strains contain neither *tdh* nor *trh* [98]. The *tdh* genes are located in a *pathogenicity island* on Ch2 [91]. In general, environmental and food isolates do not contain tdh and trh ([73] and [74]).

Gram-negative bacteria possess a fascinating injection apparatus called secretion systems. These allow bacteria to inject various substances into the cells of *Eukarya.* As is the case for many bacterial virulence factors, especially those associated autochthonous pathogens, such as the Vibrios; microbiologists understand how these factors function in human pathogenesis but do not know the function of these factors in the environment. To date, six secretion systems (T1SS, T2SS, T3SS, T4SS, T5SS and T6SS) have been described. In 2003, the sequenced genome of *V. parahaemolyticus* RIMD2210633 was shown to contain two different T3SS genes – T3SS1 and T3SS2 [91]. T3SS1 is located in a pathogenicity island on Ch1, T3SS2 is in a Ch2 pathogenicity island, and it is now known that there are two different T3SS2 – T3SS2α and T3SS2β [103].

Makino et al. [91] identified at least 50 other genes in *V. parahaemolyticus* that may be involved with pathogenesis. Some of these additional virulence factors include urease [112], attachment mechanisms [60, 130], ToxR [30], and RTX toxin [91].

Vibrio vulnificus

One of the more recently discovered pathogenic Vibrios [46, 123], *V. vulnificus*, causes serious diseases in humans and is thought by some Vibrio biologists to be the most virulent bacterium now known (death rate >50% in patients at risk [104]; the overall case fatality rate reported by CDC [97] for 1992–1997 was 39%). *V. vulnificus* is thought to cause three human diseases: primary septicemia (caused by the ingestion of raw or undercooked shellfish – especially oysters), gastroenteritis (caused by the ingestion of raw or undercooked shellfish), and wound infections (caused by contact with water, barnacles, fish barbs, and other marine objects). The literature documenting simple gastroenteritis is scant and, accordingly, some experts question this disease having a *V. vulnificus* etiology (J.D. Oliver, personal communication). Primary septicemia is a rapidly developing disease that can result in death in less than 48 h after consuming seafood containing the bacteria. The bacteria move from the intestinal tract into the blood stream and, from there, set up serious infections in tissues, especially the extremities. Wound infections also most frequently involve extremities and will cause bullas (Fig. 4.1). When the infection

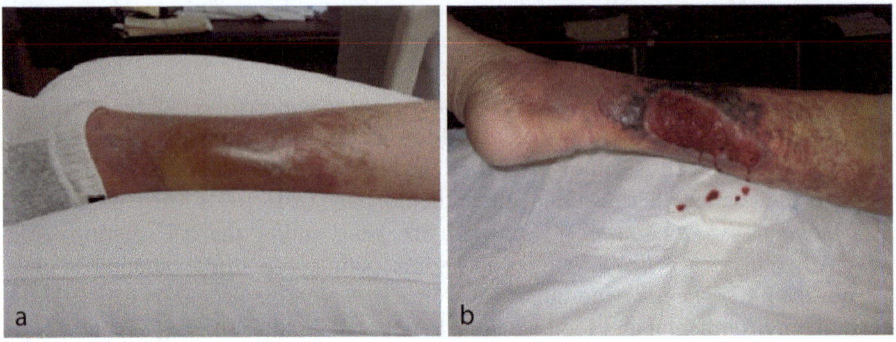

Fig. 4.1 Wound infection of unknown etiology, caused by a minor scratch from a barnacle. Within 36 h the scratch had developed into a bula (**a**). The bula was debrided (**b**) and did not progress into necrotizing fasciitis (Grimes DJ, Ekenna O unpublished data)

progresses into deep tissue and begins to cause death of that tissue, the disease is called necrotizing fasciitis. The infection shown in Fig. 4.1 did not develop into necrotizing fasciitis but is a good example of what often happens when patients are placed on aggressive antibiotic therapy to prevent progression of the infection. In this case, the patient was given three different antibiotics (clindamycin, levafloxacin, and doxycycline) prior to culture of the lesion, and these antibiotics probably caused the etiological agent to become nonculturable, a phenomenon observed and discussed by others [55, 126].

In general, *V. vulnificus* infections in healthy individuals are often not serious, but this is not always the case. Serious and fatal infections can occur in all patients and especially in those at risk; risk factors include: preexisting liver dysfunction or disease, diabetes, alcoholism, poor circulation, and immunosuppressive drug therapy. It is imperative that such individuals at risk are properly counseled about contact with seafood and seawater, so that they do not become infected with any *Vibrio* capable of causing disease in humans.

The virulence factors associated with *V. vulnificus* are not well understood. Known factors include *V. vulnificus* hemolysin (VVH) and RT [104]. There are two genotypes of *V. vulnificus* – clinical (C) and environmental (E) – and only C causes disease in humans. All strains of *V. vulnificus* have powerful iron sequestration ability (siderophores) which allows them to out-compete other species (including humans and fish) for essential iron. Capsular polysaccharide (CPS) formation is important and only encapsulated strains are virulent [104]. In addition, endotoxin (the lipoidal moiety of LPS) is very important and may be the most important cause of shock and death from *V. vulnificus* infections [104]. The complete genomic analysis of *V. vulnificus* YJ016 revealed the presence of RTX genes and they along with siderophore genes are located on Ch2 [21]. Type IV pilins, used as adherence mechanisms by many *Gammaproteobacteria* including *V. cholerae*, are a consistent feature of *V. vulnificus* and function in biofilm formation, attachment to epithelial cells, and possibly in the colonization of oysters [109, 110]. Other putative virulence factors have been described but their role in pathogenesis is unclear.

Aeromonas

Members of the genus *Aeromonas* are primarily freshwater bacteria that cause disease in both humans and aquatic animals. The most common isolates from human clinical specimens are *A. hydrophila, A. caviae, and A. sobria*, with *A. hydrophila* being the most common. Although *Aeromonas* spp. are frequently isolated from estuaries and the coastal ocean, they may not be truly autochthonous, and human disease from *Aeromonas* does not normally result from seawater or seafood. In fact, there is controversy about the ability of *Aeromonas*, a well-known pathogen of fish and amphibians, to cause disease in humans [48]. If one accepts the literature in support of human disease, it is believed that most human disease occurs in the form of gastroenteritis resulting from the ingestion of fish, shellfish, red meats, and contaminated water [48]. *Aeromonas* gastroenteritis presents either as a cholera-like disease with watery stools or a dysenteric-like illness that can include bloody stools [48]. Wound infections can also result from contact with contaminated water, either freshwater, seawater or brackish water [14, 75, 144].

Three recent reports of *Aeromonas* infections are illustrative of this seafood-borne and waterborne pathogen. In the aftermath of the 2004 tsunami that devastated eight countries and caused an estimated 225,000 deaths, the Thai Ministry of Public Health quickly began meeting health care needs in Thailand. From December 26, 2004 (the day of impact) to January 11, 2005, 1,237 cases of acute diarrhea, 356 wound infections, 177 febrile illnesses, and 156 respiratory illnesses were reported to the MOPH; only two deaths (both from aspiration pneumonia) resulted [14]. The most common isolates from the wound infections were *Proteus* spp., *Klebsiella* spp., *Pseudomonas* spp., *Enterobacter* spp. *E. coli*, and *A. hydrophila* (two isolates); surprisingly, no Vibrios were isolated, as was the case with Hurricanes Katrina and Rita.

Eighty-two strains of presumptive *Aeromonas* spp. were isolated from 250 frozen freshwater fish (Tilapia, *Oreochromis niloticus niloticus*) intended for human consumption and purchased in local markets in Mexico City. The isolates were identified with standard molecular techniques (16 S rRNA) and 88.3% were placed in two species – *A. salmonicida* (67.5%) and *A. bestiarum* (20.9%). The remaining isolates were identified as *A. veronii* (5.2%), *A. encheleia* (3.9%), and *A. hydrophila* (2.6%). The authors noted that this was one of the first major *Aeromonas* studies conducted in Mexico, and further noted that their results demonstrated the need for concern over putative pathogens with antimicrobial resistance and known virulence factors being present in food meant for human consumption [9].

Finally, a study of the prevalence of *A. hydrophila* in marketed seafood (fish and prawns) was conducted in land-locked city in South India by Vivekanandhana et al. [145]. Random samples of seafood (536 fish and 278 prawns) were collected from several vendors in a popular seafood market, and fish showing visible spoilage, injury or disease were avoided. Overall, 180 fish samples (33.6%) and 49 prawn samples (17.6%) contained *A. hydrophila*. The authors attributed the incidence to temperature abuse, fly contamination from a nearby sewage treatment plant, and the ability of *A. hydrophila* to grow at refrigerator temperatures (4–7°C); and they further noted that *A. hydrophila* is a pathogen of emerging importance [145].

Aeromonas Soft-Tissue Wound Infection
In late summer of 2010, a 7-year-old boy suffered a large and complex laceration injury to the right calf while recreational boating on a coastal river. He was brought within 2 h to the hospital, after initial first aid in the field. The wound was thoroughly washed and cleansed at surgery and closed with over 30 fine stitches. He was discharged the next day in stable condition and on oral antibiotics.

He failed to take the prescribed antibiotics because of nausea and vomiting (no diarrhea or fever). On readmission 6 days later, the wound was infected. Wound culture showed a rapid growth of a gram-negative rod (GNR) that was beta-hemolytic on blood agar, and also grew on MacConkey and chocolate agars. It was oxidase positive, mucoid, catalase positive, and motile.

The organism was confirmed to be *Aeromonas hydrophila*. It was found to be sensitive to second and third-generation cephalosoprins, quinolones, tetracycline, trimethoprim/sulfamethoxazole, and aminoglycosides. He responded well to intravenous ceftriaxone (a third-generation cephalosporin), local wound care, and later to applied skin graft to the injured calf. He made a full recovery.

The infection was most likely caused by wound contamination from the freshwater (river), and occurred because of a combination of factors: premature surgical closure of wound, and inability of the patient to take prescribed oral antibiotics to which the organism was sensitive.

Edwardsiella

There are three species of *Edwardsiella*, and the one that causes human disease is the opportunistic pathogen *E. tarda*. *E. tarda* causes a Salmonellosis-type enteritis in humans and typically derives from freshwater and freshwater animals (e.g., pet turtles). Infections from marine sources are unknown. *E. ictaluria* is a serious fish pathogen, often associated with septicemia in catfish (especially in aquaculture of catfish), but it is not known to cause disease in humans or marine fishes.

Yersinia

Several *Yersinia* spp. cause diseases in humans (Bubonic plague or Black Death, pseudotuberculosis, enteritis, extraintestinal complications) and in fish, including marine fish (salmonids). However, none of the marine fish pathogens are known to cause human disease and will therefore not be discussed here.

Brucella

Long known to cause disease in terrestrial animals and humans, these zoonotic bacteria are now known to exist in the ocean. The classic *Brucella* spp., *B. abortus*, *B. melitensis* and *B. suis*, cause brucellosis or undulant fever in humans and domestic animals. In domestic animals, the disease outcome is often abortion because the bacteria prefer to metabolize mesoerythritol which is found in the uterus and fetus of animals but not in humans. In humans, symptoms are general and include fever, chills, malaise, with heavy sweating, and high fever at night.

The three classic *Brucella* species have not been reported in marine mammals. Instead, it appears that marine mammals carry their own *Brucella* spp. [71], and they have caused three naturally acquired cases of human disease. Two cases were reported in Peru in 2003 (Sohn et al. 2003) and one in New Zealand [95]. The three *Brucella* human isolates were characterized with several molecular methods, and all three were found to share a common genotype with previously reported marine mammal *Brucella* spp. [152]. Representing 173 animals and one human patient were analyzed using a molecular method called multiple loci VNTR (variable number of tandem repeats), and the authors targeted 16 genetic loci (MLVA-16) that had been shown to be highly descriptive for *Brucella* spp. [93]. The study included two new species isolated from marine mammals, *B. ceti* and *B. pinnipedialis*, and concluded that these two species cluster into three distinct clades. Interestingly, the three isolates described by Whatmore et al. [152] did not cluster within the three clades but were, however, closely linked to the three marine mammal groups [93].

Enterococcus

Members of the genus *Enterococcus* are largely commensal colonizing organisms of the gastrointestinal tract of humans and warm-blooded animals and are commonly recovered in their feces [49]. These organisms are gram-positive facultative anaerobes that do not form spores but are capable of survival and growth in a wide variety of environmental conditions. These include tolerance of temperatures ranging from 10°C to 45°C, pH from 4.5 to 9.0 and high sodium chloride concentrations [63]. Although *Enterococcus* species have been found in many different marine and freshwater environments [85, 92, 124, 143, 158] as well as being associated with processed and fresh fish and seafood [31, 70, 99, 116, 154], these organisms are not usual pathogens for fish or marine mammals. *Enterococci* are known to be introduced into these environments by sewage contamination from known point sources, such as sewage treatment plants, and are used as indicator organisms for the probable presence of disease-producing pathogens in marine waters [147]. It is unlikely that point sources are the sole contributor of these organisms to an aquatic environment. Domestic and wild animals, water runoff from storms or agricultural sources, wind-driven sediment resuspension events, and humans utilizing the waters have all been shown to contribute to the presence of *Enterococcus* species in aquatic environments. ([1, 33, 42, 43, 107]; Rebarchik DM, Grimes DJ unpublished data).

The principal pathogenic *Enterococcus* in humans, *Enterococcus feacalis* and *E faecium*, are among the *Enterococcus* species isolated from aquatic environments; however, *Enterococci* found in marine settings have not been linked directly to the onset of human enterococcal infections. In general, these organisms are primarily associated with serious, often fatal, nosocomial infections, including postsurgical wound infections, endocarditis, urinary tract infections, and sepsis; and they are currently emerging common pathogens [49]. Enterococci lack significant virulence factors associated with disease but are intrinsically resistant to many antibiotics currently in use. These bacteria, especially *E. fecium*, are known to easily acquire antibiotic resistance genes from other microorganisms encountered in their environment. In addition to its importance as an indicator organism, the significance of *Enterococcus* in a marine-water setting is the increased likelihood that the organisms will be exposed to other microorganisms from which they might acquire antibiotic resistance genes, thus adding to the difficulty of treating an already challenging infection.

Streptococcus

Numerous species of fish are susceptible to infection by members of the genus *Streptococcus*. Although these infections are not common, when they do occur, it is often in an aquaculture setting and can be responsible for significant mortality and large economic costs. One species that is responsible for such infections, *Streptococcus iniae*, is a primary pathogen for fish that can also infect humans. This organism was first isolated from infected freshwater dolphins in 1996 [113] and has subsequently been associated with sporadic infections in multiple fish species [41, 77, 80, 115]. It was recognized as a human pathogen in the mid-1990s with several documented infections in North America [11, 150], and later in Japan [84]. *S. iniae* infections in humans present as fever and cellulitis, often with bacterimia, and can be treated with intravenous penicillin and gentamicin [84].

S. agalactiae (Lancefield group B) is a significant human pathogen especially in newborn infants where it can cause sepsis, pneumonia, and meningitis; and in pregnant women where it is associated with urinary tract infections. This organism has been linked to disease outbreaks and some massive kills in several fish species [121]. Investigations performed to type the bacteria isolated from infected fish and environmental samples indicated that sewage contamination was a likely source for the infections in fish [121, 125]. There is no evidence linking fish or a marine environment with human disease.

Listeria

The genus *Listeria* consists of six species including two that are recognized as pathogens. *L. monocytogenes* is an important human pathogen, and *L. ivanovii* is an animal pathogen that may very rarely infect humans but is not associated with marine related infections. *L. monocytogenes* are short, motile, gram-positive rods that may appear as coccobaccili. They are hardy organisms that grow as facultative anaerobes in a wide range of temperatures (from 1°C to 45°C), pH (pH 4.3 to pH 9), and they tolerate high salt concentrations. They are ubiquitous organisms that are commonly found in soil and water, on vegetation, and decaying matter and excreted in feces of humans and animals [89]. *L. monocytogenes* is the causative agent of listeriosis, a serious but rare infection caused by eating food contaminated with the bacteria. Multiple types of food have been associated with *Listeria* infections, classically soft cheeses made from unpasteurized milk, meat and processed meat products, and fish and shell fish [87]. Food may become contaminated before, during, or after preparation; and the usual measures for prevention of growth of contaminating organisms, low temperatures, extremes of pH, and high salt, are ineffective against *Listeria*. Included among the marine sources implicated in food contamination are crab meats and dips, lobster and shell fish, and many varieties of fish, especially those smoked or processed [6, 36, 81–83]. In a recent report from the CDC addressing the incidence of infection with pathogens that are commonly transmitted via a food borne route, of the greater than that 17,000 culture-proven infections, Listeria accounted for less than 1% of cases with an incidence of 0.34 per 100,000 [18].

Listeriosis occurs primarily in pregnant women, newborn infants, elderly patients, and patients who are immunocompromised, and in all but the newborns; the infections result in an initial mild influenza-like illness that may progress to sepsis and meningitis. In pregnant women, there is an increased risk of miscarriage, and in newborns the infection is associated with a high mortality, up to 40%, and long-term side effects [7, 120]. Adults and children acquire the disease after ingestion of contaminated foods, whereas newborns acquire infecting organisms either transplacentally or via an ascending route during labor and delivery. Onset of the disease varies with population and exposure routes, and it may be from days to weeks. The mortality rates for infections in these populations are high, up to 25%. *Listeria* are facultative intracellular pathogens, a characteristic that contributes to the severity of listeriosis. After being phagocytosed, the bacterium utilizes unique virulence factors to spread from cell to cell without an extracellular state, thus evading the humoral immune response. Included among these factors are internalins (inlA, inlB, inlC) that facilitate attachment to the cells. Once inside the cell, listerolysin O and phospholipase C enzymes act to release the bacteria to the cytosol where the bacteria utilizes a protein, ActA, to coordinate the assembly of actin into a "tail" that propels the bacterium across one cell and into the next. *Listeria* infections can be treated with common antibiotics, such as ampicillin, ciprofloxacin, linezolid and azithromicin.

Mycobacterium

The genus *Mycobacterium* consists of numerous aerobic, nonmotile, non-spore-forming, acid-fast rods that occur widely in nature. These bacteria range from soil-dwelling saprophytes to pathogens of humans and animals. Of the greater than 20 human pathogenic *Mycobacterium* species, few have been associated with infections from or in a marine environment [155]. *Mycobacterium marinum* (synonym *M balnei*) is the primary aquatic pathogen and is a free living organism found throughout the world in both fresh and marine water environments. It was first discovered as a pathogen for fish, causing skin nodules and ulcers, in the mid-1920s and has since been recognized as a natural pathogen for other *ectotherms*, such as frogs. Since the early 1950s [88] it has been recognized as a cause of human infections first described as causing superficial skin lesions, nodules referred to as "swimming pool granulomas," in children who swam in contaminated pools. Like most other human pathogenic Mycobacteria, *M marinum* grows slowly, having a typical incubation period of 2–4 weeks before cutaneous lesions appear, but occasionally may progress rapidly. Unlike other human pathogenic Mycobacteria, it optimally grows at lower temperatures, 30–33°C, and is usually inhibited at 37°C, helping to explain the typical location of the lesions on the extremities and the usual lack of systemic involvement.

Currently *M marinum* is the most common cause of atypical *Mycobacterium* infection in humans with a reported incidence of 0.27 cases per 100,000 adults [128] Infection with these organisms can occur at any age, but it usually occurs in adults who handle fresh- or saltwater fish or fish tanks, and is now rarely associated with swimming pools, as proper construction and chlorination has eliminated this source [40]. Human exposure now primarily comes through inoculation of existing skin abrasions while handling fish or their tanks, or directly through fish bites or puncture wounds from fins. Local trauma is an important factor in establishing *M marinum* infections and their sequelea. Infections obtained after inoculation of an existing abrasion or a direct puncture manifests as a localized nodule or granuloma at the site of bacterial entry that may develop into an ulcer or progress to involve nearby lymph nodes, sporotrichotic lymphangitis. In healthy individuals it rarely progresses to involve bones, joints, or other systemic sites. Immunocompromised states increase the risk for becoming infected and can be associated with more aggressive systemic disease [111, 151]. Diagnosis and treatment are often delayed because of a lack of suspicion for mycobacterial involvement versus more common bacterial pathogens and are contributed to by the long incubation period.

Treatment for *M marinum* is driven by the severity of the infection [122] and ranges from oral monotherapy with minocycline, clarithromycin, doxycycline, ciprofloxacin, and trimethoprim-sulfamethoxazole for superficial cutaneous infections with susceptible organisms to combination therapies for drug resistant strains. Severe infections, including those with a sporotrichoid distribution pattern, generally require combination therapy with rifampicin and ethambutol. Surgical debridement is not usually recommended, however, other alternative topical therapies, such as cryotherapy, X-ray therapy, electrodesiccation, photodynamic therapy, and local hyperthermic therapy can be effective.

Allochthonous Pathogens

Escherichia

The first species of this genus, *Bacterium coli*, was first isolated in the late 1800s and it was proposed as an indicator of fecal contamination of water. The isolate was renamed *Escherichia coli* in 1919 and today remains in use as the fecal indicator recommended by the USEPA for freshwater [142]. Unfortunately, *E.coli* and the enterococci do not correlate with indigenous pathogens such as the vibrios [29] and enteric viruses (see s.no. 12, J. Woods, Waterborne Disease of the Ocean, Enteric Viruses, this volume). In addition to its worldwide use as a fecal indicator, *E. coli* is the most common cause of urinary tract disease in humans (ca. 90% of human UTIs): certain strains cause gastroenteritis of various degrees of severity (e.g., STEC or the Shiga toxin-producing *E. coli* O157 usually derived from meats and produce) and nonpathogenic strains such as *E. coli* K12 that laid the early foundation for much of what is known about metabolism and enzymology.

Despite extensive documentation of *E. coli* causing human disease from the consumption of contaminated raw and undercooked foods, oral–fecal transmission in public facilities such as nurseries and day care centers and contaminated drinking water, transmission from seafood and seawater is rare and does not occur in the USA with great frequency. The overall foodborne STEC O157 incidence in the USA for 2009 was 0.99 per 100,000 population, and the STEC non-O157 incidence was 0.57 [18] for many years. This disease agent will not be further discussed in this chapter, and readers interested in this important pathogen are encouraged to peruse other literature.

Shigella

Shigella spp. cause a significant incidence of foodborne disease worldwide, but they are not often acquired from the ocean. Some strains produce the powerful Shiga toxin which, like CTX, is a lysogenic conversion product. In the USA, shigellosis (bacillary dysentery) has a fairly high incidence (3.99 per 100,000) as a foodborne agent of disease [18].

Salmonella

The Salmonellae are important pathogens but are much like *E. coli* and *P. aeruginosa*, with regard to their importance as marine pathogens. There are two species of *Salmonella*, *S. enterica* and S. *bongori*, and these two species are comprised of many serovars and subspecies. Both *Salmonella enterica* serotype Typhi (formerly *S. typhi*) and the other gastroenteritis-causing *Salmonella* spp. are human pathogens that historically were frequently acquired from the ingestion of contaminated seafood (especially filter-feeding bivalves, such as raw or undercooked oysters) and seawater. With the advent of fecal indicator monitoring of seafood and seawater, refrigeration of seafood, sanitary surveys (especially surveys of molluscan shellfish beds), and sewage treatment in developed nations, the origin of these diseases from the ocean declined significantly. Today, diseases caused by Salmonellae are still frequent worldwide, most commonly caused by *S. enterica* subspp. *enterica*, and they are almost always foodborne in both developed and developing countries (see http://www.who.int/mediacentre/factsheets/fs139/en/; [97]). In the USA, CDC reported a Salmonella foodborne disease incidence of 15.9 per 100,000 population; this is the highest incidence of any foodborne disease, but there was no breakdown on food type, i.e., seafood incidence was not given [18].

Non-typhoid salmonellosis in humans is most commonly gastroenteritis although complications, such as septicemia, can occur; deaths are rare if the patients are kept hydrated and placed on appropriate antibiotic treatment. Virulence is largely determined by pathogenicity islands, and the non-typhoid Salmonellae contain at least 12 of these genetically mobile elements [65]. Human salmonellosis is usually acquired from food, although contaminated water can also serve as a vehicle for transmission. The Food and Agriculture Organization (FAO) of the United Nations recently published a report on the control of *Salmonella* in sustainable aquaculture, and it contains a very nice review of occurrence and survival in the aquatic environment [44].

The incidence of salmonellosis deriving from seafood in the USA is low but is probably far underreported. DePaola et al. [34] recently provided evidence that, even though the reported incidence is low, the potential for acquisition in the USA certainly exists. They conducted a 2-year study of market oysters collected twice each month from retail establishments (restaurants and raw bars, seafood markets, wholesale dealers) in nine states. In all, FDA collected 397 samples representing 258 establishments. Salmonella was detected in 8.6% of the market oysters, a rate only exceeded by *V. parahaemolyticus* and *V. vulnificus* [34].

In the FAO report [44], it was noted that many studies have shown *Salmonella* serotype Sneftenberg to be the major serotype in marine environments and raw seafood worldwide. The report further noted that *Salmonella* spp. have been isolated from many marine mammals.

Morganella

Human disease from *Morganella morganii* is common (postoperative and other nosocomial) but these infections rarely emanate from the ocean. However, *M. morganii is* often associated with the decomposition of seafood; and if such seafood is consumed, scombroid fish poisoning can result. Scombroid results from histamine build up (and possibly buildup of other vasoactive amines) in the seafood as a result of histidine decarboxylation during the seafood spoilage process. This disease is a true food poisoning or intoxication, as opposed to a food infection, e.g., Salmonellosis and *Vibrio* gastroenteritis. Scombroid is probably caused by several enteric bacteria, including *Proteus*, spp. *Klebsiella pneumoniae, Hafnia alvei, Enterobacter* spp., *Serratia* spp., and *Citrobacter freundii;* in addition, *V. alginolyticus, Aeromonas* spp., and *Photobacterium* spp. are also histamine formers; and all have been isolated from spoiled fish [139]. The incidence of scombroid in the USA is thought to be common (http://www.fda.gov/Food/FoodSafety/FoodborneIllness/FoodborneIllnessFoodbornePathogens NaturalToxins/BadBugBook/ucm070823.htm), and worldwide incidence is also common. However, because scombroid is not a reportable disease, documented cases are very low, e.g., only 103 incidents involving 4 people were reported from 1968 to 1980 (see above FDA URL). The most common fish involved with this intoxication are tuna, bonito, mackerel, and *mahi mahi*, and once the amines are formed in the meat, neither cooking, canning, or freezing lowers toxicity. Onset of this intoxication is rapid (often 30 min) and symptoms include a tingling or burning sensation in the mouth, rash on the upper body, and a drop in blood pressure; nausea, vomiting, and diarrhea may also present, and hospitalization may be required.

Pseudomonas

The most common human pathogen in this genus is *Pseudomonas aeruginosa*, and, like the Vibrios, this species is metabolically very diverse. *P. aeruginosa* is a very common cause of death in third-degree burn patients, it is a common nosocomial agent of disease, is frequently resistant to most clinically useful antibiotics, it is a cause of urinary tract infections, and it is a common (and often lethal) complication of cystic fibrosis. There have been numerous reports of human *P. aeruginosa* infections occurring from various types of freshwater contact (e.g., swimming pools, whirlpools, hot tubs, atomizers), but this literature is far too extensive to be summarized here. Although *P. aeruginosa* is frequently isolated from the coastal ocean [58], the authors are not aware of any literature documenting that disease caused by *P. aeruginosa* came from the ocean. It is primarily a freshwater bacterium [114].

Campylobacter

Campylobacter spp. are frequently isolated from healthy cattle, chickens, and birds [48], and they are also associated with several foods (unpasteurized milk, poultry, shellfish, fruits, and vegetables), freshwater ponds, and streams contaminated with fecal material [69]; Campylobacters are not normally isolated from seawater. While a few reports of seawater isolations exist [3, 58], most marine isolates come from shellfish [69]. The association with shellfish is similar to that of *Salmonella*, in that shellfish acquire *Campylobacter* spp., and usually *C. jejuni*, from filter-feeding in water contaminated with fecal material [2]. The incidence of foodborne *Campylobacter* disease in the USA is 13.02 per 100,000, second only to Salmonella [18]. However, seafood-borne disease is rare [2, 69].

Staphylococcus

The genus *Staphylococcus* is made up of at least 40 species of gram-positive, facultative anaerobic organisms that are found throughout the world. Most of these organisms exist as commensal colonizing organisms of animals and humans, but may also be found in soil, on surfaces and in untreated water. They are hardy organisms that grow in the presence of bile salts and NaCl (up to 6.5%), and they can survive on many types of surfaces for extended periods of time making them a challenge to eliminate in public environments, such as gyms, prisons and hospitals. *Staphylococcal* species are differentiated from other important gram-positive organisms by the presence of the enzyme catalase, and they are differentiated amongst themselves by the presence of the enzyme coagulase which is present in the more clinically relevant organisms.

Among the organisms that make up this genus are a large variety of coagulase-negative staphylococci (CoNS) that are the primary commensal colonizers of humans. CoNS are pathogenic primarily for compromised populations only, such as preterm infants or persons with implanted prosthetic devices. *Staphylococcus aureus* is coagulase positive, and this species is the principle pathogen associated with human infection. In humans, *S. aureus* strains are opportunistic pathogens that may colonize, without infecting, up to 40% of the population [24, 79, 146], but may occasionally gain access to the host, evade the immune response, and causes disease [52, 53]. In addition, these organisms have acquired resistance to most of the antibiotics used against them making treatment of infections challenging.

Most infections caused by *S. aureus* are limited to the cutaneous tissues and are caused by a person's own colonizing organisms. However, these bacteria are also capable of causing serious, life threatening systemic disease. In fact, *S. aureus* including the methicillin resistant *S. aureus*, MRSA, are among the leading causes of *nosocomial* infections [10, 78]. Furthermore, MRSA have emerged as significant causes of community as well as hospital-associated infections [38, 78].

The diseases caused by *S. aureus* in humans and animals are often produced through the action of specific toxins or virulence factors that different bacterial isolates can produce. Specific toxins are associated with particular syndromes, such as toxic shock syndrome toxin and scalded skin syndrome. Many *S. aureus* are also capable of producing and secreting toxins responsible for staphylococcal food poisoning, termed enterotoxins, that only need to be ingested to cause intoxication and do not require the continued presence of the bacteria for disease. Consumption of seafood contaminated with *S. aureus* producing enterotoxins leads to staphylococcal food poisoning. Contamination of the food products often comes during processing as is seen in Listeriosis; however, organisms in the water and associated with marine life as seen in Peter the Great Bay, Japan, and Nha Trang Bay, South China, seas may also contribute to human disease [5].

S. aureus including drug resistant MRSA and CoNS have been isolated at recreational beaches, from marine ccc and temperate environments [5, 19, 50, 57, 133]. Adults and toddlers in diapers have been shown to shed *S. aureus* and the indicator organism *Enterococcus* into recreational marine waters and sand [118]. Persons using these recreational beaches may transmit and receive these organisms from the environment [42, 51, 131]. A retrospective epidemiological/ microbiological monitoring study performed in Hawaii in subtropical marine waters and beaches found that persons were four times more likely to have staphylococcal skin infections if they had a history of seawater contact [20]. Whether there is a correlation of the microbial load in these environments and increased infections is yet to be verified; however, a recent study performed at a South Florida recreational beach did show a correlation between the average number of bathers in the water and the presence of *S. aureus* [119].

S. aureus and MRSA have also been isolated from marine mammals, including bottlenose dolphins (*Tursiops truncates*), seals, and walruses (blowholes, gastric fluids, fecal and anal cultures), both in captivity and in the wild, and have been associated with both colonization and disease [45, 105, 129]. In marine mammals in captivity, it is likely that the source of *S. aureus* and MRSA are colonized human handlers. The source of these organisms for the non-captive animals is not clear; however, colonized wild mammals were primarily identified in locations associated with human recreational use in the estuarine waters of Charleston, SC and Indian River Lagoon, FL [100, 129]. It has been suggested that some of the marine mammals, such as bottlenose dolphins, might serve as sentinels for transfer of resistant organisms from humans and animals into this environment, or simply indicate that the antibiotics are reaching this environment. To date, there are no confirmed cases of human infection from colonized or infected marine mammals.

***Staphylococcus aureus* and MRSA at Recreational Beaches**

Since the early 1990s, investigators in Hawaii have isolated *S. aureus* from the waters used for recreation and suspected that exposure of bathers to the organisms in this environment might put them at risk for staphylococcal infections. Recent investigations have also isolated *S. aureus* and methicillin resistant, MRSA, from recreational marine beaches at multiple locations, including Hawaii, Puget Sound, California, and South Florida. The sources of the bacteria in these environments are likely multiple and not yet completely appreciated. *S. aureus* are not known to have a marine reservoir; however, humans and other mammals that may be present at the beach are known to be colonized with the bacteria and are potential sources. In fact, humans have been shown to readily shed their colonizing methicillin sensitive *S. aureus* (MSSA) and MRSA into marine waters and sand.

A recent study completed at a South Florida recreational beach collected MSSA and MRSA from ambient water, water nearby bathers and sand and evaluated the bacteria present in these environments for their potential to be associated with infection by determining the virulence factors they could produce. The study showed that 30–37% of water samples had *S. aureus*; however, it was unable to show a correlation between exposure to *S. aureus* and reported illness. The majority, greater than 97%, of *S. aureus* found at this location were MSSA that carried few virulence factors known to be associated with infection. However; the MRSA isolated from this location were similar to the MRSA found in the community that are known to have the potential to cause serious infections. The lack of association between exposure to *S. aureus* and illness in this study, at this location, and the lack of an adequate number of participants (sample size) to establish an association with an organism that is present in only present 37% of samples overall is explained in part by the low percentage of potentially virulent MRSA. The populations and concentrations of MSSA and MRSA at other more crowded, recreational beaches would likely be as different as the human populations utilizing these beaches. Bathers exposed to greater numbers of more virulent organisms could be at increased risk for infections. Further studies are required to establish the true risk to bathers exposed to MSSA and especially MRSA in these settings.

General Treatment Principles

Many of the marine or ocean-dwelling microorganisms important in human disease are gram-negative bacteria. The diseases caused by these microorganisms run the spectrum, from septicemia (e.g., *V. vulnificus*), to gastroenteritis, wound infections, ear infections, and eye infections. It is not surprising that treatment is also varied from syndrome to syndrome.

Cholera (*V. cholera*) is the classic example of a cause of noninflammatory severe gastroenteritis (rice water stools). In cholera, the diarrhea is toxin induced. The pathogenesis of diarrhea in other organisms may be inflammatory (e.g., *Salmonella, Campylobacter* or *Shigella*). The nature or mechanism of the diarrhea may affect the primary choice of, or decision regarding, treatment.

As a general rule, severe diarrheal diseases require fluid replacement, given either by mouth (oral rehydration) or by intravenous means as the primary mode of treatment, while inflammatory diarrheas require, in addition to fluids, also antimicrobial therapy. The mode of fluid delivery will depend on how sick the patient is, the availability of which treatment, and the capacity or resources available to deliver the fluid replacement to the patient.

Sometimes, antimicrobial treatment may be detrimental in severe diarrhea. A good example is the bloody diarrhea caused by *E. coli* 0157:H7, which may lead to hemolytic-uremic syndrome, especially in children exposed to both the toxin-producing *E. coli*, as well as antibiotic therapy [156]. In such cases, withdrawal of antimicrobial therapy may sometimes help prevent further complications of renal failure [108]. Treatment in this circumstance is entirely supportive.

For toxin-producing diarrheas, antibody binding in situ presents an attractive and elegant option, but this type of therapy has not yet been developed (for all practical purposes) for most diarrheal diseases seen in the clinical setting [101].

Septicemia (e.g., due to *V. vulnificus*) requires aggressive management in the intensive care unit (ICU) setting, surgical debridement where necessary, as well as antimicrobial therapy. Severe wound infections like necrotizing fasciitis require primary surgical debridement, antimicrobial treatment, and sometimes hyperbaric oxygen therapy.

Most patients with severe wound infections or septicemia do not typically die or have other adverse outcomes due to or as a consequence of antimicrobial resistance, but complications result because they present too late to the hospital, have devitalized tissues that were not promptly debrided, or for other host factor reasons.

In our experience, most *Vibrio* organisms seen in clinical practice on the Gulf Coast are sensitive to the third-generation cephalosporins (sometimes also to second generation cephalosporins), quinolones, tetracyclines, and aminoglycosides. The same goes for *Aeromonas hydrophila*, with similar susceptibilities as above, in addition to usual sensitivity to trimethorprim/sulfamethoxazole.

Acute diarrheal illnesses require antimicrobial therapy usually only for a short period (5 days is typically enough); while septicemia (e.g., typhoid fever and *Vibrio* sepsis) would require longer therapy (2 weeks or longer), depending on the complications.

Severe wound infections of the necrotizing fasciitis type often require multiple surgical debridements, in addition to antimicrobial therapy and local wound care. In addition, skin grafting or plastic surgery is often required to cover defective skin or tissue.

The key to successful treatment of all of these disease entities is timeliness in starting treatment. The earlier appropriate therapy is started, the better the clinical outcome. The later treatment is started, the more complications one can expect.

Adjustments in antimicrobial therapy can and should be made after in vitro antimicrobial susceptibility studies are available. However, prompt treatment must be started very early, empirically (best guess or educated guess), before the laboratory susceptibility reports are available.

The Vibrios have predictable antimicrobial susceptibilities, more so than *Salmonella, E. coli,* or *Shigella.* Antimicrobial resistance to enteric pathogens reflects the pattern of use of antibiotics in a given environment, as well as the ease of antimicrobial drug availability and abuse in the area or locale where the infection was acquired. Typically for *Aeromonas*, resistance to ampicillin-like agents and first-generation cephalosporins is common; these agents should therefore not be used to treat infections due to *Aeromonas.*

In general, a gram-negative bacillus found in coastal or ocean water (outside and far away from a sewage drainage site) is likely to be free-living and, therefore, is more likely to be sensitive to multiple antimicrobial agents. Organisms causing disease acquired through human-to-human or foodborne transmission (e.g., *Salmonella*), on the other hand, are more likely to have been previously exposed to antibiotics (e.g., in animals or food products). Infections acquired through such contacts may therefore be more resistant to antimicrobials than free-living ocean, river, or brackish water bacteria [4].

The sensitivity of human-to-human or animal-to-human transmitted gram-negative bacteria (*E. coli, Salmonella, Shigella, Pseudomonas,* etc.) will usually reflect the pattern of local prevailing antimicrobial use in that community. The local hospital's antibiogram should provide the initial guide in the choice of empiric therapy, with necessary adjustments made after in vitro susceptibility studies are available.

Antimicrobial therapy for Brucellosis is often prolonged (up to 6 weeks) in order to prevent relapse. Often, combination therapy that includes a tetracycline plus rifampin, or an aminoglycoside, is required and recommended for complete cure of this debilitating disease [127, 132].

Acute gastroenteritis caused by food poisoning (e.g., staphylococcal preformed heat-stable enterotoxin) is often rapid in onset (within 1–6 h of food ingestion) and is also self-limited. The symptoms of severe nausea and vomiting occur usually within 1–6 h, and are usually over in less than 24 h [13]. Antimicrobial therapy is usually not required.

Treatment Considerations for Gram-Positive Organisms

Enterococci lack significant virulence factors associated with disease but are intrinsically resistant to many antibiotics currently in use. These bacteria are known to easily acquire antibiotic resistance genes from other microorganisms encountered in their environment. Treatment is guided by the determined antibiotic sensitivities of the infecting organisms and may be prolonged.

Streptococcus iniae infections usually present after exposure to fish with fever and cellulitis, often with bacterimia, and can be treated with intravenous penicillin and gentamicin [84]. *S. agalactiae* (Lancefield group B) remains sensitive to penicillin.

Listeria infections can be treated with common antibiotics, such as ampicillin, ciprofloxacin, linezolid and azithromicin. The delivery method of antibiotic is determined by the severity of diease.

Treatment for *Mycobacterium marinum* is driven by the severity of the infection [122] and ranges from oral monotherapy with minocycline, clarithromycin, doxycycline, ciprofloxacin, and trimethoprim-sulfamethoxazole for superficial cutaneous infections with susceptible organisms to combination therapies for drug-resistant strains. Severe infections, including those with a sporotrichoid distribution pattern, generally require combination therapy with rifampicin and ethambutol. Surgical debridement is not usually recommended however other alternative topical therapies such as cryotherapy, X-ray therapy, electrodesiccation, photodynamic therapy, and local hyperthermic therapy can be effective.

The majority of infections by *Staphylococcus aureus* are cutaneous infections limited to skin and soft tissues. Minor skin infections are usually treated with topical antibiotics, such as a nonprescription triple-antibiotic mixture or mupirocin. In some cases, oral antibiotics may be given for more severe skin infections. If abscesses are present, surgical drainage may be required and for smaller abscesses may be curative. More serious and life-threatening systemic infections are treated with intravenous antibiotics. The choice of antibiotic depends on the susceptibility of the particular staphylococcal strains, as determined by culture results in the laboratory. MRSA from the community may be sensitive to several antibiotics effective against MSSA; however, hospital-associated MRSA are usually resistant multiple antibiotics and may be challenging to treat. Vancomycin remains the drug of choice for multidrug resistant MRSA.

Future Directions

Clearly, most of what is known about waterborne human pathogens is based on laboratory and clinical observations, and this is especially true for autochothonous bacteria. The diseases and metabolic capabilities of these pathogens can be described in great detail, and in many cases the genomes of these bacteria have been completely sequenced. However, until their role or niche in the ocean and in freshwater habitats is fully investigated, it will not be possible to fully understand their ability to cause disease in humans. Accordingly, scientists need to continue asking the question, "how does this microorganism live in the ocean and yet invade humans to cause disease?"

Bibliography

Primary Literature

1. Abdelzaher AM, Wright ME, Ortega C, Solo-Gabriele HM, Miller G, Elmir S, Newman X, Shih P, Bonilla JA, Bonilla TD, Palmer CJ, Scott T, Lukasik J, Harwood VJ, McQuaig S, Sinigalliano C, Gidley M, Plano LR, Zhu X, Wang JD, Fleming LE (2010) Presence of

pathogens and indicator microbes at a non-point source subtropical recreational marine beach. Appl Environ Microbiol 76:724–732
2. Abeyta C, Deeter FG, Kaysner CA, Stott RF, Wekell MM (1993) *Campylobacter jejuni* in a Washington state shellfish growing bed associated with illness. J Food Prot 56:323–325
3. Alonso JL, Alonso MA (1993) Presence of *Campylobacter* in marine waters of Valencia, Spain. Water Res 27:1559–1562
4. Baya AM, Brayton PR, Brown VL, Grimes DJ, Russek-Cohen E, Colwell RR (1986) Coincident plasmids and antimicrobial resistance in marine bacteria isolated from polluted and unpolluted Atlantic Ocean samples. Appl Environ Microbiol 51:1285–1292
5. Beleneva IA (2011) Incidence and characteristics of Staphylococcus aureus and Listeria monocytogenes from the Japan and South China seas. Mar Pollut Bull 62(2):382–387
6. Boerlin P, Boerlin-Petzold F et al (1997) Typing Listeria monocytogenes isolates from fish products and human listeriosis cases. Appl Environ Microbiol 63(4):1338–1343
7. Bortolussi R (2008) Listeriosis: a primer. CMAJ 179(8):795–797
8. Brock TD (1999) Robert Koch, a life in medicine and bacteriology. ASM Press, Washington, DC
9. Castro-Escarpulli G, Figuerasb MJ, Aguilera-Arreolaa G, Solerb L, Ferna´ndez-Rendo´na E, Aparicioa GO, Guarrob J, Chaco´n MR (2003) Characterisation of *Aeromonas* spp. isolated from frozen fish intended for human consumption in Mexico. Int J Food Microbiol 84:41–49
10. Centers for Disease Control and Prevention (1999) National Nosocomial Infections Surveillance system report: data from 1997–1999, Atlanta, GA
11. CDC (1996) Invasive infection due to *Streptococcus iniae* – Ontario, 1995–1996. Morb Mortal Wkly Rep 45:650–653
12. CDC (1998) Outbreak of *Vibrio parahaemolyticus* infections associated with eating raw oysters – Pacific Northwest, 1997. MMWR 47:457–462
13. CDC (2001) Diagnosis and management of foodborne illnesses: a primer for physicians. MMWR 50(RR02):1–69
14. CDC (2005) Rapid health response, assessment, and surveillance after a tsunami – Thailand, 2004–2005. MMWR 54:61–64
15. CDCb (2006) Vibrio parahaemolyticus infections associated with consumption of raw shellfish – three states, 2006. MMWR 55:1–2
16. CDCa (2006) Two cases of toxigentic Vibrio cholerage O1 infection after Hurricanes Katrina and Rita – Louisiana, October 2005. MMWR 55(02):31–32
17. CDC (2009) Preliminary FoodNet data on the incidence of infection with pathogens transmitted commonly through food – 10 states, 2008. MMWR 58:333–337
18. CDC (2010) Preliminary foodnet data on the incidence of infection with pathogens transmitted commonly through food – 10 states, 2009. MMWR 59:418–422
19. Charoenca N, Fujioka R (1993) Assessment of staphylococcus bacteria in Hawaii marine recreational waters. Water Sci Technol 27:283–289
20. Charoenca N, Fujioka RS (1995) Association of staphylococcal skin infections and swimming. Water Sci Technol 31:11–17
21. Chen C-Y, Wu K-M, Chang Y-C, Chang C-H, Tsai H-C, Liao T-L, Liu Y-M, Chen H-J, Shen AB-T, Li J-C et al (2003) Comparative genome analysis of *Vibrio vulnificus*, a marine pathogen. Cold Spring Harbor Laboratory Press, pp 2577–2587
22. Colwell RR, West PA, Maneval D, Remmers EF, Elliot EL, Carlson NE (1984) Ecology of the pathogenic vibrios in Chesapeake Bay. In: Colwell RR (ed) Vibrios in the environment. Wiley, New York
23. Chowdhury NR, Chakraborty S, Ramamurthy T, Nishibuchi M, Yamasaki S, Takeda Y, Nair GB (2000) Molecular evidence of clonal Vibrio parahaemolyticus pandemic strains. Emerg Infect Dis 6:631–636
24. Cole AM, Tahk S, Oren A, Yoshioka D, Kim YH, Park A, Ganz T (2001) Determinants of *Staphylococcus aureus* nasal carriage. Clin Diagn Lab Immunol 8:1064–1069

25. Colwell RR (1996) Global climate and infectious disease: the cholera paradigm. Science 274:2025–2031
26. Colwell RR, Huq A, Islam MS, Aziz KMA, Yunus M, Khan NH, Mahmud A, Sack RB, Nair GB, Chakraborty J, Sack DA, Russek-Cohen E (2003) Reduction of cholera in Bangladeshi villages by simple filtration. Proc Natl Acad Sci 100:1051–1055
27. Colwell RR (2006) A global and historical perspective of the genus *Vibrio*. In: Thompson FL, Austin B, Swings J (eds) The biology of Vibrios. ASM Press, Washington, DC, pp 3–11
28. Constantin de Magny G, Murtugudde R, Sapiano MRP, Nizam A, Brown CW, Busalacchi AJ, Yunus M, Nair GB, Gil AI, Lanata CF, Calkins J, Manna B, Rajendran K, Bhattacharya MK, Huq A, Sack RB, Colwell RR (2008) Environmental signatures associated with cholera epidemics. Proc Natl Acad Sci 105:17676–17681
29. Cook DW, Ruple AD (1989) Indicator bacteria and Vibrionaceae multiplication in post-harvest shellstock oysters. J Food Prot 52:343–349
30. Croci L, Suffredini E, Cozzi L, Paniconi M, Ciccaglioni G, Colombo MM (2007) Evaluation of different polymerase chain reaction methods for the identification of *Vibrio parahaemolyticus* strains isolated by cultural methods. J AOAC Int 90:1588–1597
31. Dalgaard P, Vancanneyt M et al (2003) Identification of lactic acid bacteria from spoilage associations of cooked and brined shrimps stored under modified atmosphere between 0 degrees C and 25 degrees C. J Appl Microbiol 94(1):80–89
32. DePaola A, Kaysner CA, Bowers J, Cook DW (2000) Environmental investigations of *Vibrio parahaemolyticus* in oysters after outbreaks in Washington, Texas, and New York (1997 and 1998). Appl Environ Microbiol 66:4649–4654
33. DePaola A, Nordstrom JL, Bowers JC, Wells JG, Cook DW (2003) Seasonal abundance of total and pathogenic *Vibrio parahaemolyticus* in Alabama oysters. Appl Environ Microbiol 69:1521–1526
34. DePaola A, Jones JL, Woods J, Burkhardt W III, Calci KR, Krantz JA, Bowers JC, Kasturi K, Byars RH, Jacobs E, Williams-Hill D, Nabe K (2010) Bacterial and viral pathogens in live oysters: 2007 United States survey. Appl Environ Microbiol 76:2754–2768
35. DeLong EF, Pace NR (2000) Environmental diversity of bacteria and archaea. Syst Biol 50:470–478
36. Destro MT (2000) Incidence and significance of Listeria in fish and fish products from Latin America. Int J Food Microbiol 62(3):191–196
37. Dibrov P (2005) The sodium cycle in *Vibrio cholerae*: riddles in the dark. Biochemistry (Mosc) 70:150–153
38. Diep BA, Carleton HA et al (2006) Roles of 34 virulence genes in the evolution of hospital- and community-associated strains of methicillin-resistant *Staphylococcus aureus*. J Infect Dis 193(11):1495–1503
39. Eckburg PB, Lepp PW, Relman DA (2003) Archaea and their potential role in human disease. Infect Immun 71:591–596
40. Edelstein H (1994) Mycobacterium marinum skin infections. Report of 31 cases and review of the literature. Arch Intern Med 154(12):1359–1364
41. Eldar A, Bejerano Y, Bercovier H (1994) *Streptococcus shiloi* and *Streptococcus difficile*: two new streptococcal species causing a meningoencephalitis in fish. Curr Microbiol 28:139–143
42. Elmir SM, Wright ME, Abdelzaher A, Solo-Gabriele HM, Fleming LE, Miller G, Rybolowik M, Peter Shih MT, Pillai SP, Cooper JA, Quaye EA (2007) Quantitative evaluation of bacteria released by bathers in a marine water. Water Res 41:3–10
43. Elmir SM, Shibata T, Solo-Gabriele HM, Sinigalliano CD, Gidley ML, Miller G, Plano LR, Kish J, Withum K, Fleming LE (2009) Quantitative evaluation of enterococci and Bacteroidales released by adults and toddlers in marine water. Water Res 43:4610–4616
44. FAO (2010) FAO expert workshop on the application of biosecurity measures to control *Salmonella* contamination in sustainable aquaculture. Food and Agriculture Organization of the United Nations, Rome (ISBN 978-92-5-106553-2)

45. Faires MC, Gehring E et al (2009) Methicillin-resistant Staphylococcus aureus in marine mammals. Emerg Infect Dis 15(12):2071–2072
46. Farmer JJ (1980) Revival of the name *Vibrio vulnificus*. Int J Syst Bacteriol 30:656
47. Farmer JJI, Janda M, Brenner FW, Cameron DN, Birkhead KM (2005) Genus 1. *Vibrio*. In: Brenner DJ, Krieg NR, Staley JT, Garrity GM (eds) Bergey's manual of systematic bacteriology, vol 2, The proteobacteria, Part B. The gammaproteobacteria. Springer, New York, pp 494–546
48. FDA (2009) Bad bugs book: foodborne pathogenic microorganisms and natural toxins handbook. http://www.fda.gov/Food/FoodSafety/FoodborneIllness/FoodborneIllnessFoodbornePathogensNaturalToxins/BadBugBook/ucm070523.htm
49. Fisher K, Phillips C (2009) The ecology, epidemiology and virulence of Enterococcus. Microbiology 155(Pt 6):1749–1757
50. Fleisher JM, Fleming LE et al (2010) The BEACHES Study: health effects and exposures from non-point source microbial contaminants in subtropical recreational marine waters. Int J Epidemiol 39(5):1291–1298
51. Fleming L, Solo Gabriel H et al (2008) Final report on the pilot epidemiologic assessment of microbial indicators for monitoring recreational water quality in marine sub/tropical environments. The NSF NIEHS OHH Center, Rosenstiel School of Marine and Atmospheric Sciences, University of Miami, Miami
52. Foster TJ (2004) The *Staphylococcus aureus* "superbug". J Clin Invest 114(12):1693–1696
53. Foster TJ (2005) Immune evasion by staphylococci. Nat Rev Microbiol 3(12):948–958
54. Fujino T, Okuno Y, Nakada D, Aoyama A, Fukai K, Mukai T, Ueho T (1953) On the bacteriological examination of shirasu-food poisoning. Med J Osaka Univ 4:299–304
55. Gefen O, Mumcuoglu M, Engel G, Balaban NQ (2008) Single-cell protein induction dynamics reveals a period of vulnerability to antibiotics in persister bacteria. Proc Natl Acad Sci 105:6145–6149
56. Gil AL, Louis VR, Rivera ING, Lipp E, Huq A, Lanata CF, Taylor DN, Russek-Cohen E. Choopun N, Sack RB, Colwell RR (2004) Occurance and distribution of *Vibrio cholerae* in the coastal environment of Peru. Environ Microbiol 6:699–706
57. Goodwin KD, Pobuda M (2009) Performance of CHROMagar Staph aureus and CHROMagar MRSA for detection of Staphylococcus aureus in seawater and beach sand–comparison of culture, agglutination, and molecular analyses. Water Res 43:4802–4811
58. Grimes DJ, Singleton FL, Stemmler J, Palmer LM, Brayton P, Colwell RR (1984) Microbiological effects of wastewater effluent discharge into coastal waters of Puerto Rico. Water Res 18:613–619
59. Grimes DJ, Mills AL, Nealson KH (2000) The importance of viable but nonculturable bacteria in biogeochemistry. In: Colwell RR, Grimes DJ (eds) Nonculturable microorganisms in the environment. ASM Press, Washington, DC, pp 209–227
60. Guvener ZT, McCarter LL (2003) Multiple regulators control capsular polysaccharide production in *Vibrio parahaemolyticus*. J Bacteriol 185:5431–5441
61. Hara-Kudo Y, Saito S, Ohtsuka K, Yamasaki S, Yahiro S, Nishio T, Iwade Y, Otomo T, Konuma H, Tanaka H, Nakagawa H, Sugiyama K, Sugita-Konishi Y, Kumagai S (2010) Decreasing *Vibrio parahaemolyticus* infections and analysis of seafood contamination in Japan. Vibrios in the Environment-2010, Biloxi, MS (Abstracts)
62. Harth E, Matsuda L, Hernández C, Rioseco ML, Romero J, González-Escalona N, Martínez-Urtaza J, Espejo RT (2009) Epidemiology of *Vibrio parahaemolyticus* outbreaks, southern Chile. Emerg Infect Dis 15:163–168
63. Harwood VJ, Whitlock J, Withington V (2000) Classification of antibiotic resistance patterns of indicator bacteria by discriminant analysis: use in predicting the source of fecal contamination in subtropical waters. Appl Environ Microbiol 66:3698–3704
64. He H, Wang Q, Sheng L, Liu Q, Zhang Y (2010) Functional characterization of *Vibrio alginolyticus* twin-arginine translocation system: Its roles in biofilm formation, extracellular protease activity, and virulence towards fish. Curr Microbiol. doi:10.1007/s00284-010-9844-6

65. Hensel M (2004) Evolution of pathogenicity islands of *Salmonella enterica*. Int J Med Microbiol 294:95–102
66. Ho PL, Tang WM, Lo KS, Yuen KY (1998) Necrotizing fasciitis due to *Vibrio alginolyticus* following an injury inflicted by a stingray. Scand J Infect Dis 30:192–193
67. Honda T, Iida T (1993) The pathogenicity of *Vibrio parahaemolyticus* and the role of thermostable direct haemolysin and related haemolysins. Rev Med Microbiol 4:106–113
68. Huq A, Sack RB, Nizam A, Longini IM, Balakrish Nair G, Ali A, Morris JG Jr, Khan MNH, Siddique AK, Yunus M, Albert MJ, Sack DA, Colwell RR (2005) Critical factors influencing the occurrence of *Vibrio cholerae* in the environment of Bangladesh. Appl Environ Microbiol 71:4645–4654
69. Jacobs-Reitsma W (2000) *Campylobacter* in the food supply. In: Nachamkin I, Blaser MJ (eds) Campylobacter, 2nd edn. ASM Press, Washington, DC, pp 467–480
70. Jaffres E, Sohier D et al (2009) Study of the bacterial ecosystem in tropical cooked and peeled shrimps using a polyphasic approach. Int J Food Microbiol 131(1):20–29
71. Jensen AE, Cheville NF, Thoen CO, MacMillan AP, Miller WG (1999) Genomic fingerprinting and development of a dendrogram for *Brucella* spp. isolated from seals, porpoises, and dolphins. J Vet Diagn Invest 11:152–157
72. Jia A, Woo NY, Zhang XH (2010) Expression, purification, and characterization of thermolabile hemolysin (TLH) from *Vibrio alginolyticus*. Dis Aquat Organ 90:121–127
73. Johnson CN, Flowers AR, Young VC, Gonzalez-Escalona N, DePaola A, Grimes DJ (2009) Genetic relatedness among *tdh*$^{+}$ and *trh*$^{+}$ *Vibrio parahaemolyticus* cultured from Gulf of Mexico oysters (*Crassostrea virginica*) and surrounding water and sediment. Microb Ecol 57:437–443
74. Johnson CN, Flowers AR, Noriea NF III, Zimmerman AM, Bowers J, DePaola A, Grimes DJ (2010) Relationships between environmental factors and pathogenic vibrios in the northern Gulf of Mexico. Appl Environ Microbiol 76:7076–7084
75. Joseph SW, Daily OP, Hunt WS, Seidler RJ, Allen DA, Colwell RR (1979) *Aeromonas* primary wound infection of a diver in polluted waters. J Clin Microbiol 10:46–49
76. Kaper JB, Morris JG, Levine MM (1995) Cholera. Clin Microbiol Rev 8:48–86
77. Kitao T, Aoki T, Sakoh R (1981) Epizootic caused by beta-hemolytic *Streptococcus* species in cultured freshwater fish. Fish Pathol 19:173–180
78. Klevens RM, Morrison MA, Nadle J, Petit S, Gershman K, Ray S, Harrison LH, Lynfield R, Dumyati G, Townes JM, Craig AS, Zell ER, Fosheim GE, McDougal LK, Carey RB, Fridkin SK (2007) Invasive methicillin-resistant Staphylococcus aureus infections in the United States. JAMA 298:1763–1771
79. Kluytmans J, van Belkum A, Verbrugh H (1997) Nasal carriage of *Staphylococcus aureus*: epidemiology, underlying mechanisms, and associated risks. Clin Microbiol Rev 10:505–520
80. Kusuda R (1992) Bacterial fish diseases in marineculture in Japan with special emphasis on streptococcosis. Isr J Aquacult 44:140
81. Lappi VR, Ho A et al (2004) Prevalence and growth of Listeria on naturally contaminated smoked salmon over 28 days of storage at 4 degrees C. J Food Prot 67(5):1022–1026
82. Lappi VR, Thimothe J et al (2004) Longitudinal studies on Listeria in smoked fish plants: impact of intervention strategies on contamination patterns. J Food Prot 67(11):2500–2514
83. Lappi VR, Thimothe J et al (2004) Impact of intervention strategies on Listeria contamination patterns in crawfish processing plants: a longitudinal study. J Food Prot 67(6):1163–1169
84. Lau SK, Woo PC et al (2003) Invasive Streptococcus iniae infections outside North America. J Clin Microbiol 41(3):1004–1009
85. Lauková A, Juris P (1997) Distribution and characterization of *Enterococcus* species in municipal sewages. Microbios 89:73–80
86. Li XC, Xiang ZY, Xu XM, Yan WH, Ma JM (2009) Endophthalmitis Caused by *Vibrio alginolyticus*. J Clin Microbiol 47:3379–3381

87. Lianou A, Sofos JN (2007) A review of the incidence and transmission of Listeria monocytogenes in ready-to-eat products in retail and food service environments. J Food Prot 70(9):2172–2198
88. Linell F, Norden A (1954) Mycobacterium balnei, a new acid-fast bacillus occurring in swimming pools and capable of producing skin lesions in humans. Acta Tuberc Scand Suppl 33:1–84
89. Liu D (ed) (2008) Handbook of *Listeria monocytogenes*. CRC Press, Boca Raton
90. Louis VR, Russek-Cohen E, Choopun N, Rivera ING, Gangle B, Jiang SC, Rubin A, Patz JA, Huq A, Colwell RR (2003) Predictability of *Vibrio cholerae* in Chesapeake Bay. Appl Environ Microbiol 69:2773–2785
91. Makino K, Oshima K, Kurokawa K, Yokoyama K, Uda T, Tagomori K, Iijima Y, Najima M, Nakano M, Yamashita A, Kubota Y, Kimura S, Yasunaga T, Honda T, Shinagawa H, Hattori M, Iida T (2003) Genome sequence of *Vibrio parahaemolyticus* a pathogenic mechanism distinct from that of *V. cholerae*. Lancet 361:743–749
92. Manero A, Blanch AR (1999) Identification of *Enterococcus* spp. with a biochemical key. Appl Environ Microbiol 65:4425–4430
93. Maquart M, LeFleche P, Foster G, Tryland M, Ramisse F, Djønne B, Al Dahouk S, Jacques I, Neubauer H, Walravens K, Godfroid J, Cloeckaert A, Vergnaud G (2009) MLVA-16 typing of 295 marine mammal *Brucella* isolates from different animal and geographic origins identifies 7 major groups within *Brucella ceti* and *Brucella pinnipedialis*. BMC Microbiol 9:145. doi:10.1186/1471-2180-9-145
94. McCarter LL (2006) Motility and chemotaxis. In: Thompson FL, Austin B, Swings J (eds) The biology of Vibrios. ASM Press, Washington, DC, pp 115–132
95. McDonald WL, Jamaludin R, Mackereth G, Hansen M, Humphrey S, Short P, Taylor T, Swingler J, Dawson CE, Whatmore AM, Stubberfield E, Perrett LL, Simmons G (2006) Characterisation of a *Brucella* sp. strain as a marine-mammal type despite isolation from a patient with spinal osteomyelitis in New Zealand. J Clin Microbiol 44:4363–4370
96. McLaughlin JB, DePaola A, Bopp CA, Martinek KA, Napolilli NP, Allison CG, Murray SL, Thompson EC, Bird MM, Middaugh JP (2005) Outbreak of *Vibrio parahaemolyticus* gastroenteritis associated with Alaskan oysters. N Engl J Med 353:1463–1470
97. Mead PS, Slutsker L, Dietz V, McCaig LF, Bresee JS, Shapiro C, Griffin PM, Tauxe RV (1999) Food-related illness and death in the United States. Emerg Infect Dis 5:607–625
98. Meador CE, Parsons MM, Bopp CA, Gerner-Smidt P, Painter JA, Vora GJ (2007) Virulence gene- and pandemic group-specific marker profiling of clinical *Vibrio parahaemolyticus* isolates. J Clin Microbiol 45:1133–1139
99. Mejlholm O, Kjeldgaard J et al (2008) Microbial changes and growth of Listeria monocytogenes during chilled storage of brined shrimp (Pandalus borealis). Int J Food Microbiol 124(3):250–259
100. Morris PJ, Johnson WR et al (2011) Isolation of culturable microorganisms from free-ranging bottlenose dolphins (*Tursiops truncatus*) from the southeastern United States. Vet Microbiol 148(2–4):440–447
101. Mukherjee J, Chios K, Fishwild D, Hudson D, O'Donnell S, Rich SM, Donohue-Rolf A, Tzipori S (2002) Human Stx2-specific monoclonal antibodies prevent systemic complications of *Escherichia coli* O157:H7 infection. Infect Immun 70(2):612–619
102. Nishibuchi M (2006) Miscellaneous human pathogens. In: Thompson FL, Austin B, Swings J (eds) The biology of Vibrios. ASM Press, Washington, DC, pp 367–381
103. Noriea NF III, Johnson CN, Griffitt KJ, Grimes DJ (2010) Distribution of type III secretion systems in *Vibrio parahaemolyticus* from the Northern Gulf of Mexico. J Appl Microbiol 109:953–962
104. Oliver JD (2006) *Vibrio vulnificus*. In: Thompson FL, Austin B, Swings J (eds) The biology of Vibrios. ASM Press, Washington, DC, pp 349–366
105. O'Mahony R, Abbott Y et al (2005) Methicillin-resistant Staphylococcus aureus (MRSA) isolated from animals and veterinary personnel in Ireland. Vet Microbiol 109(3–4): 285–296

106. O'Neill KR, Jones SH, Grimes DJ (1992) Seasonal incidence of *Vibrio vulnificus* in the Great Bay Estuary of New Hampshire and Maine. Appl Environ Microbiol 58:3257–3262
107. Oshiro R, Fujioka R (1995) Sand, soil, and pigeon droppings – sources of indicator bacteria in the waters of Hanauma Bay, Oahu, Hawaii. Water Sci Technol 31:251–254
108. Panos GZ, Betsi GI, Falagas ME (2006) Systematic review: are antibiotics detrimental or beneficial for the treatment of patients with *Escherichia coli* O157:H7 infection? Aliment Pharmacol Ther 24(5):731–742
109. Paranjpye RN, Strom MS (2005) A *Vibrio vulnificus* Type IV pilin contributes to biofilm formation, adherence to epithelial cells, and virulence. Infect Immun 73:1411–1422
110. Paranjpye RN, Johnson AB, Baxter AE, Strom MS (2007) Role of Type IV pilins in persistence of *Vibrio vulnificus* in *Crassostrea virginica* oysters. Appl Environ Microbiol 73:5041–5044
111. Parenti DM, Symington JS et al (1995) Mycobacterium kansasii bacteremia in patients infected with human immunodeficiency virus. Clin Infect Dis 21(4):1001–1003
112. Park K-S, Iida T, Yamaichi Y, Oyagi T, Yamamoto K, Honda T (2000) Genetic characterization of DNA region containing the *trh* and *ure* genes of *Vibrio parahaemolyticus*. Infect Immun 68:5742–5748
113. Peir GB, Madin SH (1976) *Streptococcus iniae* sp. nov., a beta-hemolytic streptococcus isolated from an Amazon freshwater dolphin, *Inia geoffrensis*. Int J Syst Bacteriol 26:545–553
114. Pellett S, Bigley DV, Grimes DJ (1983) Distribution of *Pseudomonas aeruginosa* in a riverine ecosystem. Appl Environ Microbiol 45:328–332
115. Perera RP, Johnson SK, Collins MD, Lewis DH (1994) *Streptococcus iniae* associated with mortality of *Tilapia nilotica* and *T. aurea* hybrids. J Aquat Anim Health 6:335–340
116. Petersen A, Dalsgaard A (2003) Species composition and antimicrobial resistance genes of Enterococcus spp, isolated from integrated and traditional fish farms in Thailand. Environ Microbiol 5(5):395–402
117. Phillips AMB, DePaolo A, Bowers J, Ladner S, Grimes DJ (2007) An evaluation of the use of remotely sensed parameters for prediction of incidence and risk associated with *Vibrio parahaemolyticus* in Gulf Coast oysters (*Crassostrea virginica*). J Food Prot 70:879–884
118. Plano LR, Garza AC et al (2011) Shedding of Staphylococcus aureus and methicillin-resistant Staphylococcus aureus from adult and pediatric bathers in marine waters. BMC Microbiol 11(1):5
119. Plano LRW, Shibata T et al (2011) Characterization of *Staphylococcus aureus* and community associated MRSA at a recreational marine beach in South Florida. J Antimicrob Chemother (in review)
120. Posfay-Barbe KM, Wald ER (2009) Listeriosis. Semin Fetal Neonatal Med 14(4):228–233
121. Qasem JA, Sameer A-Z, Salwa A-M, Samee A-A, Ahmed A-M, Al-Sharifi Faisal A-S (2008) Molecular investigation of *Streptococcus agalactiae* isolates from environmental samples and fish specimens during a massive fish kill in Kuwait Bay. Pakistan J Biol Sci 11:2500–2504
122. Rallis E, Koumantaki–Mathioudaki E (2007) Treatment of Mycobacterium marinum cutaneous infections. Expert Opin Pharmacother 8(17):2965–2978
123. Reichelt JL, Bauman P, Bauman L (1979) Study of genetic relationships among marine species *Beneckea* and *Photobacterium* by means of DNA/DNA hybridization. Arch Microbiol 1110:101–120
124. Rice EW, Messer JW et al (1995) Occurrence of high-level aminoglycoside resistance in environmental isolates of enterococci. Appl Environ Microbiol 61(1):374–376
125. Rivas AL, Gonzalez RN, Wiedmann M, Bruce JL, Cole EM et al (1997) Diversity of *Streptococcus agalactiae* and *Staphylococcus aureus* ribotypes recovered from New York dairy herds. Amer J Vet Res 58:482–487

126. Rivers B, Steck TR (2001) Viable but nonculturable uropathogenic bacteria are present in the mouse urinary tract following urinary tract infection and antibiotic therapy. Urol Res 29:60–66
127. Roushan MRH, Mohraz M, Hajiahmadi M, Ramzani A, Valayati AA (2006) Efficacy of Gentamicin plus Doxycycline versus Streptomycin plus Doxycycline in the treatment of Brucellosis in humans. Clin Infect Dis 42:1075–1080
128. Saubolle M (1989) Nontuberculous mycobacteria as agents in human disease in the United States. Clin Microbiol Newslett 11:113–117
129. Schaefer AM, Goldstein JD et al (2009) Antibiotic-resistant organisms cultured from Atlantic bottlenose dolphins (*Tursiops truncatus*) inhabiting estuarine waters of Charleston, SC and Indian River Lagoon, FL. Ecohealth 6(1):33–41
130. Shime-Hattori A et al (2006) Two type IV pili of *Vibrio parahaemolyticus* play different roles in biofilm formation. FEMS Microbiol Lett 264:89–97
131. Sinigalliano CD, Fleisher JM et al (2010) Traditional and molecular analyses for fecal indicator bacteria in non-point source subtropical recreational marine waters. Water Res 44(13):3763–3772
132. Skalsky K, Yahav D, Bishara J, Pitlik S, Leibovici L, Paul M (2008) Treatment of human brucellosis: systematic review and meta-analysis of randomized trials. BMJ 336(7646):701–704
133. Soge OO, Meschke JS, No DB, Roberts MC (2009) Characterization of methicillin resistant Staphylococcus aureus and methicillin-resistant coagulase negative Staphylococcus spp isolated from US West coast public marine beaches. J Antimicrob Chemother 64:1148–1155
134. Sun B, Zhang XH, Tang X, Wang S, Zhong Y, Chen J, Austin B (2007) A single residue change in *Vibrio harveyi* hemolysin results in the loss of phospholipase and hemolytic activities and pathogenicity for turbot (*Scophthalmus maximus*). J Bacteriol 189:2575–2579
135. Takeda T, Peina Y, Ogawa A, Dohi S, Abe H, Nair GB, Pal SC (1991) Detection of heat-stable enterotoxin in a cholera toxin gene-positive strain of *Vibrio cholerae* O1. FEMS Microbiol Lett 64:23–27
136. Tauxe R, Seminario L, Tapia R, Libel M (1994) The Latin American epidemic. In: Wachsmuth IK, Blake PA, Olsvik O (eds) *Vibrio cholerae* and cholera: molecular to global perspectives. American Society for Microbiology, Washington, DC
137. Thompson FL, Hoste B, Vandemeulebroecke K, Swings J (2003) Reclassification of *Vibrio hollisae* as *Grimontia hollisae* gen. nov., comb. Nov. Int J Syst Evol Microbiol 53:1615–1617
138. Tokuda H, Unemoto T (1982) Characterization of the respiration-dependant Na^+ pump in the marine bacterium *Vibrio alginolyticus*. J Biol Chem 257:10007–10014
139. Tsai YH, Kung H-F, Lee T-M, Lin G-T, Hwang D-F (2004) Histamine-related hygienic quantities and bacteria found in popular commercial scombroid fish fillets in Taiwan. J Food Prot 67:407–412
140. Uh Y, Park JS, Hwang GY, Jang IH, Yoon KJ, Park HC, Hwang SO (2001) *Vibrio alginolyticus* acute gastroenteritis: a report of two cases. Clin Microbiol Infect 7:104–106
141. Ulitzer S (1975) The mechanism of swarming of *Vibrio alginolyticus*. Arch Microbiol 104:67–71
142. USEPA (2000) Improved enumeration methods for the recreational water quality indicators: Enterococci and *Escherichia coli*. EPA/821/R-97/004, Washington, DC
143. Valdivia E, Martin-Sanchez I et al (1996) Incidence of antibiotic resistance and sex pheromone response among enterococci isolated from clinical human samples and from municipal waste water. J Appl Bacteriol 81(5):538–544
144. Vally H, Whittle A, Cameron S, Dowse GK, Watson T (2004) Outbreak of *Aeromonas hydrophila* wound infections associated with mud football. Clin Infect Dis 38:1084–1089
145. Vivekanandhana G, Hathab AAM, Lakshmanaperumalsamy P (2005) Prevalence of *Aeromonas hydrophila* in fish and prawns from the seafood market of Coimbatore, South India. Food Microbiol 22:133–137

146. Von Eiff C, Becker K, Machka K, Stammer H, Peters G (2001) Nasal carriage as a source of Staphylococcus aureus bacteremia. Study Group. N Engl J Med 344:11–16
147. Wade TJ, Pai N et al (2003) Do U.S. Environmental Protection Agency water quality guidelines for recreational waters prevent gastrointestinal illness? A systematic review and meta-analysis. Environ Health Perspect 111(8):1102–1109
148. Waldor MK, Mekalanos JJ (1996) Lysogenic conversion by a filamentous phage encoding cholera toxin. Science 272:1910–1914
149. Wang SX, Zhang XH, Zhong YB, Sun BG, Chen JX (2007) Genes encoding the *Vibrio harveyi* haemolysin (VHH)/thermolabile haemolysin (TLH) are widespread in vibrios. Wei Sheng Wu Xue Bao 47:874–881
150. Weinstein MR, Litt M, Kertesz DA, Wyper P, Rose D, Coulter M, McGeer A, Facklam R, Ostach C, Willey BM, Borczyk A, Low DE et al (1997) Invasive infections due to a fish pathogen, Streptococcus iniae. N Engl J Med 337:589–594
151. Wendt JR, Lamm RC et al (1986) An unusually aggressive Mycobacterium marinum hand infection. J Hand Surg Am 11(5):753–755
152. Whatmore AM, Dawson CE, Groussaud P, Koylass MS, King AC, Shankster SJ, Sohn AH, Probert WS, McDonald WL (2008) Marine mammal *Brucella* genotype associated with zoonotic infection. Emerg Infect Dis 14:517–518
153. WHO (World Health Organization) (2010) Cholera, 2009. Wkly Epidemiol Rec 85:293–308
154. Wilson IG, McAfee GG (2002) Vancomycin-resistant enterococci in shellfish, unchlorinated waters, and chicken. Int J Food Microbiol 79(3):143–151
155. Wolinsky E (1992) Mycobacterial diseases other than tuberculosis. Clin Infect Dis 15(1):1–10
156. Wong CS, Jelacic S, Habeeb RL, Watkins SL, Tarr PI (2000) The risk of the hemolytic-uremic syndrome after antibiotic treatment of *Escherichia coli* O157:H7 infections. N Engl J Med 342(26):1930–1936
157. Yakushi T, Maki S, Homma M (2004) Interaction of PomB with the third transmembrane segment of PomA in the Na^+-driven polar flagellum of *Vibrio alginolyticus*. J Bacteriol 186:5281–5291
158. Yamahara KM, Walters SP et al (2009) Growth of enterococci in unaltered, unseeded beach sands subjected to tidal wetting. Appl Environ Microbiol 75(6):1517–1524
159. Zimmerman AM, Rebarchik DM, Flowers AR, Williams JL, Grimes DJ (2009) *Escherichia coli* detection using mTEC agar and fluorescent antibody direct viable counting on coastal recreational water samples. Lett Appl Microbiol 49:478–483

Books and Reviews

Belkin S, Colwell RR (eds) (2005) Pathogenic microorganisms in the sea. Kluwer/Plenum, New York

Colwell RR, Grimes DJ (eds) (2000) Nonculturable microorganisms in the environment. ASM Press, Washington, DC

Walsh PJ, Smith SL, Fleming LE, Solo-Gabriele HM, Gerwick WH (eds) (2008) Oceans and human health: risk and remedies from the sea. Elsevier, St. Louis

Chapter 5
Infectious Disease Modeling

Angela R. McLean

Glossary

Basic reproductive number	A summary parameter that encapsulates the infectiousness of an infectious agent circulating in a population of hosts.
Host	An organism that acts as the environment within which an infectious agent replicates.
Infectious agent	A microorganism that replicates inside another organism.
Pathogen	An infectious agent that damages its host.
Variant	One of several types of an infectious agent, often closely related to and sometimes evolved from other variants under consideration.

Definition of the Subject

Infectious disease models are mathematical descriptions of the spread of infection. The majority of infectious disease models consider the spread of infection from one host to another and are sometimes grouped together as "mathematical epidemiology." A growing body of work considers the spread of infection within an

This chapter, which has been modified slightly for the purposes of this volume, was originally published as part of the Encyclopedia of Sustainability Science and Technology edited by Robert A. Meyers. DOI:10.1007/978-1-4419-0851-3

A.R. McLean (✉)
Zoology Department, Institute of Emerging Infections, University of Oxford,
South Parks Road, Oxford OX1 3PS, UK
e-mail: angela.mclean@zoo.ox.ac.uk

P. Kanki and D.J. Grimes (eds.), *Infectious Diseases: Selected Entries from the Encyclopedia of Sustainability Science and Technology*,
DOI 10.1007/978-1-4614-5719-0_5, © Springer Science+Business Media New York 2013

individual, often with a particular focus on interactions between the infectious agent and the host's immune responses. Such models are sometimes grouped together as "within-host models." Most recently, new models have been developed that consider host–pathogen interactions at two levels simultaneously: both within-host dynamics and between-host transmissions. Infectious disease models vary widely in their complexity, in their attempts to refer to data from real-life infections and in their focus on problems of an applied or more fundamental nature. This entry will focus on simpler models tightly tied to data and aimed at addressing well-defined practical problems.

Introduction

Why is it that smallpox was eradicated in 1979 [1] but measles, once scheduled for eradication by the year 2000, still kills over a hundred thousand children each year [2]? Both diseases can be prevented with cheap, safe, and effective vaccines which probably induce lifelong immunity, and neither virus has an environmental or animal reservoir.

One way to address this question is to consider the comparative ease of spread of the two infections. A useful parameter that summarizes this ease of spread is the "basic reproductive number" always denoted as R_0. The definition of the basic reproductive number is the number of secondary infections caused during the entire duration of one infection if all contacts are susceptible (i.e., can be infected). The concept has widespread currency in the literature on infectious disease models with varying degrees of affection [3]. There is no question that it has been a useful, simple rule of thumb for characterizing how easily an infection can spread [4]. Furthermore, the simplest of calculations relate R_0 to the degree of intervention needed to bring an infection under control and, eventually, eradicate it. The relationship between the basic reproductive number and disease control arises from the simple fact that if each infectious person causes less than one secondary case, then the number of infections must fall. If it is always true, even when there is no infection circulating, that each new case causes less than one secondary case, then the infection will die out. This simple observation leads to a straightforward calculation for the proportion of a population that must be vaccinated in order to achieve eradication, p_c:

$$p_c = 1 - \frac{1}{R_0} \tag{5.1}$$

This relationship arises from the fact that if $(R_0 - 1)$ out of the R_0 people a case might have infected have been vaccinated, then each case, in a population vaccinated to that degree, will cause less than one secondary case. For example, say $R_0 = 10$, if 9/10 of the population are successfully immune following vaccination, then infection cannot spread. Thus, in general, $p_c = (R_0 - 1)/R_0$, as stated in Eq. 5.1. If the fraction of

Table 5.1 The basic reproductive number, R_0, and the critical vaccination proportion for eradication, p_c, for measles and smallpox

Infection	Place	Time	R_0	p_c (%)
Smallpox	India	1970s	3	67
Measles	India	1970s	15	93
Measles	UK	1960s	15	93

the population that are successfully vaccinated is greater than p_c, then each case will cause, on average, less than one case, and infection cannot spread.

Comparing estimated values for R_0 for measles and smallpox and inferred values for the proportion that need to be vaccinated to ensure eradication (Table 5.1) leads to a simple answer to the question why has smallpox been eradicated, but not measles? Smallpox, with a basic reproductive number around three, was eradicated with vaccination coverage of around 67%. The higher R_0 for measles, nearer to 15, requires vaccination coverage close to 95% to ensure eradication. Many parts of the world remain unable to achieve such high coverage; measles remains suppressed by tremendous efforts at vaccination but is not yet eradicated.

These calculations are so straightforward that they can be made without recourse to any formal modeling. However, embedding these ideas inside a formal modeling framework has proven very useful. The next section describes the simplest applicable model.

The SIR Model

The "plain vanilla" model of mathematical epidemiology is called the SIR model because it splits the host population into three groups:

- The susceptible (S) can be infected if exposed
- The infectious (I) are both infected and infectious to others
- The recovered (R) are no longer infectious and are immune to further infection

The SIR model's structure then consists of a set of assumptions about how people flow into, out of, and between these three groups. Those assumptions can be represented graphically as in Fig. 5.1.

The assumptions of the SIR model with vaccination are the following: People are born at a constant rate B, and a proportion p of them are vaccinated at birth. Vaccinated newborns are immune for life and so they join the recovered class. Unvaccinated newborns enter the susceptible class. Susceptibles are infected at a per capita rate proportional to I, the number of infectious people in the population. This gives rise to a transfer from the susceptible to the infectious class at rate βIS. Susceptibles are also subject to a per capita background death rate μ. Infectious people recover into the recovered class I at per capita rate γ or die at the per capita background death rate μ. Recovered individuals are immune for the rest of their lives, so the only exit from the recovered class is at the per capita background death rate μ.

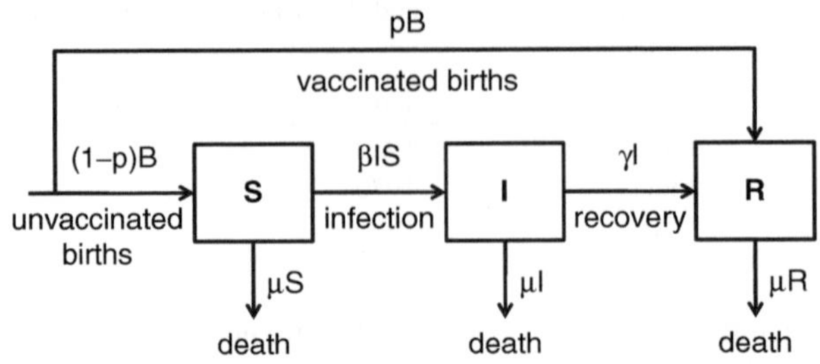

Fig. 5.1 The SIR model in graphical form. The host population is divided into three groups, and transitions of people between those groups are described. Those transitions represent the five processes: birth, vaccination, death, infection, and recovery

These assumptions can be written in several different forms of equations, for example, difference equations, ordinary differential equations, or stochastic differential equations. The difference equation form is as follows:

$$S(t+1) = S(t) + (1-p)\,B - \beta I(t)S(t) - \mu S(t) \tag{5.2}$$

$$I(t+1) = I(t) + \beta I(t)S(t) - \gamma I(t) - \mu I(t) \tag{5.3}$$

$$R(t+1) = R(t) + pB + \gamma I(t) - \mu R(t) \tag{5.4}$$

This difference equation form is particularly easy to handle numerically and can be straightforwardly solved in a spreadsheet. Fig. 5.2a shows the solutions to Eqs. 5.2–5.4 over 50 years with a 1-week timestep. Parameters are set so that an infection with a basic reproductive number of 5 and a 1-week duration of infection is spreading in a population of 100,000 individuals. The figure illustrates how this model shows damped oscillations towards a stable state. The same is true for the ODE version of this model.

This model is useful for understanding the impact of vaccination. In Fig. 5.2b, the solutions to Eqs. 5.2–5.4 are shown when vaccination at birth is introduced 10 years into the model run. With a basic reproductive number of 5, Eq. 5.1 tells us that vaccination of over 80% of newborns will lead to eradication. This is exemplified in the pink line where 90% vaccination leads to no further cases. Vaccination coverage below this threshold value reduces the numbers of cases and increases the inter-epidemic period but does not lead to eradication. Notice the very long inter-epidemic period at 70% vaccination. This phenomen occurs when vaccine coverage levels are close to but do not achieve the critical coverage level. Under these circumstances, it takes a very long time to accumulate enough susceptibles to trigger the first epidemic after vaccination is introduced. It may therefore appear as though eradication has been achieved even though vaccination coverage is below the critical level. This phenomenon, named "the honeymoon period," [5] was first described in modeling studies and later identified in field data [6].

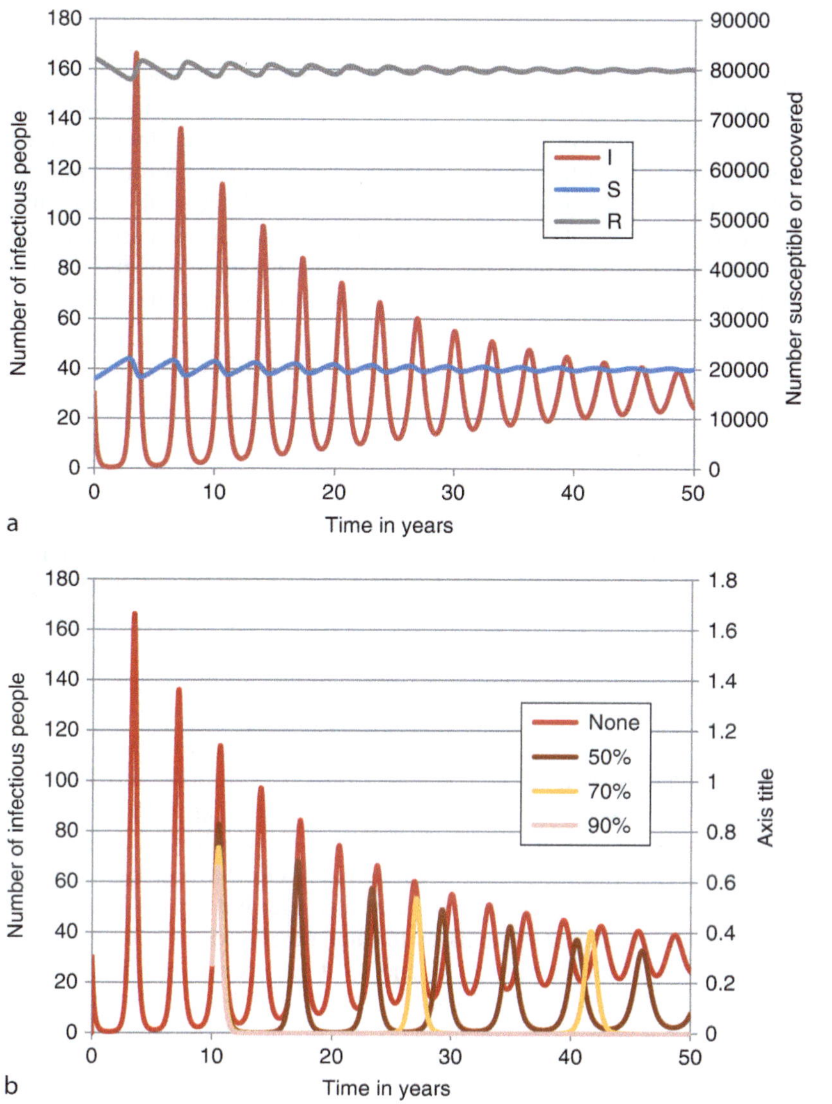

Fig. 5.2 Numerical solutions to the SIR difference equation model. Infection circulates in a population of 100,000 individuals, with an expectation of life at birth of 50 years. The infectious period is 1 week, and the basic reproductive number is 5. This gives the following model parameters: B = 38 per week, $\mu = 0.00385$ per person per week, $\gamma = 1$ per person per week, and $\beta = 0.00005$ per infected per susceptible per week. (**a**) shows damped oscillations in all three classes after an initial perturbation of 20% of the susceptible class into the recovered class. In (**b**), vaccination of 50%, 70%, or 90% of newborns is introduced at time 10 years. With $R_0 = 5$, the critical vaccination proportion $p_c = 0.8$. Vaccination coverage above this level (at 90%) leads to eradication

The ability to identify target vaccination levels predicted to lead to disease eradication has been widely influential in policy circles [7]. Models with the same fundamental structure as the SIR model are used to set targets for vaccination coverage in many settings [8]. Similar models are also used to understand the likely impact of different interventions of other sorts, for example, drug treatment [9] or measures for social distancing [10]. However, models for informing policy need to explore more of the wrinkles and complexities of the real world than are acknowledged in the simple equations of the SIR model. The next section describes some of the types of host heterogeneity that have been explored in making versions of the SIR model that aim to be better representations of the real world.

Host Heterogeneity

There are many aspects of host heterogeneity that have bearing on the transmission and impact of infections. Two of the most important are host age and spatial distribution. In this section, the modeling of these two types of host heterogeneity is introduced with reference to two specific infections: rubella and foot-and-mouth disease.

Rubella is a directly transmitted viral infection that usually causes mild disease when contracted during childhood. However, infection of a woman during early pregnancy can lead to serious birth defects for her unborn child. The set of consequent conditions is labeled "congenital rubella syndrome" or CRS. Because vaccination acts to extend the time between epidemics (Fig. 5.1b), it also acts to increase the average age at infection. This sets up a complex trade-off when introducing rubella vaccination to a community. On the one hand, vaccinated girls are protected from catching rubella at any age, but on the other hand, the girls who remain unvaccinated are likely to catch rubella when they are older, more likely to be in their childbearing years and so at greater risk of CRS. This means that vaccination with low coverage can actually lead to more CRS, and only when coverage levels get above a certain level do the benefits of vaccinating the community outweigh the costs. Calculating where that level lies then becomes an important public health question.

Because age is such an important component of the risks associated with rubella infection, models of this system need to take account of host age. The relevant versions of Eqs. 5.2–5.4 are difference equations with two independent variables, age (a) and time (t):

$$S(a+1,t+1) = S(a,t) - S(a,t)[\Sigma_{a'}\beta(a,a')I(a,t)] - \mu(a)S(a,t) \tag{5.5}$$

$$\begin{aligned} I(a+1,t+1) = {} & I(a,t)+S(a,t)[\Sigma_{a'}\beta(a,a')I(a,t)] \\ & - \gamma(a)I(a,t) - \mu(a)I(a,t) \end{aligned} \tag{5.6}$$

$$R(a+1, t+1) = R(a, t) + \gamma(a)I(a, t) - \mu(a)R(a, t) \tag{5.7}$$

Notice how these equations, by taking account of age as well as time, allow consideration of several different kinds of age dependence. Firstly, Eq. 5.6 calculates the number of cases of infection *of a given age* over time. Since the main consideration in balancing up the pros and cons of rubella vaccination is the number of cases in women of childbearing age, this is an essential model output. Secondly, the per capita rate at which susceptibles become infected depends on their age and on the age of all the infected people. This model is thus able to take account of the complexities of family, school, and working life which drive people of different ages to age-dependent patterns of mixing. Thirdly, the recovery rate $\gamma(a)$ and, more importantly, the background death rate $\mu(a)$ can both be made to depend on age. Since a fixed per capita death rate is a particularly bad approximation of human survival, this is another important advance on models without age structure.

Models with age structure akin to that presented in Eqs. 5.5–5.7 have been essential components of the planning of rubella vaccination strategies around the world [11, 12]. A model as simple as these equations would never be used for formulating policy; furthermore, most age-structured models use the continuous time and age versions and so have the structure of partial differential equations. Nevertheless, Eqs. 5.5–5.7 illustrate the fundamentals of how to include age in an epidemiological model.

The spatial distribution of hosts is another important aspect of their heterogeneity. If the units of infection are sessile (e.g., plants), the assumption that all hosts are equally likely to contact each other becomes particularly egregious and models that acknowledge the spatial location of hosts more important. One example of units of infection that do not move is farms. If trade between farms has been halted because of a disease outbreak, then disease transmission between farms is likely to be strongly dependent upon their location. This was the case during the 2001 foot-and-mouth disease epidemic in the UK, and spatial models of that epidemic are nice examples of how to explicitly include the distance between hosts in a model epidemic.

On February 19, 2001, a vet in Essex reported suspected cases of foot-and-mouth disease (FMD) in pigs he had inspected at an abattoir. FMD is a highly infectious viral disease of cloven-hoofed animals. Because of its economic and welfare implications for livestock, FMD had been eradicated from Western Europe. The FMD outbreak that unfolded in the UK over the ensuing months had a huge impact with millions of farm animals killed and major economic impact in the countryside as tourism was virtually shut down.

There was heated debate about the best way to control the spread of infection from farm to farm. FMD virus is so very infectious that no attempt was made to control its spread within a farm. Once infection of livestock on a farm was detected, all susceptible animals were slaughtered. Mathematical models of the spread of this epidemic thus treat each farm as a unit of infection, and, as before, farms can be categorized as susceptible, infectious, etc. The best of these models [13] keeps track of every single farm in the United Kingdom, characterizing farms by their location and the number of sheep and cattle they hold. The model classifies farms into four groups: susceptible, incubating, infectious, or slaughtered. As in all epidemic models, the heart of the model is the per capita rate at which susceptible farms become infected – the so-called force of infection. Because this FMD model is an individual-based, stochastic simulation, it is not possible to write out its equations in a simple form as before, but the probability of infection for a single farm can easily be written.

Suppose all farms in the UK are listed and indexed with i. Then p_i, the probability that an individual farm i becomes infected during one unit of time, is:

$$p_i = \beta_i \Sigma_{\text{all infectious farms} j} \tau_j K(d_{ij}) \quad (5.8)$$

where β_i is the susceptibility of farm i, determined by the number of sheep and cows it holds; τ_j is the infectiousness of farm j, also determined by the number of sheep and cows it holds; and K (d_{ij}) is a function of the distance between the pair of farms i and j which determines how quickly infectiousness falls off with increasing distance. K is known as the "infection kernel." In the FMD example, the infection kernel was estimated from contact tracing data on farms that were sources of infection and their secondary cases. This observed relationship shows a very sharp falling off of infectiousness, with a farm just 2 km distant being less than tenfold as infectious to a susceptible farm than one that is adjacent.

This section describes just two of the possible heterogeneities that are often included when making models of the spread of epidemics. There is almost no end to how complex an epidemiological model can become. However, it is very easy for complex models to outstrip the data available to calculate their parameters. In some cases, this can mean that models become black boxes concealing ill-informed guesswork, rather than prisms unveiling the implications of well-sourced and well-understood data.

Within-Host Dynamics

Mathematical models can also be used to investigate the dynamics of events that unfold within infected hosts. In these models, the units of study are often infected cells and immune cells responding to infection. As with epidemiological models, there is a wide range of modeling styles: Some models detail many different interacting components; others make a virtue of parsimony in their description of within-host interactions. In this section, a simple model of the within-host evolution of HIV is used to illustrate how pared-down, within-host models of infection can address important practical questions.

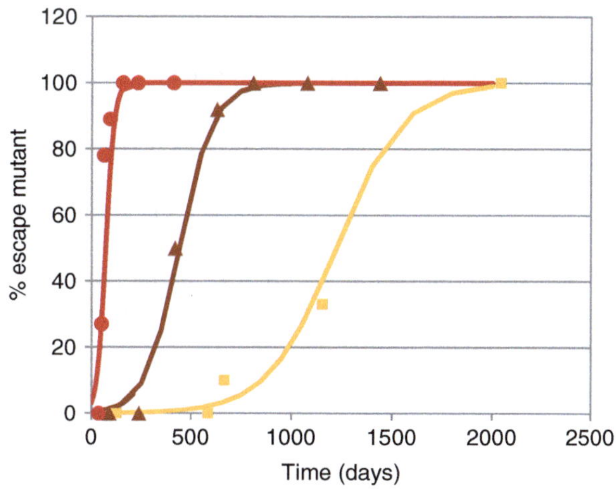

Fig. 5.3 The outgrowth of HIV CTL escape mutants through time. Data sets from three different patients (reviewed in [19]) are shown as *red*, *brown*, and *yellow* symbols. Equation 5.12 is fitted to these data, yielding rates of outgrowth, c−(a − â), of 0.048 (*red*), 0.012 (*brown*), and 0.006 (*yellow*)

Several trials of prophylactic HIV vaccines have shown little or no effect [14–16], and understanding why these vaccines failed is a major research priority [17]. A quantitative description of the interaction between HIV and host immune cells would be an asset to such understanding. For one component of host immunity – the cytotoxic T cell (CTL) response – such a description can be derived. The question is how effective are host CTL responses at killing HIV-infected cells? Not how many CTLs are present, nor which cytokines they secrete, but how fast do they kill HIV-infected cells?

During the course of a single infection, HIV evolves to escape from the selection pressure imposed by host CTLs [18]. In this process, new HIV variants emerge that are not recognized by the host CTLs. These variants are called "CTL escape mutants." These CTL escape mutants can be seen to grow out in hosts who mount relevant CTL responses (Fig. 5.3) and to revert in hosts who do not. The rate of reversion in hosts without relevant CTL responses reflects the underlying fitness cost of the mutation. The rate of outgrowth in hosts who do mount relevant CTL responses is a balance between the efficacy of those responses and the fitness cost of the mutations. These costs and benefits need to be examined in the context of the underlying rate of turnover of HIV-infected cells. All this can be represented in a two-line mathematical model [19].

Let x be the number of host cells infected with "wild-type" virus – that is, virus that can be recognized by the relevant host CTL responses. Let y be the number of host cells infected with escape mutant virus. The model then consists of a pair of ordinary differential equations describing the growth rate of each population of infected cells. The wild-type population grows at rate a, is killed by the CTL response in question at rate c, and is killed by all other processes at rate b. The escape mutant population grows at rate â (â < a, reflecting the underlying fitness cost of the mutation) and is killed by all other processes at rate b. Escape-mutant-infected cells are not killed by the CTL response in question because of the presence of the escape mutation in the viral genome. These assumptions give rise to the following pair of linear ordinary differential equations:

$$x' = ax - bx - cx \tag{5.9}$$

$$y' = \hat{a}y - by \tag{5.10}$$

The observed quantity, call it p, is the fraction of virus that is of the escape mutant type; $p = y/(x + y)$. Simple application of the quotient rule for differentiation yields the single differential equation

$$p' = (c - (a - \hat{a}))p(1 - p) \tag{5.11}$$

with solution:

$$p = (k \exp(-(c - (a - \hat{a}))t) + 1)^{-1} \tag{5.12}$$

where for p_0, the fraction escaped at time 0:

$$k = \frac{(1 - p_0)}{p_0} \tag{5.13}$$

It is straightforward to fit the analytic expression (12) to data on the outgrowth of escape mutants to obtain estimates of the quantity $c - (a - \hat{a})$. Figure 5.3 shows fitted curves with estimates of $c - (a - \hat{a})$ of 0.048, 0.012, and 0.006. The quantity of interest is the parameter c – the rate at which CTL kills cells infected with wild-type virus. Fortunately, independent estimates of the fitness cost of the escape mutation $(a - \hat{a})$ are available. The median of several such observations yields $(a - \hat{a}) = 0.005$ [19]. Taken together and combined with further data, the inference is that on average, a single CTL response kills infected cells at rate 0.02 per day.

The half-life of an HIV-infected cell is about 1 day. This figure was itself derived from the application of elegantly simple models to data on the post-treatment dynamics of HIV [20, 21]. If a single CTL response kills infected cells at rate 0.02 per day and their overall death rate is one, then just 2% of the death of infected cells can be attributed to killing by one CTL response. Patients will typically mount many responses – but probably not more than a dozen. This analysis shows that even though CTL responses are effective enough to drive viral evolution, they are, in quantitative terms, very weak. A vaccine to protect against HIV infection would have to elicit immune responses that are manyfold stronger than the natural responses detected in ongoing infection. This simple, model-based observation greatly helps understand why the vaccines trialed so far have failed.

Multilevel Models

The models discussed so far deal either with events inside individuals or with transmission amongst individuals (people or farms) in a population. Some questions require simultaneous consideration of events at both levels of organization. This is particularly true for questions about the evolution of infectious agents as their evolution proceeds within individual hosts, but they are also transmitted between hosts. Models that capture events at both the within-host and between-host levels are fairly recent additions to the literature on infectious disease modeling. Here, they are illustrated with two examples, a set of models that consider the emergence of a zoonotic infection in humans and a model of the within-host evolution and between-host transmission of HIV.

Emerging infections are a continuing threat to human well-being. The pandemics of SARS in 2003 and H1N1 swine flu in 2009 illustrated how quickly a new infectious agent spreads around the world. Neither of these was as devastating as some predicted, but the continuing pandemic of HIV is ample proof that emerging infectious diseases can have devastating consequences for human communities. Many novel emerging infections arise as zoonoses – that is, infections that cross from animals into humans [22]. To become a successful emerging infection of humans – that is, one that spreads widely amongst people – is a multi-step process [23]. First, the pathogen must cross the species barrier into people, then it must transmit between people, and finally, it must transmit *efficiently enough* that epidemics arise. This latter step amounts to having a basic reproductive number, R_0, that is greater than 1. The emerging infections mentioned already, SARS, swine flu, and HIV, have transited all these steps. But there are other zoonoses that transmit to humans without emerging as epidemics or pandemics. For example, simian foamy virus, a retrovirus that is endemic in most old-world primates [24], can be detected in people who work with primates [25] or hunt them [26]. There is no record of any human-to-human transmission, implying that this zoonosis only completes the first step in becoming an emerging infection. Other infections, whilst spreading from person to person, still do not cause epidemics because that spread is insufficiently efficient. An example of such an infection is the newly discovered arenavirus from Southern Africa called "Lujo virus" [27]. This virus caused a small outbreak in the autumn of 2008. Very dramatically, four out of the five known cases died, but with five cases and just four transmission events, the basic reproductive number stayed below one, and there was no epidemic.

Acquiring $R_0 > 1$ is thus an important threshold that zoonoses must breach before they can become emerging infections. Antia and colleagues [28] developed an elegant model of the within-host evolution and between-host transmission of a zoonotic infection that initially has $R_0 < 1$, but through within-host adaptation in humans can evolve to become efficient enough at transmitting from one human to another that R_0 increases above 1 and epidemics become possible. They developed a multi-type branching process model of the transmission and evolution of a zoonosis. They found that the probability of emergence depends very strongly on the basic reproductive number of the pathogen as it crosses into humans. This is because, even when $R_0 < 1$, short chains of transmission are still possible (as exemplified with Lujo virus described above). During ongoing infections in humans, the zoonosis has opportunities to evolve towards higher transmissibility. The higher its initial R_0, the more opportunities there are for such ongoing evolution and hence for emergence.

This model of the emergence of a novel infection has been extended by other authors to address questions about the interpretation of surveillance data [29] and the role of host heterogeneity in the process of emergence [30]. These extensions confirm the original finding that the transmission efficiency (R_0) of the introduced variant (and any intermediate variants) is a very important driver of the probability of emergence. Kubiak and colleagues explored the emergence of a novel infection in populations split into several communities, with commuters acting to join those communities together. They found that most communities are sufficiently interconnected to show no effect of spatial distribution on the emergence process, even a small number of commuters being sufficient to successfully transmit any novel pathogen between settlements. Thus, although many zoonotic events happen in isolated parts of the world, unless they are really cut off from urban centers, that isolation offers little barrier to the transmission of newly emerged infections.

HIV emerged as a human infection sometime during the end of the 1800s and the early 1900s [31]. It was only recognized as a new human infection in the 1980s when cases of immunodeficiency in young Americans were unusual enough to warrant investigation [32]. As discussed above, during the course of a single infection, HIV is able to adapt to escape from the selection pressures imposed by its host's immune response. HIV variants that cannot be recognized by current host CTLs are termed "CTL escape mutants." These mutants yield important information about the strength of the immune responses that they evade. However, since they were shown to transmit from one host to another, their status has been raised to potential drivers of evolutionary change across the global HIV pandemic [33, 34].

Different hosts respond to different parts of HIV's proteins (known as epitopes). For CTL responses, it is the host class 1 human leukocyte antigen (HLA) type that determines which epitopes are recognized. When CTL escape mutants are transmitted into a host who does not make immune response to that epitope, the mutations are no longer advantageous, and the virus can revert to the wild type [35]. Global change in the prevalence of CTL escape mutants is therefore driven by three parallel processes: the selection of escape mutants in some hosts, transmission between hosts, and reversion of escape mutants in other hosts. Once again, this is a process that takes place across multiple levels of organization, evolution and reversion of escape mutations within infected hosts, and transmission between hosts.

Fryer and colleagues [36] developed a multilevel model of the three processes of within-host evolution, within-host reversion, and between-host transmission. The model is a version of the so-called SI model which is a simplified version of the SIR model presented above which does not allow recovery. The model allows heterogeneity in hosts and in the infecting virus so that there are hosts who do and do not

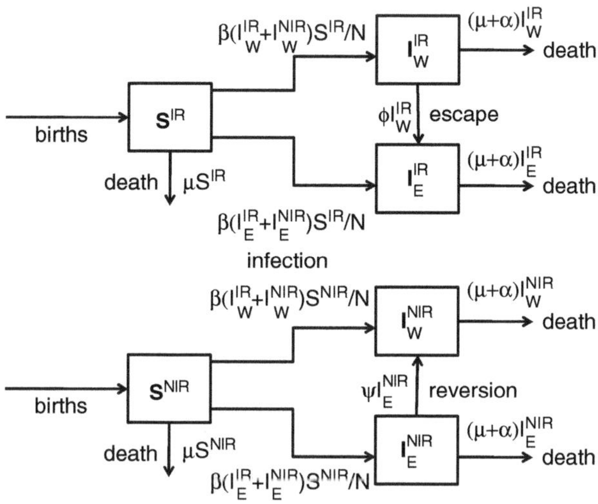

Fig. 5.4 A model of the within-host evolution and between-host transmission of HIV escape mutants [36]. Hosts are divided into two types: immune responders (superscript IR) and nonimmune responders (superscript NIR). There are also two variants of virus, wild type (subscript WT) and escape mutant (subscript E). Hosts are either susceptible, S, or infectious, I, and the type of virus with which they are infected is denoted by the subscript. Rates of infection are determined by the number of people infectious with each virus type. Immune responding hosts infected with the wild-type virus drive immune escape at per capita rate ϕ, whilst nonimmune responding hosts infected with escape mutant virus drive reversion at per capita rate ψ. All hosts are prone to per capita death rate μ, and infected hosts have an additional death rate α

mount immune responses to a given epitope and there are viruses that do and do not have escape mutations in that epitope. This model is represented in Fig. 5.4. As in the SIR model described in section "The SIR Model," the rate at which susceptibles become infected is determined by the number of infectious people present. However, in this model, because it represents the spread of a sexually transmitted disease, it is the proportion of hosts who are infectious that drives new infections. Furthermore, there are now two virus types circulating – wild type and escape mutant. Within-host adaptation allows hosts who do mount immune responses to the epitope to drive the evolution of escape mutants, and conversely, hosts who do not mount such responses can drive the reversion of escape mutant viruses back to the wild type.

This model's behavior is easy to understand. The total numbers of susceptible and infectious people simply follow the well-characterized SI model. Figure 5.5a shows total cases through time. The total epidemic goes through three phases: an initial exponential growth, a saturation phase, and then settling to a long-term equilibrium. Figure 5.5b shows the proportion of all cases that are escape mutants through time. Not surprisingly, faster escape rates and slower reversion rates lead to higher prevalence of escape mutants. Less intuitive are the following characteristics of Fig. 5.5b. Whilst the epidemic is in its exponential growth phase, so long as

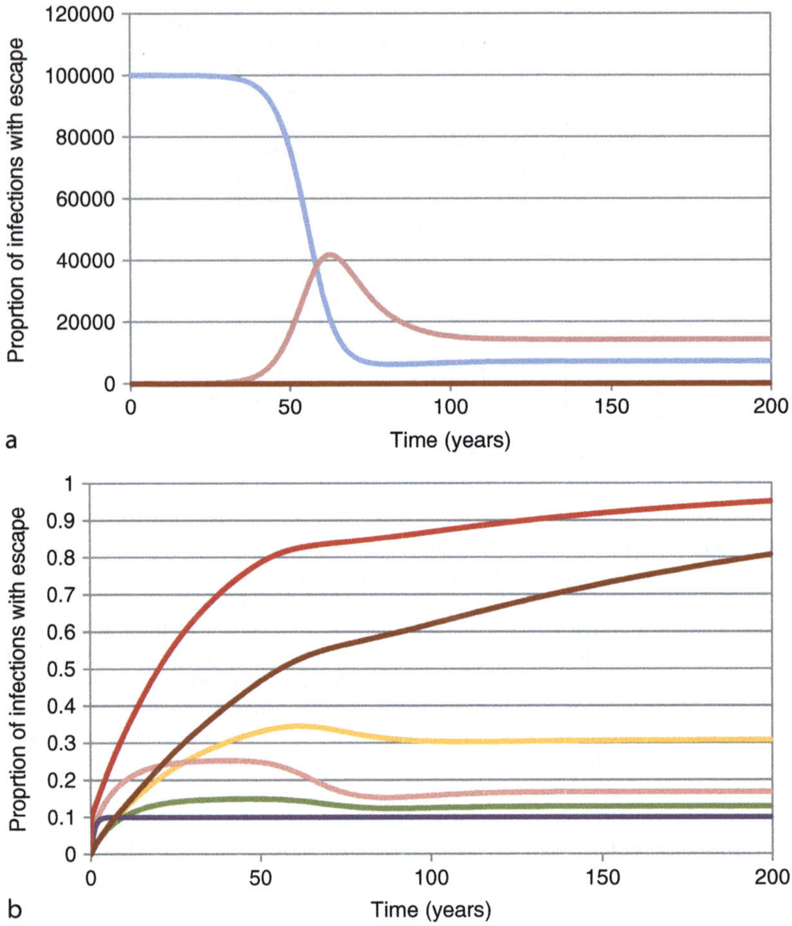

Fig. 5.5 Predictions of a model of within-host evolution and between-host transmission of HIV. (**a**) shows total numbers of susceptible (*blue*) and infectious (*red*) people through time. (**b**) shows the proportion of infections that are with escape mutant virus for a range of escape and reversion rates. The mean times to escape and reversion for each curve are as follows: *red* – escape 1 month, reversion never; *brown* – escape 5 years, reversion never; *yellow* – escape 5 years, reversion 50 years; *pink* – escape 1 year, reversion 10 years; *green* – escape 5 years, reversion 10 years; *mauve* – escape 1 year, reversion 1 year

reversion rates are reasonably fast (say once in 10 years or faster), the prevalence of escape is expected to stabilize quite quickly. However, this is not a long-term equilibrium, and as the total epidemic turns over, the escape prevalence shifts again. For an epitope that escapes fast but reverts at an intermediate rate, this leads to a substantial drop in the prevalence of escape. Secondly, fixation of escape variants only occurs if they never revert, and even then fixation takes a very long time – much longer than it takes for the underlying epidemic to equilibrate. Thirdly, the predicted dynamics and equilibrium are very sensitive to the reversion rate when that is slow. Notice the big difference, in the long term, between no reversion (brown line) and average time to reversion of 50 years (yellow line).

As well as predicting the future spread of escape mutations for different rates of escape and reversion, this model can be used to infer escape and reversion rates from data on their current prevalence. This exercise reveals a surprisingly slow average rate of escape. Across 26 different epitopes, the median time to escape was over 8 years. There is close agreement between rates of escape inferred using this model and those estimated from a longitudinal cohort study. These slow rates are in marked contrast to the general impression given by a large number of case reports in which escape is described as occurring during the first year of infection. However, a collection of case reports is a poor basis upon which to estimate an average rate of escape.

These are just two examples from the new family of infectious disease models that encapsulate processes at multiple levels of organization. As data on pathogen evolution continues to accrue, this approach will doubtlessly continue to yield new insights.

Future Directions

It seems likely that infectious diseases will continue to trouble both individuals and communities. Whilst technological advances in new drugs, new vaccines, and better methods for surveillance will undoubtedly assist with the control of infection, several trends in society pull in the opposite direction. Chief amongst these is a growing population, and second is increasing population density as more and more people live in towns and cities. What can infectious disease modeling do to help?

Models can help in two different ways. The first is to assist the understanding of systems that are intrinsically complicated. Many different interacting populations, events that occur on multiple timescales, and systems with multiple levels of organization can all be better understood when appropriate models are used as an organizing principle and a tool for formal analysis. Sometimes, the problem is that there is not enough data. A systematic description can be very revealing in searching for which new data are most needed. There are also situations where the problem is a deluge of data. In these circumstances, well-constructed models provide a useful organizing scheme with which to interrogate those data.

The second use of models is as representations of well-understood systems used as tools for comparing different intervention strategies. The model of the farm-to-farm spread of FMD described at section "Host Heterogeneity" is a fine example of this use of modeling. It includes enough detail to be a useful tool for comparing different interventions, but is still firmly rooted in available data so does not rest on large numbers of untested assumptions.

Bibliography

Primary Literature

1. Fenner F (1982) A successful eradication campaign. Global eradication of smallpox. Rev Infect Dis 4(5):916–930
2. Moss WJ, Griffin DE (2006) Global measles elimination. Nat Rev Microbiol 4:900–908
3. Farrington CP, Kanaan MN, Gay NJ (2001) Estimation of the basic reproduction number for infectious diseases from age-stratified serological survey data. J R Stat Soc Ser C Appl Stat 50:251–292
4. Anderson RM, May RM (1991) Infectious diseases of humans. Oxford University Press, Oxford
5. McLean AR, Anderson RM (1988) Measles in developing countries. Part II: the predicted impact of mass vaccination. Epidemiol Infect 100:419–442
6. McLean AR (1995) After the honeymoon in measles control. Lancet 345(8945):272
7. Babad HE et al (1995) Predicting the impact of measles vaccination in England and Wales: model validation and analysis of policy options. Epidemiol Infect 114:319–344
8. McLean AR (1992) Mathematical modelling of the immunisation of populations. Rev Med Virol 2:141–152
9. Arinaminpathy N, McLean AR (2009) Logistics for control for an influenza pandemic. Epidemics 1(2):83–88
10. Longini IM et al (2005) Containing pandemic influenza at the source. Science 309(5737):1083–1087
11. Anderson RM, Grenfell BT (1986) Quantitative investigations of different vaccination policies for the control of congenital rubella syndrome (CRS) in the United Kingdom. J Hyg 96:305–333, Cambridge University Press
12. Metcalf CJE et al (2010) Rubella metapopulation dynamics and importance of spatial coupling to the risk of congenital rubella syndrome in Peru. J R Soc Interface 8(56):369–376
13. Keeling MJ et al (2001) Dynamics of the 2001 UK foot and mouth epidemic: stochastic dispersal in a heterogeneous landscape. Science 294(5543):813–817
14. Flynn NM et al (2005) Placebo-controlled phase 3 trial of a recombinant glycoprotein 120 vaccine to prevent HIV-1 infection. J Infect Dis 191(5):654–665
15. Buchbinder SP et al (2008) Efficacy assessment of a cell-mediated immunity HIV-1 vaccine (the Step Study): a double-blind, randomised, placebo-controlled, test-of-concept trial. Lancet 372(9653):1881–1893
16. Rerks-Ngarm S et al (2009) Vaccination with ALVAC and AIDSVAX to prevent HIV-1 infection in Thailand. N Engl J Med 361(23):2209–2220, Epub 2009 Oct 20
17. Council of the Global HIV Vaccine Enterprise (2010) The 2010 scientific strategic plan of the global HIV vaccine enterprise. Nat Med 16(9):981–989
18. Phillips RE et al (1991) Human immunodeficiency virus genetic variation that can escape cytotoxic T cell recognition. Nature 354:453–459
19. Asquith B et al (2006) Inefficient cytotoxic T lymphocyte-mediated killing of HIV-1-infected cells in vivo. PLoS Biol 4:e90
20. Wei X et al (1995) Viral dynamics in human immunodeficiency virus type 1 infection. Nature 373(6510):117–122
21. Ho DD et al (1995) Rapid turnover of plasma virions and CD4 lymphocytes in HIV-1 infection. Nature 373(6510):123–126
22. Woolhouse MEJ (2002) Population biology of emerging and re-emerging pathogens. Trends Microbiol 10(10):s3–s7
23. Wolfe ND et al (2007) Origins of major human infectious diseases. Nature 447:279–283
24. Meiering CD, Linial ML (2001) Historical perspective of foamy virus epidemiology and infection. Clin Microbiol Rev 14:165–176

25. Switzer WM et al (2004) Frequent simian foamy virus infection in persons occupationally exposed to nonhuman primates. J Virol 78(6):2780–2789
26. Wolfe ND et al (2004) Naturally acquired simian retrovirus infections in central African hunters. Lancet 363:932–937
27. Briese T et al (2009) Genetic detection and characterization of Lujo virus, a new hemorrhagic fever–associated arenavirus from Southern Africa. PLoS Pathog 5(5):e1000455. doi:10.1371/journal.ppat.1000455
28. Antia R et al (2004) The role of evolution in the emergence of infectious diseases. Nature 426:658–661
29. Arinaminpathy N, McLean AR (2009) Evolution and emergence of novel human infections. Proc R Soc B 273:3075–3083
30. Kubiak R et al (2010) Insights into the evolution and emergence of a novel infectious disease. PLoS Comput Biol 6(9):e10000947
31. Worobey M et al (2008) Direct evidence of extensive diversity of HIV-1 in Kinshasa by 1960. Nature 455(7213):661–664
32. CDC (1981) Kaposi's sarcoma and *Pneumocystis pneumonia* among homosexual men – New York City and California. MMWR 30:305–308
33. Kawashima Y et al (2009) Adaptation of HIV-1 to human leukocyte antigen class I. Nature 458:641–645
34. Goulder PJ et al (2001) Evolution and transmission of stable CTL escape mutations in HIV infection. Nature 412:334–338
35. Leslie AJ et al (2004) HIV evolution: CTL escape mutation and reversion after transmission. Nat Med 10:282–289
36. Fryer HR et al (2010) Modelling the evolution and spread of HIV immune escape mutants. PLoS Pathog 6(11):e1001196

Books and Reviews

Anderson RM, May RM (1991) Infectious diseases of humans. Oxford University Press, Oxford
Keeling MJ, Rohani R (2008) Modeling infectious diseases in humans and animals. Princeton University Press, Princeton
Nowak MA, May RM (2000) Virus dynamics. Oxford University Press, Oxford

Chapter 6
Infectious Diseases, Climate Change Effects on

Matthew Baylis and Claire Risley

Glossary

Climate	The weather averaged over a long time or, succinctly, climate is what you expect, weather is what you get!
El Niño Southern Oscillation (ENSO)	A climate phenomenon whereby, following reversal of trade winds approximately every 4–7 years, a vast body of warm water moves slowly west to east across the Pacific, resulting in "an El Niño" event in the Americas and leading to a detectable change to climate (mostly disruption of normal rainfall patterns) across 70% of the earth's surface.
Emerging disease	An infection or disease that has recently increased in incidence (the number of cases), severity (how bad the disease is), or distribution (where it occurs).
Endemic stability	The counter-intuitive situation where the amount of disease rises as the amount of infection falls, such that controlling infection can exacerbate the problem.
Infection	The body of a host having been invaded by microorganisms (mostly viruses, bacteria, fungi, protozoa, and parasites).

This chapter, which has been modified slightly for the purposes of this volume, was originally published as part of the Encyclopedia of Sustainability Science and Technology edited by Robert A. Meyers. DOI:10.1007/978-1-4419-0851-3

M. Baylis (✉) • C. Risley
LUCINDA Group, Institute of Infection and Global Health, University of Liverpool, Leahurst Campus, Neston, Cheshire CH64 7TE, UK
e-mail: matthew.baylis@liverpool.ac.uk; claire.risley@liverpool.ac.uk

P. Kanki and D.J. Grimes (eds.), *Infectious Diseases: Selected Entries from the Encyclopedia of Sustainability Science and Technology*, DOI 10.1007/978-1-4614-5719-0_6, © Springer Science+Business Media New York 2013

Infectious disease	A pathology or disease that results from infection. Note that many diseases are not infectious and not all infections result in disease.
Intermediate host	A host in which a parasite undergoes an essential part of its lifecycle before passing to a second host, and where this passing is passive, that is, not by direct introduction into the next host (see *vector*).
Vector	Usually, an arthropod that spreads an infectious pathogen by directly introducing it into a host. For diseases of humans and animals, the most important vectors are flies (like mosquitoes, midges, sandflies, tsetse flies), fleas, lice, and ticks. Aphids are important vectors of diseases in plants. In some instances, other means of carriage of pathogens, such as human hands, car wheels, etc., are referred to as vectors.
Vector competence	The proportion of an arthropod vector population that can be infected with a pathogen.
Zoonosis	An infection of animals that can spread to, and cause disease in, humans (plural, zoonoses).

Definition of the Subject and Its Importance

Infectious diseases of humans continue to present a significant burden to our health, disproportionately so in the developing world. Infectious diseases of livestock affect their health and welfare, are themselves important causes of human disease and, exceptionally, can threaten our food security. Wildlife infections again present a zoonotic risk to humans, but additionally, such diseases may threaten vulnerable populations and be a cause of extinction and biodiversity loss. Wild populations are inherently more susceptible to environmental change, largely lacking any human protective influence that domesticated species and human populations may benefit from.

Many infectious diseases of humans and farmed or wild animals are influenced by weather or climate, affecting where or when disease occurs, or how severe outbreaks are, and it is therefore likely that future climate change, whether human caused or natural, will have an impact on future disease burdens. Understanding the processes involved may enable prediction of how disease burdens will change in the future and, therefore, allow mitigative or adaptive measures to be put in place.

While climate change will likely be an important cause of change in some infectious diseases in the future, there are other disease drivers which will also change over similar time scales and which may exacerbate or counteract any effects of climate change. Assessment of the future importance of climate change as an influence over future disease burdens must therefore be considered alongside other causes of change.

Introduction

The impact of infectious diseases of humans and animals seems as great now as it was a century ago. While many disease threats have disappeared or dwindled, at least in the developed world, others have arisen to take their place. Important infectious diseases of humans that have emerged in the last 30 years, for a range of reasons, include Acquired immune deficiency syndrome (AIDS), variant Creutzfeldt-Jakob disease (vCJD), multidrug resistant tuberculosis, severe acute respiratory syndrome (SARS), *E. coli* O157, avian influenza, swine flu, West Nile fever, and Chikungunya [1, 2]. The same applies to diseases of animals: Indeed, all but one of the aforementioned human diseases have animal origins – they are zoonoses – and hence the two subjects of human and animal disease, usually studied separately by medical or veterinary scientists, are intimately entwined.

What will be the global impact of infectious diseases at the end of the twenty-first century? Any single disease is likely to be affected by many factors that cannot be predicted with confidence, including changes to human demography and behavior, new scientific or technological advances including cures and vaccines, pathogen evolution, livestock management practices and developments in animal genetics, and changes to the physical environment. A further, arguably more predictable, influence is climate change.

Owing to anthropogenic activities, there is widespread scientific agreement that the world's climate is warming at a faster rate than ever before [3], with concomitant changes in precipitation, flooding, winds, and the frequency of extreme events such as El Niño. Innumerable studies have demonstrated links between infectious diseases and climate, and it is unthinkable that a significant change in climate during this century will not impact on at least some of them.

How should one react to predicted changes in diseases ascribed to climate change? The answer depends on the animal populations and human communities affected, whether the disease changes in severity, incidence, or spatiotemporal distribution and, of course, on the direction of change: Some diseases may spread but others may retreat in distribution. It also depends on the relative importance of the disease. If climate change is predicted to affect mostly diseases of relatively minor impact on human society or global biodiversity/ecosystem function, while the more important diseases are refractory to climate change's influence, then our concerns should be tempered.

To understand climate change's effects on infectious diseases in the future it is necessary to first understand how climate affects diseases today. This entry begins by first presenting examples of climate's effects on diseases of humans and livestock today and, from the understanding gained, then describes the processes by which climate change might affect such diseases in the future. Diseases of wildlife are important, to some extent, for different reasons to those of humans and livestock, and are therefore considered separately. The relative importance of climate change as a disease driver, compared to other forces, is considered, with examples provided of where climate change both is, and is not, the major force. Finally, the future prospects and the uncertainties surrounding them are considered.

Weather, Climate, and Disease

Many diseases are affected directly or indirectly by weather and climate. Remarkably, no systematic surveys of links between diseases and weather/climate seem to exist and, therefore, it is not possible to indicate whether these diseases represent a minority or majority.

The associations between diseases and weather/climate fall broadly into three categories. The associations may be *spatial*, with climate affecting the distribution of a disease; *temporal* with weather or climate affecting the timing of outbreaks; or they may relate to the *intensity* of an outbreak. Temporal associations can be further broken into at least two subcategories: *seasonal*, with weather or climate affecting the seasonal occurrence of a disease, and *interannual*, with weather or climate affecting the timing, or frequency of years in which outbreaks occur. Here a selection of these associations is presented, which is by no means exhaustive but is, rather, intended to demonstrate the diversity of effects. Furthermore, the assignment of diseases into the different categories should not be considered hard-and-fast as many diseases could come under more than one heading.

Spatial

- Schistosomosis is an important cause of human mortality and morbidity in Africa and, to a lesser extent, in Asia. The disease is caused by species of *Schistosoma* trematode parasite, for which water-living snails are intermediate hosts. The distribution of suitable water bodies is therefore important for its distribution. However, there must also be suitable temperature: In China, *Oncomelania hupensis* snail intermediate hosts cannot live north of the January 0°C isotherm (the "freezing line") while *Schistosoma japonicum* only develops within the snail at temperatures above 15.4°C. Shistosomosis risk in China is therefore restricted to the warmer southeastern part of the country [4].
- Diseases transmitted by tsetse flies (sleeping sickness, animal trypanosomosis) and ticks (such as anaplasmosis, babesiosis, East Coast fever, heartwater) impose a tremendous burden on African people and their livestock. Many aspects of the vectors' life cycles are sensitive to climate, to the extent that their spatial distributions can be predicted accurately using satellite-derived proxies for climate variables [5].
- Mosquitoes (principally *Culex* and *Aedes*) transmit several viruses of birds that can also cause mortality in humans and horses. Examples are West Nile

fever (WNF) and the viral encephalitides such as Venezuelan, western, and eastern equine encephalitis (VEE, WEE, and EEE, respectively) [6]. The spatial distributions of the mosquito vectors are highly sensitive to climate variables.

Temporal-Seasonal

All of the previous examples of spatial associations between diseases and climate can also be classified as temporal-seasonal, as the effects of climate on the seasonal cycle of the intermediate hosts (snails, tsetse flies, and mosquitoes, respectively) also determines in part the seasonal cycle of disease. There are other diseases where the associations can be described as seasonal-temporal.

- Salmonellosis is a serious food-borne disease caused by *Salmonella* bacteria, most often obtained from eggs, poultry, and pork. Salmonellosis notification rates in several European countries have been shown to increase by about 5–10% for each 1°C increase in ambient temperature [7]. Salmonellosis notification is particularly associated with high temperatures during the week prior to consumption of infected produce, implicating a mechanistic effect via poor food handling.
- Foot-and-mouth disease (FMD) is a highly contagious, viral infection of cloven-footed animals, including cattle, sheep, and pigs. Most transmission is by contact between infected and susceptible animals, or by contact with contaminated animal products. However, FMD can also spread on the wind. The survival of the virus is low at relative humidity (RH) below 60% [8], and wind-borne spread is favored by the humid, cold weather common to temperate regions. In warmer drier regions, such as Africa, wind-borne spread of FMD is considered unimportant [9].
- Peste des petits ruminants (PPR) is an acute, contagious, viral disease of small ruminants, especially goats, which is of great economic importance in parts of Africa and the Near East. It is transmitted mostly by aerosol droplets between animals in close contact. However, the appearance of clinical PPR is often associated with the onset of the rainy season or dry cold periods [10], a pattern that may be related to viral survival. The closely related rinderpest virus survives best at low or high relative humidity, and least at 50–60% [11].
- Several directly transmitted human respiratory infections, including those caused by rhinoviruses (common colds) and seasonal influenza viruses (flu) have, in temperate countries, seasonal patterns linked to the annual temperature cycle. There may be direct influences of climate, such as the effect of humidity on survival of the virus in aerosol [12], or indirect influences via, for example, seasonal changes in the strength of the human immune system or more indoor crowding during cold weather [13].

Temporal-Interannual

The previous examples of spatial associations between diseases and climate, which were further categorized as temporal-seasonal, can also be classified as temporal-interannual, as the effects of climate on the intermediate hosts (snails, tsetse flies, and mosquitoes, respectively) will determine in part the risk or scale of a disease outbreak in a given year. There are other diseases where the associations can be described as seasonal-interannual.

- Anthrax is an acute infectious disease of most warm-blooded animals, including humans, with worldwide distribution. The causative bacterium, *Bacillus anthracis* forms spores able to remain infective for 10–20 years in pasture. Temperature, relative humidity, and soil moisture all affect the successful germination of anthrax spores, while heavy rainfall may stir up dormant spores. Outbreaks are often associated with alternating heavy rainfall and drought, and high temperatures [14].
- Cholera, a diarrheal disease which has killed tens of millions of people worldwide, is caused by the bacterium *Vibrio cholerae*, which lives amongst sea plankton [15]. High temperatures causing an increase in algal populations often precede cholera outbreaks. Disruption to normal rainfall helps cholera to spread further, either by flooding, leading to the contamination of water sources, such as wells, or drought which can make the use of such water sources unavoidable. Contaminated water sources then become an important source of infection in people.
- Plague is a flea-borne disease caused by the bacterium *Yersinia pestis*; the fleas' rodent hosts bring them into proximity with humans. In Central Asia, large scale fluctuations in climate synchronize the rodent population dynamics over large areas [16], allowing population density to rise over the critical threshold required for plague outbreaks to commence [17].
- African horse sickness (AHS), a lethal infectious disease of horses, is caused by a virus transmitted by *Culicoides* biting midges. Large outbreaks of AHS in the Republic of South Africa over the last 200 years are associated with the combination of drought and heavy rainfall brought by the warm phase of the El Niño Southern Oscillation (ENSO) [18].
- Rift Valley Fever (RVF), an important zoonotic viral disease of sheep and cattle, is transmitted by *Aedes* and *Culex* mosquitoes. Epizootics of RVF are associated with periods of heavy rainfall and flooding [19–21] or, in East Africa, with the combination of heavy rainfall following drought associated with ENSO [20, 22]. ENSO-related floods in 1998, following drought in 1997, led to an epidemic of RVF (and some other diseases) in the Kenya/Somalia border area and the deaths of more than 2000 people and two-thirds of all small ruminant livestock [23]. Outbreaks of several other human infections, including malaria and dengue fever have, in some parts of the world, been linked to ENSO events.

Intensity

In addition to associations between climate and the spatial and temporal distributions of disease outbreaks, there are some examples of associations between climate and the intensity or severity of the disease that results from infection. It is theoretically possible, for example, that climate-induced immunosuppression of hosts may favor the multiplication of some microparasites (viruses, bacteria, rickettsia, fungi, protozoa), thereby increasing disease severity or, alternatively, reduce the disease-associated immune response to infection, thereby reducing disease severity.

However, the clearest examples pertain to macroparasites (helminth worms) which, with the notable exception of *Strongyloides* spp., do not multiply within the host. Disease severity is therefore directly correlated with the number of parasites acquired at the point of infection or subsequently, and in turn this is frequently associated with climate, which affects both parasite survival and seasonal occurrence.

- Fasciolosis, caused by the *Fasciola* trematode fluke, is of economic importance to livestock producers in many parts of the world and also causes disease in humans. In sheep, severe pathology, including sudden death, results from acute fasciolosis which occurs after ingestion of more than 2,000 metacercariae (larval flukes) of *Fasciola hepatica* at pasture, while milder pathology associated with subacute and chronic fasciolosis occurs after ingestion of 200–1,000 metacercariae [24]. Acute fasciolosis is therefore most common in places or in years when rainfall and temperature favor the survival of large numbers of metacercariae.
- *Nematodirus battus* is a nematode parasite of the intestine of lambs. Eggs deposited in the feces of one season's lambs do not hatch straightaway; instead, they build up on the pasture during summer and remain as eggs over winter, not hatching until temperatures the following spring exceed 10°C [25]. Once the mean daily temperature exceeds this threshold the eggs hatch rapidly, leading to a sharp peak in the number of infective larvae on the pasture. If this coincides with the new season's lambs grazing on the pasture, there is likely to be a large uptake of larvae and severe disease, called nematodiriasis. If, however, there is a warm spell early in the year, the peak in larvae on pasture may occur while lambs are still suckling rather than grazing, such that fewer larvae are ingested and the severity of nematodiriasis is reduced.

Climate Change and Disease

There is a substantial scientific literature on the effects of climate change on health and disease, but with strong focus on human health and vector-borne disease [5, 26–42]. By contrast, the effects of climate change on diseases spread by other means, or animal diseases in general, have received comparatively little attention

[43–48]. Given the global burden of diseases that are not vector-borne, and the contribution made by animal diseases to poverty in the developing world [49], attention to these areas is overdue.

The previous section demonstrates the range of climate influences upon infectious disease. Such influences are not the sole preserve of vector-borne diseases: Food-borne, water-borne, and aerosol-transmitted diseases are also affected. A common feature of non-vector-borne diseases affected by climate is that the pathogen or parasite spends a period of time outside of the host, subject to environmental influence. Examples include the infective spores of anthrax; FMD viruses in temperate regions; the *Salmonella* bacteria that contaminate food products; the cholera-causing *vibrio* bacteria in water; and the moisture- and temperature-dependent survival of the parasites causing schistosomosis and fasciolosis.

By contrast, most diseases transmitted directly between humans (for example, human childhood viruses, sexually transmitted diseases (STDs), tuberculosis) have few or no reported associations with climate. This is also the case for animal infections such as avian influenza, bovine tuberculosis, brucellosis, Newcastle's disease of poultry, and rabies. Clear exceptions are the viruses that cause colds and seasonal flu in humans, and PPR in small ruminants; these viruses are spread by aerosol between individuals in close contact but are nevertheless sensitive to the effects of ambient humidity and possibly temperature.

The influence of climate on diseases that are not vector-borne appears to be most frequently associated with the timing (intra- or interannual) of their occurrence rather than their spatial distribution. There are examples of such diseases that occur only in certain parts of the world (for example, PPR) but most occur worldwide. By contrast, the associations of vector-borne diseases with climate are equally apparent in time and space, with very few vector-borne diseases being considered a risk worldwide. This is a reflection of the strong influence of climate on both the spatial and temporal distributions of intermediate vectors. If there are exceptions to this rule, the vectors are likely to be lice or fleas, with lives so intimately associated with humans or animals that they are relatively protected from climate's influences.

In the scientific literature, many processes have been proposed by which climate change might affect infectious diseases. These processes range from the clear and quantifiable to the imprecise and hypothetical. They may affect pathogens directly or indirectly, the hosts, the vectors (if there is an intermediate host), epidemiological dynamics, or the natural environment. A framework for how climate change can affect the transmission of pathogens between hosts is shown in Fig. 6.1. Only some of the processes can be expected to apply to any single infectious disease.

Effects on Pathogens

Higher temperatures resulting from climate change may increase the rate of development of certain pathogens or parasites that have one or more lifecycle stages outside their human or animal host. This may shorten generation times and,

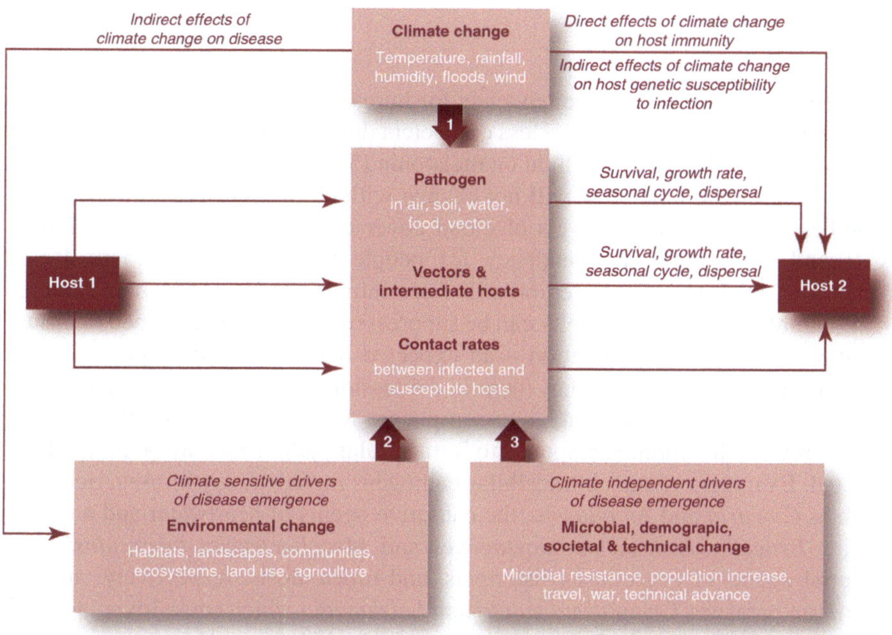

Fig. 6.1 A schematic framework of the effects of climate change on the transmission of diseases of humans and animals. Climate change can act directly on pathogens in a range of external substrates, or their vectors and intermediate hosts, thereby affecting the processes of survival, growth, seasonality, and dispersal. It can also directly affect hosts themselves or the contact rates between infected and susceptible individuals. Climate change can have indirect effects on disease transmission via its effects on the natural or anthropogenic environment, and via the genetics of exposed populations. Environmental, demographic, social, and technical change will also happen independently of climate change and have as great, or much greater, influence on disease transmission than climate change itself. The significance of climate change as a driver of disease will depend on the scale of *arrow 1*, and on the relative scales of *arrows 2* and *3*

possibly, increase the total number of generations per year, leading to higher pathogen population sizes [48]. Conversely, some pathogens are sensitive to high temperature and their survival may decrease with climate warming.

Phenological evidence indicates that spring is arriving earlier in temperate regions [50]. Lengthening of the warm season may increase or decrease the number of cycles of infection possible within 1 year for warm or cold-associated diseases respectively. Arthropod vectors tend to require warm weather so the infection season of arthropod-borne diseases may extend. Some pathogens and many vectors experience significant mortality during cold winter conditions; warmer winters may increase the likelihood of successful overwintering [41, 48].

Pathogens that are sensitive to moist or dry conditions may be affected by changes to precipitation, soil moisture, and the frequency of floods. Changes to winds could affect the spread of certain pathogens and vectors.

Effects on Hosts

A proposed explanation for the tendency for human influenza to occur in winter is that the human immune system is less competent during that time; attributable to the effects of reduced exposure to light on melatonin [51] or vitamin D production [52]. The seasonal light/dark cycle will not change with climate change, but one might hypothesize that changing levels of cloud cover could affect exposure in future. A second explanation, the tendency for people to congregate indoors during wintertime, leads to a more credible role for a future influence of climate change.

Mammalian cellular immunity can be suppressed following heightened exposure to ultraviolet B (UV-B) radiation – an expected outcome of stratospheric ozone depletion [53, 54]. In particular, there is depression of the number of T helper 1 lymphocytes, cells which stimulate macrophages to attack pathogen-infected cells and, therefore, the immune response to intracellular pathogens may be particularly affected. Examples of such intracellular pathogens include many viruses, rickettsia (such as *Cowdria* and *Anaplasma*, the causative agents of heartwater and anaplasmosis), *Brucella, Listeria monocytogenes* and *Mycobacterium tuberculosis*, the bacterial agents of brucellosis, listeriosis, and tuberculosis, respectively, and the protozoan parasites *Toxoplasma gondii* and *Leishmania* which cause toxoplasmosis and visceral leishmanosis (kala-azar), respectively, in humans [55].

A third host-related effect worthy of consideration is genetic resistance to disease. Some human populations and many animal species have evolved a level of genetic resistance to some of the diseases to which they are commonly exposed. Malaria presents a classic example for humans, with a degree of resistance to infection in African populations obtained from heterozygosity for the sickle-cell genetic trait. Considering animals, wild mammals in Africa may be infected with trypanosomes, but rarely show signs of disease; local Zebu cattle breeds, which have been in the continent for millennia, show some degree of trypanotolerance (resistance to disease caused by trypanosome infection); by contrast, recently introduced European cattle breeds are highly susceptible to trypanosomosis. In stark contrast, African mammals proved highly susceptible to rinderpest which swept through the continent in the late nineteenth century, and which they had not previously encountered. It seems unlikely that climate change will directly affect genetic or immunologic resistance to disease in humans or animals. However, significant shifts in disease distributions driven by climate change pose a greater threat than simply the exposure of new populations. Naïve populations may, in some cases, be particularly susceptible to the new diseases facing them.

Certain diseases show a phenomenon called *endemic stability*. This occurs when the severity of disease is less in younger than older individuals, when the infection is common or endemic and when there is life-long immunity after infection. Under these conditions most infected individuals are young, and experience relatively mild disease. Counter-intuitively, as endemically stable infections become rarer, a higher proportion of cases are in older individuals (it takes longer, on average, to acquire infection) and the number of cases of severe disease rises. Certain tick-borne diseases of livestock in Africa, such as anaplasmosis, babesiosis, and cowdriosis, show a degree of endemic stability [56], and it has been proposed to occur for some human diseases like malaria [57]. If climate change drives such diseases to new areas, nonimmune individuals of all ages in these regions will be newly exposed, and outbreaks of severe disease could follow.

Effects on Vectors

Much has already been written about the effects of climate change on invertebrate disease vectors. Indeed, this issue, especially the effects on mosquito vectors, has dominated the debate so far. It is interesting to bear in mind, however, that mosquitoes are less significant as vectors of animal disease than they are of human disease (Table 6.1). Mosquitoes primarily, and secondarily lice, fleas, and ticks, transmit between them a significant proportion of important human infections. By contrast, biting midges, brachyceran flies (e.g., tabanids, muscids, myiasis flies, hippoboscids), ticks, and mosquitoes (and, in Africa, tsetse) all dominate as vectors of livestock disease. Therefore, a balanced debate on the effects of climate change on human and animal disease must consider a broad range of vectors.

There are several processes by which climate change might affect disease vectors. First, temperature and moisture frequently impose limits on their distribution. Often, low temperatures are limiting because of high winter mortality, or high temperatures because they involve excessive moisture loss. Therefore, cooler regions which were previously too cold for certain vectors may begin to allow them to flourish with climate change. Warmer regions could become even warmer and yet remain permissive for vectors if there is also increased precipitation or humidity. Conversely, these regions may become less conducive to vectors if moisture levels remain unchanged or decrease, with concomitant increase in moisture stress.

For any specific vector, however, the true outcome of climate change will be significantly more complex than that outlined above. Even with a decrease in future moisture levels, some vectors, such as certain species of mosquito, could become more abundant, at least in the vicinity of people and livestock, if the response to warming is more water storage and, thereby, the creation of new breeding sites. One of the most important vectors of the emerging Chikungunya virus (and to a lesser extent dengue virus) is the Asian tiger mosquito (*Aedes albopictus*) which is a container breeder and therefore thrives where humans store water. Equally, some vectors may be relatively insensitive to direct effects of climate change, such as muscids which breed in organic matter or debris, and myiasis flies which breed in hosts' skin.

Changes to temperature and moisture will also lead to increases or decreases in the abundance of many disease vectors. This may also result from a change in the frequency of extreme weather events such as ENSO. Outbreaks of several biting midge and mosquito-borne diseases, for example, have been linked to the occurrence of ENSO [18, 22, 59–62] and mediated, at least in part, by increase in the vector population size in response to heavy rainfall, or rainfall succeeding drought, that ENSO sometimes brings [18, 22]. Greater intra- or interannual variation in rainfall, linked or unlinked to ENSO, may lead to an increase in the frequency or scale of outbreaks of such diseases.

Table 6.1 The major diseases transmitted by arthropod vectors to humans and livestock (Adapted from [58])

Vector	Diseases of humans	Diseases of livestock
Phthiraptera (Lice)	Epidemic typhus Trench fever Louse-borne relapsing fever	
Reduvidae (Assassin bugs)	Chagas' disease	
Siphonaptera (Fleas)	Plague Murine typhus Q fever Tularaemia	Myxomatosis
Psychodidae (Sand flies)	Leishmanosis Sand fly fever	Canine leishmanosis Vesicular stomatitis
Culicidae (Mosquitoes)	Malaria Dengue Yellow fever West Nile Filiariasis Encephalitides (WEE, EEE, VEE, Japanese encephalitis, Saint Louis encephalitis) Rift Valley fever	West Nile fever Encephalitides Rift Valley fever Equine infectious anemia
Simulidae (Black flies)	Onchocercosis	Leucocytozoon (birds) Vesicular stomatitis
Ceratopogonidae (Biting midges)		Bluetongue African horse sickness Akabane Bovine ephemeral fever
Glossinidae (Tsetse flies)	Trypanosomosis	Trypanosomosis
Tabanidae (Horse flies)	Loiasis	Surra Equine infectious anemia *Trypanosoma vivax*
Muscidae (Muscid flies)	Shigella *E. coli*	Mastitis Summer mastitis Pink-eye (IBK)
Muscoidae, Oestroidae (Myiasis-causing flies)	Bot flies	Screwworm Blow fly strike Fleece rot
Hippoboscoidae (Louse flies, keds)		Numerous protozoa
Acari (Mites)	Chiggers Scrub typhus (tsutsugamushi) Scabies	Mange Scab Scrapie?
Ixodidae (Hard ticks) Argasidae (Soft ticks)	Human babesiosis Tick-borne encephalitis Tick fevers Ehrlichiosis Q fever Lyme disease	Babesiosis East coast fever (Theileriosis) Louping ill African Swine Fever Ehrlichiosis Q fever

(continued)

Table 6.1 (continued)

Vector	Diseases of humans	Diseases of livestock
	Tick-borne relapsing fever Tularaemia	Heartwater Anaplasmosis Borreliosis Tularaemia

The ability of some insect vectors to become or remain infected with pathogens (their vector competence) varies with temperature [63, 64]. In addition to this effect on vector competence, an increase in temperature may alter the balance between the lifespan of an infected vector, its frequency of feeding, and the time necessary for the maturation of the pathogen within it. This balance is critical, as a key component of the risk of transmission of a vector-borne disease is the number of blood meals taken by a vector between the time it becomes infectious and its death [65]. Accordingly, rising ambient temperature can increase the risk of pathogen transmission by shortening the time until infectiousness in the vector and increasing its feeding frequency at a faster rate than it shortens the vector's lifespan, such that the number of feeds taken by an infectious vector increases. However, at even higher temperatures this can reverse [66] such that the number of infectious feeds then decreases relative to that possible at lower temperatures. This point is extremely important, as it means that the risk of transmission of vector-borne pathogens does not uniformly increase with rising temperature, but that it can become too hot and transmission rates decrease. This effect will be most important for short-lived vectors such as biting midges and mosquitoes [30].

Lastly, there may be important effects of climate change on vector dispersal, particularly if there is a change in wind patterns. Wind movements have been associated with the spread of epidemics of many *Culicoides*- and mosquito-borne diseases [67–72].

Effects on Epidemiological Dynamics

Climate change may alter transmission rates between hosts by affecting the survival of the pathogen or the intermediate vector, but also by other, indirect, forces that may be hard to predict with accuracy. Climate change may influence human demography, housing, or movement or be one of the forces that leads to changes in future patterns of international trade, local animal transportation, and farm size. All of these can alter the chances of an infected human or animal coming into contact with a susceptible one. For example, a series of droughts in East Africa between 1993 and 1997 resulted in pastoral communities moving their cattle to graze in areas normally reserved for wildlife. This resulted in cattle infected with a mild lineage of rinderpest transmitting disease both to other cattle and to susceptible wildlife, causing severe disease, for example, in buffalo, lesser kudu, and impala, and devastating certain populations [73].

Indirect Effects

No disease or vector distribution can be fully understood in terms of climate only. The supply of suitable hosts, the effects of co-infection or immunological cross-protection, the presence of other insects competing for the same food sources or breeding sites as vectors [74], and parasites and predators of vectors themselves, could have important effects [48]. Climate change may affect the abundance or distribution of hosts or the competitors/predators/parasites of vectors and influence patterns of disease in ways that cannot be predicted from the direct effects of climate change alone.

Equally, it has been argued that climate change-related disturbances of ecological relationships, driven perhaps by agricultural changes, deforestation, the construction of dams, and losses of biodiversity, could give rise to new mixtures of different species, thereby exposing hosts to novel pathogens and vectors and causing the emergence of new diseases [40]. A possible "example in progress" is the reemergence in the UK of bovine tuberculosis, for which the badger (*Meles meles*) is believed to be a carrier of the causative agent, *Mycobacterium bovis*. Farm landscape, such as the density of linear features like hedgerows, is a risk factor for the disease, affecting the rate of contact between cattle and badger [75]. Climate change will be a force for modifying future landscapes and habitats, with indirect and largely unpredictable effects on diseases.

Other Drivers of Disease

The future disease burden of humans and animals will depend not only on climate change and its direct and indirect effects on disease, but also on how other drivers of disease change over time. Even for diseases with established climate links, it may be the case that in most instances these other drivers will prove to be more important than climate. A survey of 335 events of human disease emergence between 1940 and 2004 classified the underlying causes into 12 categories [2]. One of these, "climate and weather," was only listed as the cause of ten emergence events. Six of these were non-cholera *Vibrio* bacteria which cause poisoning via shellfish or exposure to contaminated seawater; the remaining four were a fungal infection and three mosquito-borne viruses. The other 11 categories included, however, "land use changes" and "agricultural industry changes," with 36 and 31 disease emergence events, respectively, and both may be affected by climate change. The causes of the remaining 77% of disease emergence events – "antimicrobial agent use," "international travel and commerce," "human demography and behavior," "human susceptibility to infection," "medical industry change," "war and famine," "food industry changes," "breakdown of public health," and "bushmeat" – would be expected to be either less or not influenced by climate change. Hence, climate change's indirect effects on human or animal disease may exceed its direct effects, while drivers unsusceptible to climate change may be more important still at determining our disease future.

Climate Change and Disease in Wildlife

Wildlife disease is important for different reasons to those which make disease in humans and domestic animals important. It has the potential for endangerment of wildlife and can be a source of zoonoses and livestock disease.

Wildlife Disease as a Cause of Endangerment

Disease in wild populations may limit or cause extreme fluctuations in population size [76] and reduce the chances of survival of endangered or threatened species [77]. Indeed, disease can be the primary cause of extinction in animals or be a significant contributory factor toward it. For example:

- The Christmas Island rat, *Rattus macleari*, is believed to have been extinct by 1904. There is molecular evidence that this was caused by introduction of murine trypanosomes apparently brought to the island by black rats introduced shortly before 1904 [78].
- Similarly, the last known Po'o-uli bird (*Melamprosops phaeosoma*) in Hawaii died from avian malaria brought by introduced mosquitoes [79].
- Canine Distemper in the Ethiopian Wolf (*Canis simensis*) has brought about its decline [80].
- Devil facial tumor disease, an aggressive nonviral transmissible parasitic cancer, continues to threaten Tasmanian Devil (*Sarcophilus harrisii*) populations [81].
- The white nose fungus G*eomyces destructans* is decimating bat populations in Northeastern US states and is currently spreading in Europe [82]. This is perhaps the most recent emergence of a disease of concern to wildlife endangerment.

Although disease can cause endangerment and extinction, its importance relative to other causes is uncertain. A review of the causes of endangerment and extinction in the International Union for Conservation of Nature (IUCN) red list of plant and animal species found that disease was implicated in a total of 254 cases, some 7% of the total examined [83]. Although the other factors may be more important overall, disease remains an important cause of endangerment and extinction for certain animal groups. A contender for the single issue of greatest current conservation concern is the epidemic of the chytrid fungus *Batrachochytrium dendrobatidis* in amphibians. With a broad host range and high mortality, this pathogen is likely to be wholly or partly responsible for all recent amphibian extinction events, which, remarkably, comprise 29% of all extinctions attributable to disease since the year 1500 [77].

Wildlife Disease as a Source of Infections for Humans and Livestock

Many diseases of wildlife frequently cross the species barrier to infect humans or domestic animals [84, 85]. Closely related organisms often share diseases. A particular risk to humans is presented by diseases of primates: The human immunodeficiency viruses (HIV) that cause AIDS originated as simian immunodeficiency viruses in African monkeys and apes. Humans acquire or have acquired many other pathogens from mammals other than primates, especially those that humans choose to live close to (livestock, dogs, cats), or that choose to live close to us (rodents, bats) [86]. In addition, there are examples of human infections shared with birds (e.g., avian influenza), and reptiles and fishes (e.g., *Salmonella* spp.). Insects are frequent carriers of pathogens between vertebrate hosts, and there may even be a pathogen transmissible from plants to humans [87].

Wildlife populations are the primary source of emerging infectious diseases in humans. A search of the scientific literature published between 1940 and 2004 attempted to quantify the causes of disease introductions into human populations and found that about 72% were introduced from wildlife [2].

How Climate Change Can Influence Wildlife Disease

The effects of climate change on wildlife disease are important when the changes produced lead to increased risk of endangerment or extinction of the wildlife, or increased transmission risk to humans or domestic animals.

Climate change can increase the threat of endangerment or extinction, via reduction in population size of the wildlife host (by altering habitats, for example), or increase in pathogen range or virulence, such that the persistence of a host population is at risk, and climate change can increase the risk of disease emergence and spread to humans or livestock via change to the distribution of wildlife hosts, such that encroachment on human or livestock populations is favored.

Changes in species' distribution may arise directly under climate change as a result of an organism's requirement for particular climatic conditions or indirectly via ecological interactions with other species which are themselves affected. Climate change can cause the appearance of new assemblages of species and the disappearance of old communities [88], which can create new disease transmission opportunities or end existing ones.

Climatic factors potentially affecting wildlife disease transmission more directly include the growth rate of the pathogen in the environment or in a cold-blooded (ectothermic) wildlife host (e.g., fish, amphibian, reptile). Therefore, effects which are more marked in wildlife disease in comparison to human or livestock disease include the occurrence of ectothermic hosts, and also the vast range of potential vectors that may transmit disease. It is therefore more difficult to generalize about the effects of climate change on wildlife.

In colder climates, the parasite that causes the most severe form of human malaria, *Plasmodium falciparum* does not develop rapidly enough in its mosquito vectors for there to be sufficient transmission to maintain the parasite. Avian malaria (*Plasmodium relictum*) exhibits an elevational gradient due mainly to temperature and is subject to similar constraints [89].

An example of a wildlife pathogen constrained due to its dependence on environmental transmission is anthrax. Infective spores of anthrax bacilli can lie dormant in soils for decades, becoming dangerous when climatic conditions, particularly precipitation, favor it. It is well established that when a host population size is reduced, the pool of susceptible individuals may be too small for pathogen survival. This effect is particularly acute for host-specific diseases, such as the transmissible cancer of Tasmanian devils, *Sarcophilus harrisii*, DFTD or Devil facial tumor disease (discussed in [81]); at low host population sizes, DFTD may become extinct. By contrast, diseases with a broad host range may threaten individual species down to the last individual. As anthrax has both a broad host range and can lie dormant in the environment, it is a particular threat for species with very low numbers, and is currently a conservation consideration for many species. For example the Javan rhinoceros population is down to fewer than 60 individuals and identified as a priority for conservation (an "EDGE" species) because of its uniqueness and scarcity [90].

Climate change may have particular impact on marine animals, because of the preponderance of ectothermic animals in the sea, the multiple ways in which climate change is predicted to affect the marine environment, and the multiple stresses that marine organisms and ecosystems are already experiencing due to anthropogenic influence. Disease is an important part of this impact. For example, warming of the Pacific in the range of the oyster *Crassostrea virginica* caused range expansion of the protozoan parasite *Perkinsus marinus* probably due to a combination of increased overwinter survival, greater summer proliferation, and increased host density [91]. Coral reefs are also sensitive to at least 12 potential factors associated with climate change: [92] CO_2 concentration, sea surface temperature, sea level, storm intensity, storm frequency, storm track, wave climate, run off, seasonality, overland rainfall, and ocean and air circulation [92]. Although these factors might not all increase levels of disease, the synergism between disease, climate, and other stressors might lead to accelerating degradation of the coral reef habitat.

From a geographic perspective, there is evidence that the greatest change in ecosystems attributable to climate change is likely to be in the tropics; the second being the arctic [88]. The impacts of this change on wildlife disease and its consequences may be particularly great in these two regions, and there is evidence that it is already occurring. The tropics have the most species in imminent danger of extinction [93] while tropical coral reefs comprise much of the biodiversity of the oceans. In addition to extinction risks, tropical forests may also pose a zoonosis risk. An increase in animal-human interaction is likely in tropical forests, which have a diverse fauna subject to increasing human encroachment.

With regard to the Arctic, a model of the effect of global warming on a protostrongylid lung-dwelling nematode *Umingmakstrongylus pallikuukensis*, in Canadian Arctic muskoxen *Ovibos moschatus*, found that warming was likely to significantly influence infection, making the muskoxen more susceptible to predation [94]. Muskoxen were also subject to climate-influenced outbreaks of fatal pneumonia [95]. Indeed, the combination of climate change's effects on pathogen survival and transmission, and the stress to host species from changing environmental conditions, may have serious impact on arctic populations of fish, muskoxen, sheep, and others [96].

Dependency of Disease on Climate: The Example of Chytridiomycosis in Amphibians

In wildlife epidemiology, the host may be of equal importance to the pathogen and vector when considering the impact of climate, as wildlife may be impacted by climate in more diverse ways than humans or domestic animals, and are subject to much reduced human mitigation of those impacts. The importance of climate in *Batrachochytrium dendrobatidis* epidemiology, the cause of chytridiomycosis, and numerous amphibian extinctions is fiercely debated. Although the pathogen *B. dendrobatidis* is neither vector-borne nor has a prolonged environmental phase, it is affected by temperature because the environment of its ectothermic hosts is not kept constant when external temperature varies. It belongs to a basal group within the fungi and has a brief motile zoospore stage for dispersal, followed by the penetration of the outer layers of amphibian skin and asexual intracellular multiplication [97]. Its growth is limited by warmer temperatures, perhaps because amphibians shed their outer layers of skin more frequently in warmer temperatures [97]. Pounds et al. [98] were the first to make a connection between climate and *B. dendrobatidis*-mediated amphibian extinctions, reporting spatiotemporal associations between warming and last year of detection of frog species. The development rate of the *B. dendrobatidis* fungus depends on summer temperature [99], and its survival is dependent on winter freezing [100]. The fungus appears to cause more mortality in mountainous regions [101], yet may be limited at the upper extremes of altitude. Climate may also affect the impact of the disease due to host factors. The habitat of the golden toad *Bufo periglenes* in 1987, the last year of its existence, was much reduced due to an especially dry summer. This may have caused crowded conditions in the remaining ponds, facilitating the spread of chytridiomycosis [102].

In addition, climate may affect mortality associated with the disease. The mortality of frogs exposed to *B. dendrobatidis* spores as adults has been shown to be dependent on the condition of the frogs [103]. In a changing climate where amphibians are shifting their ranges into suboptimal areas, hosts are likely to be more susceptible to the damaging effects of *B. dendrobatidis* infection.

On the other hand, it has been argued that climate is not important in *B. dendrobatidis* outbreaks [104]. The authors contrast the climate-linked epidemic hypothesis with the hypothesis that disease outbreaks occur largely due to introduction into unexposed, naive populations, and describe spatiotemporal "waves" of declines across Central America as evidence that it is disease introduction (and not climatic variables) causing declines.

Evidence of Climate Change's Impact on Disease

Climate warming has already occurred in recent decades. If diseases are sensitive to such warming, then one might expect a number of diseases to have responded by changing their distribution, frequency, or intensity. A major difficulty, however, is the attribution of any observed changes in disease occurrence to climate change because, as shown above, other disease drivers also change over time. It has been argued that the minimum standard for attribution to climate change is that there must be known biological sensitivity of a disease or vector to climate, and that the change in disease or vector (change in seasonal cycle, latitudinal or altitudinal shifts) should be statistically associated with observed change in climate [28]. This has been rephrased as the need for there to be change in both disease/vector and climate at the same time, in the same place, and in the "right" direction [105]. Given these criteria, few diseases make the standard: Indeed, only a decade ago one group concluded that the literature lacks strong evidence for an impact of climate change on even a single vector-borne disease, let alone other diseases.

This situation has changed. One disease in particular, bluetongue, has emerged dramatically in Europe over the last decade and this emergence can be attributed to recent climate change in the region. It satisfies the right time, right place, right direction criterion [64], but in fact reaches a far higher standard: A model for the disease, with variability in time driven only by variation in climate, produces quantitative estimates of risk which fit closely with the disease's recent emergence in both space and time.

Bluetongue

Bluetongue is a viral disease of sheep and cattle. It originated in Africa, where wild ruminants act as natural hosts for the virus, and where a species of biting midge, *Culicoides imicola*, is the major vector [106]. During the twentieth century bluetongue spread out of Africa into other, warm parts of the world, becoming endemic in the Americas, southern Asia, and Australasia; in most of these places, indigenous *Culicoides* became the vectors. Bluetongue also occurred very infrequently in the far extremes of southern Europe: once in the southwest (southern Spain and Portugal, 1955–1960), and every 20–30 years in the southeast (Cyprus, 1924,

1943–46); Cyprus and Greek islands close to Turkey (1977–1978); the presence of *C. imicola* was confirmed in both areas and this species was believed to be the responsible vector. Twenty years after this last 1977–1978 outbreak, in 1998 bluetongue once again reappeared in southeastern Europe [107]. Subsequent events, however, are unprecedented.

Between 1998 and 2008 bluetongue accounted for the deaths of more than one million sheep in Europe – by far the longest and largest outbreak on record. Bluetongue has occurred in many countries or regions that have never previously reported this disease or its close relatives. There have been at least two key developments. First, *C. imicola* has spread dramatically, now occurring over much of southern Europe and even mainland France. Second, indigenous European *Culicoides* species have transmitted the virus. This was first detected in the Balkans where bluetongue occurred but no *C. imicola* could be found [108]. In 2006, however, bluetongue was detected in northern Europe (The Netherlands) from where it spread to neighboring countries, the UK and even Scandinavia. The scale of this outbreak has been huge, yet the affected countries are far to the north of any known *C. imicola* territory [109].

Recently, the outputs of new, observation based, high spatial resolution (25 km) European climate data, from 1960 to 2006 have been integrated within a model for the risk of bluetongue transmission, defined by the basic reproduction ratio R_0 [110]. In this model, temporal variation in transmission risk is derived from the influence of climate (mainly temperature and rainfall) on the abundance of the vector species, and from the influence of temperature alone on the ability of the vectors to transmit the causative virus. As described earlier, this arises from the balance between vector longevity, vector feeding frequency, and the time required for the vector to become infectious. Spatial variation in transmission risk is derived from these same climate-driven influences and, additionally, differing densities of sheep and cattle. This integrated model successfully reproduces many aspects of bluetongue's distribution and occurrence, both past and present, in Western Europe, including its emergence in northwest Europe in 2006. The model gives this specific year the highest positive anomaly (relative to the long-term average) for the risk of bluetongue transmission since at least 1960, but suggests that other years were also at much higher-than-average risk. The model suggests that the risk of bluetongue transmission increased rapidly in southern Europe in the 1980s and in northern Europe in the 1990s and 2000s.

These results indicate that climate variability in space and time are sufficient to explain many aspects of bluetongue's recent past in Europe and provide the strongest evidence to date that this disease's emergence is, indeed, attributable to changes in climate. What then of the future? The same model was driven forward to 2050 using simulated climate data from regional climate models. The risk of bluetongue transmission in northwestern Europe is projected to continue increasing up to at least 2050 (Fig. 6.2). Given the continued presence of susceptible ruminant host populations, the models suggest that by 2050, transmission risk will have increased by 30% in northwest Europe relative to the 1961–1999 mean. The risk is also projected to increase in southwest Europe, but in this case only by 10% relative to the 1961–1999 mean.

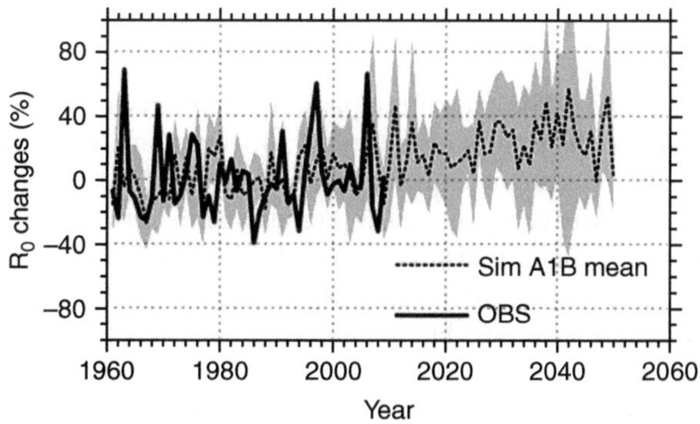

Fig. 6.2 Projections of the effect of climate change on the future risk of transmission of bluetongue in northern Europe. The *y-axis* shows relative anomalies (%) with respect to the 1961–1999 time period for the risk of bluetongue transmission, during August-October in north-west Europe, as defined by the basic reproductive ratio, R_0. R_0 was estimated from climate observations (*OBS – thick black line*), and an ensemble of 11 future climate projections (*SimA1B*), for which the *dashed line* presents the mean and the *grey* envelope the spread (Adapted from [110])

The matching of observed change in bluetongue with quantitative predictions of a climate-driven disease model provides evidence for the influence of climate change far stronger than the "same place, same time, right direction" criterion described earlier. Indeed, it probably makes bluetongue the most convincing example of a disease that is responding to climate change. In this respect, bluetongue differs remarkably from another vector-borne disease, malaria.

Malaria

Some 3.2 billion people live with the risk of malaria transmission, between 350 and 500 million clinical episodes of malaria occur each year and the disease kills at least one million people annually [111]. Of these, each year about 12 million cases and 155,000–310,000 deaths are in epidemic areas [112]. Interannual climate variability primarily drives the timing of these epidemics.

Malaria is caused by *Plasmodium* spp. parasites. Part of the parasite's life cycle takes place within anopheline mosquitoes while the remainder of the life cycle occurs within the human host. The parasite and mosquito life cycles are affected by weather and climate (mainly rain, temperature, and humidity), allowing models of the risk of malaria transmission to be driven by seasonal forecasts from ensemble prediction systems [113], thereby permitting forecasts of potential malaria outbreaks with lead times of up to 4–6 months [114, 115].

Among scientists there are contrasting views about the overall importance of climate on the transmission of malaria, and therefore on the importance of future climate change. Some argue that climate variability or change is the primary actor in any changing transmission pattern of malaria, while others suggest that any changing patterns today or in the foreseeable future are due to non-climate factors [35, 116, 117].

A key insight is that while global temperatures have risen, there has been a net reduction in malaria in the tropics over the last 100 years and temperature or rainfall change observed so far cannot explain this reduction [118]. Malaria has moved from being climate sensitive (an increasing relationship between ambient temperature and the extent of malaria transmission) in the days before disease interventions were widely available to a situation today where regions with malaria transmission are warmer than those without, but within the malaria-affected region, warmer temperatures no longer mean more disease transmission. Instead, other variables affecting malaria, such as good housing, the running of malaria control schemes, or ready access to affordable prophylaxis, now play a greater role than temperature in determining whether there are higher or lower amounts of transmission. This would suggest that the importance of climate change in discussions of future patterns of malaria transmission is likely to have been significantly overplayed.

What is clearly recognized, by all sides in the malaria and climate debate, is that mosquitoes need water to lay their eggs in, and for larval development, and that adult mosquitoes need to live long enough in an environment with high humidity and with sufficiently high temperature for transmission to be possible to the human host. Hence, while the spatial distribution of higher versus lower degrees of malaria transmission appears to have become, in a sense, divorced from ambient temperature, it seems likely that the weather plays as important a role as ever in determining when seasonal transmission will start and end. Climate change may therefore still have a role to play in malaria: not so much affecting where it occurs but, via changing rainfall patterns and mosquito numbers, when or for how long people are most at risk.

Malaria has only recently become confined to the developing world and tropics. It is less than 40 years since malaria was eradicated in Europe and the United States; and the 15°C July isotherm was the northern limit until the mid nineteenth century [119]. Changes in land use and increased living standards, in particular, acted to reduce exposure to the mosquito vector in these temperate zones, leading ultimately to the final removal of the disease. In the UK, a proportion of the reduction has been attributed to increasing cattle numbers and the removal of marshland [120]. In Finland, changes in family size, improvements in housing, changes in farming practices, and the movement of farmsteads out of villages lead to the disappearance of malaria [121], where it had formerly been transmitted indoors in winter. While future increases in temperature may, theoretically, lead to an increased risk of malaria transmission in colder climes than at present [120, 122], the much-altered physical and natural environment may preclude this risk increasing to a level that merits concern. Once again, a more important future driver of malaria risk, in the UK at least, may be the pressure to return some of our landscape to its former state, such as the reflooding of previously drained marshland.

Future Directions

Climate change is widely considered to be a major threat to human and animal health, and the viability of certain endangered species, via its effects on infectious diseases. How realistic is this threat? Will most diseases respond to climate change, or just a few? Will there be a net increase in disease burden or might as many diseases decline in impact as increase?

The answers to these questions are important, as they could provide opportunities to mitigate against new disease threats, or may provide the knowledge-base required for policy makers to take necessary action to combat climate change itself. However, both the methodology to accurately predict climate change's effects on diseases and, in most cases, the data to apply the methodology to a sufficiently wide-range of pathogens is currently lacking.

The majority of pathogens, particularly those not reliant on intermediate hosts or arthropod vectors for transmission, either do occur, or have the potential to occur, in most parts of the world already. Climate change has the capacity to affect the frequency or scale of outbreaks of these diseases: Good examples would be the frequency of food poisoning events from the consumption of meat (such as salmonellosis) or shellfish (caused by *Vibrio* bacteria).

Vector-borne diseases are usually constrained in space by the climatic needs of their vectors, and such diseases are therefore the prime examples of where climate change might be expected to cause distributional shifts. Warmer temperatures usually favor the spread of vectors to previously colder environments, thereby exposing possibly naïve populations to new diseases.

However, altered rainfall distributions have an important role to play. Many pathogens or parasites, such as those of anthrax, haemonchosis, and numerous vector-borne diseases, may in some regions be subject to opposing forces of higher temperatures promoting pathogen or vector development, and increased summer dryness leading to more pathogen or vector mortality. Theoretically, increased dryness could lead to a declining risk of certain diseases. A good example is fasciolosis, where the lymnaeid snail hosts of the *Fasciola* trematode are particularly dependent on moisture. Less summer rainfall and reduced soil moisture may reduce the permissiveness of some parts of the UK for this disease. The snail and the free-living fluke stages are, nevertheless, also favored by warmer temperatures and, in practice, current evidence is that fasciolosis is spreading in the UK [123].

One way to predict the future for disease in a specific country is to learn from countries that, today, are projected to have that country's future climate [37, 39]. At least some of thc complexity behind the multivariate nature of disease distributions should have precipitated out into the panel of diseases that these countries currently face.

For example, in broad terms, the UK's climate is predicted to get warmer, with drier summers and wetter winters, becoming therefore increasingly "mediterranean." It would seem reasonable, therefore, to predict that the UK of the future might experience diseases currently present in, or that occur periodically in, southern Europe. For humans, the best example would be leishmanosis (cutaneous and visceral) [124], while for animals, examples include West Nile fever [125], *Culicoides imicola*–transmitted bluetongue and African horse sickness [41], and canine leishmanosis [126]. The phlebotomid sandfly vectors of the latter do not currently occur in the UK, but they are found widely in southern continental Europe, including France, with recent reports of their detection in Belgium [127]. The spread of the Asian tiger mosquito into Europe and the recent transmission in Europe of both dengue fever [128] and Chikungunya [129] by this vector are further cause for alarm.

However, the contrasting examples of bluetongue and malaria – one spreading because of climate change and one retreating despite it – show that considerations which focus entirely on climate may well turn out to be false. Why are these two diseases, both vector-borne and subject to the similar epidemiological processes and temperature dependencies, so different with respect to climate change? The answer lies in the relative importance of other disease drivers. For bluetongue, it is difficult to envisage epidemiologically relevant drivers of disease transmission, other than climate, that have changed significantly over the time period of the disease's emergence [64]. Life on the farm for the midges that spread bluetongue is probably not dramatically different today from the life they enjoyed 30 years ago. Admittedly, changes in the trade of animals or other goods may have been important drivers of the increased risk of introduction of the causative viruses into Europe, but after introduction, climate change may be the most important driver of increased risk of spread.

For malaria, change in drivers other than climate, such as land use and housing, the availability of prophylaxis, insecticides and, nowadays, insecticide-treated bed nets, have played far more dominant roles in reducing malaria occurrence than climate change may have played in increasing it. Two key reasons, then, for the difference between the two diseases are, first, that life for the human hosts of malaria has changed more rapidly than that of the ruminant hosts of bluetongue, and second, the human cost of malaria was so great that interventions were developed; while the (previously small) economic burden of bluetongue did not warrant such effort and our ability to combat the disease 5 years ago was not very different from that of 50 years before. The very recent advent of novel inactivated vaccines for bluetongue may now be changing this situation.

This entry began by asking whether climate change will affect most diseases or just a few. The examples of malaria and bluetongue demonstrate that a better question may be as follows: Of those diseases that are sensitive to climate change, how many are relatively free from the effects of other disease drivers such that the pressures brought by a changing climate can be turned into outcomes?

Bibliography

Primary Literature

1. Morens DM, Folkers GK, Fauci AS (2004) The challenge of emerging and re-emerging infectious diseases. Nature 430:242–249
2. Jones KE, Patel NG, Levy MA, Storeygard A, Balk D, Gittleman JL, Daszak P (2008) Global trends in emerging infectious diseases. Nature 451:990–993

3. IPCC (2001) Climate change 2001: the scientific basis. Intergovernmental Panel on Climate Change, Cambridge
4. Zhou XN, Yang GJ, Yang K, Wang XH, Hong QB, Sun LP, Malone JB, Kristensen TK, Bergquist NR, Utzinger J (2008) Potential impact of climate change on schistosomiasis transmission in China. Am J Trop Med 78:188–194
5. Rogers DJ, Packer MJ (1993) Vector-borne diseases, models, and global change. Lancet 342:1282–1284
6. Weaver SC, Barrett ADT (2004) Transmission cycles, host range, evolution and emergence of arboviral disease. Nat Rev Microbiol 2:789–801
7. Kovats RS, Edwards SJ, Hajat S, Armstrong BG, Ebi KL, Menne B (2004) The effect of temperature on food poisoning: a time-series analysis of salmonellosis in ten European countries. Epidemiol Infect 132:443–453
8. Donaldson AI (1972) The influence of relative humidity on the aerosol stability of different strains of foot-and-mouth disease virus suspended in saliva. J Gen Virol 15:25–33
9. Sutmoller P, Barteling SS, Olascoaga RC, Sumption KJ (2003) Control and eradication of foot-and-mouth disease. Virus Res 91:101–144
10. Wosu LO, Okiri JE, Enwezor PA (1992) Optimal time for vaccination against peste des petits ruminants (PPR) disease in goats in the humid tropical zone in southern Nigeria. In: Rey B, Lebbie SHB, Reynolds L (eds) Small ruminant research and development in Africa: proceedings of the first biennial conference of the African small ruminant research network. International Laboratory for Research in Animal Diseases (ILRAD), Nairobi
11. Anderson J, Barrett T, Scott GR (1996) Manual of the diagnosis of Rinderpest. Food and Agriculture Organization of the United Nations, Rome
12. Soebiyanto RP, Adimi F, Kiang RK (2010) Modeling and predicting seasonal influenza transmission in warm regions using climatological parameters. PLoS ONE 5:e9450
13. Lowen AC, Mubareka S, Steel J, Palese P (2007) Influenza virus transmission is dependent on relative humidity and temperature. PLoS Pathog 3:e151
14. Parker R, Mathis C, Looper M, Sawyer J (2002) Guide B-120: anthrax and livestock. Cooperative Extension Service, College of Agriculture and Home Economics, University of New Mexico, Las Cruces
15. Eiler A, Johansson M, Bertilsson S (2006) Environmental influences on *Vibrio* populations in northern temperate and boreal coastal waters (Baltic and Skagerrak Seas). Appl Environ Microbiol 72:6004–6011
16. Kausrud KL, Viljugrein H, Frigessi A, Begon M, Davis S, Leirs H, Dubyanskiy V, Stenseth NC (2007) Climatically driven synchrony of gerbil populations allows large-scale plague outbreaks. Proc R Soc B Biol Sci 274:1963–1969
17. Davis S, Begon M, De Bruyn L, Ageyev VS, Klassovskiy NL, Pole SB, Viljugrein H, Stenseth NC, Leirs H (2004) Predictive thresholds for plague in Kazakhstan. Science 304: 736–738
18. Baylis M, Mellor PS, Meiswinkel R (1999) Horse sickness and ENSO in South Africa. Nature 397:574
19. Davies F, Linthicum K, James A (1985) Rainfall and epizootic Rift Valley fever. Bull World Health Org 63:941–943
20. Linthicum KJ, Anyamba A, Tucker CJ, Kelley PW, Myers MF, Peters CJ (1999) Climate and satellite indicators to forecast Rift Valley fever epidemics in Kenya. Science 285:397–400
21. Linthicum KJ, Bailey CL, Davies FG, Tucker CJ (1987) Detection of Rift Valley fever viral activity in Kenya by satellite remote sensing imagery. Science 235:1656–1659
22. Anyamba A, Linthicum KJ, Mahoney R, Tucker CJ, Kelley PW (2002) Mapping potential risk of Rift Valley fever outbreaks in African savannas using vegetation index time series data. Photogramm Eng Remote Sens 68:137–145
23. Little PD, Mahmoud H, Coppock DL (2001) When deserts flood: risk management and climatic processes among East African pastoralists. Clim Res 19:149–159

24. Behm CA, Sangster NC (1999) Pathology, pathyophysiology and clinical aspects. In: Dalton JP (ed) Fasciolosis. CAB International, Wallingford, pp 185–224
25. Christie MG (1962) On hatching of *Nematodirus battus*, with some remarks on *N. filicollis*. Parasitology 52:297
26. Githeko AK, Lindsay SW, Confalonieri UE, Patz JA (2000) Climate change and vector-borne diseases: a regional analysis. Bull World Health Org 78:1136–1147
27. Hay SI, Cox J, Rogers DJ, Randolph SE, Stern DI, Shanks GD, Myers MF, Snow RW (2002) Climate change and the resurgence of malaria in the East African highlands. Nature 415:905–909
28. Kovats RS, Campbell-Lendrum DH, McMichael AJ, Woodward A, Cox JS (2001) Early effects of climate change: do they include changes in vector-borne disease? Philos Trans R Soc Lond B 356:1057–1068
29. Kovats RS, Haines A, Stanwell-Smith R, Martens P, Menne B, Bertollini R (1999) Climate change and human health in Europe. Br Med J 318:1682–1685
30. Lines J (1995) The effects of climatic and land-use changes on insect vectors of human disease. In: Harrington R, Stork NE (eds) Insects in a changing environment. Academic Press, London, pp 157–175
31. McMichael AJ, Githeko AK (2001) Human health (Chapter 9). In: McCarthy OFCJJ, Leary NA, Dokken DJ, White KS (eds) Climate change 2001: impacts, adaptation, and vulnerability: contribution of working group II to the third assessment report of the Intergovernmental Panel on Climate Change. Cambridge University Press, Cambridge, pp 453–485
32. Patz JA, Kovats RS (2002) Hotspots in climate change and human health. Br Med J 325:1094–1098
33. Randolph SE (2004) Evidence that climate change has caused 'emergence' of tick- borne diseases in Europe? Int J Med Microbiol 293:5–15
34. Reeves WC, Hardy JL, Reisen WK, Milby MM (1994) Potential effect of global warming on mosquito-borne arboviruses. J Med Entomol 31:323–332
35. Reiter P, Thomas CJ, Atkinson PM, Hay SI, Randolph SE, Rogers DJ, Shanks GD, Snow RW, Spielman A (2004) Global warming and malaria: a call for accuracy. Lancet Infect Dis 4:323–324
36. Rogers DJ, Randolph SE (2000) The global spread of malaria in a future, warmer world. Science 289:1763–1766
37. Rogers DJ, Randolph SE, Lindsay SW, Thomas CJ (2001) Vector-borne diseases and climate change. Department of Health, London
38. Semenza JC, Menne B (2009) Climate change and infectious diseases in Europe. Lancet Infect Dis 9:365–375
39. Sutherst RW (1998) Implications of global change and climate variability for vector-borne diseases: generic approaches to impact assessments. Int J Parasitol 28:935–945
40. WHO (1996) Climate change and human health. World Health Organisation, Geneva
41. Wittmann EJ, Baylis M (2000) Climate change: effects on *Culicoides*-transmitted viruses and implications for the UK. Vet J 160:107–117
42. Zell R (2004) Global climate change and the emergence/re-emergence of infectious diseases. Int J Med Microbiol 293:16–26
43. Baylis M, Githeko AK (2006) State of science review: the effects of climate change on infectious diseases of animals. Office of Science and Innovation, London
44. Cook G (1992) Effect of global warming on the distribution of parasitic and other infectious diseases: a review. J R Soc Med 85:688–691
45. Gale P, Adkin A, Drew T, Wooldridge M (2008) Predicting the impact of climate change on livestock disease in Great Britain. Vet Rec 162:214–215
46. Gale P, Drew T, Phipps LP, David G, Wooldridge M (2009) The effect of climate change on the occurrence and prevalence of livestock diseases in Great Britain: a review. J Appl Microbiol 106:1409–1423
47. Harvell CD, Kim K, Burkholder JM, Colwell RR, Epstein PR, Grimes DJ, Hofmann EE, Lipp EK, Osterhaus A, Overstreet RM, Porter JW, Smith GW, Vasta GR (1999) Review:

marine ecology – emerging marine diseases – climate links and anthropogenic factors. Science 285:1505–1510
48. Harvell CD, Mitchell CE, Ward JR, Altizer S, Dobson AP, Ostfeld RS, Samuel MD (2002) Ecology – climate warming and disease risks for terrestrial and marine biota. Science 296:2158–2162
49. Perry BD, Randolph TF, McDermott JJ, Sones KR, Thornton PK (2002) Investing in animal health research to alleviate poverty. International Livestock Research Institute, Nairobi
50. Walther GR, Post E, Convey P, Menzel A, Parmesan C, Beebee TJC, Fromentin JM, Hoegh-Guldberg O, Bairlein F (2002) Ecological responses to recent climate change. Nature 416:389–395
51. Dowell SF (2001) Seasonal variation in host susceptibility and cycles of certain infectious diseases. Emerg Infect Dis 7:369–374
52. Cannell JJ, Vieth R, Umhau JC, Holick MF, Grant WB, Madronich S, Garland CF, Giovannucci E (2006) Epidemic influenza and vitamin D. Epidemiol Infect 134:1129–1140
53. Aucamp PJ (2003) Eighteen questions and answers about the effects of the depletion of the ozone layer on humans and the environment. Photochem Photobiol Sci 2:9–24
54. de Gruijl FR, Longstreth J, Norval M, Cullen AP, Slaper H, Kripke ML, Takizawa Y, van der Leun JC (2003) Health effects from stratospheric ozone depletion and interactions with climate change. Photochem Photobiol Sci 2:16–28
55. Jankovic D, Liu ZG, Gause WC (2001) Th1- and Th2-cell commitment during infectious disease: asymmetry in divergent pathways. Trends Immunol 22:450–457
56. Eisler MC, Torr SJ, Coleman PG, Machila N, Morton JF (2003) Integrated control of vector-borne diseases of livestock – pyrethroids: panacea or poison? Trends Parasitol 19:341–345
57. Coleman PG, Perry BD, Woolhouse MEJ (2001) Endemic stability – a veterinary idea applied to human public health. Lancet 357:1284–1286
58. Mullen G, Durden L (2002) Medical and veterinary entomology. Academic, Orlando
59. Gagnon AS, Bush ABG, Smoyer-Tomic KE (2001) Dengue epidemics and the El Nino Southern Oscillation. Clim Res 19:35–43
60. Gagnon AS, Smoyer-Tomic KE, Bush ABG (2002) The El Nino Southern Oscillation and malaria epidemics in South America. Int J Biometeorol 46:81–89
61. Hales S, Weinstein P, Souares Y, Woodward A (1999) El Niño and the dynamics of vector borne disease transmission. Environ Health Perspect 107:99–102
62. Kovats RS (2000) El Nino and human health. Bull World Health Org 78:1127–1135
63. Kramer LD, Hardy JL, Presser SB (1983) Effect of temperatures of extrinsic incubation on the vector competence of *Culex tarsalis* for western equine encephalomyelitis virus. Am J Trop Med 32:1130–1139
64. Purse BV, Mellor PS, Rogers DJ, Samuel AR, Mertens PPC, Baylis M (2005) Climate change and the recent emergence of bluetongue in Europe. Nat Rev Microbiol 3:171–181
65. Macdonald G (1955) The measurement of Malaria transmission. Proc R Soc Med Lond 48:295–302
66. De Koeijer AA, Elbers ARW (2007) Modelling of vector-borne disease and transmission of bluetongue virus in North-West Europe. In: Takken W, Knols BGJ (eds) Emerging pests and vector-borne diseases in Europe. Wageningen Academic, Wageningen, pp 99–112
67. Sellers RF (1992) Weather, *Culicoides*, and the distribution and spread of bluetongue and African horse sickness viruses. In: Walton TE, Osburn BI (eds) Bluetongue, African horse sickness and related Orbiviruses. CRC Press, Boca Raton, pp 284–290
68. Sellers RF, Maarouf AR (1991) Possible introduction of epizootic hemorrhagic disease of deer virus (serotype 20) and bluetongue virus (serotype 11) into British Columbia in 1987 and 1988 by infected *Culicoides* carried on the wind. Can J Vet Res 55:367–370
69. Sellers RF, Pedgley DE (1985) Possible windborne spread to Western Turkey of bluetongue virus in 1977 and of Akabane virus in 1979. J Hyg Camb 95:149–158
70. Sellers RF, Pedgley DE, Tucker MR (1977) Possible spread of African horse sickness on the wind. J Hyg Camb 79:279–298
71. Sellers RF, Pedgley DE, Tucker MR (1978) Possible windborne spread of bluetongue to Portugal, June-July 1956. J Hyg Camb 81:189–196

72. Sellers RF, Pedgley DE, Tucker MR (1982) Rift Valley fever, Egypt 1977: disease spread by windborne insect vectors? Vet Rec 110:73–77
73. Kock RA, Wambua JM, Mwanzia J, Wamwayi H, Ndungu EK, Barrett T, Kock ND, Rossiter PB (1999) Rinderpest epidemic in wild ruminants in Kenya 1993–97. Vet Rec 145:275–283
74. Davis AJ, Jenkinson LS, Lawton JH, Shorrocks B, Wood S (1998) Making mistakes when predicting shifts in species range in response to global warming. Nature 391:783–786
75. White PCL, Brown JA, Harris S (1993) Badgers (*Meles meles*), cattle and bovine tuberculosis (*Mycobacterium bovis*) – a hypothesis to explain the influence of habitat on the risk of disease transmission in southwest England. Proc R Soc Lond B 253:277–284
76. Tompkins DM, Dobson AP, Arneberg P, Begon ME, Cattadori IM, Greenman JV, Heesterbeek JAP, Hudson PJ, Newborn D, Pugliese A, Rizzoli AP, Rosa R, Rosso F, Wilson K (2001) Parasites and host population dynamics. In: Hudson PJ, Dobson AP (eds) Ecology of wildlife diseases. Oxford University Press, Oxford, pp 45–62
77. Smith KF, Sax DF, Lafferty KD (2006) Evidence for the role of infectious disease in species extinction and endangerment. Conserv Biol 20:1349–1357
78. Wyatt KB, Campos PF, Gilbert MTP, Kolokotronis SO, Hynes WH, DeSalle R, Daszak P, MacPhee RDE, Greenwood AD (2008) Historical mammal extinction on christmas island (Indian ocean) correlates with introduced infectious disease. PLoS ONE 3(11):e3602
79. Freed LA, Cann RL, Goff ML, Kuntz WA, Bodner GR (2005) Increase in avian malaria at upper elevation in Hawai'i. Condor 107:753–764
80. Haydon DT, Laurenson MK, Sillero-Zubiri C (2002) Integrating epidemiology into population viability analysis: managing the risk posed by rabies and canine distemper to the Ethiopian wolf. Conserv Biol 16:1372–1385
81. McCallum H, Jones M (2006) To lose both would look like carelessness: Tasmanian devil facial tumour disease. PLoS Biol 4:1671–1674
82. Frick WF, Pollock JF, Hicks AC, Langwig KE, Reynolds DS, Turner GG, Butchkoski CM, Kunz TH (2010) An emerging disease causes regional population collapse of a common North American bat species. Science 329:679–682
83. Smith KF, Acevedo-Whitehouse K, Pedersen AB (2009) The role of infectious diseases in biological conservation. Anim Conserv 12:1–12
84. Daszak P, Cunningham AA, Hyatt AD (2000) Wildlife ecology – emerging infectious diseases of wildlife – threats to biodiversity and human health. Science 287:443–449
85. Gortazar C, Ferroglio E, Hofle U, Frolich K, Vicente J (2007) Diseases shared between wildlife and livestock: a European perspective. Eur J Wildl Res 53:241–256
86. Wolfe ND, Dunavan CP, Diamond J (2007) Origins of major human infectious diseases. Nature 447:279–283
87. van der Riet FD (1997) Diseases of plants transmissible between plants and man (phytonoses) exist. Med Hypotheses 49:359–361
88. Williams JW, Jackson ST, Kutzbacht JE (2007) Projected distributions of novel and disappearing climates by 2100 AD. Proc Natl Acad Sci USA 104:5738–5742
89. Benning TL, LaPointe D, Atkinson CT, Vitousek PM (2002) Interactions of climate change with biological invasions and land use in the Hawaiian islands: modeling the fate of endemic birds using a geographic information system. Proc Natl Acad Sci USA 99:14246–14249
90. Isaac NJB, Turvey ST, Collen B, Waterman C, Baillie JEM (2007) Mammals on the EDGE: conservation priorities based on threat and phylogeny. PLoS ONE 2(3):e296
91. Ford SE, Smolowitz R (2007) Infection dynamics of an oyster parasite in its newly expanded range. Mar Biol 151:119–133
92. Sokolow S (2009) Effects of a changing climate on the dynamics of coral infectious disease: a review of the evidence. Dis Aquat Organ 87:5–18
93. Ricketts TH, Dinerstein E, Boucher T, Brooks TM, Butchart SHM, Hoffmann M, Lamoreux JF, Morrison J, Parr M, Pilgrim JD, Rodrigues ASL, Sechrest W, Wallace GE, Berlin K, Bielby J, Burgess ND, Church DR, Cox N, Knox D, Loucks C, Luck GW, Master LL, Moore R, Naidoo R, Ridgely R, Schatz GE, Shire G, Strand H, Wettengel W, Wikramanayake E (2005) Pinpointing and preventing imminent extinctions. Proc Natl Acad Sci USA 102:18497–18501

94. Kutz SJ, Hoberg EP, Polley L, Jenkins EJ (2005) Global warming is changing the dynamics of Arctic host-parasite systems. Proc R Soc Lond B 272:2571–2576
95. Ytrehus B, Bretten T, Bergsjo B, Isaksen K (2008) Fatal pneumonia epizootic in musk ox (*Ovibos moschatus*) in a period of extraordinary weather conditions. EcoHealth 5:213–223
96. Bradley M, Kutz SJ, Jenkins E, O'Hara TM (2005) The potential impact of climate change on infectious diseases of Arctic fauna. Int J Circumpolar Health 64:468–477
97. Berger L, Hyatt AD, Speare R, Longcore JE (2005) Life cycle stages of the amphibian chytrid *Batrachochytrium dendrobatidis*. Dis Aquat Organ 68:51–63
98. Pounds JA, Bustamante MR, Coloma LA, Consuegra JA, Fogden MPL, Foster PN, La Marca E, Masters KL, Merino-Viteri A, Puschendorf R, Ron SR, Sanchez-Azofeifa GA, Still CJ, Young BE (2006) Widespread amphibian extinctions from epidemic disease driven by global warming. Nature 439:161–167
99. Ribas L, Li MS, Doddington BJ, Robert J, Seidel JA, Kroll JS, Zimmerman LB, Grassly NC, Garner TWJ, Fisher MC (2009) Expression profiling the temperature-dependent amphibian response to infection by *Batrachochytrium dendrobatidis*. PLoS ONE 4(12):e8408
100. Gleason FH, Letcher PM, McGee PA (2008) Freeze tolerance of soil chytrids from temperate climates in Australia. Mycol Res 112:976–982
101. Fisher MC, Garner TWJ, Walker SF (2009) Global emergence of *Batrachochytrium dendrobatidis* and amphibian chytridiomycosis in space, time, and host. Annu Rev Microbiol 63:291–310
102. Pounds JA, Crump ML (1994) Amphibian declines and climate disturbance – the case of the golden toad and the harlequin frog. Conserv Biol 8:72–85
103. Garner TWJ, Walker S, Bosch J, Leech S, Rowcliffe JM, Cunningham AA, Fisher MC (2009) Life history tradeoffs influence mortality associated with the amphibian pathogen *Batrachochytrium dendrobatidis*. Oikos 118:783–791
104. Lips KR, Diffendorfer J, Mendelson JR, Sears MW (2008) Riding the wave: reconciling the roles of disease and climate change in amphibian declines. PLoS Biol 6:441–454
105. Rogers DJ, Randolph SE (2003) Studying the global distribution of infectious diseases using GIS and RS. Nat Rev Microbiol 1:231–237
106. Mellor PS, Boorman J, Baylis M (2000) Culicoides biting midges: their role as arbovirus vectors. Annu Rev Entomol 45:307–340
107. Mellor PS, Wittmann EJ (2002) Bluetongue virus in the Mediterranean basin 1998–2001. Vet J 164:20–37
108. Purse BV, Nedelchev N, Georgiev G, Veleva E, Boorman J, Denison E, Veronesi E, Carpenter S, Baylis M, Mellor PS (2006) Spatial and temporal distribution of bluetongue and its Culicoides vectors in Bulgaria. Med Vet Entomol 20:335–344
109. Mellor PS, Carpenter S, Harrup L, Baylis M, Mertens PPC (2008) Bluetongue in Europe and the Mediterranean basin: history of occurrence prior to 2006. Prev Vet Med 87:4–20
110. Guis H, Caminade C, Calvete C, Morse AP, Tran A, Baylis M (2011) Modelling the effects of past and future climate on the risk of bluetongue emergence in Europe. J Roy Soc Interface (in press)
111. WHO (2005) World Malaria report, rollback Malaria programme. World Health Organisation, Geneva
112. Worrall E, Rietveld A, Delacollette C (2004) The burden of malaria epidemics and cost-effectiveness of interventions in epidemic situations in Africa. Am J Trop Med 71:136–140
113. Palmer TN, Alessandri A, Andersen U, Cantelaube P, Davey M, Delecluse P, Deque M, Diez E, Doblas-Reyes FJ, Feddersen H, Graham R, Gualdi S, Gueremy JF, Hagedorn R, Hoshen M, Keenlyside N, Latif M, Lazar A, Maisonnave E, Marletto V, Morse AP, Orfila B, Rogel P, Terres JM, Thomson MC (2004) Development of a European multimodel ensemble system for seasonal-to-interannual prediction (DEMETER). Bull Am Meteorol Soc 85:853–872
114. Morse AP, Doblas-Reyes FJ, Hoshen MB, Hagedorn R, Palmer TN (2005) A forecast quality assessment of an end-to-end probabilistic multi-model seasonal forecast system using a malaria model. Tellus Ser A 57:464–475

115. Jones AE, Morse AP (2010) Application and validation of a seasonal ensemble prediction system using a dynamic malaria model. J Clim 23:4202–4215
116. Lafferty KD (2009) The ecology of climate change and infectious diseases. Ecology 90:888–900
117. Epstein P (2010) The ecology of climate change and infectious diseases: comment. Ecology 91:925–928
118. Gething PW, Smith DL, Patil AP, Tatem AJ, Snow RW, Hay SI (2010) Climate change and the global malaria recession. Nature 465:342–344
119. Reiter P (2008) Global warming and malaria: knowing the horse before hitching the cart. Malar J 7:S3
120. Kuhn KG, Campbell-Lendrum DH, Armstrong B, Davies CR (2003) Malaria in Britain: past, present, and future. Proc Natl Acad Sci USA 100:9997–10001
121. Hulden L (2009) The decline of malaria in Finland – the impact of the vector and social variables. Malar J 8:94
122. Lindsay SW, Hole DG, Hutchinson RA, Richards SA, Willis SG (2010) Assessing the future threat from vivax malaria in the United Kingdom using two markedly different modelling approaches. Malar J 9:70
123. Pritchard GC, Forbes AB, Williams DJL, Salimi-Bejestani MR, Daniel RG (2005) Emergence of fasciolosis in cattle in East Anglia. Vet Rec 157:578–582
124. Dujardin JC, Campino L, Canavate C, Dedet JP, Gradoni L, Soteriadou K, Mazeris A, Ozbel Y, Boelaert M (2008) Spread of vector-borne diseases and neglect of leishmaniasis, Europe. Emerg Infect Dis 14:1013–1018
125. Gould EA, Higgs S, Buckley A, Gritsun TS (2006) Potential arbovirus emergence and implications for the United Kingdom. Emerg Infect Dis 12:549–555
126. Shaw SE, Langton DA, Hillman TJ (2008) Canine leishmaniosis in the UK. Vet Rec 163:253–254
127. Depaquit J, Naucke TJ, Schmitt C, Ferté H, Léger N (2005) A molecular analysis of the subgenus *Transphlebotomus* Artemiev, 1984 (Phlebotomus, Diptera, Psychodidae) inferred from ND4 mtDNA with new northern records of *Phlebotomus mascittii* Grassi, 1908. Parasitol Res 95:113–116
128. La Ruche G, Souares Y, Armengaud A, Peloux-Petiot F, Delaunay P, Despres P, Lenglet A, Jourdain F, Leparc-Goffart I, Charlet F, Ollier L, Mantey K, Mollet T, Fournier JP, Torrents R, Leitmeyer K, Hilairet P, Zeller H, Van Bortel W, Dejour-Salamanca D, Grandadam M, Gastellu-Etchegorry M (2010) First two autochthonous dengue virus infections in metropolitan France, September 2010. Euro Surveill 15:2–6
129. Eitrem R, Vene S (2008) Chikungunya fever-a threat for Europeans. A review of the recent literature. Parasitol Res 103:S147–S148

Books and Reviews

International Panel on Climate Change (2007) Climate change 2007: impacts, adaptation and vulnerability. Cambridge University Press, Cambridge

Chapter 7
Infectious Diseases, Vibrational Spectroscopic Approaches to Rapid Diagnostics

Jeremy D. Driskell and Ralph A. Tripp

Glossary

Chemometrics	A term to describe the use of multivariate statistics used to extract chemical information.
Fourier-transform infrared spectroscopy (FTIR)	A specific technique for acquiring IR absorption spectra in which all wavelengths are simultaneously measured.
Infrared spectroscopy	An absorption-based vibrational spectroscopic technique which primarily probes non-polar bonds.
Polymerase chain reaction (PCR)	An enzymatic method for amplifying a specific nucleic acid sequence.
Raman spectroscopy	A scattering vibrational spectroscopic technique which primarily probes polar bonds.
Surface-enhanced Raman spectroscopy (SERS)	A technique used to amplify Raman scattered signal via adsorption to a nanometer-scale metallic surface.
Vibrational (molecular) spectroscopy	A general term for the use of light to probe vibrations in a sample as a means of determining chemical composition and structure.

This chapter, which has been modified slightly for the purposes of this volume, was originally published as part of the Encyclopedia of Sustainability Science and Technology edited by Robert A. Meyers. DOI:10.1007/978-1-4419-0851-3

J.D. Driskell (✉) • R.A. Tripp
Department of Infectious Diseases, College of Veterinary Medicine, Animal Health Research Center, University of Georgia, Athens, GA 30602, USA
e-mail: jdriske@IllinoisState.edu; ratripp@uga.edu; ratripp@gmail.com

P. Kanki and D.J. Grimes (eds.), *Infectious Diseases: Selected Entries from the Encyclopedia of Sustainability Science and Technology*,
DOI 10.1007/978-1-4614-5719-0_7, © Springer Science+Business Media New York 2013

Definition of the Subject and Its Importance

Importance of Rapid Diagnosis of Infectious Diseases

Infectious diseases are a major burden on human health with the World Health Organization (WHO) reporting that infectious diseases are responsible for one in ten deaths in the world's richest nations. The impact of infectious diseases is even greater in poorer regions of the world where six of every ten deaths are caused by a spectrum of infectious diseases that include bacteria, viruses, parasites and fungi. These infectious agents can further be described as classical pathogens, e.g., tuberculosis and malaria, seasonal epidemics, e.g., influenza and rhinoviruses, emerging infectious disease, e.g., highly pathogenic avian influenza and hemorrhagic fevers, or global pandemics such as the most recent outbreak of novel H1N1 influenza virus. Central to the management of each of these diseases are diagnostics. Early and rapid detection of an infectious agent is not only imperative to prevent the spread of disease, but it is also an essential first step to identify appropriate therapeutics that target the disease, as well as to overcome inappropriate administration of ineffective drugs that may drastically lead to drug-resistant pathogens such as methicillin-resistant *Staphylococcus aureus* (MRSA). This is just a succinct example which highlights the importance of diagnostic testing; however, the sections that follow discuss the current status of diagnostics and introduce an emerging approach to diagnostics based on vibrational spectroscopy which has tremendous potential to significantly advance the field.

Introduction

Classical Culture-Based Diagnostics

Despite the importance of diagnostic tests for infectious diseases, relatively few technological advances have supplanted classical microbiological approaches, e.g., in vitro culture, as a diagnostic standard. Clinical laboratories routinely rely on selective and chromogenic growth media to identify bacterial agents. For example, an αβ−chromogenic medium, which includes two substrates, has been developed to selectively isolate *Salmonella* spp. with 100% sensitivity and 90.5% specificity [1]. More recently chromogenic media have been developed to identify *Staphylococcus aureus* and distinguish methicillin-resistant strains (MRSA) [2, 3]. Culture-based diagnostics provide a method for definitive identification of many bacteria, and the tests are relatively inexpensive; however, the identification process has generally low throughput and substantial time is required for diagnostics. Typically, culture requires 24–72 h to allow the bacteria to grow while slow-growing organisms such as mycobacteria require substantially longer (6–12 week) incubation times.

Obviously, this is not ideal as the time frame can delay patient treatment. It should also be noted that not all pathogens can be cultured in a laboratory environment, thus the technique cannot be universally applied. There are several additional drawbacks of culture-based diagnostic methods including the requirement for species specific reagents, appropriate culture and storage environments, and labor intensive procedures.

Antibody-Based Diagnostics

Antigen detection and serology are common approaches used in clinical laboratories as alternative or complementary tools to culture-based detection. Common to both of these methods is the use of antibodies either to directly label and detect the antigen or to capture the host response, e.g., antibody responses to infection. Typically, an enzyme-linked immunosorbent assay (ELISA) is employed for antigen detection in diagnostic laboratories. As a first step, ELISA requires an unknown amount of antigen in a sample to be specifically, via a capture antibody in a sandwich assay format, or nonspecifically, via adsorption, immobilized to a solid phase such as a microtiter plate. After removing excess antigen, a known amount of detection antibody specific to the pathogen is then introduced to bind any immobilized antigen. The detection antibody is either directly labeled with an enzyme, or in an additional step, detected with an enzyme-labeled secondary antibody. After removal of excess reagent, a substrate is introduced to react with the enzyme producing a quantifiable color change. A slight modification of this approach replaces the enzyme with a fluorophor for fluorescence-based readout, eliminating the final substrate incubation step. A similar approach is taken for ELISA-based serological assays in which a known amount of purified antigen is immobilized onto the solid phase and incubated with serum to detect the presence of antibodies. While the procedure requires multitudinous steps, reagents, and substantial labor, ELISA is considered rapid relative to culture-based diagnostics as in many cases the assay can be completed within several hours. ELISA-based assays continue to be an integral part of laboratory diagnostics, but in their original form they are limited to the laboratory.

Lateral flow immunoassays, also called dipstick assays, immunochromatography, sol particle immunoassays, or rapid diagnostic tests, have been developed to overcome many limitations of ELISAs by eliminating the complex multi-step procedure, reducing labor, and allowing field or point-of-care testing. Lateral flow assays, like ELISAs, utilize pathogen-specific antibodies for the direct detection of antigen or detection of antibody response. However, for the case of lateral-flow assays the labeled detection antibody, capture antibody, and control reagents are dried on a prefabricated carrier strip. By design, these assays overcome diffusion-limited kinetics to exploit the rapid kinetics of antibody-antigen recognition [4, 5] to yield results in 10–20 min. Thus, because of the "reagentless" nature and rapid results, these assays are well suited for field use and resource-poor regions where reagent storage and test sites are severely limited. It should be noted, however, that these benefits are at the loss of quantitative information and often a lower threshold of detection.

Numerous lateral flow immunoassays have been developed commercially for clinical diagnostics. Several competing manufacturers offer rapid diagnostic tests for influenza virus in which a conserved antigen is detected in a lateral flow assay format. Some detect influenza A and influenza B without distinction of the subtypes, e.g., QuickVue Influenza Test (Quidel), others detect and differentiate A and B strains, e.g., QuickVue Influenza A + B Test (Quidel) and 3 M™ Rapid Detection Flu A + B Test, and some only identify A or B strains, e.g., SAS Influenza A Test, (SA Scientific). Similarly, commercial rapid diagnostic tests are available for detection of a conserved protein for rotavirus A, e.g., IVD Rotavirus A Testing Kit. Not all lateral flow assays are designed for antigen detection; a rapid diagnostic test developed for leptospirosis diagnosis target anti-*Leptospira* IgM antibodies [6].

Despite continued advancements in antibody-based diagnostics these platforms will always be limited by the need for species-specific reagents, i.e., antibodies where the assays can only perform with the sensitivity and specificity inherent to the antibody. For example, commercial lateral flow assays for influenza only provide 50–70% sensitivity and 90–95% specificity with respect to culture-based diagnosis [7]. While the lateral flow assays may be performed rapidly, their low sensitivity may preclude early diagnosis due to low levels or unsustained levels of antigen through disease available for detection. Moreover, serological-based assays developed to detect agent-specific antibodies require that the infection elicit a detectable sustained immune response before the assay can be performed, a feature which substantially delays diagnosis.

Molecular Diagnostics

Nucleic acid and sequence-based methods for diagnostics offer significant advantages over conventional culture- and antibody-based diagnostics with regard to sensitivity, specificity, speed, cost, and portability. Central to molecular diagnostics is the use of a complementary nucleic acid probe that hybridizes to a unique species-specific region of the infectious agents RNA or DNA. While several molecular platforms have been developed for infectious diseases diagnostics, e.g., fluorescence in situ hybridization (FISH), polymerase chain reaction (PCR) is the most commonly employed molecular method for diagnostics. PCR is a method of amplifying targeted segments of nucleic acid by several orders of magnitude to facilitate detection. In principal, complementary primer sequences are used to hybridize to a target nucleic acid sequence. A thermostable DNA polymerase, e.g., Taq polymerase, is then employed to extend the primer sequence. Thermal control facilitates extension, melting, and annealing, and via temperature-controlled cycling, the number of target sequences increases exponentially with each cycle. Amplification of the target sequence can be read out in an ethidium bromide-stained agarose gel or in real-time via cleavage of a fluorescent tag from the primer during the extension step. Appropriate selection of the primers provides extremely specific detection, while the amplification of the target nucleic acid provides excellent sensitivity.

PCR has been demonstrated to be sensitive to single-copy numbers of DNA/RNA targets. In these most sensitive PCR assays, primers are chosen to fully complement a region of the target sequence. Perfect complement probes are also ideal for maximizing the assay specificity to a particular infectious agent. However, in practice, genetic mutations, particularly prevalent in RNA viruses such as influenza, can render a primer/probe set ineffective for diagnosis. Thus, degenerate probes are sometimes chosen to encompass some genetic diversity at the expense of sensitivity.

Multiplexed PCR utilizes multiple primer/probe sets that target different pathogens. Multiplexed assays are implemented when the sample size is limited, preventing multiple individual singleplex PCR analyses, and the clinician is unable to determine the most likely causative agent based on early clinical presentation. Multiplexed PCR assays are not quantitative due to target competition for reagents, are typically less sensitive than singleplex assays, and because of increased reagents, are more expensive to perform than singleplex assays. Moreover, multiple PCR products cannot be simultaneously read out by fluorescence, thus microarray analysis or electrophoresis to identify PCR products of different lengths is required to detect multiple PCR products. However, breakthroughs in multiplexed detection and quantitation are forthcoming [8–10].

Thus, for detection and diagnosis of many diseases such as viruses, PCR offers many advantages over classical methods of diagnostics, and its role will continue to expand in clinical diagnostic laboratories. However, there are challenges associated with PCR. For pathogens in which culture and microscopy can be used, molecular diagnostics are not the most cost effective. For example, the cost of a commercial PCR assay for tuberculosis ranges from $40 to $80, whereas microscopy and culture can be implemented for $1 and $5, respectively. Another consideration is how to interpret PCR results. Due to the sensitivity afforded by PCR, extremely low levels of infectious agent can be detected which may be below clinically relevant thresholds for disease presentation. Thus, quantitative PCR rather than qualitative PCR is typically more informative when correlating to clinical diagnosis. PCR assays developed in the laboratory are not always translational to analysis of clinical samples. In general, PCR cannot be performed directly on biological fluids such as blood because compounds such as hemoglobin, lactoferrin, heme, and IgG inhibit amplification [11–13]. Therefore, DNA and RNA are extracted as a first step, prior to PCR, but inefficiency of extraction kits often lead to a decrease in analytical performance compared to laboratory cultures [14, 15]. Moreover, isolation of nucleic acids is time consuming and technically challenging unless automated, which requires expensive equipment and reagents. A final consideration is the need for temperature-sensitive reagents, thermocyclers, skilled workers, and a clean laboratory environment to prevent contamination leading to false-negative results. While tremendous efforts are focused on PCR automation, incorporation of microfluidics [16, 17], and isothermal amplification [18–21], e.g., loop-mediated isothermal amplification (LAMP), to overcome these challenges, the current status of PCR precludes widespread use of PCR diagnostics in point-of-care settings and developing nations.

Limitations of Classical and Conventional Diagnostics

As discussed above, diagnostics are essential to healthcare and disease management. While there is merit in further advancing current diagnostic methods, there are shortcomings for each approach. As noted above, culture is sensitive, but the length of time prevents rapid and early diagnosis. Antibody-based techniques are often limited in sensitivity, and like molecular diagnostics, are expensive, and require species-specific detection reagents. It is likely that any improvements in these methods to address these challenges will be incremental; however there are several next generation diagnostics that offer potential paradigm shifting approaches to detection and diagnoses that are currently being investigated. One important area is the use of molecular or vibrational spectroscopy for "whole-organism" fingerprinting. This innovative approach to diagnostics promises to be rapid, specific, and truly reagentless.

Spectroscopy-Based Diagnostics

Vibrational spectroscopy includes a number of nondestructive analytical techniques which provide molecular information about the chemical makeup, e.g., functional groups, of a sample. Subtle changes in the frequency of a particular functional group vibration, e.g., group frequency, provides additional details of chemical structure, local environment surrounding the bond, bond angle, length, geometry, and conformation. These attributes of vibrational spectroscopy have led to the development of vibrational spectroscopic approaches to generate whole-organisms fingerprints to serve as unique biochemical signatures for pathogen identification. Unique to this approach of infectious agent identification is rapid readout, and perhaps most importantly, there is no need for species-specific reagents or other reagents of any kind. Three of the most developed vibrational spectroscopic techniques include infrared absorption spectroscopy, Raman scattering spectroscopy, and surface-enhanced Raman spectroscopy (SERS). These three methods, as well as their development for diagnosing infectious diseases are described in detail below.

Infrared Spectroscopy

Infrared spectroscopy is an absorption technique in which the sample is irradiated most commonly with mid-infrared light with wavelengths in the range of

2.5–50 μm. Photons of appropriate energy to bring about a transition from one vibrational state to an excited vibrational state are absorbed by the analyte. Selection rules govern which vibrations are allowed to absorb IR photons, providing chemical and structural information. These rules require a net change in the dipole moment of the molecule as a result of the vibration. Thus, IR absorption spectra are dominated by asymmetrical vibrations. With consideration to these selection rules, proteins and nucleic acids (building blocks of bacteria, viruses, etc.) are excellent absorbers of IR radiation making IR spectroscopy an ideal tool for characterization of infectious agents.

The chemical complexity and sheer size of bacteria and viruses tend to produce complex spectra with broad and overlapping bands. Yet subtle changes in band shape, slight shifts in band peak positions, and variation in relative band intensities provide significant insight into chemical structure. Thus, careful evaluation of the full spectral fingerprint of whole-organisms, rather than analysis of single peaks common to small molecule studies, can lead to identification and classification of microorganisms.

IR spectroscopy as a technique for whole-organism fingerprinting dates back to 1952 [22]. In this early study, Stevenson and Boulduan showed that the IR spectra for *Escherichia coli*, *Pseudomonas aeruginosa*, *Bacillus subtilis*, and six other species of cultured bacteria are unique to each species. In addition to species identification, six strains of *Bacterium tularense* were spectroscopically differentiated. This initial work did not immediately translate to the diagnostic applications of IR spectroscopy. Two major breakthroughs that did not occur for several decades after the original findings were essential to further increase the utility of IR for whole-organism fingerprinting. First, prior to the 1980s when Fourier transform infrared spectroscopy (FTIR) was introduced, dispersive instruments were typically used. Dispersive instruments do not provide the speed or analytical performance required for IR-based diagnostics. FTIR instruments provide much better signal-to-noise spectra with improved spectral resolution, and are acquired in less time. These advantages provided by FTIR were essential to accurately analyze these complex biological spectra by distinguishing subtle changes in spectral band shape and to rapidly collect data. Additional developments in the methods of data analysis were also essential to move IR whole-organism fingerprinting forward. The large number of variables, i.e., wavelengths, each containing relevant information, inherent to IR spectra and rather subtle changes in intensity prevented traditional single-peak analysis for bacterial identification. Moreover, visual inspection of the spectra by the analyst is too labor intensive, prone to operator error, and unrealistic for large numbers of spectra and/or organisms from which to identify. The introduction of chemometrics, i.e., multivariate statistics applied to spectroscopy, led to the continued interest and advancement of IR-based diagnostics. These methods, including principal component analysis [23, 24], discriminant analysis, multiple regression analysis [25], and artificial neural networks [26, 27], function to simplify the high dimensional dataset by identifying the most significant variables with the ultimate goal of sample identification or quantification.

FTIR has received the greatest attention with respect to vibrational spectroscopic techniques over the last 2 decades and has been developed to the point that it can be considered an established method for the identification of both bacterial species and strains [28–33]. Researchers continue to investigate laboratory cultures of bacteria in an effort to standardize sampling protocols so that spectral databases can be shared among laboratories and to standardize the methods of analysis including spectral preprocessing, feature selection, and classification algorithms. As the method has matured, and while it continues to be tweaked and standardized, FTIR is now being applied to the analysis of clinical specimens. For example, FTIR has been successfully used to analyze clinical sputum samples collected from cystic fibrosis patients [26]. This FTIR capability is important as historically, lung infection caused by *Pseudomonas aeruginosa*, *Staphylococcus aureus*, and *Haemophilus influenzae* were the major causes of morbidity and mortality in cystic fibrosis patients; however, the number of emerging nonfermenting species is on the rise [34], and many of these species are closely related and not appropriately identified using typical clinical diagnostics and microbiological approaches [35]. Using a FTIR spectral library and an artificial neural network built for pathogen identification, the results from the FTIR method were compared to conventional microbiology detection methods. A two-tiered ANN classification scheme was built in which the top-level network identified *P. aeruginosa*, *S. maltophilia*, *Achromobacter xylosoxidans*, *Acinetobacter* spp., *R. pickettii*, and *Burkholderia cepacia* complex (BCC) bacteria. The second-level network differentiated among four species of BCC, *B. cepacia*, *B. multivorans*, *B. cenocepacia*, and *B. stabilis*. Ultimately, this method resulted in identification success rates of 98.1% and 93.8% for the two ANN levels, respectively. However, before this optimized method was established, the research highlighted three important considerations. First, not all bacterial isolates produce poly-β-hydroxybutyric acid (PHB) which contributes to the IR spectra and confounds classification. To overcome this, each isolate was cultured on TSA medium and harvested after 5 h of growth, prior to the expression of PHB. This step enriches the bacteria for analysis and eliminates interference from PHB. Second, flagella or pilus fibers were determined to contribute to spectral heterogeneity. Vigorous vortexing and subsequent centrifugation removes the fibers to significantly improve spectral reproducibility and classification results (Fig. 7.1). Third, the classification algorithm significantly affects the classification results. The authors show that hierarchical clustering algorithms (HCA) discriminate between reference and clinical strains rather than based on bacterial identity. Advanced methods, such as ANN, that determine spectral variables that vary only as a result of the bacteria was necessary to correctly classify according to strain. This example work demonstrates the power of IR-based diagnostics, but suggests that these methods may require problem-specific standardization of experimental protocols and data analysis.

These groundbreaking efforts to develop IR for bacterial analysis have led to the realization that spectroscopic methods have advantages for exploring detection of other pathogens. For example, FTIR has been employed for the distinction of yeast and fungi with success [36, 37]. More recently IR has been investigated as a method

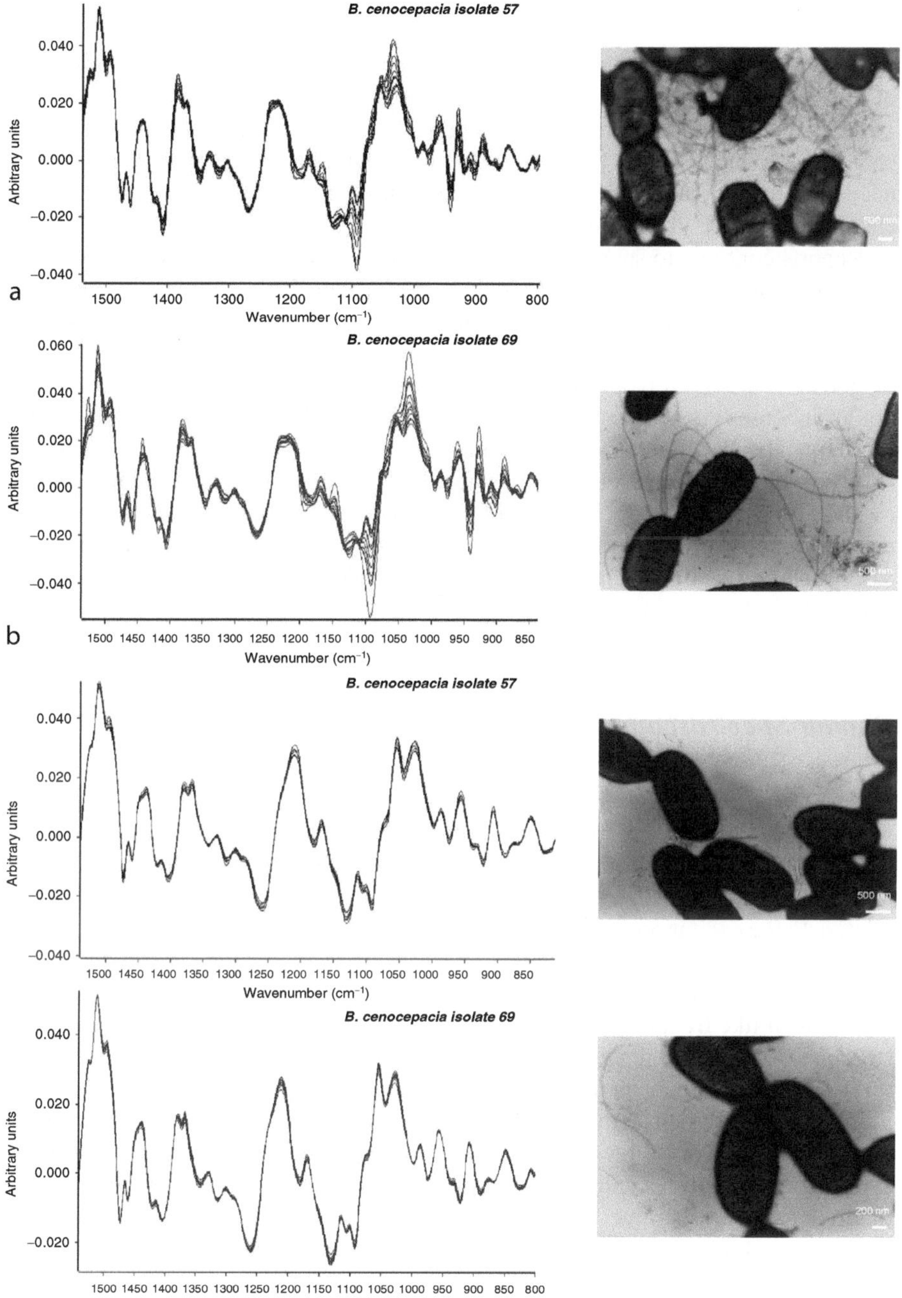

Fig. 7.1 Vector-normalized first-derivative spectra of two *B. cenocepacia* clinical isolates (isolates 57 and 69) in the 1,500–800 cm^{-1} range. (**a**) The heterogeneity of 15 replicate measurements for each strain in the spectral ranges of 1,200–900 cm^{-1} and 1,500–1,300 cm^{-1} and the corresponding micrographs obtained by TEM are shown. (**b**) Vector-normalized first-derivative spectra measured after vortexing of similar cells at the maximum intensity for 15 min and subsequent centrifugation at 8,000 × g for 5 min to separate the cells from free pilus appendages in the supernatants. Micrographs of the cells obtained by TEM after they were vortexed without centrifugation show the small fragments of pili or fibers suspended in the supernatants (From [26])

to detect viral infections, although current experiments are limited to viral infection of cells in culture [38–41]. While mock-infected and herpes simplex virus type 1-infected Vero cells are readily distinguished via IR, infection-induced spectral changes are inconsistent [39, 40]. Thus, substantially more research effort is necessary to standardize protocols and correlate the spectral response to the biochemical response upon infection.

Reports continue to support the utility of FTIR-based diagnostics in the clinical laboratory, but there are certain limitations to consider. First, water is a particularly strong absorber of IR light. Thus, care must be taken to completely dehydrate the sample prior to data acquisition. This obviously does not prevent IR-based diagnostics, it is merely an inconvenience. Second, IR absorption spectroscopy is not an inherently sensitive method and trace levels of a pathogen are not readily apparent. Hence, clinical samples will likely require a culture step to generate sufficient biomass for IR analysis. As noted above, this sample enrichment can be as short as 5 h, and with IR data acquisition on the order of minutes, the total analysis time is still more rapid, less labor intensive, and more informative in many cases than conventional diagnostic methods and does not require species-specific reagents.

Raman Spectroscopy

Raman spectroscopy is a scattering technique, in which the sample is irradiated with a monochromatic light source, almost always a laser. The majority of the scattered photons are elastically scattered and maintain the same frequency as the excitation source; however, a small fraction of the photons are shifted in frequency relative to the excitation source. The difference in the energy between the excitation and inelastically, i.e., Raman, scattered photons correspond to the energy necessary to bring about a transition from one vibrational state to an excited vibrational state. Thus, much like IR spectroscopy, Raman spectra provide insight into the chemical structure, local environment, geometry, and conformation of the sample and can serve as a whole-organism fingerprinting method. Selection rules also govern which vibrations are Raman active. These rules require a change in the polarizability during the vibration to be Raman active. Thus, Raman spectra are dominated by symmetrical vibrations and the technique is often seen as a complementary rather than competing technique with IR spectroscopy. However, for application to the analysis of biological materials and whole-organism fingerprinting methods, Raman offers many inherent advantages over IR spectroscopy.

Because of the selection rules, the main chain and aromatic side chains of peptides rather than aliphatic side chains are probed via Raman scattering in contrast to IR. Raman bands of nucleic acids are limited to heterocyclic bases or phosphodiester groups making up the backbone. Raman bands are narrower and less likely to overlap, thus the spectra are much less complicated compared to IR spectra because of the many more nonsymmetric vibrations that are possible.

Another major advantage of Raman is that water does not interfere since its vibrations do not fit the selection rules criteria. This is an extremely important consideration when analyzing biological samples which are endemic to aqueous environments. Other advantages of Raman include the flexibility to analyze samples in any state, e.g., gas, liquid, or solid, and the ability to analyze small sample volumes and masses because of the tight focus of incident laser light (square microns) compared to the incident IR beams (square centimeters).

Viruses were the first infectious agent analyzed by Raman spectroscopy, although not in a diagnostic capacity [42]. In this first work, Raman spectroscopy was used to probe the RNA and protein structure upon viral packaging. In the 1970s, Raman spectroscopy suffered from poor sensitivity due to instrument limitations. The first evaluation of Raman spectroscopy for pathogen detection was not until 1987 when spectra were collected for five species of bacteria including *E. coli*, *P. fluorescens*, *S. epidermidis*, *B. subtilis*, and *E. cloacae* [43]. To overcome the limited sensitivity of the instruments at the time, an ultraviolet laser was used for excitation to enhance spectral features of RNA, DNA, tyrosine, and tryptophan via resonance Raman. Unique spectra were observed for each bacterium, although analysis relied on visual interpretation since chemometrics had not been implemented for spectral analysis yet. UV Raman instruments, while producing the requisite sensitivity for pathogen analysis, is quite expensive and non-resonant vibrations are not observed which results in a significant loss in information that is valuable for differentiation.

Despite the recognized benefits of Raman-based diagnostics, particularly when compared to conventional and IR-based diagnostics, instrumentation has limited the maturation of Raman-based diagnostics. After development of UV Raman for pathogen detection [43–46], Fourier transform Raman (FT-Raman) instruments were introduced for microbiological studies which increased instrument sensitivity [47, 48]. Raman instruments have now evolved to include NIR lasers to reduce fluorescence from biological and NIR-sensitive CCD detectors. These modern instruments have only been developed in this decade to fully explore the potential of Raman as a diagnostic technique [49–55]. Thus Raman-based whole-organism fingerprinting is less developed than IR-based methods and examples are generally limited to the analysis of laboratory cultures.

In an early study, Maquelin et al. [54] utilized Raman spectroscopy to directly analyze five bacterial strains, including three strains of *Staphylococcus* spp., *E. coli*, and *E. faecium*, on solid culture medium. The flexibility in sample type afforded by Raman spectroscopy allowed direct measurement on the culture plate that would not be possible using IR spectroscopy. The background Raman spectrum resulting from the culture medium was subtracted from those spectra collected from the bacterial microcolonies. Hierarchical cluster analysis yielded two major groupings, one consisting of the three *Staphylococcus* strains and one consisting of the *E. coli* and *E. faecium*. The *E. coli* and *E. faecium* spectra clearly grouped according to species within the latter subcluster while spectra in the *Staphylococcus* subcluster grouped according to strain. While chemometric analysis of these spectra collected from same-day cultures yielded a successful classification rate of 100% for external validation samples, combined data collected from 3 days dropped the accuracy to 83% for classification of two *S. aureus* strains (ATCC 29213 and UHR 28624). However, these two strains are extremely similar and in general the results demonstrate the utility of Raman-based diagnostics.

The most rigorous evaluation of Raman spectroscopy for reagentless detection and identification of pathogens was performed in collaboration with a US government laboratory. In this work, a comprehensive library of Raman spectra has been established for over 1,000 species, including 281 CDC category A and B biothreats, 146 chemical threats, 310 environmental interferents, and numerous others [52]. Spectral signatures were collected using Raman chemical imaging spectroscopy (RCIS) [56]. RCIS technology combines digital imaging and Raman spectroscopy. Digital imaging automatically discriminates against background particulates and identifies regions of interest on a sample platform that are then targeted for Raman analysis. Sample analysis is faster and completely automated using this approach. Two commercially available instruments were tested, one in the laboratory (ChemImage Corp., Falcon) and the other in the field (ChemImage Corp., Eagle). To test the robustness of the Raman spectral library and classification scheme, blinded samples containing one of four *Bacillus* strains were analyzed and identified. The predictive performance ranged from 89.4% to 93.1% for these closely related bacteria. It was concluded that key to the success of this diagnostic approach is the extensiveness of the spectral library. There are many more bacterial phenotypes than genotypes, and it has been found that Raman fingerprints correlate with cell phenotype, thus an all-inclusive library must contain spectra for each bacterial strain grown under different conditions and at different stages of development. In a subsequent study untrained personnel at the Armed Forces Institute of Pathology evaluated 14 bacteria to generate a spectral library and sent 20 blinded samples to ChemImage for external validation in which all 20 samples were correctly identified. This comprehensive study is the first to establish the true utility of automated Raman-based diagnostics carried out off-site by untrained personnel. However, these samples were prepared in water, cell culture media, or spiked nasal swabs, none of which are truly clinical samples.

An early study to evaluate clinical samples for *Acinetobacter* by Raman spectroscopy and compare the results with an established diagnostic method were among the first showing the power and speed of Raman-based detection [55]. In this study, 25 *Acinetobacter* isolates from five hospitals in three countries were analyzed using selective amplification of restriction fragments (AFLP), an established molecular technique for typing bacteria strains. Dendograms resulting from the hierarchical cluster analysis of Raman and AFLP fingerprints for the isolates were generated and compared (Fig. 7.2). Both dendograms resulted in five clusters that separate the strains according to the five outbreaks, with the exception of one Basildon isolate RUH 3242 which clustered with isolates from Venlo in the Raman-based dendogram. Overall results from Raman fingerprinting of these clinical isolates were very similar to those obtained for established methods, but with the advantage of faster analysis and less complicated procedures.

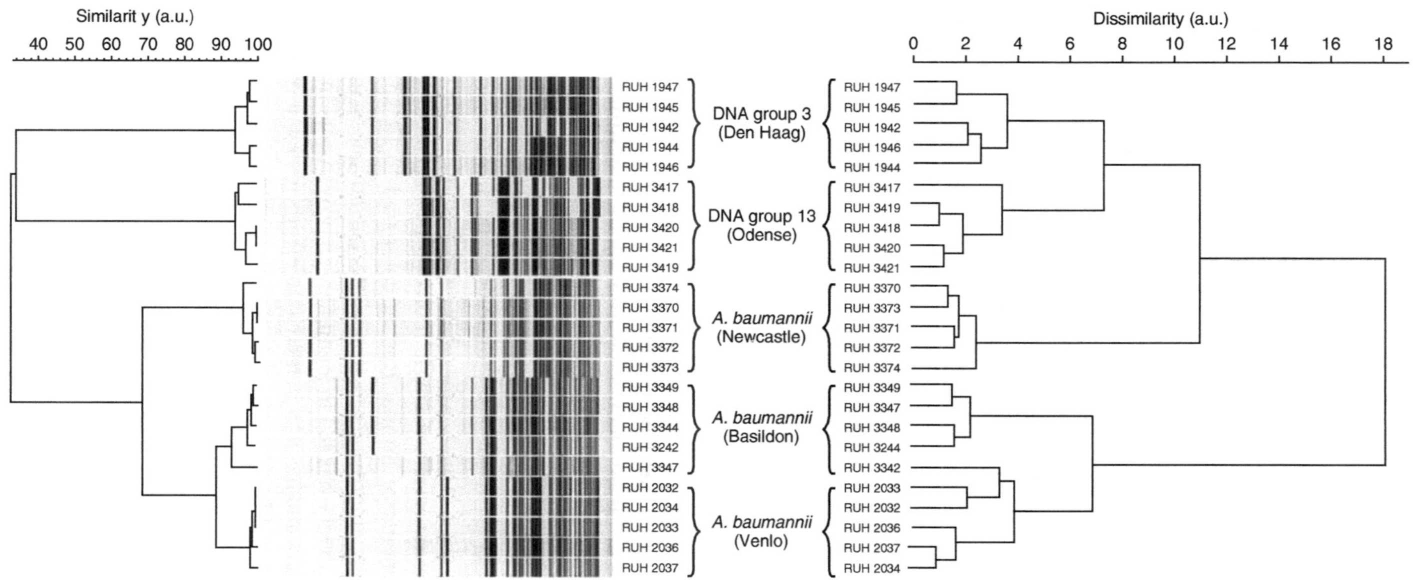

Fig. 7.2 Dendrograms resulting from the hierarchical cluster analysis of (*left*) AFLP analysis and (*right*) Raman analysis of the isolates. The asterisk marks the strain RUH 3242 misclassified via Raman fingerprinting (From [55])

Despite the advancement of Raman spectroscopy instrumentation and methods for pathogen fingerprinting, Raman is still often limited by poor sensitivity. Only $\sim$1 in 10^6–10^8 photons are inelastically scattered as the vast majority are elastically scattered. This means that high quality spectra with the requisite signal-to-noise can take minutes to acquire. While this may not be a limitation in laboratory experiments, or developmental stages in research, it prohibits its usefulness in clinical diagnostic laboratories which analyze hundreds to thousands of samples per day. Thus, there is great interest in enhancing the Raman signal. One such method is to excite the sample with a frequency that resonates with an electronic transition, so called resonance Raman spectroscopy. For biological samples, this requires UV lasers for excitation, and as noted above, is cost prohibitive to widespread adoption of this method. Moreover, chemical information is lost when performing resonance Raman which would likely reduce classification accuracy of closely related pathogens. An alternative method to amplify Raman scattering is surface-enhanced Raman spectroscopy (SERS). SERS has received a great deal of attention, particularly with respect to whole-organism fingerprinting and is the subject of the next section.

SERS

Surface-enhanced Raman spectroscopy is a technique in which the Raman signal of a sample is significantly amplified via adsorption onto a metallic nanostructured surface. A laser excitation frequency is selected such that it is in resonance with the collective oscillation of the conduction electrons in the nanostructures, i.e., surface plasmon resonance. When resonance conditions are met, the local electromagnetic field experienced by molecules in close proximity to the surface is significantly increased to yield rather large enhancements in the Raman scattering. While the signal enhancement is substrate and sample dependent, typical enhancements are on the order of 10^4–10^{14} with respect to normal Raman intensities, with several studies reporting the detection of single molecules using this technique [57, 58]. SERS offers the benefits of normal Raman compared to IR spectroscopy while providing a markedly improved sensitivity. Recent advances in nanofabrication methods and SERS theory has led to significant improvements in SERS substrates in the last several years and has driven increased efforts to develop SERS for whole-organism fingerprinting [59–78].

The major focus of whole-organism fingerprinting via SERS has been on bacteria identification [51, 64–74, 77, 78]. Most of these studies report differentiation among bacteria species, with many demonstrating discrimination of different strains of the same species. However, there are several inconsistencies that have been noted by researchers, particularly in the earlier studies. For example, Grow et al. found SERS spectra for strains that belong to the same species were sometimes less similar than spectra collected from different species [65], and Jarvis and Goodacre observed similar spectra for the same bacteria using different preparations of silver nanoparticles, but noted subtle changes in signal intensities among nanoparticle batches [68]. These discrepancies evident in these early studies highlight the primary challenge of SERS-based diagnostics, i.e., the enhancing substrate. The SERS signal is highly dependent on the enhancing substrate, thus a reliable means of fabricating nanostructured materials is vital to the success of SERS-based diagnostics.

Several research laboratories have analyzed and published SERS spectra for both *Bacillus subtilis* and *E. coli*; however, each reported incongruent spectral fingerprints [67, 68, 71, 72]. The experimental protocols, however, varied among each study. For example, in two different reports Jarvis et al. used two different chemical synthesis preparations to generate colloidal silver, citrate reduction [67] and borohydride reduction [51], to serve as the SERS substrate. The SERS spectra were drastically different in each study. It is well known that spectra are dependent on the enhancing nanostructure, e.g., material, size, shape, interparticle spacing, etc., but given the same final nanostructure similar spectra were expected. The authors attributed the differences to the effect of diverse chemistries used to prepare each silver colloid [79]. However, it should be noted that different excitation sources, 7 nm and 785 nm, were employed in the two studies. For normal Raman, the Raman shifts should be independent of the excitation source, thus spectral fingerprinting should not be affected by the choice of the laser. SERS spectra, however, can be influenced by the excitation source because of the requisite pairing of the excitation frequency and plasmon resonance of the substrate. Therefore, it is perhaps more probable that spectral differences observed by Jarvis et al. are due to greater signal enhancement for the 7 nm excitation source rather than due to differences in chemical preparation of the colloidal silver. This interpretation is supported by a study in which a third variation in experimental parameters was implemented utilizing citrate-reduced silver colloid but acquired spectra with a 647 nm laser [71]. Results from this study closely resembled the results for *B. subtilis* obtained by Jarvis et al. employing borohydride reduced silver nanoparticles and 7 nm excitation. Collectively, these studies also demonstrate the need for procedural consistency.

In a pivotal study, scientists at a US Army research laboratory evaluated the SERS signatures for many bacteria using a standardized sampling protocol and instrumentation. To date, three SERS substrates were directly compared using the standardized protocol: silver nanoparticles, silver film over nanospheres (FONS), and commercially available Klarite. Interaction between substrate and bacteria vary significantly as visualized with electron microscopy which likely results in different spectral fingerprints. Moreover the signal intensities varied significantly among the substrates reflecting differences in enhancing quality. Details of these experiments are approved for public release as a technical report (ARL-TR-4957).

In another key study, SERS and Raman fingerprints were directly compared to assess the advantages of SERS analysis [72]. Raman and SERS spectra were collected for several bacteria, including four strains of *Bacillus*, *S. typhimurium*, and *E. coli*. As noted above, the substrate is a critical factor in SERS analysis, and in this study aggregated gold nanoparticle films were grown in-house and established as a reliable means of substrate preparation for acquisition of repeatable spectra. As anticipated, SERS yielded much greater signal-to-noise spectra compared to normal Raman. The study also identified two unexpected benefits of SERS. Normal Raman signal for *Bacillus* species was overwhelmed by native fluorescence of the sample; however, in the SERS analysis, the metal substrate functioned to quench the fluorescence component in addition to enhancing the Raman signal. It was also observed that normal Raman spectra are more complex than SERS spectra. This is explained by the fact that bulk Raman interrogates all components throughout the entire bacterium equally, while the distance-dependence of SERS enhancement preferentially probes the region of the bacterium closest to the metal substrate and bands for the internal components are not detected. Fortunately, most chemical variation among bacterial strains and species are expressed on the cell surface, thus greater spectral differences are observed among SERS spectra of different samples than compared to bulk Raman spectra. This added advantage is exemplified by greater discrimination of bacteria when utilizing SERS spectra as compared to Raman spectra [72].

A number of novel nanofabrication methods have recently emerged for producing SERS substrates with the potential for addressing the issues noted above due to substrate heterogeneity. These include electron beam lithography [80, 81], nanosphere lithography [82–84], a template method [85–88], oblique angle vapor deposition (OAD) [89–91], and a proprietary wet-etching technology used to produce commercially available Klarite (D3 technologies). It should be noted, however, that with the exception of OAD and Klarite, these fabrication methods are not adaptable to large-scale production due to the complexity of the fabrication procedure. Not only is it likely that these substrates will lead to significant advances in SERS-diagnostics of bacteria, the use of OAD and Klarite substrates has already lead to successful application to virus identification [59, 60, 62, 63, 75, 76].

In the most recent investigation of SERS-based viral fingerprinting, eight strains of rotavirus were analyzed [63]. These isolates were recovered from clinical fecal samples and propagated in MA104 cells and represent the 5 G and 3 P genotypes responsible for the most severe infections. Unique SERS fingerprints were acquired for each strain when adsorbed onto OAD-fabricated silver nanorods. Representative spectra for each strain and negative control, as well as the difference spectra which subtract out the background cell lysate signal are displayed in Fig. 7.3. Classification algorithms based on partial least squares discriminant analysis were constructed to identify the samples according to (1) rotavirus positive or negative, (2) P4, P6, or P8 genotype, (3) G1, G2, G3, G4, or G9 genotype, or (4) strain. Respectively, these four classification models resulted in 100%, 98–100%, 96–100%, and 100% sensitivity and 100%, 100%, 99–100%, and 99–100% specificity.

Compilation and critical analysis of reports to date demonstrate the potential of Raman-based diagnostics and its advantages over IR, normal Raman spectroscopy, and convention diagnostic methods, but also highlight the need for standardization. The challenge in the future is standardization of substrates and sampling protocols since background can “quench” signal from the analyte. For example, blood

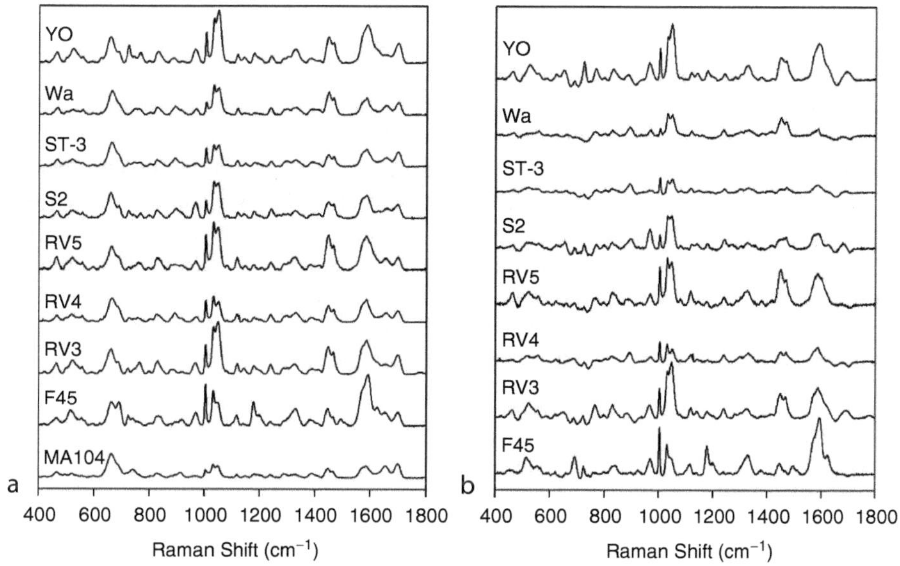

Fig. 7.3 (**a**) Average SERS spectra for eight strains of rotavirus and the negative control (MA104 cell lysate). Spectra were baseline corrected, normalized to the band at 633 cm^{-1}, and offset for visualization. (**b**) Difference SERS spectra for eight strains after subtraction of MA104 spectrum (From [63])

analysis requires sample processing to remove some competing elements [92], yet SERS spectra highly dependent on the sample pretreatment procedure as remaining chemical species will also contribute signal and degrade the performance of matching in spectral library databases. The outlook of SERS is not a question of spectral quality and reproducibility in a controlled environment, the question is how to control the environment across laboratories.

Future Directions

The future of spectroscopic-based diagnostics is bright as demonstrated by the many studies cited and discussed above. In addition to the success found in these studies, areas of improvement have also been identified. An important area of potential development is the methods used for statistical analysis. Well-established algorithms such as PCA, HCA, and discriminant analysis continue to provide high predictive accuracy, but recent examples have shown that more creative and novel approaches such as artificial neural networks, "bar-coding" [70], or innovative uses of PLS [59] can further improve the predictive value. A revolution in instrumentation is also occurring. Vibrational spectroscopy has recently filled niches in quality control of pharmaceuticals and raw materials as well as identification of chemical threats. The nature of these applications and explosion in interest have driven the instrumentation industry to invest in the development and production of high-quality yet affordable handheld instruments for mobile, on-site analysis. This market-driven commercialization, in effect, is paving the way for point-of-care, mobile, and cost-effective spectroscopy-based diagnostics. The most important factor for widespread realization of spectroscopic diagnosis will be the emergence of a universal protocol for sampling, and for the case of SERS, a standard substrate. The accepted protocol must then be used to build a spectral database covering a variety of phenotypes and developmental stages as illustrated above. Implementing a standard practice is crucial for the success of the technique, but once developed this technology has the potential to become the first and immediate response to clinical cases in which infection is suspected.

Bibliography

Primary Literature

1. Perry JD, Ford M, Taylor J, Jones AL, Freeman R et al (1999) ABC medium, a new chromogenic agar for selective isolation of *Salmonella* spp. J Clin Microbiol 37:766–768
2. Hedin G, Fang H (2005) Evaluation of two new chromogenic media, CHROMagar MRSA and S. aureus ID, for identifying Staphylococcus aureus and screening methicillin-resistant S-aureus. J Clin Microbiol 43:4242–4244
3. Perry JD, Davies A, Butterworth LA, Hopley ALJ, Nicholson A et al (2004) Development and evaluation of a chromogenic agar medium for methicillin-resistant *Staphylococcus aureus*. J Clin Microbiol 42:4519–4523
4. Nygren H, Stenberg M (1985) Kinetics of antibody-binding to surface-immobilized antigen – influence of mass-transport on the enzyme-linked immunosorbent-assay (ELISA). J Colloid Interface Sci 107:560–566
5. Nygren H, Werthen M, Stenberg M (1987) Kinetics of antibody-binding to solid-phase-immobilized antigen – effect of diffusion rate limitation and steric interaction. J Immunol Methods 101:63–71
6. Gussenhoven GC, vanderHoorn M, Goris MGA, Terpstra WJ, Hartskeerl RA et al (1997) LEPTO dipstick, a dipstick assay for detection of Leptospira-specific immunoglobulin M antibodies in human sera. J Clin Microbiol 35:92–97
7. Gordon A, Videa E, Saborio S, Lopez R, Kuan G et al (2010) Diagnostic accuracy of a rapid influenza test for pandemic influenza A H1N1. Plos One 5:e10364
8. Erdman DD, Weinberg GA, Edwards KM, Walker FJ, Anderson BC et al (2003) GeneScan reverse transcription-PCR assay for detection of six common respiratory viruses in young children hospitalized with acute respiratory illness. J Clin Microbiol 41:4298–4303
9. Lassauniere R, Kresfelder T, Venter M (2010) A novel multiplex real-time RT-PCR assay with FRET hybridization probes for the detection and quantitation of 13 respiratory viruses. J Virol Methods 165:254–260
10. Li H, McCormac MA, Estes RW, Sefers SE, Dare RK et al (2007) Simultaneous detection and high-throughput identification of a panel of RNA viruses causing respiratory tract infections. J Clin Microbiol 45:2105–2109
11. Abu al-Soud W, Radstrom P (2001) Purification and characterization of PCR-inhibitory components in blood cells. J Clin Microbiol 39:485–493

12. Abu Al-Soud W, Radstrom P (2001) Effects of amplification facilitators on diagnostic PCR in the presence of blood, feces, and meat. J Clin Microbiol 38:4463–4470
13. Widjojoatmodjo MN, Fluit AC, Torensma R, Verdonk G, Verhoef J (1992) The magnetic immunopolymerase chain-reaction assay for direct detection of salmonellae in fecal samples. J Clin Microbiol 30:3195–3199
14. Chui LW, King R, Lu P, Manninen K, Sim J (2004) Evaluation of four DNA extraction methods for the detection of *Mycobacterium avium* subsp *paratuberculosis* by polymerase chain reaction. Diagn Microbiol Infect Dis 48:39–45
15. McOrist AL, Jackson M, Bird AR (2002) A comparison of five methods for extraction of bacterial DNA from human faecal samples. J Microbiol Methods 50:131–139
16. Kaigala GV, Hoang VN, Stickel A, Lauzon J, Manage D et al (2008) An inexpensive and portable microchip-based platform for integrated RT-PCR and capillary electrophoresis. Analyst 133:331–338
17. Zhang NY, Tan HD, Yeung ES (1999) Automated and integrated system for high-throughput DNA genotyping directly from blood. Anal Chem 71:1138–1145
18. Aryan E, Makvandi M, Farajzadeh A, Huygen K, Bifani P et al (2010) A novel and more sensitive loop-mediated isothermal amplification assay targeting IS6110 for detection of *Mycobacterium tuberculosis* complex. Microbiol Res 165:211–220
19. Fang XE, Liu YY, Kong JL, Jiang XY (2010) Loop-mediated isothermal amplification integrated on microfluidic chips for point-of-care quantitative detection of pathogens. Anal Chem 82:3002–3006
20. Notomi T, Okayama H, Masubuchi H, Yonekawa T, Watanabe K et al (2000) Loop-mediated isothermal amplification of DNA. Nucl Acids Res 28:e63
21. Shivakoti S, Ito H, Murase T, Ono E, Takakuwa H et al (2010) Development of reverse transcription-loop-mediated isothermal amplification (RT-LAMP) assay for detection of avian influenza viruses in field specimens. J Vet Med Sci 72:519–523
22. Stevenson HJR, Bolduan OEA (1952) Infrared spectrophotometry as a means for identification of bacteria. Science 116:111–113
23. Lin MS, Al-Holy M, Al-Qadiri H, Kang DH, Cavinato AG et al (2004) Discrimination of intact and injured *Listeria monocytogenes* by Fourier transform infrared spectroscopy and principal component analysis. J Agric Food Chem 52:5769–5772
24. Ngo-Thi NA, Kirschner C, Naumann D (2003) Characterization and identification of microorganisms by FIF-IR microspectrometry. J Mole Struct 661:371–380
25. Janbu AO, Moretro T, Bertrand D, Kohler A (2008) FT-IR microspectroscopy: a promising method for the rapid identification of *Listeria* species. FEMS Microbiol Lett 278:164–170
26. Bosch A, Minan A, Vescina C, Degrossi J, Gatti B et al (2008) Fourier transform infrared spectroscopy for rapid identification of nonfermenting gram-negative bacteria isolated from sputum samples from cystic fibrosis patients. J Clin Microbiol 46:2535–2546
27. Rebuffo-Scheer CA, Schmitt J, Scherer S (2007) Differentiation of Listeria monocytogenes serovars by using artificial neural network analysis of Fourier-transformed infrared spectra. Appl Environ Microbiol 73:1036–1040
28. Bouhedja W, Sockalingum GD, Pina P, Allouch P, Bloy C et al (1997) ATR-FTIR spectroscopic investigation of *E coli* transconjugants beta-lactams-resistance phenotype. FEBS Lett 412:39–42
29. Goodacre R, Timmins EM, Rooney PJ, Rowland JJ, Kell DB (1996) Rapid identification of *Streptococcus* and *Enterococcus* species using diffuse reflectance-absorbance Fourier transform infrared spectroscopy and artificial neural networks. FEMS Microbiol Lett 140:233–239
30. Helm D, Labischinski H, Naumann D (1991) Elaboration of a procedure for identification of bacteria using Fourier-transform IR spectral libraries – a stepwise correlation approach. J Microbiol Methods 14:127–142
31. Helm D, Labischinski H, Schallehn G, Naumann D (1991) Classification of bacteria by Fourier-transform infrared-spectroscopy. J Gen Microbiol 137:69–79

32. Naumann D, Fijala V, Labischinski H, Giesbrecht P (1988) The rapid differentiation and identification of pathogenic bacteria using Fourier-transform infrared spectroscopic and multivariate statistical-analysis. J Mole Struct 174:165–170
33. Naumann D, Helm D, Labischinski H (1991) Microbiological characterizations by FT-IR spectroscopy. Nature 351:81–82
34. Ferroni A, Sermet-Gaudelus I, Abachin E, Quesne G, Lenoir G et al (2002) Use of 16 S rRNA gene sequencing for identification of nonfermenting gram-negative bacilli recovered from patients attending a single cystic fibrosis center. J Clin Microbiol 40:3793–3797
35. Miller MB, Gilligan PH (2003) Laboratory aspects of management of chronic pulmonary infections in patients with cystic fibrosis. J Clin Microbiol 41:4009–4015
36. Fischer G, Braun S, Thissen R, Dott W (2006) FT-IR spectroscopy as a tool for rapid identification and intra-species characterization of airborne filamentous fungi. J Microbiol Methods 64:63–77
37. Sandt C, Sockalingum GD, Aubert D, Lepan H, Lepouse C et al (2003) Use of Fourier-transform infrared spectroscopy for typing of *Candida albicans* strains isolated in intensive care units. J Clin Microbiol 41:954–959
38. Erukhimovitch V, Karpasasa M, Huleihel M (2009) Spectroscopic detection and identification of infected cells with Herpes viruses. Biopolymers 91:61–67
39. Erukhimovitch V, Mukmanov I, Talyshinsky M, Souprun Y, Huleihel M (2004) The use of FTIR microscopy for evaluation of herpes viruses infection development kinetics. Spectrochimica Acta Part A-Mole Biomol Spectrosc 60:2355–2361
40. Hastings G, Krug P, Wang RL, Guo J, Lamichhane HP et al (2009) Viral infection of cells in culture detected using infrared microscopy. Analyst 134:1462–1471
41. Salman A, Erukhimovitch V, Talyshinsky M, Huleihil M, Huleihel M (2002) FTIR spectroscopic method for detection of cells infected with herpes viruses. Biopolymers 67:406–412
42. Hartman KA, Clayton N, Thomas GJ (1973) Studies of virus structure by Raman spectroscopy 1. R17 virus and R17 RNA. Biochem Biophys Res Commun 50:942–949
43. Dalterio RA, Baek M, Nelson WH, Britt D, Sperry JF et al (1987) The resonance Raman microprobe detection of single bacterial-cells from a chromobacterial mixture. Appl Spectrosc 41:241–244
44. Chadha S, Manoharan R, Moenneloccoz P, Nelson WH, Peticolas WL et al (1993) Comparison of the UV resonance Raman-spectra of bacteria, bacteria-cell walls, and ribosomes excited in the deep UV. Appl Spectrosc 47:38–43
45. Ghiamati E, Manoharan R, Nelson WH, Sperry JF (1992) UV resonance Raman-spectra of bacillus spores. Appl Spectrosc 46:357–364
46. Manoharan R, Ghiamati E, Dalterio RA, Britton KA, Nelson WH et al (1990) UV resonance Raman-spectra of bacteria, bacterial-spores, protoplasts and calcium dipicolinate. J Microbiol Methods 11:1–15
47. Edwards HGM, Russell NC, Weinstein R, Wynnwilliams DD (1995) Fourier-transform Raman-spectroscopic study of fungi. J Raman Spectrosc 26:911–916
48. Williams AC, Edwards HGM (1994) Fourier-transform Raman-spectroscopy of bacterial-cell walls. J Raman Spectrosc 25:673–677
49. Choo-Smith LP, Maquelin K, van Vreeswijk T, Bruining HA, Puppels GJ et al (2001) Investigating microbial (micro) colony heterogeneity by vibrational spectroscopy. Appl Environ Microbiol 67:1461–1469
50. Huang WE, Griffiths RI, Thompson IP, Bailey MJ, Whiteley AS (2004) Raman microscopic analysis of single microbial cells. Anal Chem 76:4452–4458
51. Jarvis RM, Brooker A, Goodacre R (2004) Surface-enhanced Raman spectroscopy for bacterial discrimination utilizing a scanning electron microscope with a Raman spectroscopy interface. Anal Chem 76:5198–5202
52. Kalasinsky KS, Hadfield T, Shea AA, Kalasinsky VF, Nelson MP et al (2007) Raman chemical imaging spectroscopy reagentless detection and identification of pathogens: signature development and evaluation. Anal Chem 79:2658–2673

53. Maquelin K, Choo-Smith LP, Endtz HP, Bruining HA, Puppels GJ (2002) Rapid identification of *Candida* species by confocal Raman micro spectroscopy. J Clin Microbiol 40:594–600
54. Maquelin K, Choo-Smith LP, van Vreeswijk T, Endtz HP, Smith B et al (2000) Raman spectroscopic method for identification of clinically relevant microorganisms growing on solid culture medium. Anal Chem 72:12–19
55. Maquelin K, Dijkshoorn L, van der Reijden TJK, Puppels GJ (2006) Rapid epidemiological analysis of *Acinetobacter* strains by Raman spectroscopy. J Microbiol Methods 64:126–131
56. Schaeberle MD, Morris HR, Turner JF, Treado PJ (1999) Raman chemical imaging spectroscopy. Anal Chem 71:175A–181A
57. Kneipp K, Wang Y, Kneipp H, Perelman LT, Itzkan I et al (1997) Single molecule detection using surface-enhanced Raman scattering (SERS). Phys Rev Lett 78:1667–1670
58. Nie SM, Emory SR (1997) Probing single molecules and single nanoparticles by surface-enhanced Raman scattering. Science 275:1102–1106
59. Alexander TA (2008) Development of methodology based on commercialized SERS-active substrates for rapid discrimination of Poxviridae virions. Anal Chem 80:2817–2825
60. Alexander TA (2008) Surface-enhanced Raman spectroscopy: a new approach to rapid identification of intact viruses. Spectroscopy 23:36–42
61. Bao PD, Huang TQ, Liu XM, Wu TQ (2001) Surface-enhanced Raman spectroscopy of insect nuclear polyhedrosis virus. J Raman Spectrosc 32:227–230
62. Driskell JD, Shanmukh S, Liu YJ, Hennigan S, Jones L et al (2008) Infectious agent detection with SERS-active silver nanorod arrays prepared by oblique angle deposition. IEEE Sens J 8:863–870
63. Driskell JD, Zhu Y, Kirkwood CD, Zhao YP, Dluhy RA et al (2010) Rapid and sensitive detection of rotavirus molecular signatures using surface enhanced Raman spectroscopy. Plos One 5(4):e10222
64. Goeller LJ, Riley MR (2007) Discrimination of bacteria and bacteriophages by Raman spectroscopy and surface-enhanced Raman spectroscopy. Appl Spectrosc 61:679–685
65. Grow AE, Wood LL, Claycomb JL, Thompson PA (2003) New biochip technology for label-free detection of pathogens and their toxins. J Microbiol Methods 53:221–233
66. Guicheteau J, Christesen SD (2006) Principal component analysis of bacteria using surface-enhanced Raman spectroscopy. Proc SPIE 6218:62180G
67. Jarvis RM, Brooker A, Goodacre R (2006) Surface-enhanced Raman scattering for the rapid discrimination of bacteria. Faraday Discuss 132:281–292
68. Jarvis RM, Goodacre R (2004) Discrimination of bacteria using surface-enhanced Raman spectroscopy. Anal Chem 76:40–47
69. Laucks ML, Sengupta A, Junge K, Davis EJ, Swanson BD (2005) Comparison of Psychro-active arctic marine Bacteria and common Mesophillic bacteria using surface-enhanced Raman spectroscopy. Appl Spectrosc 59:1222–1228
70. Patel IS, Premasiri WR, Moir DT, Ziegler LD (2008) Barcoding bacterial cells: a SERS-based methodology for pathogen identification. J Raman Spectrosc 39:1660–1672
71. Pearman WF, Fountain AW (2006) Classification of chemical and biological warfare agent simulants by surface-enhanced Raman spectroscopy and multivariate statistical techniques. Appl Spectrosc 60:356–365
72. Premasiri WR, Moir DT, Klempner MS, Krieger N, Jones G et al (2005) Characterization of the surface enhanced Raman scattering (SERS) of bacteria. J Phys Chem B 109:312–320
73. Premasiri WR, Moir DT, Lawrence DZ (2005) Vibrational fingerprinting of bacterial pathogens by surface enhanced Raman scattering (SERS). Proc SPIE 5795:19–29
74. Sengupta A, Laucks ML, Davis EJ (2005) Surface-enhanced Raman spectroscopy of bacteria and pollen. Appl Spectrosc 59:1016–1023
75. Shanmukh S, Jones L, Driskell J, Zhao Y, Dluhy R et al (2006) Rapid and sensitive detection of respiratory virus molecular signatures using a silver nanorod array SERS substrate. Nano Lett 6:2630–2636

76. Shanmukh S, Jones L, Zhao Y-P, Driskell JD, Tripp RA et al (2008) Identification and classification of respiratory syncytial virus (RSV) strains by surface-enhanced Raman spectroscopy and multivariate statistical techniques. Anal Bioanal Chem 390:1551–1555
77. Yan F, Vo-Dinh T (2007) Surface-enhanced Raman scattering detection of chemical and biological agents using a portable Raman integrated tunable sensor. Sens Actuat B-Chem 121:61–66
78. Zeiri L, Bronk BV, Shabtai Y, Czege J, Efrima S (2002) Silver metal induced surface enhanced Raman of bacteria. Colloids Surf A-Physicochem Eng Aspects 208:357–362
79. Jarvis RM, Goodacre R (2008) Characterisation and identification of bacteria using SERS. Chem Soc Rev 37:931–936
80. DeJesus MA, Giesfeldt KS, Oran JM, Abu-Hatab NA, Lavrik NV et al (2005) Nanofabrication of densely packed metal-polymer arrays for surface-enhanced Raman spectrometry. Appl Spectrosc 59:1501–1508
81. Kahl M, Voges E, Kostrewa S, Viets C, Hill W (1998) Periodically structured metallic substrates for SERS. Sens Actuat B-Chem 51:285–291
82. Haynes CL, Van Duyne RP (2001) Nanosphere lithography: a versatile nanofabrication tool for studies of size-dependent nanoparticle optics. J Phys Chem B 105:5599–5611
83. Hulteen JC, Treichel DA, Smith MT, Duval ML, Jensen TR et al (1999) Nanosphere lithography: size-tunable silver nanoparticle and surface cluster arrays. J Phys Chem B 103:3854–3863
84. Jensen TR, Malinsky MD, Haynes CL, Van Duyne RP (2000) Nanosphere lithography: tunable localized surface plasmon resonance spectra of silver nanoparticles. J Phys Chem B 104:10549–10556
85. Broglin BL, Andreu A, Dhussa N, Heath JA, Gerst J et al (2007) Investigation of the effects of the local environment on the surface-enhanced Raman spectra of striped gold/silver nanorod arrays. Langmuir 23:4563–4568
86. Lombardi I, Cavallotti PL, Carraro C, Maboudian R (2007) Template assisted deposition of Ag nanoparticle arrays for surface-enhanced Raman scattering applications. Sensor Actuat B-Chem 125:353–356
87. Ruan CM, Eres G, Wang W, Zhang ZY, Gu BH (2007) Controlled fabrication of nanopillar arrays as active substrates for surface-enhanced Raman spectroscopy. Langmuir 23:5757–5760
88. Yao JL, Pan GP, Xue KH, Wu DY, Ren B et al (2000) A complementary study of surface-enhanced Raman scattering and metal nanorod arrays. Pure Appl Chem 72:221–228
89. Chaney SB, Shanmukh S, Zhao Y-P, Dluhy RA (2005) Randomly aligned silver nanorod arrays produce high sensitivity SERS substrates. Appl Phys Lett 87:31908–31910
90. Driskell JD, Shanmukh S, Liu Y, Chaney SB, Tang XJ et al (2008) The use of aligned silver nanorod arrays prepared by oblique angle deposition as surface enhanced Raman scattering substrates. J Phys Chem C 112:895–901
91. Liu YJ, Fan JG, Zhao YP, Shanmukh S, Dluhy RA (2006) Angle dependent surface enhanced Raman scattering obtained from a Ag nanorod array substrate. Appl Phys Lett 89:173134
92. Premasiri WR, Moir DT, Klempner MS, Ziegler LD (2007) Surface-enhanced Raman scattering of microorganisms. New Approaches Biomed Spectrosc 963:164–185

Books and Reviews

Carter EA, Marshall CP, Ali MHM, Ganendren R, Sorrell TC, Wright L, Lee Y-C, Chen C-I, Lay PA (2007) Infrared spectroscopy of microorganisms: characterization, identification, and differentiation. In: Kneipp K, Aroca R, Kneipp H, Wentrup-Byrne E (eds) New approaches in biomedical spectroscopy. American Chemical Society, Washington, DC, pp 64–84

Huang WE, Li M, Jarvis RM, Goodacre R, Banwart SA (2010) Shining light on the microbial world: the application of Raman microspectroscopy. In: Laskin A, Sariaslani S, Gadd GM (eds) Advances in applied microbiology, vol 70. Elsevier, San Diego, pp 153–186

Ince J, McNally A (2009) Development of rapid, automated diagnostics for infectious disease: advances and challenges. Expert Rev Med Devices 6(6):641–651

Posthuma-Trumpie G, Korf J, van Amerongen A (2009) Lateral flow (immune)assay: its strengths, weakness, opportunities and threats. A literature survey. Anal Bioanaly Chem 393:569–582

Premasiri WR, Moir DT, Klempner MS, Ziegler LD (2007) Surface-enhanced Raman scattering of microorganisms. In: Kneipp K, Aroca R, Kneipp H, Wentrup-Byrne E (eds) New approaches in biomedical spectroscopy. American Chemical Society, Washington, DC, pp 164–199

Quan P-L, Briese T, Palacios G, Lipkin WI (2008) Rapid sequence-based diagnosis of viral infection. Antiviral Res 79:1–5

Sharaf MA, Illman DL, Kowalski BR (1986) Chemometrics. Wiley, New York

Tuma R, Thomas GJ Jr (2002) Raman spectroscopy of viruses. In: Chalmers JM, Griffiths PR (eds) Handbook of vibrational spectroscopy applications in life, pharmaceutical and natural sciences, vol 5. Wiley, West Sussex, pp 3519–3535

Chapter 8
Malaria Vaccines

Christopher V. Plowe

Glossary

Adjuvant	A substance added to a vaccine to stimulate a stronger or more effective immune response.
Blood stage	The stage of the malaria parasite life cycle responsible for clinical symptoms. Vaccines that target the blood stage are intended to prevent disease and death, but they do not prevent infection and may not affect malaria transmission.
Challenge trial	Small experimental Phase 1/2 clinical trial in which healthy volunteers receive a malaria vaccine and are exposed to the bites of malaria-infected mosquitoes or injected with malaria parasites under carefully controlled conditions.
Immunogenicity	The ability of a vaccine to produce specific immune responses (usually antibodies) that recognize the vaccine antigen.
Pre-erythrocytic	Stages of the malaria parasite that are injected by a mosquito and develop in the liver before emerging into the blood where they can cause symptoms. Vaccines targeting pre-erythrocytic stages are intended

This chapter, which has been modified slightly for the purposes of this volume, was originally published as part of the Encyclopedia of Sustainability Science and Technology edited by Robert A. Meyers. DOI:10.1007/978-1-4419-0851-3

C.V. Plowe (✉)
Howard Hughes Medical Institute/Center for Vaccine Development, University of Maryland School of Medicine, 685 West Baltimore Street, Baltimore, MD 21201, USA
e-mail: cplowe@medicine.umaryland.edu

P. Kanki and D.J. Grimes (eds.), *Infectious Diseases: Selected Entries from the Encyclopedia of Sustainability Science and Technology*,
DOI 10.1007/978-1-4614-5719-0_8, © Springer Science+Business Media New York 2013

	to prevent infection altogether and, if highly effective, would also prevent disease and block transmission.
Sexual stage	The male and female forms of malaria parasites that are responsible for transmission through mosquitoes. Vaccines directed against sexual stages are intended to prevent malaria transmission.
Subunit vaccine	A vaccine based on a small portion of the organism, usually a peptide or protein.
Vaccine resistance	The ability of malaria parasites to escape strain-specific immune responses by exploiting genetic diversity to increase the frequencies of non-vaccine-type variants in a population or to evolve new diverse forms.
Whole-organism vaccine	A vaccine based on an attenuated or killed whole parasite.

Definition of the Subject

Vaccines are the most powerful public health tools mankind has created, and research toward malaria vaccines began not long after the parasite responsible for this global killer was discovered and its life cycle described more than 100 years ago. But parasites are bigger, more complicated, and wilier than the viruses and bacteria that have been conquered or controlled with vaccines, and a malaria vaccine has remained elusive. High levels of protective efficacy were achieved in crude early experiments in animals and humans using weakened whole parasites, but the results of more sophisticated modern approaches using molecular techniques have ranged from modest success to abject failure. A subunit recombinant protein vaccine that affords in the neighborhood of 25–50% protective efficacy against malaria is in the late stages of clinical evaluation in Africa. Incremental improvements on this successful vaccine are possible and worth pursuing, but the best hope for a malaria vaccine that would improve prospects for malaria eradication may lie with the use of attenuated whole parasites and powerful immune-boosting adjuvants.

Introduction

The malaria parasite is thought to have killed more human beings throughout history than any other single cause [1]. Today, along with AIDS and tuberculosis, malaria remains one of the "big three" infectious diseases, every year exacting a heavy toll on human life and health in parts of Central and South America, large regions of Asia, and throughout most of sub-Saharan Africa, where up to 90% of malaria deaths occur [2].

In the middle of the twentieth century, the availability of the long-acting insecticide dichlorodiphenyltrichloroethane (DDT) and the safe and effective anti-malarial drug chloroquine provided the basis for optimism that global eradication was possible. Although malaria was eliminated in several countries around the margins of the malaria map, factors including the emergence of DDT-resistant mosquitoes and chloroquine-resistant parasites, as well as waning economic and political support, led the campaign to stall by the late 1960s, resulting in the rapid resurgence to previous disease levels in many locations [3].

For the next 30 years, the spread of drug-resistant malaria and donor fatigue contributed to an overall lack of progress against the disease. But starting in the late 1990s, new tools, including long-lasting insecticide-impregnated nets and highly efficacious combination drug therapies, led to a wave of successes, including dramatic reductions in disease burden in some areas and complete elimination of malaria in others [4]. These success stories have stimulated a renewed sense of optimism about prospects for global eradication [5].

If it is to succeed, this nascent drive toward country-by-country elimination and possible eventual worldwide eradication of malaria will require powerful new tools, importantly including malaria vaccines that produce protective immune responses that surpass those acquired through natural exposure to malaria [6]. Successful global or regional campaigns to eradicate smallpox, polio, and measles have all relied on vaccines. Those for yellow fever, hookworm, and yaws – and for malaria – which have relied on non-vaccine measures such as vector control or drug treatment have all failed [7]. While the odds of successful global malaria eradication would be very long even with an ideal malaria vaccine, they are virtually nil without one. In the meantime, even a modestly effective vaccine could substantially reduce the continuing heavy burden of malaria-attributable disease (247 million annual cases) and death (881,000 annual deaths) [2].

After nearly a century of malaria vaccine research, today, one modestly effective vaccine based on a parasite surface protein is being tested in a large Phase 3 trial in hopes of licensure within a few years [8]. Many other vaccine candidates have fallen short and been abandoned before reaching this stage, although several are in early stages of clinical development. The chief reasons that it has taken this long to get this far are that malaria parasites replicate and propagate through an extremely complex life cycle involving vertebrate hosts and an insect vector, and that they have evolved a repertoire of mechanisms for evading both natural and vaccine-induced immunity.

This review focuses on key themes that have emerged over 75 years of malaria vaccine research and development and on a few key examples of malaria vaccines that have reached the stage of testing for efficacy in clinical trials in humans. Several recent review articles listed after the primary bibliography explore malaria immunology and preclinical vaccine development in more detail.

The Malaria Life Cycle

Malaria is a potentially fatal parasitic disease transmitted to humans and other vertebrate animals by mosquitoes. Four species of *Plasmodium* cause malaria disease primarily in humans: *P. falciparum, P. vivax, P. ovale* and *P. malariae*. A fifth, *P. knowlesi*, infects mainly nonhuman primates but was recently found also to infect and sicken humans [9], and hundreds of other malaria species infect other mammals, reptiles, and birds. Because it is responsible for most severe malaria and deaths, *P. falciparum* has been the target of most vaccine development efforts and is the main focus of this review. However, in many parts of the world, *P. vivax* is the predominant species, and it causes more severe malaria than has sometimes been appreciated [10]. A vivax malaria vaccine would be highly beneficial, especially if it interrupted transmission.

The malaria life cycle begins when the female *Anopheles* mosquito injects sporozoites from her salivary gland into the skin as she takes a blood meal (Fig. 8.1). The worm-like sporozoites – about 7 μ in length, or as long as an erythrocyte is wide – invade liver hepatocytes, each sporozoite multiplying over several days into tens of thousands of tiny (about 1 μ in diameter) merozoites packed into a single infected hepatocyte. These pre-erythrocytic stages cause no clinical signs or symptoms. A highly efficacious pre-erythrocytic vaccine would thus completely block infection, preventing parasites from reaching the blood and causing disease, and also preventing transmission.

Rupturing hepatocytes release showers of merozoites into the circulation, initiating the blood stage of malaria infection that is responsible for disease. Merozoites quickly invade erythrocytes and undergo asexual multiplication, dividing, growing, bursting from the erythrocyte, and re-invading in a periodic pattern with each cycle lasting 2 days (3 days in the case of *P. malariae*), until interrupted by host immunity, drug treatment, or death. Malaria vaccines that target the blood stage are thought of as anti-disease vaccines and would be expected to prevent or reduce clinical illness but would not prevent infection.

Some blood-stage parasites develop into male and female gametocytes. These sexual forms are taken up by the mosquito vector during a blood meal, mate to form a brief diploid stage, and then develop through further haploid stages and migrate from the gut to the salivary glands (Fig. 8.1). Each mating pair of gametocytes yields up to 1,000 infectious sporozoites, which are injected into the host to complete the transmission cycle. Vaccines targeting the sexual stages would prevent neither infection nor disease in the vaccinated individual and are thought of as transmission-blocking vaccines. Highly efficacious pre-erythrocytic or blood-stage vaccines that prevent sexual reproduction would also block transmission, so the term "transmission-blocking" need not refer exclusively to vaccines against sexual or mosquito stages of the parasite. The term "vaccines that interrupt transmission" (VIMT) encompasses all vaccines that interrupt transmission, whatever stage they target [11].

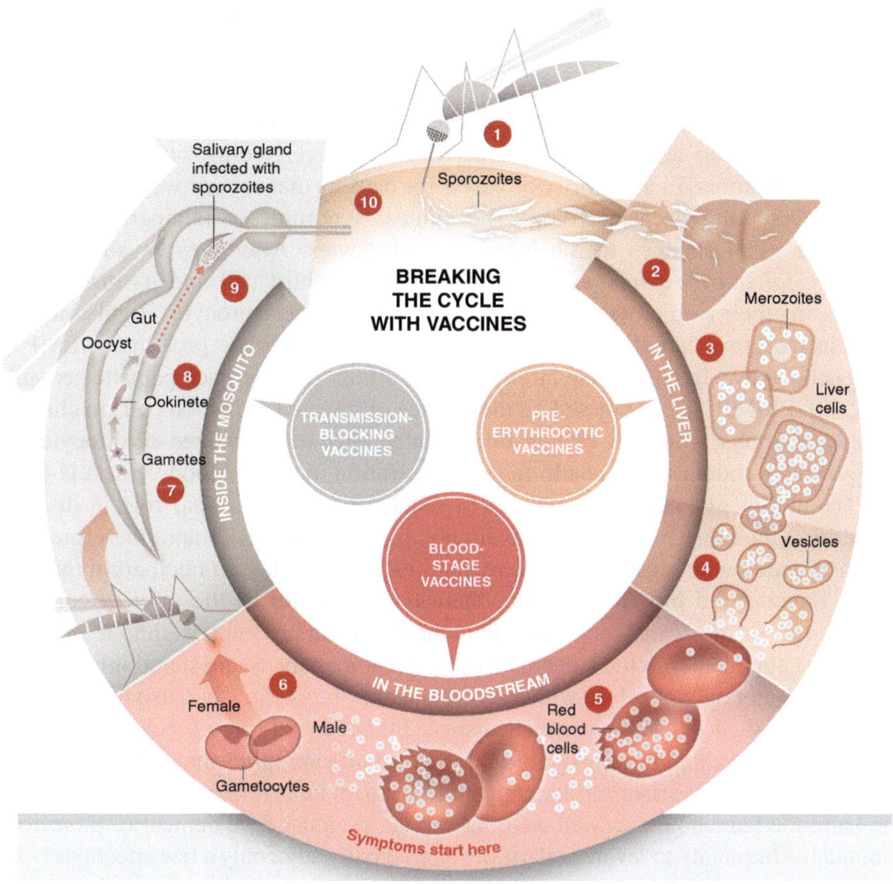

Fig. 8.1 *Life cycle of malaria and stages targeted by vaccines.* (Source: PATH – Malaria Vaccine Initiative). 1. Malaria infection begins when an infected female *Anopheles* mosquito bites a person, injecting *Plasmodium* parasites, in the form of sporozoites, into the bloodstream. 2. The sporozoites pass quickly into the human liver. 3. The sporozoites multiply asexually in the liver cells over the next 7–10 days, causing no symptoms. 4. The parasites, in the form of merozoites, burst from the liver cells. 5. In the bloodstream, the merozoites invade red blood cells (erythrocytes) and multiply again until the cells burst. Then, they invade more erythrocytes. This cycle is repeated, causing fever each time parasites break free and invade blood cells. 6. Some of the infected blood cells leave the cycle of asexual multiplication. Instead of replicating, the merozoites in these cells develop into sexual forms of the parasite, called gametocytes, that circulate in the bloodstream. 7. When a mosquito bites an infected human, it ingests the gametocytes, which develop further into mature sex cells called gametes. 8. The gametes develop into actively moving ookinetes that burrow into the mosquito's midgut wall and form oocysts. 9. Inside the oocyst, thousands of active sporozoites develop. The oocyst eventually bursts, releasing sporozoites that travel to the mosquito salivary glands. 10. The cycle of human infection begins again when the mosquito bites another person

Pathogenesis and Disease

While semi-immune individuals can be chronically infected with malaria and experience no symptoms of illness, malaria infection of a nonimmune person usually causes an acute illness characterized by fever, chills, aches, and other flu-like symptoms. In a minority of cases, for reasons that are not well understood, more severe illness can develop. The clinical syndrome of falciparum malaria originates with changes in the infected erythrocytes. After they invade, blood-stage *P. falciparum* parasites effectively hijack the host cell and its machinery, expressing their own proteins on the surface of the host erythrocyte. Highly variant protein receptors called *P. falciparum* erythrocyte membrane proteins (PfEMP1) are encoded by a large, diverse family of up to 60 *var* genes in each parasite genome [12, 13]. PfEMP1 are expressed on the surface of infected red blood cells in clumps known as knobs, which are responsible for adherence of parasitized erythrocytes to the vascular endothelium, resulting in sequestration in tissue blood vessels [14].

The ability of falciparum malaria to sequester plays a critical role in disease severity – the other human malarias do not appear to sequester and, therefore, are not associated with most of the severe manifestations seen with falciparum malaria [15]. Infected red blood cells cytoadhere and sequester in the microcirculatory compartments of organs, most notably in the brain and placenta, leading to disease, and most abundantly in the spleen, causing splenomegaly. Sequestered infected red blood cells not only interfere with microcirculatory blood flow but also hide outside the reach of host defense mechanisms. Infected red blood cells lose their deformability and compromise blood flow in small capillaries and venules [16].

The presence of variant surface antigens leads to immune responses that appear both to harm the human host as well as to lead to the eventual development of protective immunity. Immunity to severe malaria develops rapidly, after only a few infections [17], possibly due to antibody responses that protect against a relatively conserved subset of PfEMP1 variants that are associated with severe malaria. In contrast, the slow acquisition of immune protection against uncomplicated malaria over years of repeated exposure to malaria is thought to represent the accumulation of protective immune responses to a repertoire of diverse antigens, probably including both PfEMP1 and the surface proteins that are the targets of most vaccine candidates [18].

Epidemiology

The epidemiology of malaria is determined primarily by the patterns and intensity of malaria transmission, which in turn drives the prevalence of malaria infection and the incidence of different forms of malaria disease. In low-transmission settings with unstable malaria, there is a potential for epidemic disease when transmission recurs or increases as a result of reintroduction to a population not recently exposed to malaria or to changing climactic or environmental conditions that favor contact between humans and malaria-transmitting *Anopheles* mosquitoes. Outbreaks can occur when malaria-naïve populations such as transmigrants, miners, or soldiers are exposed to malaria, causing high rates of disease [19].

Depending largely on the degree of host immunity, the manifestations of malaria infection can range from completely asymptomatic parasitemia, to mild disease that can be treated on an outpatient basis with oral drugs, to acute catastrophic life-threatening illness requiring intensive care. Very young infants are thought to be protected from malaria disease by maternal antibodies and persistent hemoglobin F. While they may be infected congenitally or by mosquito bites in the first days or weeks of life, infants do not experience clinical disease until they are a few months old. Following this brief period of relative insusceptibility in early infancy, protective immunity against malaria disease is acquired through repeated exposure and is therefore related to transmission intensity.

Where malaria transmission is moderate (average of one or more infected mosquito bites per month) or high (two or more infected mosquito bites per week; up to more than one per day in some areas), the risk period for death from malaria is highest in infants and young children who are in the process of developing acquired immunity. In a typical moderate- or high-transmission setting in sub-Saharan Africa, most severe malaria is experienced by children aged less than 5 years; children aged up to 10–12 years experience frequent episodes of uncomplicated malaria; and older teenagers and adults, while still often infected, rarely experience symptoms of malaria illness. Severe anemia is more frequent in the youngest infants, while cerebral malaria tends to peak in children aged 3–4 years who have experienced previous malaria episodes, suggesting that an overly exuberant immune response contributes to the pathogenesis of cerebral malaria.

In contrast, in low-transmission settings, persons of all ages have a similar risk of infection and uncomplicated malaria, most who are infected become sick, and the risk for severe malaria persists throughout life. Semi-immune adults, although they remain susceptible to asymptomatic parasitemia, are protected against clinical malaria disease, rarely becoming ill even when persistently infected. This protective immunity is lost after a few years in the absence of exposure. Acquired immunity is also diminished in pregnancy, in that women pregnant with their first child are susceptible to severe *P. falciparum* disease from placental malaria because they lack immunity to placenta-specific cytoadherence proteins. As placental immunity develops in subsequent pregnancies, there is a reduced risk of adverse effects of malaria in pregnancy on the mother and fetus [20].

Based on these epidemiological patterns, the primary populations targeted for malaria vaccines are infants and young children in areas of moderate and high transmission who bear the greatest burden of disease and death, and women of childbearing age in these same areas. Malaria-naïve travelers and military troops would also benefit from a malaria vaccine. As more countries move toward malaria elimination and global eradication is considered [21], the general population of malaria-endemic areas may be vaccinated to drive down transmission [6, 11].

Immunity

Insights into malaria immunity come not only from studies in various animal models including birds, rodents, and nonhuman primates but also from important studies in humans, including classic passive transfer experiments [22, 23] and early studies of malaria therapy for neurosyphilis [24, 25], as well as immuno-epidemiological studies that try to identify correlates of clinical protection [26, 27]. Both humoral and cellular factors contribute to acquired immune protection against malaria. Broadly speaking, cellular immune responses are thought to be more important in controlling the pre-erythrocytic stages of malaria infection [28], and antibodies are thought to block erythrocyte invasion to suppress blood-stage infection [22, 26]. For these reasons, cellular immune responses are typically emphasized in the development of pre-erythrocytic vaccines and antibody responses in the development of blood-stage vaccines. However, despite nearly 100 years of human and animal research, the basis of protective immunity against malaria is poorly understood, and no specific immune response has been established as an essential correlate of clinical protection, complicating malaria vaccine development.

New genomic tools have the potential to improve understanding of malaria immunity and may aid in vaccine development. For example, while it has long been known that there is some degree of strain specificity to malaria immunity, the discovery of large families of genes encoding highly variable surface antigens that mediate cytoadherence and immune evasion [14] has led to models for explaining the slow acquisition of protective immunity as a process of building up a repertoire of variant-specific immune responses until protection is in place against the full range of locally prevalent variants. As next-generation sequencing technologies improve their ability to generate sequence from clinical samples and to assemble genomes and map variant sequences to a reference genome, genomic epidemiology studies that relate parasite genotypes to clinical risk and allele-specific immune responses will permit testing of the hypotheses generated by such models. In another new approach, high-density protein arrays permit serological profiling of large numbers of serum samples against thousands of recombinant malaria proteins [29]. When this protein array was used to identify *P. falciparum* proteins that were differentially recognized by the sera of children who were resistant to clinical malaria, several previously unknown antigens were identified as possibly being important in acquired immunity, providing possible new vaccine targets [30].

In addition to acquired immunity, several host genetic factors offer some degree of protection against malaria, generally not by preventing infection but by reducing the risk of clinical illness or severe disease. Sickle cell trait [31] and other hemoglobinopathies [32–34] are more prevalent in populations at risk of malaria because they afford protection against clinical malaria. Various other human genetic polymorphisms associated with the host immune response [35] and with host–parasite binding [36] have also been correlated with susceptibility to clinical malaria in genetic association studies.

Early Malaria Vaccines

In 1880, Charles Louis Alphonse Laveran, a 33-year-old French Army doctor working in Algeria, discovered motile worm-like parasites, later understood to be exflagellating male gametes, in the blood of a feverish soldier [37]. Seven years later, Ronald Ross established that malaria parasites were transmitted to birds by the bites of infected mosquitoes [38]. The tremendous public health benefit that would be provided by an effective malaria vaccine was quickly appreciated, and from the 1930s to the 1970s, malaria vaccine researchers used primarily birds (including ducklings, canaries, chickens, and turkeys) and occasionally monkeys as model systems, and inactivated or killed whole parasites or parasite extracts as vaccines, often accompanied by immune-boosting adjuvant systems. This wave of vaccine development research crested in the late 1940s as World War II ended and attention shifted to the global campaign to eradicate malaria using drugs and anti-vector methods, and only resurged in the late 1960s when it became apparent that eradication was not possible with existing tools.

Early work focused on attenuated whole-parasite vaccines. Working in India, Russell and Mohan protected chickens from mosquito challenge with *P. gallinaceum* by immunizing them with sporozoites inactivated by ultraviolet light [39]. In 1945, the Hungarian-American immunologist Jules Freund (of Freund's complete adjuvant fame) and colleagues reported that they had successfully protected ducks against intravenous challenge with the avian malaria *P. lophurae* by immunizing them with formalin-inactivated malaria-infected blood cells and an adjuvant system consisting of a lanolin-like substance, paraffin oil, and killed tubercle bacilli [40]. They used a similar vaccine formulation to protect rhesus monkeys against *P. knowlesi* challenge [41]. These pioneering studies demonstrated two important principles that remain highly relevant to contemporary malaria vaccine development efforts, namely that good protective efficacy can be achieved with whole-organism vaccines and that strong immune-boosting adjuvants can achieve levels of protection that match or exceed those acquired through repeated natural exposure.

In a thoughtful and prescient paper published in 1943 describing protection of ducklings when a bacterial toxin adjuvant was added to a killed blood-stage vaccine, Henry Jacobs briefly cited an abstract presented by W. B. Redmond at the 1939 annual meeting of the American Society of Parasitologists, writing that "Redmond. . .noted some protection against bird malaria when he vaccinated with irradiated parasites" [42]. Redmond's abstract described using a frozen killed vaccine but made no mention of inactivation by irradiation [43]. Unfortunately, Redmond never published this work, but one can speculate that he may have described using some form of a radiation-attenuated *P. lophurae* vaccine preparation in his presentation at the 1939 meeting, anticipating subsequent work using this approach, including attenuation by irradiation of both blood stages [44, 45] and later and more famously of sporozoites [46, 47].

Columbia University scientists, who were aware of "inconclusive" earlier studies reported in the German medical literature, described in 1946 their own unsuccessful attempts to protect humans against intravenous challenge with *P. vivax* using a vaccine consisting of formalin-treated and freeze-thawed blood containing 65–150 million blood-stage *P. vivax* parasites [48]. No adjuvant was used in these earliest human studies, which may partially explain the disappointing results.

Several of these historical threads came together in a series of major breakthroughs in the late 1960s and early 1970s. First, Ruth Nussenzweig and Jerome Vanderberg at New York University reported in 1967 that intravenous immunization with irradiated *P. berghei* sporozoites could protect mice against subsequent intravenous challenge with viable sporozoites [46]. This advance was quickly translated into human trials of radiation-attenuated *P. falciparum* sporozoites delivered by the bites of infected, irradiated mosquitoes [49, 50]. In these and subsequent malaria challenge trials, 90% of volunteers who were immunized with radiation-attenuated sporozoites by receiving at least 1,000 infected bites over several sessions were fully protected against infection [51]. All unvaccinated volunteers acquired malaria from the bites of non-irradiated mosquitoes. These pioneering studies, first done by University of Maryland investigators in prisoners [52], provided proof that humans could be protected against infection with deadly *P. falciparum* through immunization, and spurred the identification of specific sporozoite proteins that could serve as antigens for subunit vaccines, as described in the following sections.

At around this time, two major advances ushered in the modern era of malaria vaccine development: the advent of molecular biology and the ability to clone, sequence, and express parasite genes in heterologous expression systems such as bacteria and yeast, and the development of methods for growing *P. falciparum* parasites in continuous in vitro culture [53, 54], providing a reliable and reproducible source of malaria parasites, proteins, and genes.

Obstacles to Malaria Vaccines

Why is there still no licensed malaria vaccine after 75 years of vaccine development research and evaluation of more than 70 vaccine candidates [55] in preclinical and clinical testing? Obstacles slowing progress toward an effective malaria vaccine include the size and complexity of the parasite, its genetic diversity, the efficiency of its amplification, the incomplete and temporary nature of naturally acquired immunity, and the fact that in addition to providing protection, immune responses also contribute to pathogenesis. Furthermore, parasite material must generally be obtained from infected hosts or mosquitoes – only one species, *P. falciparum*, can be grown in continuous culture. Finally, validated immune correlates of protection are lacking, so candidate vaccines can only be down-selected by conducting costly efficacy trials in humans.

The *P. falciparum* genome has about 23 million bases of DNA organized into 14 chromosomes and about 5,000 genes [56]. This is orders of magnitude larger than the genomes of the viruses and bacteria to which vaccines have been successfully developed. This complexity and the large number of gene products provide the means and materials needed for the highly complex life cycle stages in vertebrate hosts and mosquitoes (Fig. 8.1). Moreover, mutation during mitotic reproduction in the haploid liver and blood stages and genetic recombination during the diploid sexual reproductive stages in the mosquito result in extensive genetic diversity that is driven by selection pressure from the immune system, as well as by drugs and, when they are deployed, potentially by vaccines [57]. All of this complexity and diversity greatly complicates the choice of candidate antigens for vaccine development.

At least 18 different forms of one leading blood-stage antigen [58] and more than 200 variants of another [59] have been documented in a single African village. If vaccines targeting these antigens generate immune responses that are insufficiently cross-protective, vaccines based on just one or two genetic variants are unlikely to be broadly efficacious [57]. To date, the choices of which variants of target antigens to include in malaria vaccines have not been made in consideration of the frequencies of these variants in natural populations. Careful molecular epidemiological studies are beginning to pinpoint which of the many polymorphisms in some of these antigens are the most important determinants of strain-specific natural immunity [58, 59], and this approach may help inform the design of polyvalent or chimeric vaccines that protect against diverse parasite strains [57].

Immunization with whole parasites of a given life cycle stage or with stage-specific proteins typically protects against only that life cycle stage, hence the notion of vaccines that prevent infection, disease, or transmission by targeting the different stages. While a highly efficacious pre-erythrocytic vaccine would prevent not only infection, but by doing so also prevent disease as well as transmission [6], even a single surviving sporozoite could theoretically result in a full-blown infection, severe disease, and transmission. This is because of the parasite's ability to multiply rapidly – one sporozoite gives rise to tens of thousands of merozoites emerging from the liver about a week and a half after a mosquito bite, and each merozoite multiplies roughly tenfold during the 48-h blood cycle, quickly resulting in billions of parasites circulating in the body. In reality, the rate at which parasites amplify their numbers is determined not just by the parasite's maximum reproductive capacity but also by host defenses and other factors. The ability of a partially efficacious pre-erythrocytic vaccine to reduce the risk of clinical malaria illness [60] supports the idea that there is some benefit from slowing the rate of parasite reproduction short of complete prevention of blood-stage infection – a so-called leaky vaccine, perhaps better thought of as an injectable bednet.

Most successful vaccines prevent infection or illness with pathogens that naturally result in strong and long-lasting immune protection after a single exposure. As described above, the naturally acquired protective immunity to malaria is hard won and short lived. An effective malaria vaccine would need to produce stronger immune responses more quickly than those that develop even under intense continuous natural exposure to malaria, and a vaccine intended to prevent infection will need to surpass natural immunity, which gradually protects against clinical illness but does not completely prevent infection.

Because the host immune response contributes to malaria pathogenesis, a vaccine could theoretically increase the risk of harmful inflammatory responses to subsequent infection, especially for vaccines directed against the blood stages that are responsible for pathology. One blood-stage malaria vaccine based on the *P. falciparum* merozoite surface protein 1 (MSP1) protected monkeys against lethal infection but then resulted in life-threatening anemia following subsequent exposure to malaria [61]. No such post-vaccination anemia has been observed in malaria-exposed adults [62, 63] or children [64, 65] in African trials of a different MSP1 malaria vaccine. A Phase 2 trial of a blood-stage malaria vaccine based on a different antigen, the apical membrane antigen 1 (AMA1), reported a possible increased risk of anemia in vaccinated malaria-exposed children in an unplanned post-hoc analysis [66]. No increased risk of anemia has been seen in trials with a more highly immunogenic AMA1 vaccine tested in similar populations [67, 68]. Vigilance for untoward inflammatory responses to malaria vaccines or to post-vaccination malaria infection will continue to be an important aspect of clinical malaria vaccine development, especially for blood-stage vaccines.

Pre-erythrocytic Vaccines

The vaccine furthest along in clinical development, RTS,S/AS01, targets the pre-erythrocytic circumsporozoite protein (CSP) of *P. falciparum*. The gene encoding CSP was the first *P. falciparum* gene cloned [69], and as the major surface protein coating sporozoites, CSP was immediately of great interest as a vaccine candidate. Thought to be important in sporozoite development and motility [70], CSP contains a central repeat region that elicits antibody responses [71], flanked on each side by non-repetitive regions containing T-cell epitopes [69]. Antibodies directed against the central repeat region cause the protein coat to slough off and block invasion of hepatocytes, suggesting that vaccine-induced antibodies might prevent infection [72]. Early synthetic CSP vaccines based on the repeat region using aluminum hydroxide as an adjuvant were poorly immunogenic, although a few individuals who achieved high antibody titers were protected against experimental challenge with homologous sporozoites [73, 74].

Seroepidemiological studies failed to find an association between anti-CSP antibodies and protection against infection [75], however, and the addition of adjuvants that produced higher antibody levels did not result in improved efficacy [76], leading the developers of RTS,S to include the flanking regions containing T-cell epitopes in subsequent versions of the vaccine. The final form of RTS,S is comprised of the central repeat region (R) and T-cell epitopes (T) using the hepatitis B surface antigen (S) as a carrier matrix, and co-expressed in *Saccharomyces cerevisiae* with additional S, hence "RTS,S." The clinical development of RTS,S has included progressive improvements in adjuvant systems, resulting in improved efficacy both in experimental challenges [77–79] and in clinical trials in malaria-exposed adults [80] and children [60, 81]. The current formulation includes the liposomal-based Adjuvant System AS01, which contains the immunostimulants monophosphoryl lipid A and QS21, a saponin derivative extracted from the bark of the South American soap bark tree *Quillaja saponaria*. The development of RTS,S supports the notion that strong adjuvants are a necessary component for efficacious subunit protein malaria vaccines.

RTS,S/AS01 (and its immediate forebears with oil-in-water versions of the AS0-adjuvant system) was the first malaria vaccine to demonstrate meaningful levels of clinical protection in field trials. Trials in children and infants who are naturally exposed to malaria have demonstrated efficacy against clinical disease in the range of 30–56% and up to 66% against infection, and a good record of safety and tolerability [60, 81, 82]. A large Phase 2 trial in 1,465 Mozambican children showed 26% efficacy against all malaria episodes over nearly 4 years, 32% against first or only episode of clinical malaria, and 38% in preventing severe clinical episodes [83]. After 45 months, the prevalence of parasitemia was significantly lower in vaccines compared to the control group (12% vs 19%). The magnitude of the protective effect after nearly 4 years was thus modest, but importantly, there was no evidence of a post-immunization "rebound" effect – a theoretical concern that the vaccine might interfere with the natural acquisition of protective immunity.

Based on these demonstrations of a level of efficacy that is well below levels of protection expected for vaccines against other common pathogens but rare good news for malaria vaccines, RTS,S/AS01 is currently being evaluated in a large Phase 3 trial of 16,000 children and infants in seven African countries. The cost-effectiveness of licensing and deploying a malaria vaccine with efficacy in the range of 25–50% is debated, but where the malaria burden remains high, as in much of sub-Saharan Africa, such a vaccine is likely to be sought and used. Strategies being investigated to improve on the efficacy of RTS,S/AS01 include adding antigens tested with the same adjuvant system to create a multistage, multi-antigen RTS,S-based vaccine [84] and priming with an adenovirus expressing CSP before boosting with RTS,S/AS01 [85].

Several other pre-erythrocytic vaccine candidates have progressed through various stages of preclinical and early clinical development [55], including recombinant subunit protein vaccines as well as DNA vaccines and viral vectored vaccines [81]. Even though some of these have generated seemingly good humoral and especially cellular immune responses when formulated with strong adjuvants [86], protective efficacy has not been achieved in clinical testing in humans. Prime-boost approaches using DNA vaccines or viral vectors have also resulted in improved immunogenicity including cell-mediated responses in some participants and, in some cases, in measurable delays in time to infection in experimental sporozoite challenge trials [87]. Although prime-boost vaccine strategies have failed to demonstrate meaningful protection in published clinical efficacy trials, recent unpublished reports are more promising, with about 25% sterile protection provided by DNA prime and viral vector boost using both CSP and the blood-stage antigen AMA1 (T. Richie, personal communication). Efforts to improve these approaches to get more consistent cellular immune responses and higher levels of protection may be hampered by variability in host responses to vaccination. Other novel approaches such as a self-assembling polypeptide nanoparticle CSP vaccine are showing promise in preclinical testing [88]. Mining of genomic and proteomic data has led to the identification of new pre-erythrocytic vaccine candidate proteins, some of which have shown promising results in early preclinical testing in animal models [89].

Blood-Stage Vaccines

Most blood-stage malaria vaccine candidates are based on antigens that coat the surface of the invasive merozoites and/or that are involved with the process of erythrocyte invasion, in hopes of generating antibodies that block invasion and curtail parasite replication in the blood, reducing the risk or severity of clinical illness. The merozoite surface protein 1, or MSP1, was the first and best characterized of many proteins on the merozoite surface that are being targeted for vaccine development. MSP1 undergoes cleavage into four fragments that remain on the merozoite surface as a complex. Before erythrocyte invasion, the entire MSP1 complex is shed except for the C-terminal 19-kDa fragment ($MSP1_{19}$), which remains on the surface as the merozoite enters the erythrocyte [90]. Naturally acquired antibodies to $MSP1_{19}$ inhibit erythrocyte invasion and are associated with protection from clinical malaria in field studies [91, 92], supporting its potential as a vaccine candidate. Studies of recombinant MSP1 vaccines in monkeys were encouraging [93]. An $MSP1_{19}$-based vaccine on the same adjuvant platform as RTS,S produced antibodies in Malian adults that recognized MSP1 from diverse strains of *P. falciparum* [62], but had no protective efficacy against clinical malaria in Kenyan children [65]. Comparison of the degree of homology with the vaccine strain of MSP1 sequences in the infections experienced by children in the vaccine and control groups will clarify the extent to which the genetic diversity of MSP1 accounted for this lack of efficacy.

The apical membrane antigen 1, or AMA1, resides in the apical complex of the merozoite [94] before being processed and moving to the surface as the merozoite is released from the infected erythrocyte [95], where it is thought to play a role in erythrocyte invasion [96]. Proteomic studies have shown that AMA1 is also expressed in the sporozoite stage [97], suggesting that it may play a similar role in hepatocyte invasion. People living in malaria-endemic areas produce antibodies to AMA1 that can inhibit erythrocyte invasion in vitro [98] and that are associated with protection in field studies [99]. Studies in animal models show strain specificity in the inhibitory activity of anti-AMA1 antibodies [98], and these results have been corroborated by subsequent allelic exchange experiments [100, 101].

Sequencing of the gene encoding AMA1 in samples from a single Malian village identified more than 200 unique AMA1 variants in about 500 *P. falciparum* infections [59], raising the daunting prospect that a 200-valent AMA1 vaccine might be required to achieve broad protective efficacy. However, molecular epidemiological analyses showed that a group of just eight polymorphic amino acids lying adjacent to the presumed erythrocyte-binding site on AMA1 were responsible for strain-specific naturally acquired immunity, suggesting that a vaccine comprised of as few as ten "serotypes" of AMA1 might be sufficient to protect against 80% of unique variants.

Two AMA1 vaccines have reached the stage of efficacy trials in humans. A bivalent vaccine with two different forms of AMA1 adjuvanted with aluminum hydroxide failed to provide any protection against parasitemia or clinical malaria [66], and molecular analyses of pre- and post-immunization infections turned up no evidence of strain-specific efficacy or selection of non-vaccine variants [102]. A monovalent AMA1 vaccine formulated with the same adjuvant system as that used with RTS,S was more highly immunogenic [68] in a similar population of African children. Although this vaccine did not prevent infection after experimental sporozoite challenge [103] and showed marginal overall efficacy against clinical malaria, it demonstrated strong strain-specific efficacy, reducing the risk of clinical malaria caused by parasites with AMA1 homologous to the vaccine strain by more than 60% [104]. This encouraging result suggests that it may be possible to develop a more broadly efficacious multivalent or chimeric next-generation AMA1 vaccine, and efforts are being made to do this [105–107].

The *P. falciparum* proteins that are expressed on the surface of infected erythrocytes and that mediate cytoadherence and immune evasion and contribute to pathogenesis would seem to be attractive candidates for anti-disease vaccines. These PfEMP1 proteins are encoded by a large family of diverse *var* genes [14] with up to 60 variants in each parasite genome. Designing a vaccine that would be broadly protective against such an extraordinarily polymorphic target is likely to be very difficult. One possible approach to overcome this difficulty may be the identification of conserved epitopes that are nevertheless immunogenic [108]. Because a single PfEMP1 that is somewhat less polymorphic, VAR2CSA, mediates cytoadherence in placental malaria, prospects for a PfEMP1-based pregnancy malaria vaccine may be better, and research toward this goal is underway [109].

Multistage, multi-antigen vaccines that include blood-stage components are discussed in a subsequent section.

Transmission-Blocking Vaccines

Transmission-blocking vaccines are vaccines that are specifically intended to block transmission by targeting molecules that are unique to gametocytes, the male and female forms that are taken up during a blood meal and that mate in the mosquito midgut, or that target subsequent mosquito stages. Antibodies directed against such targets are capable of blocking the development of mosquito stages, thus interrupting transmission [110]. In a rare example of a vaccine designed to target multiple species, a vaccine based on mosquito-stage proteins in both *P. falciparum* and *P. vivax* was recently shown to produce dose-dependent antibody-mediated transmission-blocking activity [111]. However, the vaccine, which was formulated with the powerful adjuvant Montanide ISA 51, was unacceptably reactogenic. With the recent renewed call for global malaria eradication [5, 21], transmission-blocking vaccines will be increasingly emphasized. In one novel and promising approach, vaccines that target mosquito molecules are being contemplated in hopes of avoiding selection pressure within the host that favors "vaccine-resistant" parasites [112].

Although not traditionally thought of as transmission-blocking vaccines, highly efficacious pre-erythrocytic vaccines that provide sterile immunity would also interrupt transmission. Because they would completely prevent infection, such vaccines would also have the important added benefit of preventing disease caused by the blood stages. In the context of malaria elimination, a highly efficacious pre-erythrocytic vaccine would thus be the product development target for "vaccines that interrupt transmission" [11].

Multistage, Multi-antigen Vaccines

Compared to the viral and bacterial human pathogens for which effective vaccines exist, malaria parasites are big and complex and elicit equally complex and multifaceted immune responses. In retrospect, it may not be surprising that so many candidate vaccines that target just a single variant of a single antigen have failed to demonstrate clinical efficacy, especially against heterologous natural challenge. Several attempts have been made to improve on the efficacy of single-antigen vaccines by developing multistage, multi-antigen vaccines. One of the earliest was SPf66, a synthetic vaccine consisting of peptides derived from the blood-stage antigen MSP1 linked by the central repeat of the pre-erythrocytic antigen CSP. While reports of efficacy in initial trials in South America generated great excitement [113], subsequent studies in Africa and Asia showed no significant protective efficacy [114–116]. In another approach that yielded disappointing results, vaccinia virus was used as a vector to express seven *P. falciparum* genes, but both immunogenicity and efficacy were limited [117]. Attempts to develop DNA vaccines with from two to as many as 15 *P. falciparum* genes [118] were likewise unsuccessful.

Based on the modest efficacy of RTS,S against clinical malaria and evidence that a blood-stage vaccine using a similar adjuvant system can produce strain-specific efficacy [104], it is reasonable to believe that it may be possible to construct a multistage, multi-antigen recombinant protein that improves on the efficacy of RTS,S [84]. However, it seems not unlikely that vaccines that target 2, 5, or even 15 of the 5,000 gene products will still fall short of the high levels of protection seen with radiation-attenuated whole-organism vaccines when delivered through the bites of infected mosquitoes [51].

Whole-Organism Vaccines

Even though live attenuated vaccines were the earliest and remain some of the best vaccines against other pathogens, and even though birds and monkeys had been protected by live attenuated malaria parasites in the earliest vaccine studies [40, 41], the protection seen with irradiated sporozoites in the early 1970s [49] was interpreted not as a direct path to a malaria vaccine but as proof that a vaccine was possible and as justification for the ensuing decades of research aimed at identifying the "right" vaccine antigen or heterologous expression system. The limited success of these Sisyphean research efforts led to reevaluation of the dogma that it would be impossible to manufacture an attenuated sporozoite vaccine in mosquitoes [51]. A radiation-attenuated, metabolically active, non-replicating sporozoite vaccine has been manufactured in and purified from aseptically raised mosquitoes [47] and was recently evaluated for safety and efficacy in an experimental sporozoite challenge trial in humans. The goal was to administer by needle essentially the same immunogen – albeit aseptic, purified, and cryopreserved – that had previously demonstrated 90% protective efficacy in the form of sterilizing immunity when delivered by the bites of at least 1,000 irradiated infected mosquitoes. When administered by intradermal or subcutaneous routes, the sporozoite vaccine did not have significant protective efficacy [119]. Contemporaneous studies show that the vaccine sporozoites are highly immunogenic in monkeys when administered intravenously but not subcutaneously [119]. The likeliest explanation for this result is thought to be that sporozoites delivered into the skin by needle injection in a comparatively large volume of fluid were unable to reach the liver with efficiency approaching that of sporozoites injected either by a mosquito probing for small blood vessels or intravenously.

The potential for this approach to yield a highly efficacious pre-erythrocytic whole-organism vaccine remains well worth pursuing. Efforts are underway to improve delivery methods to more closely approximate the probing mosquito's efficient delivery of attenuated sporozoites into the circulation, starting with the intravenous route that is clearly superior in animal models. Whole-parasite vaccine experiments in the 1940s showed that adjuvants could boost protection, and this will likely be tried with the sporozoite vaccine. Genetic attenuation of sporozoites as an alternative to radiation is also being explored [120]. In a similar "back to the future" paradigm shift, attenuated whole-parasite blood-stage vaccines are now also back on the table [121, 122] more than 60 years after this approach was shown to work in monkeys.

Future Directions

While about 30 of the more than 100 countries with malaria transmission are actively trying to eliminate malaria, it is generally agreed that global malaria eradication is not possible without either global economic development to the levels that permitted malaria elimination in the United States, Europe, and the former Soviet Union or new tools such as a highly efficacious malaria vaccine that interrupts transmission [11]. A powerfully adjuvanted pre-erythrocytic single-antigen recombinant protein vaccine, RTS,S/AS01, significantly reduces the clinical burden of malaria in African children [60, 81], but it does not prevent infection, and, somewhat surprisingly, its effect on transmission, if any, has not been reported. While the results of a field trial of a similarly adjuvanted blood-stage vaccine [104] as well as combination of RTS,S with a viral vector in a prime-boost regimen [85] provide a rationale for pursuing improvements on RTS,S and similar vaccines, the dire public health need for a highly efficacious malaria vaccine calls for more than incremental improvements. Whole-parasite vaccines may be the radically different new (and yet very old) approach that is needed. Challenges remain to produce and deliver such a vaccine, but none that are insurmountable.

Research from the 1940s and many more recent studies point to the importance of strong adjuvants for subunit as well as for whole-organism vaccines, and wide access to immunogenic and safe adjuvants will be important for accelerating malaria vaccine development. Researchers are focusing now on the painfully elusive goal of achieving a high degree of efficacy, but the ideal vaccine would be not only safe and efficacious but also thermostable and protective for a long period of time after a single immunization – characteristics that will not be easy to achieve but which might be possible through pharmaceutical technologies such as controlled release formulations.

All predictions of when a malaria vaccine will be available have been overly optimistic, but barring unforeseen setbacks, a vaccine that substantially reduces the malaria burden should be licensed within just a few years. This will be a magnificent accomplishment that will mark not the end of the road for malaria vaccine development but a critical milestone on the path leading toward the malaria vaccine that the world needs.

Acknowledgments The author is supported by the National Institute of Allergy and Infectious Diseases of the U.S. National Institutes of Health, by the Doris Duke Charitable Foundation, and by the Howard Hughes Medical Institute and wishes to thank Kirsten Lyke, Kavita Gandhi, Matthew Laurens, Mark Travassos, Thomas Richie, and Judith Epstein for critical reading of the manuscript and to acknowledge the U.S. Agency for International Development, the Walter Reed Army Institute of Research, GlaxoSmithKline Biologicals, Sanaria Inc., the U.S. Military Malaria Vaccine Program – Naval Medical Research Center, and the Malaria Research and Training Center of the University of Bamako, Mali, for collaboration on malaria vaccine trials.

Bibliography

Primary Literature

1. Garnham PCC (1966) Malaria parasites and other haemosporidia. Blackwell, Oxford
2. World Health Organization (2008) The global malaria action plan for a malaria free world. World Health Organization, Geneva
3. Rieckmann KH (2006) The chequered history of malaria control: are new and better tools the ultimate answer? Ann Trop Med Parasitol 100:647–662. doi:10.1179/136485906X112185
4. WHO (2007) United Arab Emirates certified malaria-free. Wkly Epidemiol Rec 82:30–32
5. Roberts L, Enserink M (2007) Malaria. Did they really say...eradication? Science 318:1544–1545

6. Plowe CV, Alonso P, Hoffman SL (2009) The potential role of vaccines in the elimination of falciparum malaria and the eventual eradication of malaria. J Infect Dis 200:1646–1649. doi:10.1086/646613
7. Henderson DA (1999) Lessons from the eradication campaigns. Vaccine 17(Suppl 3):S53–S55, S0264410X99002935 [pii]
8. Casares S, Brumeanu TD, Richie TL (2010) The RTS,S malaria vaccine. Vaccine 28:4880–4894. doi:10.1016/j.vaccine.2010.05.033, S0264-410X(10)00721–8 [pii]
9. Cox-Singh J, Davis TM, Lee KS, Shamsul SS, Matusop A, Ratnam S, Rahman HA, Conway DJ, Singh B (2008) *Plasmodium knowlesi* malaria in humans is widely distributed and potentially life threatening. Clin Infect Dis 46:165–171. doi:10.1086/524888
10. Price RN, Tjitra E, Guerra CA, Yeung S, White NJ, Anstey NM (2007) Vivax malaria: neglected and not benign. Am J Trop Med Hyg 77:79–87, 77/6_Suppl/79 [pii]
11. The malERA Consultative Group on Vaccines (2010) A research agenda for malaria eradication: vaccines. PLoS Med 8:e1000398
12. Baruch DI, Pasloske BL, Singh HB, Bi X, Ma XC, Feldman M, Taraschi TF, Howard RJ (1995) Cloning the *P. falciparum* gene encoding PfEMP1, a malarial variant antigen and adherence receptor on the surface of parasitized human erythrocytes. Cell 82:77–87
13. Su X, Heatwole VM, Wertheimer SP, Guinet F, Herrfeldt JA, Peterson DS, Ravetch JV, Wellems TE (1995) The large diverse gene family *var* encodes proteins involved in cytoadherence and antigenic variation of *Plasmodium falciparum*-infected erythrocytes. Cell 82:89–100
14. Smith JD, Chitnis CE, Craig AG, Roberts DJ, Hudson-Taylor DE, Peterson DS, Pinches R, Newbold CI, Miller LH (1995) Switches in expression of *Plasmodium falciparum* var genes correlate with changes in antigenic and cytoadherent phenotypes of infected erythrocytes. Cell 82:101–110
15. Miller LH, Baruch DI, Marsh K, Doumbo OK (2002) The pathogenic basis of malaria. Nature 415:673–679
16. Nash GB, O'Brien E, Gordon-Smith EC, Dormandy JA (1989) Abnormalities in the mechanical properties of red blood cells caused by *Plasmodium falciparum*. Blood 74:855–861
17. Gupta S, Snow RW, Donnelly CA, Marsh K, Newbold C (1999) Immunity to non-cerebral severe malaria is acquired after one or two infections. Nat Med 5:340–343
18. Hviid L (2005) Naturally acquired immunity to *Plasmodium falciparum* malaria in Africa. Acta Trop 95:270–275. doi:10.1016/j.actatropica.2005.06.012, S0001-706X(05)00165-8
19. Baird JK, Jones TR, Danudirgo EW, Annis BA, Bangs MJ, Basri H, Purnomo MS (1991) Age-dependent acquired protection against *Plasmodium falciparum* in people having two years exposure to hyperendemic malaria. Am J Trop Med Hyg 45:65–76
20. Duffy PE (2007) Plasmodium in the placenta: parasites, parity, protection, prevention and possibly preeclampsia. Parasitology 134:1877–1881. doi:10.1017/S0031182007000170, S0031182007000170 [pii]
21. Tanner M, de Savigny D (2008) Malaria eradication back on the table. Bull World Health Organ 86:82, S0042-96862008000200002 [pii]
22. Cohen S, McGregor IA, Carrington S (1961) Gamma-globulin and acquired immunity to human malaria. Nature 192:733–737
23. McGregor IA, Carrington S, Cohen S (1963) Treatment of East African *P. falciparum* malaria with West African human gamma-globulin. Trans R Soc Trop Med Hyg 50:170–175
24. Collins WE, Jeffery GM (1999) A retrospective examination of sporozoite- and trophozoite-induced infections with *Plasmodium falciparum* in patients previously infected with heterologous species of Plasmodium: effect on development of parasitologic and clinical immunity. Am J Trop Med Hyg 61:36–43
25. Collins WE, Jeffery GM (1999) A retrospective examination of secondary sporozoite- and trophozoite-induced infections with *Plasmodium falciparum*: development of parasitologic and clinical immunity following secondary infection. Am J Trop Med Hyg 61:20–35

26. Thomas AW, Trape JF, Rogier C, Goncalves A, Rosario VE, Narum DL (1994) High prevalence of natural antibodies against *Plasmodium falciparum* 83-kilodalton apical membrane antigen (PF83/AMA-1) as detected by capture-enzyme-linked immunosorbent assay using full-length baculovirus recombinant PF83/AMA-1. Am J Trop Med Hyg 51:730–740
27. Egan AF, Chappel JA, Burghaus PA, Morris JS, McBride JS, Holder AA, Kaslow DC, Riley EM (1995) Serum antibodies from malaria-exposed people recognize conserved epitopes formed by the two epidermal growth factor motifs of MSP1(19), the carboxy-terminal fragment of the major merozoite surface protein of *Plasmodium falciparum*. Infect Immun 63:456–466
28. Schofield L (1989) T cell immunity to malaria sporozoites. Exp Parasitol 68:357–364
29. Doolan DL, Mu Y, Unal B, Sundaresh S, Hirst S, Valdez C, Randall A, Molina D, Liang X, Freilich DA, Oloo JA, Blair PL, Aguiar JC, Baldi P, Davies DH, Felgner PL (2008) Profiling humoral immune responses to *P. falciparum* infection with protein microarrays. Proteomics 8:4680–4694
30. Crompton PD, Kayala MA, Traore B, Kayentao K, Ongoiba A, Weiss GE, Molina DM, Burk CR, Waisberg M, Jasinskas A, Tan X, Doumbo S, Doumtabe D, Kone Y, Narum DL, Liang X, Doumbo OK, Miller LH, Doolan DL, Baldi P, Felgner PL, Pierce SK (2010) A prospective analysis of the Ab response to *Plasmodium falciparum* before and after a malaria season by protein microarray. Proc Natl Acad Sci USA 107:6958–6963. doi:10.1073/pnas.1001323107, 1001323107 [pii]
31. Allison AC (1954) Protection afforded by sickle-cell trait against subtertian malarial infection. Br Med J 1:290–294
32. Kay AC, Kuhl W, Prchal J, Beutler E (1992) The origin of glucose-6-phosphate-dehydrogenase (G6PD) polymorphisms in African-Americans. Am J Hum Genet 50:394–398
33. Agarwal A, Guindo A, Cissoko Y, Taylor JG, Coulibaly D, Kone A, Kayentao K, Djimde A, Plowe CV, Doumbo O, Wellems TE, Diallo D (2000) Hemoglobin C associated with protection from severe malaria in the Dogon of Mali, a West African population with a low prevalence of hemoglobin S. Blood 96:2358–2363
34. Flint J, Hill AV, Bowden DK, Oppenheimer SJ, Sill PR, Serjeantson SW, Bana-Koiri J, Bhatia K, Alpers MP, Boyce AJ (1986) High frequencies of alpha-thalassaemia are the result of natural selection by malaria. Nature 321:744–750
35. Hill AV, Allsopp CE, Kwiatkowski D, Anstey NM, Twumasi P, Rowe PA, Bennett S, Brewster D, McMichael AJ, Greenwood BM (1991) Common west African HLA antigens are associated with protection from severe malaria. Nature 352:595–600
36. Miller LH, Mason SJ, Clyde DF, McGinniss MH (1976) The resistance factor to *Plasmodium vivax* in blacks. The Duffy- blood-group genotype, FyFy. N Engl J Med 295:302–304
37. Bruce-Chwatt LJ (1981) Alphonse Laveran's discovery 100 years ago and today's global fight against malaria. J R Soc Med 74:531–8
38. Ross R (1897) Observations on a condition necessary to the transformation of the malaria crescent. Br Med J 1:251–255
39. Russell PF, Mohan BN (1942) The immunization of fowls against mosquito-borne *Plamodium gallinaceum* by injections of serum and of inactivated homologous sporozoites. J Exp Med 76:477–495
40. Freund J, Sommer HE, Walter AW (1945) Immunization against malaria: vaccination of ducks with killed parasites incorporated with adjuvants. Science 102:200–202
41. Freund J, Thomson KJ, Sommer HE, Walter AW, Schenkein EL (1945) Immunization of rhesus monkeys against malarial infection (*P. knowlesi*) with killed parasites and adjuvants. Science 102:202–204. doi:10.1126/science.102.2643.202, 102/2643/202 [pii]
42. Jacobs HR (1943) Immunization against malaria. Increased protection by vaccination of ducklings with saline-insoluble residues of *Plasmodium lophurae* mixed with a bacterial toxin. Am J Trop Med Hyg s1-23:597–606
43. Redmond WB (1939) Immunization of birds to malaria by vaccination. J Parasitol 25:28–29
44. Bennison BE, Coatney GR (1949) Effects of X-irradiation on *Plasmodium gallinaceum* and *Plasmodium lophurae* infections in young chicks. J Natl Malar Soc 8:280–289

45. Ceithaml J, Evans EA Jr (1946) The biochemistry of the malaria parasite; the in vitro effects of x-rays upon *Plasmodium gallinaceum*. J Infect Dis 78:190–197
46. Nussenzweig RS, Vanderberg J, Most H, Orton C (1967) Protective immunity produced by the injection of x-irradiated sporozoites of *Plasmodium berghei*. Nature 216:160–162
47. Hoffman SL, Billingsley PF, James E, Richman A, Loyevsky M, Li T, Chakravarty S, Gunasekera A, Li M, Stafford R, Ahumada A, Epstein JE, Sedegah M, Reyes S, Richie TL, Lyke KE, Edelman R, Laurens M, Plowe CV, Sim BK (2010) Development of a metabolically active, non-replicating sporozoite vaccine to prevent *Plasmodium falciparum* malaria. Hum Vaccin 6:97–106, 10396 [pii]
48. Heidelberger M, Prout C, Hindle JA, Rose AS (1946) Studies in human malaria III. An attempt at vaccination of paretics against blood-borne infection with *P vivax*. J Immunol 53:109–112
49. Clyde DF, Most H, McCarthy VC, Vanderberg JP (1973) Immunization of man against sporozoite-induced falciparum malaria. Am J Med Sci 266:169–177
50. Rieckmann KH, Carson PE, Beaudoin RL, Cassells JS, Sell KW (1974) Letter: sporozoite induced immunity in man against an Ethiopian strain of *Plasmodium falciparum*. Trans R Soc Trop Med Hyg 68:258–259
51. Luke TC, Hoffman SL (2003) Rationale and plans for developing a non-replicating, metabolically active, radiation-attenuated *Plasmodium falciparum* sporozoite vaccine. J Exp Biol 206:3803–3808
52. Vanderberg JP (2009) Reflections on early malaria vaccine studies, the first successful human malaria vaccination, and beyond. Vaccine 27:2–9. doi:10.1016/j.vaccine.2008.10.028, S0264-410X(08)01414-X [pii]
53. Trager W, Jensen JB (1976) Human malaria parasites in continuous culture. Science 193:673–675
54. Haynes JD, Diggs CL, Hines FA, Desjardins RE (1976) Culture of human malaria parasites *Plasmodium falciparum*. Nature 263:767–769
55. World Health Organization (2010) Initiative for vaccine research: malaria vaccines. WHO, Geneva
56. Gardner MJ, Hall N, Fung E, White O, Berriman M, Hyman RW, Carlton JM, Pain A, Nelson KE, Bowman S, Paulsen IT, James K, Eisen JA, Rutherford K, Salzberg SL, Craig A, Kyes S, Chan MS, Nene V, Shallom SJ, Suh B, Peterson J, Angiuoli S, Pertea M, Allen J, Selengut J, Haft D, Mather MW, Vaidya AB, Martin DM, Fairlamb AH, Fraunholz MJ, Roos DS, Ralph SA, McFadden GI, Cummings LM, Subramanian GM, Mungall C, Venter JC, Carucci DJ, Hoffman SL, Newbold C, Davis RW, Fraser CM, Barrell B (2002) Genome sequence of the human malaria parasite *Plasmodium falciparum*. Nature 419:498–511
57. Takala SL, Plowe CV (2009) Genetic diversity and malaria vaccine design, testing and efficacy: preventing and overcoming 'vaccine resistant malaria'. Parasite Immunol 31:560–573. doi:10.1111/j.1365–3024.2009.01138.x, PIM1138 [pii]
58. Takala SL, Coulibaly D, Thera MA, Dicko A, Smith DL, Guindo AB, Kone AK, Traore K, Ouattara A, Djimde AA, Sehdev PS, Lyke KE, Diallo DA, Doumbo OK, Plowe CV (2007) Dynamics of polymorphism in a malaria vaccine antigen at a vaccine-testing site in Mali. PLoS Med 4:e93
59. Takala SL, Coulibaly D, Thera MA, Batchelor AH, Cummings MP, Escalante AA, Ouattara A, Traore K, Niangaly A, Djimde AA, Doumbo OK, Plowe CV (2009) Extreme polymorphism in a vaccine antigen and risk of clinical malaria: implications for vaccine development. Sci Transl Med 1:2ra5
60. Alonso PL, Sacarlal J, Aponte JJ, Leach A, Macete E, Milman J, Mandomando I, Spiessens B, Guinovart C, Espasa M, Bassat Q, Aide P, Ofori-Anyinam O, Navia MM, Corachan S, Ceuppens M, Dubois MC, Demoitie MA, Dubovsky F, Menendez C, Tornieporth N, Ballou WR, Thompson R, Cohen J (2004) Efficacy of the RTS, S/AS02A vaccine against *Plasmodium falciparum* infection and disease in young African children: randomised controlled trial. Lancet 364:1411–1420

61. Egan AF, Blackman MJ, Kaslow DC (2000) Vaccine efficacy of recombinant *Plasmodium falciparum* merozoite surface protein 1 in malaria-naive, -exposed, and/or -rechallenged *Aotus vociferans* monkeys. Infect Immun 68:1418–1427
62. Thera MA, Doumbo OK, Coulibaly D, Diallo DA, Sagara I, Dicko A, Diemert DJ, Heppner DG Jr, Stewart VA, Angov E, Soisson L, Leach A, Tucker K, Lyke KE, Plowe CV (2006) Safety and allele-specific immunogenicity of a malaria vaccine in Malian adults: results of a phase I randomized trial. PLoS Clin Trials 1:e34. doi:10.1371/journal.pctr.0010034
63. Stoute JA, Gombe J, Withers MR, Siangla J, McKinney D, Onyango M, Cummings JF, Milman J, Tucker K, Soisson L, Stewart VA, Lyon JA, Angov E, Leach A, Cohen J, Kester KE, Ockenhouse CF, Holland CA, Diggs CL, Wittes J, Heppner DG Jr (2007) Phase 1 randomized double-blind safety and immunogenicity trial of *Plasmodium falciparum* malaria merozoite surface protein FMP1 vaccine, adjuvanted with AS02A, in adults in western Kenya. Vaccine 25:176–184
64. Withers MR, McKinney D, Ogutu BR, Waitumbi JN, Milman JB, Apollo OJ, Allen OG, Tucker K, Soisson LA, Diggs C, Leach A, Wittes J, Dubovsky F, Stewart VA, Remich SA, Cohen J, Ballou WR, Holland CA, Lyon JA, Angov E, Stoute JA, Martin SK, Heppner DG (2006) Safety and reactogenicity of an MSP-1 malaria vaccine candidate: a randomized Phase Ib dose-escalation trial in Kenyan children. PLoS Clin Trials 1:e32
65. Ogutu BR, Apollo OJ, McKinney D, Okoth W, Siangla J, Dubovsky F, Tucker K, Waitumbi JN, Diggs C, Wittes J, Malkin E, Leach A, Soisson LA, Milman JB, Otieno L, Holland CA, Polhemus M, Remich SA, Ockenhouse CF, Cohen J, Ballou WR, Martin SK, Angov E, Stewart VA, Lyon JA, Heppner DG, Withers MR (2009) Blood stage malaria vaccine eliciting high antigen-specific antibody concentrations confers no protection to young children in Western Kenya. PLoS One 4:e4708
66. Sagara I, Dicko A, Ellis RD, Fay MP, Diawara SI, Assadou MH, Sissoko MS, Kone M, Diallo AI, Saye R, Guindo MA, Kante O, Niambele MB, Miura K, Mullen GE, Pierce M, Martin LB, Dolo A, Diallo DA, Doumbo OK, Miller LH, Saul A (2009) A randomized controlled phase 2 trial of the blood stage AMA1-C1/Alhydrogel malaria vaccine in children in Mali. Vaccine 27:3090–3098
67. Thera MA, Doumbo OK, Coulibaly D, Diallo DA, Kone AK, Guindo AB, Traore K, Dicko A, Sagara I, Sissoko MS, Baby M, Sissoko M, Diarra I, Niangaly A, Dolo A, Daou M, Diawara SI, Heppner DG, Stewart VA, Angov E, Bergmann-Leitner ES, Lanar DE, Dutta S, Soisson L, Diggs CL, Leach A, Owusu A, Dubois MC, Cohen J, Nixon JN, Gregson A, Takala SL, Lyke KE, Plowe CV (2008) Safety and immunogenicity of an AMA-1 malaria vaccine in Malian adults: results of a Phase 1 randomized controlled trial. PLoS One 3:e1465
68. Thera MA, Doumbo OK, Coulibaly D, Laurens MB, Kone AK, Guindo AB, Traore K, Sissoko M, Diallo DA, Diarra I, Kouriba B, Daou M, Dolo A, Baby M, Sissoko MS, Sagara I, Niangaly A, Traore I, Olotu A, Godeaux O, Leach A, Dubois MC, Ballou WR, Cohen J, Thompson D, Dube T, Soisson L, Diggs CL, Takala SL, Lyke KE, House B, Lanar DE, Dutta S, Heppner DG, Plowe CV (2010) Safety and immunogenicity of an AMA1 malaria vaccine in Malian children: results of a phase 1 randomized controlled trial. PLoS One 5:e9041. doi:10.1371/journal.pone.0009041
69. Dame JB, Williams JL, McCutchan TF, Weber JL, Wirtz RA, Hockmeyer WT, Maloy WL, Haynes JD, Schneider RD (1984) Structure of the gene encoding the immunodominant surface antigen on the sporozoite of the human malaria parasite *Plasmodium falciparum*. Science 225:593–599
70. Beier JC, Vanderburg JP (1998) Sporogonic development in the mosquito. In: Sherman IW (ed) Malaria parasite biology, pathogenesis, and protection. ASM Press, Washington, DC
71. Ballou WR, Rothbard J, Wirtz RA, Gordon DM, Williams JS, Gore RW, Schneider I, Hollingdale MR, Beaudoin RL, Maloy WL et al (1985) Immunogenicity of synthetic peptides from circumsporozoite protein of *Plasmodium falciparum*. Science 228:996–999
72. Hollingdale MR, Nardin EH, Tharavanij S, Schwartz AL, Nussenzweig RS (1984) Inhibition of entry of *Plasmodium falciparum* and *P. vivax* sporozoites into cultured cells; an in vitro assay of protective antibodies. J Immunol 132:909–913

73. Herrington DA, Clyde DF, Losonsky G, Cortesia M, Murphy JR, Davis J, Baqar S, Felix AM, Heimer EP, Gillessen D et al (1987) Safety and immunogenicity in man of a synthetic peptide malaria vaccine against *Plasmodium falciparum* sporozoites. Nature 328:257–259
74. Ballou WR, Hoffman SL, Sherwood JA, Hollingdale MR, Neva FA, Hockmeyer WT, Gordon DM, Schneider I, Wirtz RA, Young JF (1987) Safety and efficacy of a recombinant DNA *Plasmodium falciparum* sporozoite vaccine. Lancet 1:1277–1281
75. Hoffman SL, Oster CN, Plowe CV, Woollett GR, Beier JC, Chulay JD, Wirtz RA, Hollingdale MR, Mugambi M (1987) Naturally acquired antibodies to sporozoites do not prevent malaria: vaccine development implications. Science 237:639–642
76. Sherwood JA, Copeland RS, Taylor KA, Abok K, Oloo AJ, Were JB, Strickland GT, Gordon DM, Ballou WR, Bales JDJ, Wirtz RA, Wittes J, Gross M, Que JU, Cryz SJ, Oster CN, Roberts CR, Sadoff JC (1996) *Plasmodium falciparum* circumsporozoite vaccine immunogenicity and efficacy trial with natural challenge quantitation in an area of endemic human malaria of Kenya. Vaccine 14:817–827
77. Gordon DM, McGovern TW, Krzych U, Cohen JC, Schneider I, LaChance R, Heppner DG, Yuan G, Hollingdale M, Slaoui M (1995) Safety, immunogenicity, and efficacy of a recombinantly produced *Plasmodium falciparum* circumsporozoite protein-hepatitis B surface antigen subunit vaccine. J Infect Dis 171:1576–1585
78. Stoute JA, Slaoui M, Heppner DG, Momin P, Kester KE, Desmons P, Wellde BT, Garcon N, Krzych U, Marchand M (1997) A preliminary evaluation of a recombinant circumsporozoite protein vaccine against *Plasmodium falciparum* malaria. RTS, S malaria vaccine evaluation group. N Engl J Med 336:86–91. doi:10.1056/NEJM199701093360202
79. Kester KE, McKinney DA, Tornieporth N, Ockenhouse CF, Heppner DG, Hall T, Krzych U, Delchambre M, Voss G, Dowler MG, Palensky J, Wittes J, Cohen J, Ballou WR (2001) Efficacy of recombinant circumsporozoite protein vaccine regimens against experimental *Plasmodium falciparum* malaria. J Infect Dis 183:640–647
80. Bojang KA, Milligan PJ, Pinder M, Vigneron L, Alloueche A, Kester KE, Ballou WR, Conway DJ, Reece WH, Gothard P, Yamuah L, Delchambre M, Voss G, Greenwood BM, Hill A, McAdam KP, Tornieporth N, Cohen JD, Doherty T (2001) Efficacy of RTS, S/AS02 malaria vaccine against *Plasmodium falciparum* infection in semi-immune adult men in The Gambia: a randomised trial. Lancet 358:1927–1934
81. Bejon P, Lusingu J, Olotu A, Leach A, Lievens M, Vekemans J, Mshamu S, Lang T, Gould J, Dubois MC, Demoitie MA, Stallaert JF, Vansadia P, Carter T, Njuguna P, Awuondo KO, Malabeja A, Abdul O, Gesase S, Mturi N, Drakeley CJ, Savarese B, Villafana T, Ballou WR, Cohen J, Riley EM, Lemnge MM, Marsh K, von Seidlein L (2008) Efficacy of RTS, S/AS01E vaccine against malaria in children 5 to 17 months of age. N Engl J Med 359:2521–2532. doi:10.1056/NEJMoa0807381, NEJMoa0807381 [pii]
82. Alonso PL, Sacarlal J, Aponte JJ, Leach A, Macete E, Aide P, Sigauque B, Milman J, Mandomando I, Bassat Q, Guinovart C, Espasa M, Corachan S, Lievens M, Navia MM, Dubois MC, Menendez C, Dubovsky F, Cohen J, Thompson R, Ballou WR (2005) Duration of protection with RTS, S/AS02A malaria vaccine in prevention of *Plasmodium falciparum* disease in Mozambican children: single-blind extended follow-up of a randomised controlled trial. Lancet 366:2012–2018
83. Sacarlal J, Aide P, Aponte JJ, Renom M, Leach A, Mandomando I, Lievens M, Bassat Q, Lafuente S, Macete E, Vekemans J, Guinovart C, Sigauque B, Sillman M, Milman J, Dubois MC, Demoitie MA, Thonnard J, Menendez C, Ballou WR, Cohen J, Alonso PL (2009) Long-term safety and efficacy of the RTS, S/AS02A malaria vaccine in Mozambican children. J Infect Dis 200:329–336. doi:10.1086/600119
84. Heppner DG Jr, Kester KE, Ockenhouse CF, Tornieporth N, Ofori O, Lyon JA, Stewart VA, Dubois P, Lanar DE, Krzych U, Moris P, Angov E, Cummings JF, Leach A, Hall BT, Dutta S, Schwenk R, Hillier C, Barbosa A, Ware LA, Nair L, Darko CA, Withers MR, Ogutu B, Polhemus ME, Fukuda M, Pichyangkul S, Gettyacamin M, Diggs C, Soisson L, Milman J, Dubois MC, Garcon N, Tucker K, Wittes J, Plowe CV, Thera MA, Duombo OK, Pau MG,

Goudsmit J, Ballou WR, Cohen J (2005) Towards an RTS, S-based, multi-stage, multi-antigen vaccine against falciparum malaria: progress at the Walter Reed Army Institute of Research. Vaccine 23:2243–2250

85. Stewart VA, McGrath SM, Dubois PM, Pau MG, Mettens P, Shott J, Cobb M, Burge JR, Larson D, Ware LA, Demoitie MA, Weverling GJ, Bayat B, Custers JH, Dubois MC, Cohen J, Goudsmit J, Heppner DG Jr (2007) Priming with an adenovirus 35-circumsporozoite protein (CS) vaccine followed by RTS, S/AS01B boosting significantly improves immunogenicity to *Plasmodium falciparum* CS compared to that with either malaria vaccine alone. Infect Immun 75:2283–2290
86. Cummings JF, Spring MD, Schwenk RJ, Ockenhouse CF, Kester KE, Polhemus ME, Walsh DS, Yoon IK, Prosperi C, Juompan LY, Lanar DE, Krzych U, Hall BT, Ware LA, Stewart VA, Williams J, Dowler M, Nielsen RK, Hillier CJ, Giersing BK, Dubovsky F, Malkin E, Tucker K, Dubois MC, Cohen JD, Ballou WR, Heppner DG Jr (2010) Recombinant liver stage antigen-1 (LSA-1) formulated with AS01 or AS02 is safe, elicits high titer antibody and induces IFN-gamma/IL-2 CD4+ T cells but does not protect against experimental *Plasmodium falciparum* infection. Vaccine 28:5135–5144. doi:10.1016/j.vaccine.2009.08.046, S0264-410X(09)01231-6 [pii]
87. Hill AV, Reyes-Sandoval A, O'Hara G, Ewer K, Lawrie A, Goodman A, Nicosia A, Folgori A, Colloca S, Cortese R, Gilbert SC, Draper SJ (2010) Prime-boost vectored malaria vaccines: progress and prospects. Hum Vaccin 6:78–83, 10116 [pii]
88. Kaba SA, Brando C, Guo Q, Mittelholzer C, Raman S, Tropel D, Aebi U, Burkhard P, Lanar DE (2009) A nonadjuvanted polypeptide nanoparticle vaccine confers long-lasting protection against rodent malaria. J Immunol 183:7268–7277. doi:10.4049/jimmunol.0901957, jimmunol.0901957 [pii]
89. Bergmann-Leitner ES, Mease RM, de la Vega P, Savranskaya T, Polhemus M, Ockenhouse C, Angov E (2010) Immunization with pre-erythrocytic antigen CelTOS from *Plasmodium falciparum* elicits cross-species protection against heterologous challenge with *Plasmodium berghei*. PLoS One 5:e12294. doi:10.1371/journal.pone.0012294
90. Blackman MJ, Heidrich HG, Donachie S, McBride JS, Holder AA (1990) A single fragment of a malaria merozoite surface protein remains on the parasite during red cell invasion and is the target of invasion-inhibiting antibodies. J Exp Med 172:379–382
91. Egan AF, Morris J, Barnish G, Allen S, Greenwood BM, Kaslow DC, Holder AA, Riley EM (1996) Clinical immunity to *Plasmodium falciparum* malaria is associated with serum antibodies to the 19-kDa C-terminal fragment of the merozoite surface antigen, PfMSP-1. J Infect Dis 173:765–9
92. Riley EM, Allen SJ, Wheeler JG, Blackman MJ, Bennett S, Takacs B, Schonfeld HJ, Holder AA, Greenwood BM (1992) Naturally acquired cellular and humoral immune responses to the major merozoite surface antigen (PfMSP1) of *Plasmodium falciparum* are associated with reduced malaria morbidity. Parasite Immunol 14:321–37
93. Stowers AW, Cioce V, Shimp RL, Lawson M, Hui G, Muratova O, Kaslow DC, Robinson R, Long CA, Miller LH (2001) Efficacy of two alternate vaccines based *on Plasmodium falciparum* merozoite surface protein 1 in an Aotus challenge trial. Infect Immun 69:18–1546. doi:10.1128/IAI.69.3.18-1546.2001
94. Peterson MG, Marshall VM, Smythe JA, Crewther PE, Lew A, Silva A, Anders RF, Kemp DJ (1989) Integral membrane protein located in the apical complex of *Plasmodium falciparum*. Mol Cell Biol 9:3151–3154
95. Narum DL, Thomas AW (1994) Differential localization of full-length and processed forms of PF83/AMA-1 an apical membrane antigen of *Plasmodium falciparum* merozoites. Mol Biochem Parasitol 67:59–68
96. Mitchell GH, Thomas AW, Margos G, Dluzewski AR, Bannister LH (2004) Apical membrane antigen 1, a major malaria vaccine candidate, mediates the close attachment of invasive merozoites to host red blood cells. Infect Immun 72:154–158

97. Florens L, Washburn MP, Raine JD, Anthony RM, Grainger M, Haynes JD, Moch JK, Muster N, Sacci JB, Tabb DL, Witney AA, Wolters D, Wu Y, Gardner MJ, Holder AA, Sinden RE, Yates JR, Carucci DJ (2002) A proteomic view of the *Plasmodium falciparum* life cycle. Nature 419:520–526
98. Hodder AN, Crewther PE, Anders RF (2001) Specificity of the protective antibody response to apical membrane antigen 1. Infect Immun 69:3286–94
99. Polley SD, Mwangi T, Kocken CH, Thomas AW, Dutta S, Lanar DE, Remarque E, Ross A, Williams TN, Mwambingu G, Lowe B, Conway DJ, Marsh K (2004) Human antibodies to recombinant protein constructs of *Plasmodium falciparum* Apical Membrane Antigen 1 (AMA1) and their associations with protection from malaria. Vaccine 23:718–728
100. Dutta S, Lee SY, Batchelor AH, Lanar DE (2007) Structural basis of antigenic escape of a malaria vaccine candidate. Proc Natl Acad Sci USA 104:12488–12493
101. Healer J, Murphy V, Hodder AN, Masciantonio R, Gemmill AW, Anders RF, Cowman AF, Batchelor A (2004) Allelic polymorphisms in apical membrane antigen-1 are responsible for evasion of antibody-mediated inhibition in *Plasmodium falciparum*. Mol Microbiol 52:159–168
102. Ouattara A, Mu J, Takala-Harrison S, Saye R, Sagara I, Dicko A, Niangaly A, Duan J, Ellis RD, Miller LH, Su XZ, Plowe CV, Doumbo OK (2010) Lack of allele-specific efficacy of a bivalent AMA1 malaria vaccine. Malar J 9:175. doi:10.1186/1475-2875-9-175, 1475-2875-9-175 [pii]
103. Spring MD, Cummings JF, Ockenhouse CF, Dutta S, Reidler R, Angov E, Bergmann-Leitner E, Stewart VA, Bittner S, Juompan L, Kortepeter MG, Nielsen R, Krzych U, Tierney E, Ware LA, Dowler M, Hermsen CC, Sauerwein RW, de Vlas SJ, Ofori-Anyinam O, Lanar DE, Williams JL, Kester KE, Tucker K, Shi M, Malkin E, Long C, Diggs CL, Soisson L, Dubois MC, Ballou WR, Cohen J, Heppner DG Jr (2009) Phase 1/2a study of the malaria vaccine candidate apical membrane antigen-1 (AMA-1) administered in adjuvant system AS01B or AS02A. PLoS One 4:e5254
104. Thera MA, Doumbo OK, Coulibaly D, Laurens MB, Ouattara A, Kone AK, Guindo AB, Traore K, Traore I, Kouriba B, Diallo DA, Diarra I, Daou M, Dolo A, Tolo Y, Sissoko MS, Niangaly A, Sissoko M, Takala-Harrison S, Lyke KE, Wu Y, Blackwelder WC, Godeaux O, Vekemans J, Dubois M-C, Ballou WR, Cohen J, Thompson D, Dube T, Soisson L, Diggs CL, House B, Lanar DE, Dutta S, Heppner DG Jr., Plowe CV (2011) A field trial to assess a blood-stage malaria vaccine. N Engl J Med 365:1004–13
105. Remarque EJ, Faber BW, Kocken CH, Thomas AW (2008) A diversity-covering approach to immunization with *Plasmodium falciparum* apical membrane antigen 1 induces broader allelic recognition and growth inhibition responses in rabbits. Infect Immun 76:2660–2670. doi:10.1128/IAI.00170-08, IAI.00170-08 [pii]
106. Remarque EJ, Faber BW, Kocken CH, Thomas AW (2008) Apical membrane antigen 1: a malaria vaccine candidate in review. Trends Parasitol 24:74–84. doi:10.1016/j.pt.2007.12.002, S1471-4922(07)00328-5 [pii]
107. Dutta S, Dlugosz LS, Clayton JW, Pool CD, Haynes JD, Gasser RA III, Batchelor AH (2010) Alanine mutagenesis of the primary antigenic escape residue cluster, c1, of apical membrane antigen 1. Infect Immun 78:661–671. doi:10.1128/IAI.00866-09, IAI.00866-09 [pii]
108. Klein MM, Gittis AG, Su HP, Makobongo MO, Moore JM, Singh S, Miller LH, Garboczi DN (2008) The cysteine-rich interdomain region from the highly variable *Plasmodium falciparum* erythrocyte membrane protein-1 exhibits a conserved structure. PLoS Pathog 4: e1000147. doi:10.1371/journal.ppat.1000147
109. Avril M, Cartwright MM, Hathaway MJ, Hommel M, Elliott SR, Williamson K, Narum DL, Duffy PE, Fried M, Beeson JG, Smith JD (2010) Immunization with VAR2CSA-DBL5 recombinant protein elicits broadly cross-reactive antibodies to placental *Plasmodium falciparum*-infected erythrocytes. Infect Immun 78:2248–2256. doi:10.1128/IAI.00410-09, IAI.00410-09 [pii]

110. Barr PJ, Green KM, Gibson HL, Bathurst IC, Quakyi IA, Kaslow DC (1991) Recombinant Pfs25 protein of *Plasmodium falciparum* elicits malaria transmission-blocking immunity in experimental animals. J Exp Med 174:1203–1208
111. Wu Y, Ellis RD, Shaffer D, Fontes E, Malkin EM, Mahanty S, Fay MP, Narum D, Rausch K, Miles AP, Aebig J, Orcutt A, Muratova O, Song G, Lambert L, Zhu D, Miura K, Long C, Saul A, Miller LH, Durbin AP (2008) Phase 1 trial of malaria transmission blocking vaccine candidates Pfs25 and Pvs25 formulated with montanide ISA 51. PLoS One 3:e2636. doi:10.1371/journal.pone.0002636
112. Dinglasan RR, Valenzuela JG, Azad AF (2005) Sugar epitopes as potential universal disease transmission blocking targets. Insect Biochem Mol Biol 35:1–10. doi:10.1016/j.ibmb.2004.09.005, S0965-1748(04)00159-6 [pii]
113. Valero MV, Amador LR, Galindo C, Figueroa J, Bello MS, Murillo LA, Mora AL, Patarroyo G, Rocha CL, Rojas M (1993) Vaccination with SPf66, a chemically synthesised vaccine, against *Plasmodium falciparum* malaria in Colombia. Lancet 341:705–710
114. Nosten F, Luxemburger C, Kyle DE, Ballou WR, Wittes J, Wah E, Chongsuphajaisiddhi T, Gordon DM, White NJ, Sadoff JC, Heppner DG (1996) Randomised double-blind placebo-controlled trial of SPf66 malaria vaccine in children in northwestern Thailand. Shoklo SPf66 Malaria Vaccine Trial Group. Lancet 348:701–707
115. D'Alessandro U, Leach A, Drakeley CJ, Bennett S, Olaleye BO, Fegan GW, Jawara M, Langerock P, George MO, Targett GA (1995) Efficacy trial of malaria vaccine SPf66 in Gambian infants. Lancet 346:462–467
116. Alonso PL, Smith T, Schellenberg JR, Masanja H, Mwankusye S, Urassa H, Bastos de Azevedo I, Chongela J, Kobero S, Menendez C et al (1994) Randomised trial of efficacy of SPf66 vaccine against *Plasmodium falciparum* malaria in children in southern Tanzania. Lancet 344:1175–1181
117. Ockenhouse CF, Sun PF, Lanar DE, Wellde BT, Hall BT, Kester K, Stoute JA, Magill A, Krzych U, Farley L, Wirtz RA, Sadoff JC, Kaslow DC, Kumar S, Church LW, Crutcher JM, Wizel B, Hoffman S, Lalvani A, Hill AV, Tine JA, Guito KP, de Taisne C, Anders R, Ballou WR (1998) Phase I/IIa safety, immunogenicity, and efficacy trial of NYVAC-Pf7, a pox-vectored, multiantigen, multistage vaccine candidate for *Plasmodium falciparum* malaria. J Infect Dis 177:1664–1673
118. Kumar S, Epstein JE, Richie TL, Nkrumah FK, Soisson L, Carucci DJ, Hoffman SL (2002) A multilateral effort to develop DNA vaccines against falciparum malaria. Trends Parasitol 18:129–135, S1471492201022073 [pii]
119. Epstein JE, Tewari K, Lyke KE, Sim BK, Billingsley PF, Laurens MB, Gunasekera A, Chakravarty S, James ER, Sedegah M, Richman A, Velmurugan S, Reyes S, Li M, Tucker K, Ahumada A, Ruben AJ, Li T, Stafford R, Eappen AG, Tamminga C, Bennett JW, Ockenhouse CF, Murphy JR, Komisar J, Thomas N, Loyevsky M, Birkett A, Plowe CV, Loucq C, Edelman R, Richie TL, Seder RA, Hoffman SL (2011) Live attenuated malaria vaccine designed to protect through hepatic CD8+ T cell immunity. Science [Epub ahead of print]
120. VanBuskirk KM, O'Neill MT, de la Vega P, Maier AG, Krzych U, Williams J, Dowler MG, Sacci JB Jr, Kangwanrangsan N, Tsuboi T, Kneteman NM, Heppner DG Jr, Murdock BA, Mikolajczak SA, Aly AS, Cowman AF, Kappe SH (2009) Preerythrocytic, live-attenuated *Plasmodium falciparum* vaccine candidates by design. Proc Natl Acad Sci USA 106. doi:10.1073/pnas.0906387106, 13004–13009. 0906387106 [pii]
121. Pinzon-Charry A, McPhun V, Kienzle V, Hirunpetcharat C, Engwerda C, McCarthy J, Good MF (2010) Low doses of killed parasite in CpG elicit vigorous CD4+ T cell responses against blood-stage malaria in mice. J Clin Invest 120:2967–2978. doi:10.1172/JCI39222, 39222 [pii]
122. McCarthy JS, Good MF (2010) Whole parasite blood stage malaria vaccines: a convergence of evidence. Hum Vaccin 6:114–123, 10394 [pii]

Books and Reviews

Bruder JT, Angov E, Limbach KJ, Richie TL (2010) Molecular vaccines for malaria. Hum Vaccin 6:54–77

Desowitz RS (1991) The malaria capers: more tales of parasites and people, research and reality. Norton, New York

Dinglasan RR, Jacobs-Lorena M (2008) Flipping the paradigm on malaria transmission-blocking vaccines. Trends Parasitol 24:364–370

Hill AV, Reyes-Sandoval A, O'Hara G, Ewer K, Lawrie A, Goodman A, Nicosia A, Folgori A, Colloca S, Cortese R, Gilbert SC, Draper SJ (2010) Prime-boost vectored malaria vaccines: progress and prospects. Hum Vaccin 6:78–83

Hviid L (2010) The role of Plasmodium falciparum variant surface antigens in protective immunity and vaccine development. Hum Vaccin 6:84–89

Langhorne J, Ndungu FM, Sponaas AM, Marsh K (2008) Immunity to malaria: more questions than answers. Nat Immunol 9:725–732

Moorthy VS, Kieny MP (2010) Reducing empiricism in malaria vaccine design. Lancet Infect Dis 10:204–211

Richards JS, Beeson JG (2009) The future for blood-stage vaccines against malaria. Immunol Cell Biol 87:377–390

Shah S (2010) The fever: how malaria has ruled humankind for 500,000 years. Farrar, Straus and Giroux, New York

Sherwin IW (2009) The elusive malaria vaccine: miracle or mirage? ASM Press, Washington, DC

Vaughan AM, Wang R, Kappe SH (2010) Genetically engineered, attenuated whole-cell vaccine approaches for malaria. Hum Vaccin 6:107–113

Wykes M, Good MF (2007) A case for whole-parasite malaria vaccines. Int J Parasitol 37:705–712

Chapter 9
Marine and Freshwater Fecal Indicators and Source Identification

Sandra L. McLellan, Alexandria B. Boehm, and Orin C. Shanks

Glossary

Alternative fecal indicators	Fecal indicators that have not been fully validated for standard water quality methods, but show potential for increased sensitivity or specificity over current indicators.
Commensal	The general meaning of this word is sharing of food and originates from the Latin word *cum mensa*, meaning "sharing a table." In the context of bacteria and host interactions, the bacteria benefit from the host without causing harm.
Enterococci	The term "enterococci" is a general reference to members of the genus *Enterococcus*: however, in the context of water quality standards, enterococci often refers to *E. faecalis* and *E. faecium*, which can be enumerated using selective and differential media.

This chapter, which has been modified slightly for the purposes of this volume, was originally published as part of the Encyclopedia of Sustainability Science and Technology edited by Robert A. Meyers. DOI:10.1007/978-1-4419-0851-3

S.L. McLellan (✉)
School of Freshwater Sciences, University of Wisconsin-Milwaukee, Milwaukee, WI, USA
e-mail: mclellan@uwm.edu

A.B. Boehm
Environmental and Water Studies, Department of Civil and Environmental Engineering, Stanford University, Stanford, CA 94305, USA
e-mail: aboehm@stanford.edu

O.C. Shanks
National Risk Management Research Laboratory, United States Environmental Protection Agency, Cincinnati, OH, USA
e-mail: shanks.orin@epa.gov

P. Kanki and D.J. Grimes (eds.), *Infectious Diseases: Selected Entries from the Encyclopedia of Sustainability Science and Technology*,
DOI 10.1007/978-1-4614-5719-0_9,

Escherichia coli (*E. coli*)	Gram-negative bacteria found in the gastrointestinal tract of almost all warm-blooded animals. These bacteria are easily cultured and can be enumerated using selective and differential media.
Fecal indicator	A chemical or biological constituent that is found in fecal matter that can be used to demonstrate the presence of contamination.
Pathogen	A microbe or microorganism such as a bacteria, virus, fungi, prion, or protozoan that causes disease in its animal or plant host.
Polymerase chain reaction (PCR)	A scientific technique in molecular biology to amplify a single or few gene copies of a nucleic acid fragment across several orders of magnitude. Amplification results in the generation of thousands to millions of copies of a particular nucleic acid sequence.
Quantitative PCR	A technique based on PCR which simultaneously amplifies and quantifies a targeted nucleic acid molecule.

Fecal Indicator Definition

Fecal indicators are organisms or chemical constituents found in fecal material or wastewater that can be measured to demonstrate the presence of fecal pollution. Fecal waste from humans and other animals can contaminant surface waters and pose a serious threat to the environment and human health. Fecal pollution serves as a vehicle for disease transmission including pathogenic bacteria, viruses, or protozoa. Fecal waste also carries with it harmless *commensal* organisms that live in the gastrointestinal (GI) tract and are often used as fecal indicators since they are present in high numbers. The type and amount of pathogens found in fecal pollution is dependent on the host source (human, agricultural animal, wildlife) and the prevalence of illness in the host population. Therefore, employing fecal indicators that provide information about human and other animal contributions is critical for estimating the likelihood that pathogens are present and for directing remediation efforts.

Introduction

Fecal indicators when detected demonstrate the presence of fecal pollution. Fecal indicators play an important role in regulation. Governmental agencies charged with the protection of human health use indicators to assess recreational water quality. In much of the developed world, *Escherichia coli* and *enterococci* are the organisms used for this purpose. Their quantitative link to human health risk in recreational epidemiology studies has led to development of water quality criteria to limit their concentrations in the USA and worldwide.

Conventional indicator methods focus on the cultivation of *E. coli* or enterococci cells isolated from an environmental sample. Culture-based methods are inexpensive and do not require extensive laboratory training to implement. However, these methods are time consuming, requiring 18–24 h to process samples. They also have other limitations such as the inability to discriminate between different animal sources and the potential of indicator microorganisms to persist and sometimes proliferate in the environment. As new scientific discoveries provide a broader view of the different microbes or chemicals associated with fecal pollution and specific sources, new indicators are being identified. These indicators are often referred to as alternative indicators since they have not been fully validated for use for standard methods in water quality testing, but show promise to address some of the limitations associated with conventional fecal indicator approaches.

Some alternative indicators are common to all sources of fecal pollution and can be used as general fecal pollution indicators. Others are associated with a particular host or group of animals. Host-associated indicators are useful for fecal source identification approaches, which are aimed at improving estimates of potential health risk due to pathogens, or identifying major pollution sources that should be remediated. Alternative indicators may take the place of conventional indicators as technology advances. Technologies such as real-time *quantitative PCR* (qPCR), flow cytometry, and advanced chemical analyses can detect previously uncultured microbes or chemicals associated with fecal pollution.

Impact of Fecal Pollution on Coastal Waters

Coastal waters are a valuable resource. Fecal pollution of beaches is not only a threat to human health [13, 52, 98, 195, 258], but can also result in economic losses to surrounding communities [113, 197]. Within the USA, the ocean and Great Lakes coasts encompass more than 15,000 miles of coastline and are the home of economic and recreational centers and unique and rich ecosystems. Many coastal areas are stressed because of dense development and subsequent anthropogenic impacts (Fig. 9.1). Studies have shown that with increasing urbanization, there is an increase of fecal pollution in waterways [91, 151, 261]. Agricultural land use in upper reaches of watersheds also contributes to fecal pollution in tributaries that ultimately discharge into the ocean or the Great Lakes [24, 261]. Fecal pollution is the major cause of biological water quality impairment in the USA and is the primary cause of recreational beach advisories and closing [252]. Currently, fecal pollution impacts are determined by measuring fecal indicator bacteria using conventional, culture-based approaches. In 2009, there were 18,682 advisories and closures at 2,876 beaches in the USA that are routinely monitored for fecal pollution.

Fig. 9.1 Plume of river water released into Lake Michigan from the Grand River (Photo provided by Dr. Philip Roberts, Georgia Institute of Technology)

The Link Between Waterborne Disease and Fecal Pollution

Fecal pollution may contain pathogens that can cause disease in humans. To date, there are more than 150 different agents of disease that can be considered waterborne pathogens. This list grows each year as additional emerging pathogens are identified. Table 9.1 lists common waterborne pathogens and their major host reservoirs. The primary reservoir of human viruses is humans themselves because viruses by nature are host specific; however, animal viruses may also be a concern if they are able to replicate in human hosts. Recent research has identified pigs as a reservoir of hepatitis E virus [99]. Sewage may contain high concentrations of human viruses and some studies have performed surveillance of the viral diseases in the community by monitoring sewage [213, 214]. Some pathogens are predominately found in nonhuman animal hosts, but if humans become infected, person-to-person or waterborne transmission may occur.

Exposure to contaminated water and potential waterborne pathogens most notably causes enteric illness, but skin, ear and eye, or respiratory illnesses may also

Table 9.1 Common pathogens responsible for enteric illnesses

Agent		Major sources	References
Viruses[a]			Reviewed by Fong and Lipp [76]
	Enteroviruses	Human	[32, 91, 122, 125, 167, 178]
	hepatitis A	Human	[86, 91, 122]
	Adenoviruses	Human	[51, 108, 122, 124, 125, 271, 273]
	Norovirus	Human	[17, 91, 263, 273]
	Rotavirus	Human	[85, 202]
	Astrovirus	Human	[47, 194]
Bacteria			
	Pathogenic *E. coli*	Humans and animals[b]	[5, 102, 109, 171, 260]
	Shigella	Humans and animals	[118]
	Vibrio cholerae	Humans and environment	[129, 130, 147]
	Campylobacter	Animals and humans	[8, 33, 112, 183, 254, 260, 270]
	Salmonella	Animals and humans	[21, 100, 144, 254]
	Yersinia enterocolitica	Animals	[104, 205]
	Aeromonas	Environment	[192, 215]
	Plesiomonas	Environment	[11, 166, 203]
	Vibrio parahemolyticus	Environment	[55, 127, 254]
Parasites			
	Cryptosporidium parvum	Animals and humans	[90, 112, 144, 149, 270]
	Giardia lamblia	Animals and humans	[90, 112]
	Entameba histolytica	Animals and humans	[152]

[a]Humans are the predominate source of human viruses, but in some cases, transmission from animals to human is possible. For example, this is suspected to be possible for hepatitis E [99], a calicivirus in the same virus family as norovirus

[b]Animals are the primary reservoir of *E. coli* O157:H7 [46], one strain of shiga-toxin-producing *E. coli*

Table 9.2 Common pathogens responsible for respiratory illness and skin infections

Agent		Primary source	References
viruses			
	Adenovirus	Humans	[51, 108, 122, 124, 125, 271, 273]
Bacteria			
	Vibrio parahemolyticus	Environment	[55, 127, 254]
	Vibrio vulnificus	Environment	[187, 264]
	Leptospira	Animals (wildlife)	[96, 236]
	Legionella spp.	Environment	[82, 186, 236]
	Staphylococcus aureus	Humans	[48, 49, 88, 254]

For an expanded list, please see [27, 168]

occur [27, 39, 65, 141, 195, 258]. Many waterborne disease agents are passed through the fecal-oral route, so any activities that involve ingesting contaminated water present a health risk. Ingesting contaminated seafood may also result in exposure to waterborne pathogens (Table 9.2). For respiratory diseases, inhalation of water droplets or direct contact with mucus membranes can expose a person to a disease-causing agent. Direct contact of contaminated water with wounds could result in an infection.

Recreational waters are of particular concern because swimmers can come into direct contact with contaminated water. Shellfish beds can also be impacted by fecal pollution and are regularly monitored to assure that harvested shellfish has not been subjected to contamination. In the Great Lakes, nearshore coastal waters are a drinking water source to nearly 40 million people. Stringent treatment requirements provide safe drinking water, but both source water and treated drinking water are closely monitored for evidence of fecal pollution to assure that treatment protocols are adequate. For a more complete discussion of this topic, see Chaps. 3–5 of this volume.

Important Attributes of Indicators. Fecal indicators can either be general indicators of fecal pollution or associated with a particular animal source. Many watersheds and coastal waters have mixed land use; therefore, both general fecal indicators and source-associated indicators have an important role in assessing water quality. Ideally, fecal indicators should be present in high levels in fecal pollution so that they can be used as a sensitive measure of the level of contamination when diluted to small concentrations in the environment. Fecal indicators should provide information about host source contribution when possible, whether it is from humans, or different agricultural animals or wildlife. Detection methods should be relatively simple and affordable considering that much of the hands-on monitoring is done by local health departments. Methods should lend themselves to rapid testing so that beach notification can happen in a timely manner.

Clearly, no single indicator can meet all of these goals. Therefore, it is critical to have multiple indicators that can be used in concert if needed. Different indicators will behave differently in various environments, e.g., marine waters versus freshwater [9, 92, 180, 227, 229]. Certain indicators may be appropriate for investigating sources of fecal pollution, or setting remediation goals, whereas others are better suited for rapid detection for recreational water quality monitoring for any fecal pollution present. Water resource managers, public health officials, and researchers must work together to identify what information is needed and choose the most appropriate indicators. For example, *E. coli* is recommended for freshwater, but it has a very short half-life in the open waters of the Great Lakes [160]; therefore, highly persistent indicators such as *Clostridium perfringens* may be more useful for long-term monitoring [71, 146, 169]. Enterococci qPCR is being developed for rapid beach testing [93, 106], but is a general indicator and host-associated indicators within the order *Bacteroidales* may be more useful for identifying sources [31, 72, 236].

Important Attributes of Fecal Indicators
Routine monitoring: Specific for fecal pollution
Present when pathogens are present
Correlate well to illnesses
Easy to quantify
Simple and cost-effective methodology
Behave in the environment in the same manner as pathogens
No growth outside of the host environment
Amenable to rapid detection methodologies
Investigations: Specific for a source of fecal pollution
Sensitive, e.g., present in high numbers in most animals with a given source
Known or predictable ecology
Known or predictable relationship to pathogens

Detection of Conventional Indicators

Common fecal indicators that are used for water quality monitoring or recreational beaches are listed in Table 9.3. All of these indicators were originally identified as constituents of fecal pollution using selective and differential culture techniques. The earliest methods data back to late 1800s and early 1900s [14, 66] for coliform bacteria. There are two culture approaches for enumerating bacteria in water samples. The most probable number (MPN) methods involve culture-based detection in liquid broth using a series of dilutions. The dilutions in which organisms are detected can be used to calculate a statistical estimate of *enterococci* concentration for that sample. The second approach involves filtering samples through a membrane filter. The filter is transferred to solid selective media that is optimized for the growth of the target organisms and inhibitory for other organisms. Various chromogenic substrates or pH indicators can be incorporated to make the media differential for fecal indicator microorganisms. A review of conventional and novel indicators can be found in Edge and Boehm [69].

Culture-based methods continue to be widely used for detection of fecal indicator bacteria; however, the time required to obtain a result is a major limitation

Table 9.3 Historic and conventional fecal indicators

Organisms	Use	Limitations
Total coliforms	Early indicator used in surface waters, currently in use for drinking water[a] since detection of total coliforms provides information on general sanitation	Not specific for fecal pollution
Fecal coliforms	Used for recreational waters until late 1980s or 1990s in some US states, still in use as a standard for wastewater, recreational waters, and shellfish	Some fecal coliforms can grow in the environment
E. coli	Currently recommended by the USEPA for fresh recreational waters[b]. Replaced fecal coliforms as a more specific indicator of fecal pollution	More specific than fecal coliforms, but has been reported to persist and grow in the environment [7, 117, 134, 265]
Fecal streptococci	Early indicator for surface water quality	Not all are fecal specific
Enterococci	Currently recommended indicator for marine recreational waters. Replaced fecal streptococci	General indicator; some grow in the environment
Clostridium perfringens	Proposed in 1963 as an indicator of wastewater and receiving waters. Used in some European countries but not the USA	May survive for long periods in the environment.

[a]Total coliforms (TC) are used since they are a good measure of bacteriological contamination, regardless of fecal or environmental sources. New proposed rules would change the standard from TC to *E. coli*, e.g., more specific for fecal pollution
[b]Criteria are being revised; new criteria will be based on detection of enterococci by culture or qPCR (total *Bacteroides*)

of these methods for providing rapid (e.g., 4 h) results of beach water quality to assure timely public notification. Molecular methods such as qPCR can be used for detection of traditional fecal indicators [106, 182, 267].

Coliforms. Coliform bacteria are a group of bacteria that were the first indicators of fecal pollution. Coliforms are gram-negative, rod-shaped, facultatively anaerobic, non-spore-forming bacteria found in warm-blooded animals, as well as in soil, water, and vegetation. Coliforms are not a specific taxonomic group of bacteria, but are classified based on a number of characteristics. These organisms are identified by fermentation of lactose with the production of acid and gas at 35–37°C. Coliforms are also negative for cytochrome oxidase and positive for β-galactosidase. Coliforms are measured by using an MPN [16] or by enumeration of colony-forming units (CFU) using membrane filtration and selective and differential media such as MI [249]. These organisms generally are within the family *Enterobacteriaceae* and include the genera *Citrobacter*, *Escherichia*, *Enterobacter*, *Hafnia*, *Klebsiella*, and *Serratia*.

Coliform bacteria were one of the earliest indicators of water quality used in the USA, with individual states setting limits of 50–2,400 coliforms per 100 ml of water as a standard for recreation waters in the 1950s and 1960s [66].

Fecal coliforms. Fecal coliforms are a subgroup of coliforms and refer more specifically to coliforms derived from feces. Like coliforms, they are not a specific taxonomic group; they are based upon several morphological and physiological characteristics. These are defined by the same criteria as coliforms, but are thermotolerant and will grow at 44.5°C. *E. coli* is one of the major fecal coliforms found in feces, in addition to members of *Klebsiella*, *Enterobacter*, and *Citrobacter*. The designation of fecal coliforms was intended to improve specificity; however, some organisms included in this group can be found free living in the environment, most notably *Klebsiella* [42, 83, 177]. Beach water samples have also been found that have evidence of fecal coliforms that have replicated in the environment [158].

The first national water quality criterion for recreational waters was based upon fecal coliforms. In 1968, the National Technical Advisory Committee, commissioned by the Federal Water Pollution Control Administration (now referred to as the Environmental Protection Agency), determined that 400 fecal coliforms per 100 ml corresponded to an adverse GI health effect [66]. Subsequent recommendations stated that for recreational waters, within a 30-day period, the geometric mean should not exceed 200 fecal coliforms per 100 ml, and 10% of the samples should not exceed 400 fecal coliforms per 100 ml. Fecal coliforms are no longer used for recreation waters in most states, but the basis of the 1968 criteria is still used for regulating water quality of wastewater treatment plant effluents and for assessing river water quality. Fecal coliforms are also still used for shellfish testing (water overlying the reefs and oyster meats).

Escherichia coli (E. coli). *E. coli* is a fecal coliform that has been suggested to be more specific for fecal pollution than testing for the group of fecal coliforms and was recommended as an indicator for freshwater in 1986 by the United States Environmental Protection Agency (USEPA) [14, 247]. *E. coli* are present in the GI tract of most warm-blooded animals, and therefore a general indicator of fecal pollution. *E. coli* is a thermotolerant coliform that produces indole from tryptophan and it can be differentiated from other microorganisms based on β-glucuronidase activity. Selective and differential media tests for this activity using methods based on membrane filtration, modified mTEC [248], or MPN approaches such as the Colilert manufactured by IDEXX [68] are commonly used to identify *E. coli* in surface water samples. One testing methodology simultaneously detects coliforms and *E. coli* using β-galactosidase and β-glucuronidase activity, respectively, as discriminators [249]. Some epidemiology studies have shown a relationship between *E. coli* densities and GI illness [65, 195]. *E. coli* has some limitations as a fecal indicator at recreational beaches because it has been shown to persist and even grow in some aquatic environments, thereby potentially interfering with the relationship between *E. coli* and recent fecal pollution events [7, 26, 134, 265].

Enterococci. Enterococci are gram-positive cocci and are nearly universally present as commensal organisms in the intestine of human and nonhuman animal hosts. The most common species in human hosts are *E. faecalis* and *E. faecium* [58, 139].

The enterococci are a subgroup of the fecal streptococci. Fecal streptococci have also been referred to as Group D streptococci according to Lancefield serotyping. The fecal streptococci have historically been used as fecal indicators and include species from two genera: *Enterococcus* and *Streptococcus*. There are two *Streptococcus* species in the fecal streptococci group – *Streptococcus bovis* and *Streptococcus equinus* – that have been shown to survive poorly in water. Hence, in water, fecal streptococci and enterococci are thought to be equivalent [116].

In the USA and the EU, enterococci are used for monitoring marine bathing waters because epidemiology studies have linked their concentration to human health outcomes [256]. The standards are tied to approved culture-based methods for their quantification: multiple-tube fermentation, membrane filtration, and defined substrate assays. Clesceri et al. (1998) describe a multiple-tube method where azide dextrose broth is used followed by confirmation with Pfizer selective *Enterococcus* (PSE) media and brain-heart infusion broth with 6.5% NaCl. Both the EU and the USA have approved the use of defined substrate assays manufactured by IDEXX for the quantification of enterococci (Enterolert and Enterolert-E). The USEPA-approved method 1600 utilizes membrane filtration onto mEI media for quantification [250]. Studies that have compared these culture-based methods for quantifying enterococci often find the methods yield slightly different results [32].

The USEPA has developed a qPCR assay for the enumeration of enterococci which has been compared to membrane filtration results [106, 253]. Enterococci measured via qPCR often yield higher concentrations than culture-based measurements since it enumerates both live and dead bacteria [32]. Enterococci measured by qPCR have been linked to human health outcomes in epidemiology studies of marine and fresh water beaches [257–259]. Ongoing work is focused on better defining these links. As the USEPA formulated new recreational water quality criteria, qPCR for enterococci is expected to be included as a rapid method which allows beach managers and public health workers to post water quality advisories on the same day the sample is taken.

Clostridium perfringens. *C. perfringens* is member of the phylum *Firmicutes* and is a gram-positive, low GC content organism. *C. perfringens* was suggested as a potential indicator in 1963 [34], and gained acceptance in EU countries, but it was not chosen for use in the USA because it survives for long periods of time in the environment [14, 66]. Epidemiology studies report a relationship between *C. perfringens* and illness [268], while other studies found no relationship [39]. However, *C. perfringens* has been shown to be a useful fecal indicator in certain environments where other indicators are highly modulated by environmental factors. Studies in tropical waters suggested *C. perfringens* is a better indicator compared with fecal coliforms because it is a spore-forming organism and does not replicate in the environment [79, 80]. Because of its spore-forming ability, *C. perfringens* has been used as a tracer of long-term fecal pollution impacts in marine and freshwater systems [36, 54, 70, 110, 169]. *C. perfringens* has also been suggested as a good indicator in the open waters of the Great Lakes because it can serve as a conservative tracer of fecal pollution and may mimic protozoan cyst or oocyst survival [164, 169, 191].

Table 9.4 Examples of biological *alternative fecal indicators* that provide animal host information

	(Notes)	References
Organisms – 16S rRNA gene		
Bacteroidales	*Bacteroidales* associated with hosts have been identified	[25, 60, 75, 78, 133, 137, 138, 153, 165, 184]
Bifidobacterium		[35, 64, 148, 173]
Methanobrevibacterium	Member of Archaea and dominant in the GI	[128, 243–245]
Faecalibacterium		[161, 279]
Lachnospiraceae		[161, 175]
Ruminococcace		[161]
Functional genes		
esp gene in enterococci	Gene responsible for attachment on human epithelial cells	[93, 212]
Toxin genes in *E. coli*		[50, 131]
Beta-glucuronidase	Polymorphisms in this gene have been linked to different host types	[200]
Unknown genes/regions		
Cattle and human markers	Identified by genomic fragment enrichment	[220, 221]
Gull marker	Identified by subtractive hybridization	[101]
Phenotypes		
Antibiotic resistance of standard fecal indicators	Based on the theory that host exposed to antibiotics will have a higher percentage of antibiotic-resistant *E. coli* or enterococci	[97, 105, 188, 269]
Viruses		
F+ coliphage	Type I and IV associated with animals and II and III associated with humans	[114]
Bacteroides phage		[12, 126, 240]
Adenovirus, enterovirus, polyomaviruses	Viruses are host specific by nature, and therefore, detection of human viruses demonstrates human sources are present.	[6, 123, 162, 179, 193]

Alternative indicators. Ongoing research studies have identified a broad array of new potential indicators of fecal pollution. Molecular-based methods have made possible the characterization of organisms that previously were either not recognized as associated with fecal pollution, or were difficult to detect due to complex cultivation requirements. Alternative indicators may also employ unique chemical constituents. Alternative indicators are being developed as general detection of fecal pollution, such as total *Bacteroides* [53, 59], as well as source identifiers associated with a particular animal group (Table 9.4).

Different sources of fecal pollution can contribute different types and concentrations of pathogens (Table 9.1 and Table 9.2). For example, human fecal sources, particularly sewage, contain waste from a large number of people and are considered a primary source of human enteric viruses. *Cryptosporidium* may be associated with cattle waste. Fecal indicators that provide information about the source will improve our ability to estimate the health risk due to pathogens as well as direct remediation efforts to major contributing sources of fecal pollution. The development of qPCR methodology has also advanced simple presence/absence detection to quantitative estimates of fecal pollution and provides a platform for the implementation of rapid methods.

16S rRNA gene targets. Many of the alternative indicators that have been described are based on detection of the organisms based on the 16S rRNA gene sequence. This gene is highly conserved among bacteria and has been used extensively to assign taxonomy.

Bacteroidales. Members of the order *Bacteroidales* are potentially useful indicators of fecal contamination because they generally are found in high numbers in fecal material of humans and other warm-blooded animals and are unlikely to survive in the beach environment [74, 136]. Early studies identified unique sequences in the *Bacteroides* 16S rRNA gene from human and ruminant *Bacteroides* species that are associated with respective fecal pollution sources [24, 136]. Sequencing of clone libraries demonstrated that sequences of members of the broader *Bacteroidales* group, rather than exclusively *Bacteroides* spp., are amplified with primers originally targeting total *Bacteroides* spp. [60]. Subsequent studies have used taxon-specific cloning to characterize *Bacteroidales* populations within humans and different animals and have identified a broad range of host-associated genetic markers [25, 60, 75, 78, 121, 133, 137, 138, 153, 165, 184]. Since culture techniques for isolation of these anaerobic bacteria are difficult to perform, molecular techniques have been developed to amplify, detect, and in some cases quantify the 16S rRNA genes of *Bacteroides* spp. from feces and ambient water [53, 59, 133, 153, 218, 262]. Many of these assays utilize the HF183 sequence first reported by Bernhard and Field [25]. The utility of the genetic markers has been tested extensively in fecal impacted environments, including beaches [1, 37, 84, 181, 207, 216]. In addition, numerous studies report information on the distribution of these host-associated genetic markers in target and non-target populations [3, 64, 133, 138, 143, 185, 224, 225, 228], relationship to pathogens [208, 209], and the decay of these genetic markers in marine and freshwaters [20, 61, 184, 210, 261].

Bifidobacterium. This genus represents another group of GI bacteria with particular species reported to be associated with human fecal pollution including *B. adolescentis*, *B. dentium*, and *B. longum* [35, 148, 155, 173]. Several technologies targeting *Bifidobacterium* genes are reported for multiplex PCR detection [35] and qPCR [150, 154]. *Bifidobacterium* typically occur at lower concentrations than *Bacteroidales* making them harder to detect in dilute ambient water samples [219] and exhibit a rapid decay based on bench-scale survival studies [201]. Thus, the detection of a *Bifidobacterium* host-associated genetic marker in a polluted water sample suggests a recent, high concentration contamination event.

Faecalibacterium. This genus of bacteria has been reported in humans and other animals and has been suggested as a potential target for development of host-associated genetic markers [81, 161, 246, 279]. Sewage and cattle have been shown to have a high abundance of *Faecalibacterium* [161, 226]. Additional characterization of this group is needed to characterize phylotypes that are associated with specific animal sources.

Lachnospiraceae. *Lachnospiraceae* are found in high abundance in human fecal samples [57, 77, 242], sewage [161], and cattle [226]. *Lachnospiraceae* are included in the group *Clostridium coccoides* [107, 150]. The proportions of *Lachnospiraceae*, *Bacteroides*, and *Bifidobacterium* of the human microbiota vary among different animal species, and quantification of these proportions has been proposed as a method for fecal pollution source identification [81]. Additional characterizations of this group are needed to characterize phylotypes that are associated with specific animal sources [161].

Gene product targets. Molecular methods have also allowed for detection of genes that serve a functional role in the organism. In some cases, the function may be linked to specific host microbe interactions, making these genetic markers potentially good host-associated alternative indicators [221]. Genetic markers have been identified with a variety of molecular methods, including subtractive hybridization, genome fragment enrichment, and other metagenomic approaches.

Toxin genes *of* E. coli. Specific subpopulations of *E. coli* contain genes coding for toxins, including heat-labile enterotoxin (LT) and heat-stable enterotoxin (ST). *E. coli* carrying toxins are generally clonal populations that are found within certain animal reservoirs and have been suggested as host-associated indicators. Specific sequences of the STII toxin gene were found to be associated with swine, but not present in sewage or dairy farm lagoons [132]. Cattle-associated LTIIa has also been reported [50, 131]. These toxin genes have a worldwide distribution [72]. The occurrence of *E. coli* positive for STII or LTIIa can be low in agricultural animal populations, potentially limiting the use of these genes for the identification of specific animal sources.

Esp *gene*. The enterococcal surface protein (*esp*) gene is a putative virulence factor in *Enterococcus faecium* that has been shown to be associated with enterococci from human origin [212]. Because this gene occurs at a low frequency, original detection methods involved an enrichment step where DNA is extracted from enterococci grown on selective media, followed by PCR. Comparison studies have shown the *esp* gene in enterococci to correlate with other human-associated genetic markers [4, 163] and this alternative indicator has been employed in numerous field studies [275]. Newer methods employ qPCR that can directly detect the *esp* gene [2].

gyrB. The genetic locus *gyr*B is a housekeeping gene (e.g., common to all bacteria because of a central function). Similar to 16S rRNA gene loci, housekeeping genes are generally highly conserved and therefore useful for identifying specific phylotypes. One study employed qPCR targeting *gyr*B in *Bacteroides fragilis* as an indicator of human specific fecal contamination [142].

Methanogens. Methanobrevibacter smithii is a dominant Archaea in the human gut [67]. The *nif*H gene of this organism has been used as a human-associated indicator [243]. Similar assays employing the same gene in *Methanobrevibacter ruminantium* have been developed [245]. Assays for quantification of the *nif*H target have also been developed [22, 128]. An Archaea genetic marker may prove useful because it may have a different survival or ecology compared with bacterial indicators and pathogens.

Bacteroides thetaiotaomicron. *B. thetaiotaomicron* is found in high numbers in humans compared with other animals and is described as a niche organism in the human gut [274]. A genomic fragment that was generated with universal primers as a second unexpected amplicon was found to distinguish *B. thetaiotaomicron* from other animal species [241]. PCR primers specific for *B. thetaiotaomicron* were developed based on the sequence of this 547-bp genomic fragment and have been tested against a number of fecal samples from humans and nonhuman sources [45, 241]. A putative gene for complex polysaccharide degradation has also been used as a genetic marker for qPCR since the trait is hypothesized to be involved in host-associated metabolic pathway [277].

Metagenomics. The majority of host-associated genetic markers available to date target the 16S rRNA gene from a limited number of different microorganisms. Advancements in DNA sequencing and sorting technologies now allow researchers to survey the entire genome of all members of fecal microbial community. Different strategies include the use of competitive hybridization approaches [101], microarrays [145, 272], and 454 pyrosequencing [161, 226, 246]. Whole genome and community approaches vastly expand the number of candidate source-associated genetic markers and may allow for the development of even more refined source identification methods.

Viruses

F specific (F^+) coliphages. F^+ coliphage RNA coliphages have serologically distinct groups that predominate in humans (groups II and III) which are distinct from those commonly found in other animals (group I and IV) [114]. Comparison studies of different alternative indicators suggest F^+ coliphage types are reliable indicators of host sources, but the groups are not exclusive to either animal or human sources [28, 180]. Further, differential survival may influence source identification in natural waters and may need to be taken into account in interpreting source identification studies [38, 172]. However, viral indicators may correlate more closely to human viral pathogens as they may have a similar ecology in the environment.

Bacteroides *phages*. Phages infecting *B. fragilis* and *B. thetaiotaomicron* have been used as indicators of human fecal pollution [12, 126, 240]. Differential ability of host strains of *Bacteroides* to detect phages from different sources has been reported [196], as well as geographic variability. Culture methods have been developed to isolate diverse host *Bacteroides* strains [190]. In survival studies, two *B. fragilis* phages were shown to survive longer in seawater compared to MS2 coliphage [157].

Human polyomaviruses. Human polyomaviruses are widespread among human populations and have been suggested as indicators of human waste [6, 162].

Table 9.5 Chemical alternative fecal indicators

Chemical constituents
Fecal sterols
Optical brighteners
Caffeine
Personal care products and pharmaceuticals

This virus is excreted in the urine and therefore may be detected in the absence of human feces. Studies have compared detection of human polyomaviruses with detection of human *Bacteroides* HF183 genetic marker and enterococci carrying the *esp* gene and found a strong correlation [4, 163].

Chemicals. Chemical methods do not detect fecal bacteria. Instead, these methods are designed to detect chemical compounds associated with human activities or sanitary sewage. Chemical indicators may provide additional evidence as to source [87, 95]. These chemicals are often found in sewage treatment facility discharges and septic tank effluent. For example, optical brighteners are commonly found in laundry detergents and have been used to indicate the presence of human fecal pollution in environmental waters [41, 62]. Fecal sterols such as coprostanol are also reported to be associated with human fecal pollution [28, 115, 176, 239]. Other potential chemical fecal indicators include antibacterial compounds, pharmaceuticals, and caffeine [95, 278] (Table 9.5).

Quantification of Bacterial Indicators Using qPCR. Conventional or endpoint PCR allows for the selective amplification of a particular genetic marker at extremely low concentrations even in the presence of a mixture of heterologous DNA targets making it ideal for environmental applications. The final result of an endpoint PCR method is either the presence or absence of the DNA target. Even though the qualitative determination of fecal pollution in a water sample can be very useful information, researchers quickly recognized the added advantage of generating quantitative data. The ability to estimate the concentration of a DNA target in a known volume of water provides a means to investigate relationships between the concentration of a fecal indicator genetic marker and numerous factors such as illness rates in swimmers or efficiency of waste management practices.

qPCR relies on the continuous monitoring of PCR product accumulation as amplification occurs. Estimation of the concentration of a genetic marker is based on the theoretical premise that there is a log-linear relationship between the starting amount of DNA target in a reaction and the fractional thermal cycle where PCR product accumulation is first significantly detectable (Table 9.2); for review see [204]. qPCR applications designed to estimate fecal bacteria concentrations in recreational waters are gaining widespread attention due to the rapid nature of these methodologies (same day results), reports linking the occurrence of DNA targets to public health risk [106, 257, 258], and the development of host-associated fecal source identification assays [40, 133, 135, 138, 163, 185, 218, 222, 223]. However, there are many technical concerns that must be addressed before these qPCR applications are ready for implementation.

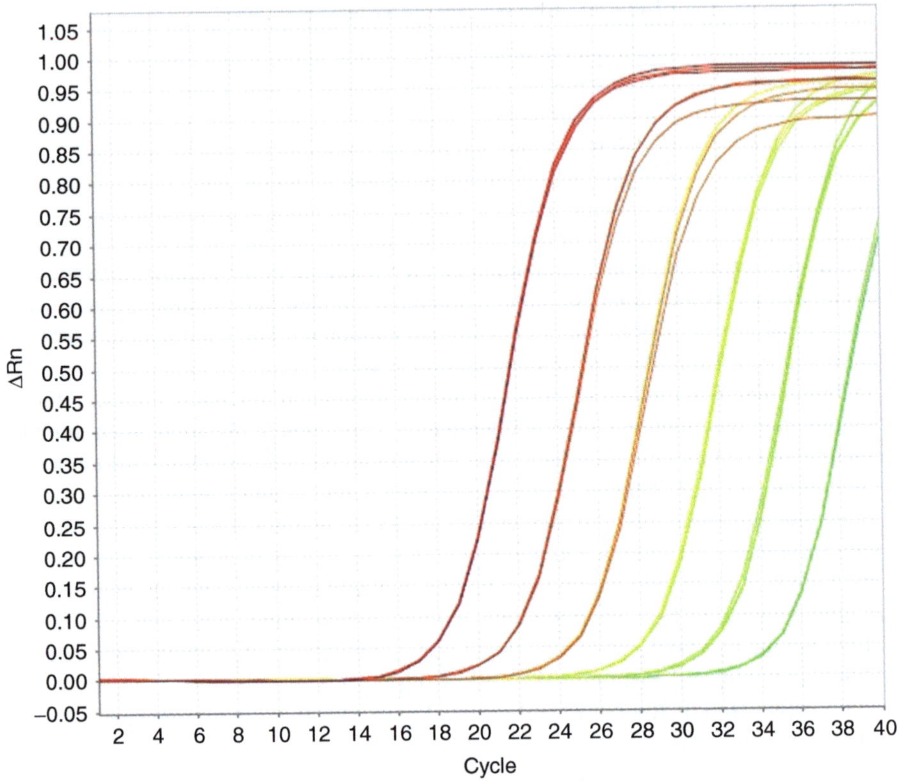

Fig. 9.2 Quantification of real-time polymerase chain reaction (qPCR) product can be achieved by observing an increase in fluorescence, indicating product formation, in relation to cycle number

It is important to recognize that a qPCR method consists of several protocols linked in succession including sample collection, sample preparation, nucleic acid purification, target amplification, and data interpretation. Each of these steps plays a critical role in the successful estimation of a DNA target concentration in an environmental sample. In addition, the extremely high level of sensitivity make qPCR methods highly susceptible to cross-contamination during field sampling, nucleic acid purification, and genetic marker amplification (Fig. 9.2). As a result, numerous studies have been conducted to address issues such as density and distribution of genetic markers in primary and secondary sources [60, 133, 199, 224, 225, 228], sample matrix interference during qPCR amplification [140, 198, 224, 255], estimating decay rates of DNA targets in ambient water [18, 23, 184, 261], loss of target DNA during nucleic acid recovery [106, 170, 238], and selection of a mathematical model to transform raw qPCR data into an estimation of concentration [230, 231].

Microbial Source Identification. Identification of the sources of fecal pollution is important for both developing remediation strategies and for estimating the likelihood of pathogen occurrence. In most cases, the source of fecal pollution in a water body of interest is originally measured because of high amounts of conventional general fecal indicators (i.e., enterococci or *E. coli*). Methods and study designs for source identification, also referred to as "microbial source tracking" (MST) or "fecal source identification" (FSI), has been reviewed extensively [72, 206, 237].

Identifying fecal pollution sources involves understanding both the physical location of the inputs and the contributing host sources. Most source identification studies begin with spatial and temporal sampling since fecal pollution sources are rarely constant and the locations of inputs are not always obvious. Following release into the environment, the ecology of fecal indicators is greatly influenced by the residence time, type of water body (e.g., marine or freshwater, oligotrophic, or nutrient rich), predation, or even potential growth by some conventional indicators [31, 236]. Therefore, it is very difficult to take one or two samples and determine the major source contributing fecal pollution to an impacted body of water.

Spatial and temporal surveys are complemented by using alternative indicators that can provide information as to the host source of fecal pollution. Often, a first tier assessment will involve distinguishing human versus nonhuman fecal pollution [89, 181]. Cross reactivity needs to be considered, along with geographic relevance of a particular indicator. The possible fecal pollution sources within the watershed need to be considered when choosing the most appropriate alternative indicators. The use of alternative indicators for microbial source identification has been reviewed extensively [20, 72, 73, 206, 211, 235, 237].

Early approaches to microbial source identification focused on library-based methods, where either phenotypic traits or genotypes of indicator bacteria were characterized from a particular source and then compared to what was found in surface waters. Methods for characterizing *E. coli* or enterococci have included antibiotic resistance, ribotyping, and repetitive extragenic palindromic PCR [43, 44, 63, 103, 105, 159, 189, 217, 269]. There are multiple complications in using library-based methods that include applicability of the library across geographic locations, specificity of *E. coli* or enterococci indicators to a particular animal host, and complex genetic relationships among these indicators [10, 72, 159, 206, 236]. Further, creating a library is expensive and multiple water samples need to be analyzed because fecal pollution inputs are usually driven by storm events and can involve multiple animal sources. Most source identification methods have moved to marker-based, or non-library dependent, approaches. Marker-based approaches involve utilizing a chemical or biological constituent that is commonly found in the fecal pollution source of interest, in high abundance so that it can be detected easily and associated with a specific human or animal source (Figs. 9.3 and 9.4).

Fig. 9.3 Stormwater outfalls introduce fecal pollution from domestic pets and wildlife into rivers. Stormwater systems can also become contaminated with human sewage from leaking sanitary sewer systems (Photos provided by Dr. Sandra McLellan, University of Wisconsin-Milwaukee)

Fig. 9.4 Large gull populations are common non-point sources of fecal pollution on beaches (Photos provided by Dr. Sandra McLellan, University of Wisconsin-Milwaukee)

Ecology of Pathogens and Indicators in the Environment

The identification of a host-associated marker of fecal pollution goes beyond microbiology. Once the fecal indicator is discharged into the environment, it becomes necessary to understand the various fate and transport mechanisms that control the concentrations of indicators and pathogens at the point of sampling.

Fate processes include dark or photo-inactivation [32], growth [111], sorption and desorption to sediments [19, 94], and grazing by zooplankton [30]. Inactivation has received the most attention of these fate processes. Although a fair amount of work has examined the interaction of pathogens and indicators with sediments, the work has primarily been focused on porous media, and simplified conditions. More work on the interactions of microbial pollutants and sediments and particles in surface waters is needed, particularly given the widespread occurrence of some indicators and pathogens in sediments and beach sands [7, 26, 56, 117, 265, 275].

Transport processes that control indicator and pathogen transport in surface waters include advection and dispersion of waterborne organisms. These processes are fairly well understood [174] and once determined in a particular surface water, they can be used to model microbial pollution. The resuspension and deposition of sediment-bound organisms is more complicated. Some work has examined these processes for *E. coli* [119, 120] and fecal coliforms [234] in streams and lagoons. Yamahara et al. [275] present a conceptual model for how enterococci in beach sands are suspended into the water column. A better mechanistic understanding of how organisms in the sediment or sand are transported into the water column is warranted.

Of the fate and transport processes described above, perhaps the most important to consider when choosing an indicator for microbial source identification is the time scale of inactivation and its tendency to sorb to sediments. For example, if the goal is enterococci source identification for designing remediation strategies, then ideally, the persistence of the genetic marker will mirror that of enterococci. A health-protective goal may be to have no feces present in a water body. If this is the case, then a source identifier with very long-persistence may be needed. A source identifier that interacts strongly with sediments may be problematic as it may allow sediments to become a secondary, environmental source of the marker. Generally, sediments are believed to be a protective environment for microorganisms, particularly bacteria, where they may persist or even grow [276]. Future work on source identifiers will need to document the importance of sorption and interactions with sediments in general.

Estimating Risk of Pathogen Exposure Using Fecal Indicators

Using fecal indicators to link the presence of fecal pollution to waterborne disease risk is challenging. The types of pathogens that might be present will depend primarily on the source of fecal pollution. For example, sanitary sewer discharges (human sources) may contain high levels of human viruses, whereas wildlife is less likely to carry human viruses, but may contain protozoan and bacteria that can infect humans. Comprehensive models that integrate data from several research fields such as occurrence of pathogens in fecal sources, dose–response relationships, source identifier decay behaviors, acceptable health risk, and route of transmission can be used to estimate risk and are termed quantitative microbial risk assessment (QMRA) [15, 232, 233]. The type of pathogen present will also depend on the prevalence of the disease-causing agent within the population at the time of contamination. Many human viruses are seasonal, and protozoans such as *Cryptosporidium* are prevalent during certain times of the year, such as spring when calves can shed high concentrations of this microorganism.

Factors that Diminish the Relationship Between Indicators and Pathogens
Seasonality of certain pathogens
Rate of infection in the host reservoir (herd, human population)
Differential decay
Differences in transport
Differences in sedimentation rates and partitioning to soil, sand, and sediments

Epidemiological Studies

Epidemiology studies have been conducted around the world to understand the correlative relationship between indicator concentrations and human health. The studies that have been conducted to date, and their methodologies, are summarized by Boehm and Soller (see Recreational Water Risk: Pathogens and Fecal Indicators. Most of the studies have focused on the health effects of recreational exposure to human fecal contamination from publicly owned treatment work discharges. These studies generally show a statistically significant correlation between enterococci and GI illness [256] in marine waters and *E. coli* and GI illness at freshwater beaches. Epidemiology studies are the cornerstone of the USA and EU water quality criteria and directives [31]. Acceptable illness rates are anchored to concentrations of indicator organisms in order to set acceptable contaminant levels. In the USA, 19 illnesses per 1,000 people is the acceptable illness level for marine water recreation, and in freshwater, the acceptable level is 8 illnesses per 1,000 people.

There are several important knowledge gaps in the understanding of how fecal contamination in recreational waters affects human health [31]. Few studies have documented the human health effects from exposure to nonhuman sources of fecal contamination including, but not limited to, bird and dog feces and urban and agricultural runoff [52, 98, 156]. A review of these studies suggests the relationship between indicator concentration and recreational waterborne illness risks is equivocal. Current studies with QMRA are trying to more fully understand the risks for exposure to animal feces in recreational waters [233].

Fecal Indicator Applications. There are numerous applications for fecal indicators and indicators need to be chosen that best serve a specific purpose or goal.

One primary purpose of an indicator is to evaluate the public health risk for recreational water. In this case, general indicators may be employed since beach managers will need to know if fecal pollution is present and at what level. Since the presence of pathogens is highly dependent on the source of fecal pollution, adequate protection of public health will depend on assuming that the indicators are derived from sources that carry the highest pathogen burden. Rapid detection of a fecal indicator is more important than the level of information provided by the indicator since water quality can change rapidly in the beach environment [29]. Ultimately, the source of fecal pollution needs to be identified and remediated to remove the health risk.

Fecal indicators also serve as important tools for sanitary survey practices and for prioritizing remediation strategies. While daily monitoring with a general indicator such as enterococci or *E. coli* will provide information on the extent of fecal pollution, the source needs to be identified in order to take corrective actions. Both extensive mapping of the physical location of fecal pollution inputs (where is it coming from?) and determination of the host sources (is it human or nonhuman sources?) are necessary. Host-associated alternative indicators are best suited for these applications.

Source identifiers can also be used to evaluate the success of best management practices and influence of many green infrastructure efforts in agriculture and urban run-off settings. For example, the installation of tile drainage systems or constructed wetlands is commonly used to control the flow of agricultural waste across the landscape during rain events. Host-associated methods provide an excellent metric for estimating the efficiency of these waste management practices.

Rapid Methods for Indicators

Recreational water quality monitoring has traditionally relied upon culture-based methods and therefore test results are not available to the public until, at the earliest, the following day. It is well established that water quality can change in a matter of hours [29, 266]. A high priority for beach managers is to utilize rapid testing methods, many of which are based on qPCR of fecal indicators. Studies have compared different rapid methods [93]. New water quality criteria that are being formulated by the USEPA are expected to include rapid methods for enterococci using qPCR.

BEACH Act Legislation. The Beaches Environmental Assessment and Coastal Health (BEACH) Act of 2000 is an amendment to the Federal Water Pollution Control Act (commonly known as the Clean Water Act). This legislation required states and tribes to adopt new or revised water quality standards by 2004. It also required the USEPA to publish new or revised criteria for pathogens and pathogen indicators. The BEACH Act authorized appropriations for states and tribes to develop and implement water quality monitoring and public notification programs at recreational beaches. The USEPA has identified scientific gaps that need to be filled in order to develop improved water quality criteria [251].

Future Directions

The identification of host-associated source identifiers represents the first step toward the successful implementation of a fecal indicator method. Several additional steps must be taken to complete the method development phase including method optimization, design of appropriate laboratory controls, and defining a data interpretation model. After method development, it is necessary to define the operational parameters of the method. In the case of qPCR, this might include factors such as generation of a calibration curve, defining the range of quantification, precision, and limit of detection. The next step is to characterize the robustness of the method by measuring specificity, host distribution of the source identifier, abundance of source identifier in target group, describing fate and transport mechanisms, establishing links to general fecal indicators, pathogens, and public health outcomes. Once the operational parameters and robustness of the method are adequately described, a multiple laboratory validation study should be conducted to address issues of reproducibility, variability between laboratories, normalization of results, standardization of controls, minimum requirements to establish laboratory efficiency, and requirements for laboratory training. It is important to note that this list is not comprehensive. There may be additional steps required depending on the intended use of the method.

Rapidly advancing technologies will provide new opportunities to expand the number and types of fecal indicators. Next-generation sequencing technologies have increased our capacity to analyze whole microbial communities, rather than single organisms. Advancing technologies will also allow for more detailed analyses of the dynamics of fecal indicators in the environment. Further, more sensitive, specific, and rapid detection strategies are needed to improve monitoring programs for devising pollution remediation strategies and for the protection of public health.

Acknowledgments We would like to thank Dr. Alford Dufour for insightful discussion and for providing historical information on the use of traditional indicators and evolution of water quality criteria. The US Environmental Protection Agency, through its Office of Research and Development, funded and managed, or partially funded and collaborated in, the research described herein. It has been subjected to the Agency's peer and administrative review and has been approved for external publication. Any opinions expressed in this entry are those of the author(s) and do not necessarily reflect the views of the Agency; therefore, no official endorsement should be inferred. Any mention of trade names or commercial products does not constitute endorsement or recommendation for use.

Bibliography

1. Ahmed W, Powell D, Goonetilleke A, Gardner T (2008) Detection and source identification of faecal pollution in non-sewered catchment by means of host-specific molecular markers. Water Sci Technol 58:579–586

2. Ahmed W, Stewart J, Gardner T, Powell D (2008) A real-time polymerase chain reaction assay for quantitative detection of the human-specific enterococci surface protein marker in sewage and environmental waters. Environ Microbiol 10:3255–3264
3. Ahmed W, Stewart J, Powell D, Gardner T (2008) Evaluation of Bacteroides markers for the detection of human faecal pollution. Lett Appl Microbiol 46:237–242
4. Ahmed W, Goonetilleke A, Powell D, Chauhan K, Gardner T (2009) Comparison of molecular markers to detect fresh sewage in environmental waters. Water Res 43:4908–4917
5. Ahmed W, Sawant S, Huygens F, Goonetilleke A, Gardner T (2009) Prevalence and occurrence of zoonotic bacterial pathogens in surface waters determined by quantitative PCR. Water Res 43:4918–4928
6. Ahmed W, Wan C, Goonetilleke A, Gardner T (2010) Evaluating sewage-associated JCV and BKV polyomaviruses for sourcing human fecal pollution in a coastal river in Southeast Queensland, Australia. J Environ Qual 39:1743–1750
7. Alm EW, Burke J, Spain A (2003) Fecal indicator bacteria are abundant in wet sand at freshwater beaches. Water Res 37:3978–3982
8. Alonso JL, Alonso MA (1993) Presence of *Campylobacter* in marine waters of Valencia. Spain Water Res 27:1559–1562
9. Anderson KL, Whitlock JE, Harwood VJ (2005) Persistence and differential survival of fecal indicator bacteria in subtropical waters and sediments. Appl Environ Microbiol 71:3041–3048
10. Anderson MA, Whitlock JE, Harwood VJ (2006) Diversity and distribution of *Escherichia coli* genotypes and antibiotic resistance phenotypes in feces of humans, cattle, and horses. Appl Environ Microbiol 72:6914–6922
11. Arai T, Ikejima N, Itoh T, Sakai S, Shimada T, Sakazaki R (1980) A survey of Plesiomonas shigelloides from aquatic environments, domestic animals, pets and humans. J Hyg (Lond) 84:203–211
12. Araujo RM, Puig A, Lasobras J, Lucena F, Jofre J (1997) Phages of enteric bacteria in fresh water with different levels of faecal pollution. J Appl Microbiol 82:281–286
13. Arnone RD, Walling JP (2007) Waterborne pathogens in urban watersheds. J Water Health 5:149–162
14. Ashbolt NJ, Grabow OK, Snozzi M (2001) Indicators of microbial water quality. In: Fewtrell L, Bartram J (eds) Water quality: guidelines, standards, and health. IWA Publishing, London
15. Ashbolt NJ, Schoen ME, Soller JA, Roser DJ (2010) Predicting pathogen risks to aid beach management: the real value of quantitative microbial risk assessment (QMRA). Water Res 44:4692–4703
16. Association, A. P. H. (1999) Standard methods for the examination of water and wastewater, 18th edn. American Public Health Association, Washington, DC
17. Aw TG, Gin KY, Ean Oon LL, Chen EX, Woo CH (2009) Prevalence and genotypes of human noroviruses in tropical urban surface waters and clinical samples in Singapore. Appl Environ Microbiol 75:4984–4992
18. Bae S, Wuertz S (2009) Rapid decay of host-specific fecal Bacteroidales cells in seawater as measured by quantitative PCR with propidium monoazide. Water Res 43:4850–4859
19. Bai S, Lung WS (2005) Modeling sediment impact on the transport of fecal bacteria. Water Res 39:5232–5240
20. Balleste E, Bonjoch X, Belanche LA, Blanch AR (2010) Molecular indicators used in the development of predictive models for microbial source tracking. Appl Environ Microbiol 76:1789–1795
21. Baudart J, Lemarchand K, Brisabois A, Lebaron P (2000) Diversity of *Salmonella* strains isolated from the aquatic environment as determined by serotyping and amplification of the ribosomal DNA spacer regions. Appl Environ Microbiol 66:1544–1552
22. Baums IB, Goodwin KD, Kiesling T, Wanless D, Diaz MR, Fell JW (2007) Luminex detection of fecal indicators in river samples, marine recreational water, and beach sand. Mar Pollut Bull 54:521–536

23. Bell A, Layton AC, McKay L, Williams D, Gentry R, Sayler GS (2009) Factors influencing the persistence of fecal Bacteroides in stream water. J Environ Qual 38:1224–1232
24. Bernhard AE, Field KG (2000) Identification of nonpoint sources of fecal pollution in coastal waters by using host-specific 16 S ribosomal DNA genetic markers from fecal anaerobes. Appl Environ Microbiol 66:1587–1594
25. Bernhard AE, Field KG (2000) A PCR assay to discriminate human and ruminant feces on the basis of host differences in Bacteroides-Prevotella genes encoding 16 S rRNA. Appl Environ Microbiol 66:4571–4574
26. Beversdorf LJ, Bornstein-Forst SM, McLellan SL (2007) The potential for beach sand to serve as a reservoir for *Escherichia coli* and the physical influences on cell die-off. J Appl Microbiol 102:1372–1381
27. Bienfang PK, Defelice SV, Laws EA, Brand LE, Bidigare RR, Christensen S, Trapido-Rosenthal H, Hemscheidt TK, McGillicuddy DJ, Anderson DM, Solo-Gabriele HM, Boehm AB, Backer LC (2011) Prominent human health impacts from several marine microbes: history, ecology, and public health implications. Int J Microbiol 2011:152815
28. Blanch AR, Belanche-Munoz L, Bonjoch X, Ebdon J, Gantzer C, Lucena F, Ottoson J, Kourtis C, Iversen A, Kuhn I, Moce L, Muniesa M, Schwartzbrod J, Skraber S, Papageorgiou GT, Taylor H, Wallis J, Jofre J (2006) Integrated analysis of established and novel microbial and chemical methods for microbial source tracking. Appl Environ Microbiol 72:5915–5926
29. Boehm AB (2007) Enterococci concentrations in diverse coastal environments exhibit extreme variability. Environ Sci Technol 41:8227–8232
30. Boehm AB, Keymer DP, Shellenbarger GG (2005) An analytical model of enterococci inactivation, grazing, and transport in the surf zone of a marine beach. Water Res 39:3565–3578
31. Boehm AB, Ashbolt NJ, Colford JM Jr, Dunbar LE, Fleming LE, Gold MA, Hansel JA, Hunter PR, Ichida AM, McGee CD, Soller JA, Weisberg SB (2009) A sea change ahead for recreational water quality criteria. J Water Health 7:9–20
32. Boehm AB, Yamahara KM, Love DC, Peterson BM, McNeill K, Nelson KL (2009) Covariation and photoinactivation of traditional and novel indicator organisms and human viruses at a sewage-impacted marine beach. Environ Sci Technol 43:8046–8052
33. Bolton F, Surman SB, Martin K, Wareing DR, Humphrey TJ (1999) Presence of *Campylobacter* and *Salmonella* in sand from bathing beaches. Epidemiol Infect 122:7–13
34. Bonde GJ (1963) Bacterial indicators of water pollution. A study of quantitative estimation. Teknisk Forlag, Copenhagen
35. Bonjoch X, Balleste E, Blanch AR (2004) Multiplex PCR with 16 S rRNA gene-targeted primers of bifidobacterium spp. to identify sources of fecal pollution. Appl Environ Microbiol 70:3171–3175
36. Bothner MH, Takada H, Knight IT, Hill RT, Butman B, Farrington JW, Colwell RR, Grassle JF (1994) Sewage contamination in sediments beneath a deep-ocean dump site off New-York. Mar Environ Res 38:43–59
37. Bower PA, Scopel CO, Jensen ET, Depas MM, McLellan SL (2005) Detection of genetic markers of fecal indicator bacteria in Lake Michigan and determination of their relationship to *Escherichia coli* densities using standard microbiological methods. Appl Environ Microbiol 71:8305–8313
38. Brion GM, Meschke JS, Sobsey MD (2002) F-specific RNA coliphages: occurrence, types, and survival in natural waters. Water Res 36:2419–2425
39. Cabelli VJ, Dufour AP, McCabe LJ, Levin MA (1982) Swimming-associated gastroenteritis and water quality. Am J Epidemiol 115:606–616
40. Caldwell JM, Raley ME, Levine JF (2007) Mitochondrial multiplex real-time PCR as a source tracking method in fecal-contaminated effluents. Environ Sci Technol 41:3277–3283

41. Cao Y, Griffith JF, Weisberg SB (2009) Evaluation of optical brightener photodecay characteristics for detection of human fecal contamination. Water Res 43:2273–2279
42. Caplenas NR, Kanarek MS (1984) Thermotolerant non-fecal source Klebsiella pneumoniae: validity of the fecal coliform test in recreational waters. Am J Public Health 74:1273–1275
43. Carson CA, Shear BL, Ellersieck MR, Asfaw A (2001) Identification of fecal *Escherichia coli* from humans and animals by ribotyping. Appl Environ Microbiol 67:1503–1507
44. Carson CA, Shear BL, Ellersieck MR, Schnell JD (2003) Comparison of ribotyping and repetitive extragenic palindromic-PCR for identification of fecal *Escherichia coli* from humans and animals. Appl Environ Microbiol 69:1836–1839
45. Carson CA, Christiansen JM, Yampara-Iquise H, Benson VW, Baffaut C, Davis JV, Broz RR, Kurtz WB, Rogers WM, Fales WH (2005) Specificity of a Bacteroides thetaiotaomicron marker for human feces. Appl Environ Microbiol 71:4945–4949
46. Center for Disease Control (2011) posting date. *Escherichia coli* O157:H7, General information
47. Chapron CD, Ballester NA, Fontaine JH, Frades CN, Margolin AB (2000) Detection of astroviruses, enteroviruses, and adenovirus types 40 and 41 in surface waters collected and evaluated by the information collection rule and an integrated cell culture-nested PCR procedure. Appl Environ Microbiol 66:2520–2525
48. Charoenca N, Fujioka RS (1993) Assessment of *Staphylococcus* bacteria in Hawaii's marine recreational waters. Water Sci Technol 27:283–289
49. Charoenca N, Fujioka RS (1995) Association of staphylococcal skin infections and swimming. Water Sci Technol 31:11–17
50. Chern EC, Tsai YL, Olson BH (2004) Occurrence of genes associated with enterotoxigenic and enterohemorrhagic *Escherichia coli* in agricultural waste lagoons. Appl Environ Microbiol 70:356–362
51. Choi S, Jiang SC (2005) Real-time PCR quantification of human adenoviruses in urban rivers indicates genome prevalence but low infectivity. Appl Environ Microbiol 71:7426–7433
52. Colford JM Jr, Wade TJ, Schiff KC, Wright CC, Griffith JF, Sandhu SK, Burns S, Sobsey M, Lovelace G, Weisberg SB (2007) Water quality indicators and the risk of illness at beaches with nonpoint sources of fecal contamination. Epidemiology 18:27–35
53. Converse RR, Blackwood AD, Kirs M, Griffith JF, Noble RT (2009) Rapid QPCR-based assay for fecal Bacteroides spp. as a tool for assessing fecal contamination in recreational waters. Water Res 43:48–4837
54. Davies CM, Long JA, Donald M, Ashbolt NJ (1995) Survival of fecal microorganisms in marine and freshwater sediments. Appl Environ Microbiol 61:1888–1896
55. DePaola A, Hopkins LH, Peeler JT, Wentz B, McPhearson RM (1990) Incidence of *Vibrio parahaemolyticus* in U.S. coastal waters and oysters. Appl Environ Microbiol 56:2299–2302
56. Desmarais TR, Solo-Gabriele HM, Palmer CJ (2002) Influence of soil on fecal indicator organisms in a tidally influenced subtropical environment. Appl Environ Microbiol 68:1165–1172
57. Dethlefsen L, Huse S, Sogin ML, Relman DA (2008) The pervasive effects of an antibiotic on the human gut microbiota, as revealed by deep 16 S rRNA sequencing. PLoS Biol 6:e280
58. Devriese LLA, van de Kerckhove A, Kilpper-Baelz R, Schleifer K (1987) Characterization and identification of *Enterococcus* species isolated from the intestines of animals. Int J Syst Bacteriol 37:257–259
59. Dick LK, Field KG (2004) Rapid estimation of numbers of fecal Bacteroidetes by use of a quantitative PCR assay for 16 S rRNA genes. Appl Environ Microbiol 70:5695–5697
60. Dick LK, Bernhard AE, Brodeur TJ, Santo Domingo JW, Simpson JM, Walters SP, Field KG (2005) Host distributions of uncultivated fecal Bacteroidales bacteria reveal genetic markers for fecal source identification. Appl Environ Microbiol 71:3184–3191
61. Dick LK, Stelzer EA, Bertke EE, Fong DL, Stoeckel DM (2010) Relative decay of Bacteroidales microbial source tracking markers and cultivated *Escherichia coli* in freshwater microcosms. Appl Environ Microbiol 76:3255–3262

62. Dickerson JW Jr, Hagedorn C, Hassall A (2007) Detection and remediation of human-origin pollution at two public beaches in Virginia using multiple source tracking methods. Water Res 41:3758–3770
63. Dombek PE, Johnson LK, Zimmerley ST, Sadowsky MJ (2000) Use of repetitive DNA sequences and the PCR To differentiate *Escherichia coli* isolates from human and animal sources. Appl Environ Microbiol 66:2572–2577
64. Dorai-Raj S, O'Grady J, Colleran E (2009) Specificity and sensitivity evaluation of novel and existing Bacteroidales and Bifidobacteria-specific PCR assays on feces and sewage samples and their application for microbial source tracking in Ireland. Water Res 43:4980–4988
65. Dufour AP (1984) Bacterial indicators of recreational water quality. Can J Public Health 75:49–56
66. Dufour AP, Schaub S (2007) The evolution of water quality criteria in the United Sates, 1922–2003. In: Wymer LJ (ed) Statistical framework for recreational water quality monitoring. Wiley, New York
67. Eckburg PB, Bik EM, Bernstein CN, Purdom E, Dethlefsen L, Sargent M, Gill SR, Nelson KE, Relman DA (2005) Diversity of the human intestinal microbial flora. Science 308:1635–1638
68. Eckner KF (1998) Comparison of membrane filtration and multiple-tube fermentation by the colilert and enterolert methods for detection of waterborne coliform bacteria, *Escherichia coli*, and enterococci used in drinking and bathing water quality monitoring in southern Sweden. Appl Environ Microbiol 64:3079–3083
69. Edge TA, Boehm AB (2011) Classical and molecular methods to measure fecal indicator bacteria. In: Sadowsky MJ, Whitman RL (eds) The fecal indicator bacteria. ASM Press, Washington, DC
70. Edwards DD, McFeters GA, Venkatesan MI (1998) Distribution of *Clostridium perfringens* and fecal sterols in a benthic coastal marine environment influenced by the sewage outfall from McMurdo Station, Antarctica. Appl Environ Microbiol 64:2596–2600
71. Emerson DJ, Cabelli VJ (1982) Extraction of *Clostridium perfringens* spores from bottom sediment samples. Appl Environ Microbiol 44:1144–1149
72. Field KG, Samadpour M (2007) Fecal source tracking, the indicator paradigm, and managing water quality. Water Res 41:3517–3538
73. Field KG, Bernhard AE, Brodeur TJ (2003) Molecular approaches to microbiological monitoring: fecal source detection. Environ Monit Assess 81:313–326
74. Fiksdal L, Maki JS, LaCroix SJ, Staley JT (1985) Survival and detection of Bacteroides spp., prospective indicator bacteria. Appl Environ Microbiol 49:148–150
75. Fogarty LR, Voytek MA (2005) Comparison of bacteroides-prevotella 16 S rRNA genetic markers for fecal samples from different animal species. Appl Environ Microbiol 71:5999–6007
76. Fong T-T, Lipp EK (2005) Enteric viruses of humans and animals in aquatic environments: health risks, detection, and potential water quality assessment tools. Microbiol Mol Biol Rev 69:357–371
77. Frank DN, St Amand AL, Feldman RA, Boedeker EC, Harpaz N, Pace NR (2007) Molecular-phylogenetic characterization of microbial community imbalances in human inflammatory bowel diseases. Proc Natl Acad Sci USA 104:13780–13785
78. Fremaux B, Gritzfeld J, Boa T, Yost CK (2009) Evaluation of host-specific Bacteroidales 16 S rRNA gene markers as a complementary tool for detecting fecal pollution in a prairie watershed. Water Res 43:4838–4849
79. Fujioka RS (2001) Monitoring coastal marine waters for spore-forming bacteria of faecal and soil origin to determine point from non-point source pollution. Water Sci Technol 44:181–188
80. Fung DYC, Fujioka R, Vijayavel K, Sato D, Bishop D (2007) Evaluation of Fung double tube test for *Clostridium perfringens* and Easyphage test for F-specific RNA coliphages as rapid screening tests for fecal contamination in recreational waters of Hawaii (vol 15, pg 217, 2007). J Rapid Meth Aut Mic 15:411–411

81. Furet JP, Firmesse O, Gourmelon M, Bridonneau C, Tap J, Mondot S, Dore J, Corthier G (2009) Comparative assessment of human and farm animal faecal microbiota using real-time quantitative PCR. FEMS Microbiol Ecol 68:351–362
82. Gast RJ, Moran D, Dennett MR, Wurtsbaugh WA, Amaral-Zettler LA (2011) Amoebae and *Legionella pneumophila* in saline environments. J Water Health 9:37–52
83. Gauthier F, Neufeld JD, Driscoll BT, Archibald FS (2000) Coliform bacteria and nitrogen fixation in pulp and paper mill effluent treatment systems. Appl Environ Microbiol 66:5155–5160
84. Gawler AH, Beecher JE, Brandao J, Carroll NM, Falcao L, Gourmelon M, Masterson B, Nunes B, Porter J, Rince A, Rodrigues R, Thorp M, Walters JM, Meijer WG (2007) Validation of host-specific Bacteriodales 16 S rRNA genes as markers to determine the origin of faecal pollution in Atlantic Rim countries of the European Union. Water Res 41:3780–3784
85. Gerba CP, Rose JB, Haas CN, Crabtree KD (1996) Waterborne rotavirus: a risk assessment. Water Res 30:2929–2940
86. Gersberg RM, Rose MA, Robles-Sikisaka R, Dhar AK (2006) Quantitative detection of hepatitis A virus and enteroviruses near the United States-Mexico border and correlation with levels of fecal indicator bacteria. Appl Environ Microbiol 72:7438–7444
87. Gilpin B, James T, Nourozi F, Saunders D, Scholes P, Savill M (2003) The use of chemical and molecular microbial indicators for faecal source identification. Water Sci Technol 47:39–43
88. Goodwin KD, Pobuda M (2009) Performance of CHROMagar Staph aureus and CHROMagar MRSA for detection of *Staphylococcus aureus* in seawater and beach sand – comparison of culture agglutination, and molecular analyses. Water Res 43:4802–4811
89. Gourmelon M, Caprais MP, Mieszkin S, Marti R, Wery N, Jarde E, Derrien M, Jadas-Hecart A, Communal PY, Jaffrezic A, Pourcher AM (2010) Development of microbial and chemical MST tools to identify the origin of the faecal pollution in bathing and shellfish harvesting waters in France. Water Res 44:4812–4824
90. Graczyk TK, Sunderland D, Tamang L, Lucy FE, Breysse PN (2007) Bather density and levels of *Cryptosporidium* Giardia, and pathogenic microsporidian spores in recreational bathing water. Parasitol Res 101:1729–1731
91. Griffin DW, Gibson CJ 3rd, Lipp EK, Riley K, Paul JH 3rd, Rose JB (1999) Detection of viral pathogens by reverse transcriptase PCR and of microbial indicators by standard methods in the canals of the Florida Keys. Appl Environ Microbiol 65:4118–4125
92. Griffin DW, Lipp EK, McLaughlin MR, Rose JB (2001) Marine recreation and public health microbiology: quest for the ideal indicator. Bioscience 51:817–825
93. Griffith JF, Cao Y, McGee CD, Weisberg SB (2009) Evaluation of rapid methods and novel indicators for assessing microbiological beach water quality. Water Res 43:4900–4907
94. Grimes DJ (1975) Release of sediment-bound fecal coliforms by dredging. Appl Microbiol 29:109–111
95. Haack SK, Duris JW, Fogarty LR, Kolpin DW, Focazio MJ, Furlong ET, Meyer MT (2009) Comparing wastewater chemicals, indicator bacteria concentrations, and bacterial pathogen genes as fecal pollution indicators. J Environ Qual 38:248–258
96. Haake DA, Dundoo M, Cader R, Kubak BM, Hartskeerl RA, Sejvar JJ, Ashford DA (2002) Leptospirosis, water sports, and chemoprophylaxis. Clin Infect Dis 34:E40–E43
97. Hagedorn C, Robinson SL, Filtz JR, Grubbs SM, Angier TA, Reneau RB Jr (1999) Determining sources of fecal pollution in a rural Virginia watershed with antibiotic resistance patterns in fecal streptococci. Appl Environ Microbiol 65:5522–5531
98. Haile RW, Witte JS, Gold M, Cressey R, McGee C, Millikan RC, Glasser A, Harawa N, Ervin C, Harmon P, Harper J, Dermand J, Alamillo J, Barrett K, Nides M, Wang G (1999) The health effects of swimming in ocean water contaminated by storm drain runoff. Epidemiology 10:355–363

99. Halbur PG, Kasorndorkbua C, Gilbert C, Guenette D, Potters MB, Purcell RH, Emerson SU, Toth TE, Meng XJ (2001) Comparative pathogenesis of infection of pigs with hepatitis E viruses recovered from a pig and a human. J Clin Microbiol 39:918–923
100. Haley BJ, Cole DJ, Lipp EK (2009) Distribution, diversity, and seasonality of waterborne salmonellae in a rural watershed. Appl Environ Microbiol 75:1248–1255
101. Hamilton MJ, Yan T, Sadowsky MJ (2006) Development of goose- and duck-specific DNA markers to determine sources of *Escherichia coli* in waterways. Appl Environ Microbiol 72:4012–4019
102. Harrison S, Kinra S (2004) Outbreak of *Escherichia coli* O157 associated with a busy beach. Commun Dis Public Health 7:47–50
103. Hartel PG, Summer JD, Hill JL, Collins JV, Entry JA, Segars WI (2002) Geographic variability of *Escherichia coli* ribotypes from animals in Idaho and Georgia. J Environ Qual 31:1273–1278
104. Harvey S, Greenwood JR, Pickett MJ, Mah RA (1976) Recovery of *Yersinia enterocolitica* from streams and lakes of California. Appl Environ Microbiol 32:352–354
105. Harwood VJ, Whitlock J, Withington V (2000) Classification of antibiotic resistance patterns of indicator bacteria by discriminant analysis: use in predicting the source of fecal contamination in subtropical waters. Appl Environ Microbiol 66:3698–3704
106. Haugland RA, Siefring SC, Wymer LJ, Brenner KP, Dufour AP (2005) Comparison of Enterococcus measurements in freshwater at two recreational beaches by quantitative polymerase chain reaction and membrane filter culture analysis. Water Res 39:559–568
107. Hayashi H, Sakamoto M, Kitahara M, Benno Y (2006) Diversity of the Clostridium coccoides group in human fecal microbiota as determined by 16 S rRNA gene library. FEMS Microbiol Lett 257:202–207
108. He J, Jiang S (2005) Quantification of enterococci and human adenoviruses in environmental samples by real-time PCR. Appl Environ Microbiol 71:2250–2255
109. Higgins JA, Belt KT, Karns JS, Russell-Anelli J, Shelton DR (2005) *tir-* and *stx*-positive *Escherichia coli* in stream waters in a metropolitan area. Appl Environ Microbiol 71:2511–2519
110. Hill RT, Straube WL, Palmisano AC, Gibson SL, Colwell RR (1996) Distribution of sewage indicated by *Clostridium perfringens* at a deep-water disposal site after cessation of sewage disposal. Appl Environ Microbiol 62:1741–1746
111. Hipsey MR, Antenucci JP, Brookes JD (2008) A generic, process-based model of microbial pollution in aquatic systems. Water Resour Res 44:26
112. Horman A, Rimhanen-Finne R, Maunula L, von Bonsdorff C-H, Torvela N, Heikinheimo A, Hanninen M-L (2004) *Campylobacter* spp., *Giardia* spp., *Cryptosporidium* spp., Noroviruses, and indicator organisms in surface water in southwestern Finland, 2000–2001. Appl Environ Microbiol 70:87–95
113. Hou D, Rabinovici SJM, Boehm AB (2006) Enterococci predictions from partial least squares regression models in conjunction with a single-sample standard improve the efficacy of beach management advisories. Environ Sci Technol 40:1737–1743
114. Hsu FC, Shieh YS, van Duin J, Beekwilder MJ, Sobsey MD (1995) Genotyping male-specific RNA coliphages by hybridization with oligonucleotide probes. Appl Environ Microbiol 61:3960–3966
115. Hussain MA, Ford R, Hill J (2010) Determination of fecal contamination indicator sterols in an Australian water supply system. Environ Monit Assess 165:147–157
116. International Organization for Standardization (2000) Water quality – Detection and enumeration of intestinal enterococci ISO 7899–2:000
117. Ishii S, Ksoll WB, Hicks RE, Sadowsky MJ (2006) Presence and growth of naturalized *Escherichia coli* in temperate soils from Lake Superior watersheds. Appl Environ Microbiol 72:612–621

118. Ishii S, Yan T, Shivley DA, Byappanahalli MN, Whitman RL, Sadowsky MJ (2006) *Cladophora* (Chlorophyta) spp. harbor human bacterial pathogens in nearshore water of Lake Michigan. Appl Environ Microbiol 72:4545–4553
119. Jamieson R, Joy DM, Lee H, Kostaschuk R, Gordon R (2005) Transport and deposition of sediment-associated *Escherichia coli* in natural streams. Water Res 39:2665–2675
120. Jamieson RC, Joy DM, Lee H, Kostaschuk R, Gordon RJ (2005) Resuspension of sediment-associated *Escherichia coli* in a natural stream. J Environ Qual 34:581–589
121. Jeter SN, McDermott CM, Bower PA, Kinzelman JL, Bootsma MJ, Goetz GW, McLellan SL (2009) Bacteroidales diversity in ring-billed gulls (Laurus delawarensis) residing at Lake Michigan beaches. Appl Environ Microbiol 75:1525–1533
122. Jiang SC, Chu W (2004) PCR detection of pathogenic viruses in southern California urban rivers. J Appl Microbiol 97:17–28
123. Jiang S, Noble R, Chu W (2001) Human adenoviruses and coliphages in urban runoff-impacted coastal waters of Southern California. Appl Environ Microbiol 67:179–184
124. Jiang SC, Nobel R, Chu W (2001) Human adenoviruses and coliphage in urban runoff-impacted coastal waters of southern California. Appl Environ Microbiol 67:179–184
125. Jiang SC, Chu W, He JW (2007) Seasonal detection of human viruses and coliphage in Newport Bay, California. Appl Environ Microbiol 73:6468–6474
126. Jofre J, Blasi M, Bosch A, Lucena F (1989) Occurrence of bacteriophages infecting Bacteroides-Fragilis and other viruses in polluted marine-sediments. Water Sci Technol 21:15–19
127. Johnson CN, Flowers AR, Noriea NF III, Zimmerman AM, Bowers JC, DePaola A, Grimes DJ (2010) Relationships between environmental factors and pathogenic Vibrios in the Northern Gulf of Mexico. Appl Environ Microbiol 76:7076–7084
128. Johnston C, Ufnar JA, Griffith JF, Gooch JA, Stewart JR (2010) A real-time qPCR assay for the detection of the nifH gene of Methanobrevibacter smithii, a potential indicator of sewage pollution. J Appl Microbiol 109:1946–1956
129. Keymer DP, Miller MC, Schoolnik GK, Boehm AB (2007) Genomic and phenotypic diversity of coastal *Vibrio cholerae* is explained by environmental factors. Appl Environ Microbiol 73:3705–3714
130. Keymer DP, Lam L, Boehm AB (2009) Biogeographic patterns in genomic diversity among a large collection of *Vibrio cholerae* isolates. Appl Environ Microbiol 75:1658–1666
131. Khatib LA, Tsai YL, Olson BH (2002) A biomarker for the identification of cattle fecal pollution in water using the LTIIa toxin gene from enterotoxigenic *Escherichia coli*. Appl Microbiol Biotechnol 59:97–104
132. Khatib LA, Tsai YL, Olson BH (2003) A biomarker for the identification of swine fecal pollution in water, using the STII toxin gene from enterotoxigenic *Escherichia coli*. Appl Microbiol Biotechnol 63:231–238
133. Kildare BJ, Leutenegger CM, McSwain BS, Bambic DG, Rajal VB, Wuertz S (2007) 16 S rRNA-based assays for quantitative detection of universal, human-, cow-, and dog-specific fecal Bacteroidales: a Bayesian approach. Water Res 41:3701–3715
134. Kinzelman J, McLellan SL, Daniels AD, Cashin S, Singh A, Gradus S, Bagley R (2004) Non-point source pollution: determination of replication versus persistence of *Escherichia coli* in surface water and sediments with correlation of levels to readily measurable environmental parameters. J Water Health 2:103–114
135. Kirs M, Smith DC (2007) Multiplex quantitative real-time reverse transcriptase PCR for F+ –specific RNA coliphages: a method for use in microbial source tracking. Appl Environ Microbiol 73:808–814
136. Kreader CA (1995) Design and evaluation of Bacteroides DNA probes for the specific detection of human fecal pollution. Appl Environ Microbiol 61:1171–1179
137. Lamendella R, Domingo JW, Oerther DB, Vogel JR, Stoeckel DM (2007) Assessment of fecal pollution sources in a small northern-plains watershed using PCR and phylogenetic analyses of Bacteroidetes 16 S rRNA gene. FEMS Microbiol Ecol 59:651–660

138. Layton A, McKay L, Williams D, Garrett V, Gentry R, Sayler G (2006) Development of Bacteroides 16 S rRNA gene TaqMan-based real-time PCR assays for estimation of total, human, and bovine fecal pollution in water. Appl Environ Microbiol 72:4214–4224
139. Layton BA, Walters SP, Lam LH, Boehm AB (2010) Enterococcus species distribution among human and animal hosts using multiplex PCR. J Appl Microbiol 109:539–547
140. Leach MD, Broschat SL, Call DR (2008) A discrete, stochastic model and correction method for bacterial source tracking. Environ Sci Technol 42:524–529
141. Leclerc H, Schwartzbrod L, Dei-Cas E (2002) Microbial agents associated with waterborne diseases. Crit Rev Microbiol 28:371–409
142. Lee CS, Lee J (2010) Evaluation of new gyrB-based real-time PCR system for the detection of *B. fragilis* as an indicator of human-specific fecal contamination. J Microbiol Meth 82:311–318
143. Lee DY, Weir SC, Lee H, Trevors JT (2010) Quantitative identification of fecal water pollution sources by TaqMan real-time PCR assays using *Bacteroidales* 16 S rRNA genetic markers. Appl Microbiol Biotechnol 88:1373–1383
144. Lemarchand K, Lebaron P (2003) Occurrence of *Salmonella* spp. and *Cryptosporidium* spp. in a French coastal watershed: relationship with fecal indicators. FEMS Microbiol Lett 218:203–209
145. Lemarchand K, Masson L, Brousseau R (2004) Molecular biology and DNA microarray technology for microbial quality monitoring of water. Crit Rev Microbiol 30:145–172
146. Leung HD, Chen G, Sharma K (2005) Effect of detached/re-suspended solids from sewer sediment on the sewage phase bacterial activity. Water Sci Technol 52:147–152
147. Lipp EK, Huq A, Colwell RR (2002) Effects of global climate on infectious disease: the cholera model. Clin Microbiol Rev 15:757–770
148. Lynch PA, Gilpin BJ, Sinton LW, Savill MG (2002) The detection of Bifidobacterium adolescentis by colony hybridization as an indicator of human faecal pollution. J Appl Microbiol 92:526–533
149. Mac Kenzie WR, Hoxie NJ, Proctor ME, Gradus MS, Blair KA, Peterson DE, Kazmierczak JJ, Addiss DG, Fox KR, Rose JB et al (1994) A massive outbreak in Milwaukee of cryptosporidium infection transmitted through the public water supply. N Engl J Med 331:161–167
150. Malinen E, Rinttila T, Kajander K, Matto J, Kassinen A, Krogius L, Saarela M, Korpela R, Palva A (2005) Analysis of the fecal microbiota of irritable bowel syndrome patients and healthy controls with real-time PCR. Am J Gastroenterol 100:373–382
151. Mallin MA, Williams KE, Esham EC, Lowe RP (2000) Effect of human development on bacteriological water quality in coastal watersheds. Ecol Appl 10:1047–1056
152. Marshall MM, Naumovitz D, Ortega Y, Sterling CR (1997) Waterborne protozoan pathogens. Clin Microbiol Rev 10:67–85
153. Matsuki T, Watanabe K, Fujimoto J, Miyamoto Y, Takada T, Matsumoto K, Oyaizu H, Tanaka R (2002) Development of 16 S rRNA-gene-targeted group-specific primers for the detection and identification of predominant bacteria in human feces. Appl Environ Microbiol 68:5445–5451
154. Matsuki T, Watanabe K, Fujimoto J, Kado Y, Takada T, Matsumoto K, Tanaka R (2004) Quantitative PCR with 16 S rRNA-gene-targeted species-specific primers for analysis of human intestinal bifidobacteria. Appl Environ Microbiol 70:167–173
155. Matsuki T, Watanabe K, Fujimoto J, Takada T, Tanaka R (2004) Use of 16 S rRNA gene-targeted group-specific primers for real-time PCR analysis of predominant bacteria in human feces. Appl Environ Microbiol 70:7220–7228
156. McBride GB, Salmond CE, Bandaranayake DR, Turner SJ, Lewis GD, Till DG (1998) Health effects of marine bathing in New Zealand. Int J Environ Health Res 8:173–189
157. McLaughlin MR, Rose JB (2006) Application of *Bacteroides fragilis* phage as an alternative indicator of sewage pollution in Tampa Bay, Florida. Estuar Coast 29:246–256

158. McLellan SL, Daniels AD, Salmore AK (2001) Clonal populations of thermotolerant Enterobacteriaceae in recreational water and their potential interference with fecal *Escherichia coli* counts. Appl Environ Microbiol 67:4934–4938
159. McLellan SL, Daniels AD, Salmore AK (2003) Genetic characterization of *Escherichia coli* populations from host sources of fecal pollution by using DNA fingerprinting. Appl Environ Microbiol 69:2587–2594
160. McLellan SL, Hollis EJ, Depas MM, Van Dyke M, Harris J, Scopel CO (2007) Distribution and fate of *Escherichia coli* in Lake Michigan following contamination with urban stormwater and combined sewer overflows. J Great Lakes Res 33:566–580
161. McLellan SL, Huse SM, Mueller-Spitz SR, Andreishcheva EN, Sogin ML (2010) Diversity and population structure of sewage-derived microorganisms in wastewater treatment plant influent. Environ Microbiol 12:378–392
162. McQuaig SM, Scott TM, Harwood VJ, Farrah SR, Lukasik JO (2006) Detection of human-derived fecal pollution in environmental waters by use of a PCR-based human polyomavirus assay. Appl Environ Microbiol 72:7567–7574
163. McQuaig SM, Scott TM, Lukasik JO, Paul JH, Harwood VJ (2009) Quantification of human polyomaviruses JC Virus and BK Virus by TaqMan quantitative PCR and comparison to other water quality indicators in water and fecal samples. Appl Environ Microbiol 75:3379–3388
164. Medema GJ, Bahar M, Schets FM (1997) Survival of *Cryptosporidium parvum, Escherichia coli*, faecal enterococci and *Clostridium perfringens* in river water: influence of temperature and autochthonous microorganisms. Water Sci Technol 35:249–252
165. Mieszkin S, Yala JF, Joubrel R, Gourmelon M (2010) Phylogenetic analysis of *Bacteroidales* 16 S rRNA gene sequences from human and animal effluents and assessment of ruminant faecal pollution by real-time PCR. J Appl Microbiol 108:974–984
166. Miller WA, Miller MA, Gardner IA, Atwill ER, Byrne BA, Jang S, Harris M, Ames J, Jessup D, Paradies D, Worcester K, Melli A, Conrad PA (2006) *Salmonella* spp., *Vibrio* spp., *Clostridium perfringens*, and *Plesiomonas shigelloides* in marine and freshwater invertebrates from coastal California ecosystems. Microb Ecol 52:198–206
167. Mocé-Llivina L, Lucena F, Jofre J (2005) Enteroviruses and bacteriophages in bathing waters. Appl Environ Microbiol 71:6838–6844
168. Moe C (2002) Waterborne transmission of infectious agents. In: Hurst CJ, Crawfod RL, Knudsen GR, McInerney MJ, Stetzenbach LD (eds) Manual of environmental microbiology, 2nd edn. ASM Press, Washington, DC, pp 184–204
169. Mueller-Spitz SR, Stewart LB, Klump JV, McLellan SL (2010) Freshwater suspended sediments and sewage are reservoirs for enterotoxin-positive *Clostridium perfringens*. Appl Environ Microbiol 76:5556–5562
170. Mumy KL, Findlay RH (2004) Convenient determination of DNA extraction efficiency using an external DNA recovery standard and quantitative-competitive PCR. J Microbiol Methods 57:259–268
171. Muniesa M, Jofre J, Garcia-Aljaro C, Blanch AR (2006) Occurrence of *Escherichia coli* O157:H7 and other enterohemorrhagic *Escherichia coli* in the environment. Environ Sci Technol 40:7141–7149
172. Muniesa M, Payan A, Moce-Llivina L, Blanch AR, Jofre J (2009) Differential persistence of F-specific RNA phage subgroups hinders their use as single tracers for faecal source tracking in surface water. Water Res 43:1559–1564
173. Nebra Y, Bonjoch X, Blanch AR (2003) Use of *Bifidobacterium dentium* as an indicator of the origin of fecal water pollution. Appl Environ Microbiol 69:2651–2656
174. Nevers MB, Boehm AB (2010) Modeling fate and transport of fecal bacteria in surface water. In: Sadowsky MJ, Whitman RL (eds) The fecal indicator bacteria. ASM Press, Washington, DC

175. Newton RJ, VandeWalle JL, Borchardt MA, Gorelick MH, McLellan SL (2011) *Lachnospiraceae* and *Bacteroidales* alternative fecal indicators reveal chronic human sewage contamination in an urban harbor. Appl Environ Microbiol 77:6972–6981
176. Nichols PD, Leeming R, Rayner MS, Latham V, Ashbolt NJ, Turner C (1993) Comparison of the abundance of the fecal sterol soprostanol and fecal bacerial groups in inner-shlef waters and sediments near Sydney, Australia. J Chromatogr 643:189–195
177. Niemela SI, Vaatanen P (1982) Survival in lake water of *Klebsiella pneumoniae* discharged by a paper mill. Appl Environ Microbiol 44:264–269
178. Noble RT, Fuhrman JD (2001) Enteroviruses detected by reverse transcriptase polymerase chain reaction from the coastal waters of Santa Monica Bay, California: low correlation to bacterial indicator levels. Hydrobiologia 460:175–183
179. Noble RT, Allen SM, Blackwood AD, Chu W, Jiang SC, Lovelace GL, Sobsey MD, Stewart JR, Wait DA (2003) Use of viral pathogens and indicators to differentiate between human and non-human fecal contamination in a microbial source tracking comparison study. J Water Health 1:195–207
180. Noble RT, Moore DF, Leecaster MK, McGee CD, Weisberg SB (2003) Comparison of total coliform, fecal coliform, and enterococcus bacterial indicator response for ocean recreational water quality testing. Water Res 37:1637–1643
181. Noble RT, Griffith JF, Blackwood AD, Fuhrman JA, Gregory JB, Hernandez X, Liang X, Bera AA, Schiff K (2006) Multitiered approach using quantitative PCR to track sources of fecal pollution affecting Santa Monica Bay, California. Appl Environ Microbiol 72:1604–1612
182. Noble RT, Blackwood AD, Griffith JF, McGee CD, Weisberg SB (2010) Comparison of rapid quantitative PCR-based and conventional culture-based methods for enumeration of *Enterococcus* spp. and *Escherichia coli* in recreational waters. Appl Environ Microbiol 76:7437–7443
183. Obiri-Danso K, Jones K (2000) Intertidal sediments as reservoirs for hippurate negative campylobacters, salmonellae and faecal indicators in three EU recognized bathing waters in North West England. Water Res 34:519–527
184. Okabe S, Shimazu Y (2007) Persistence of host-specific *Bacteroides-Prevotella* 16 S rRNA genetic markers in environmental waters: effects of temperature and salinity. Appl Microbiol Biotechnol 76:935–944
185. Okabe S, Okayama N, Savichtcheva O, Ito T (2007) Quantification of host-specific Bacteroides-Prevotella 16 S rRNA genetic markers for assessment of fecal pollution in freshwater. Appl Microbiol Biotechnol 74:890–901
186. Palmer CJ, Tsai YL, Paszko-Kolva C, Mayer C, Sangermano LR (1993) Detection of *Legionella* species in sewage and ocean water by polymerase chain reaction, direct fluorescent- antibody, and plate culture methods. Appl Environ Microbiol 59:3618–3624
187. Panicker G, Myers ML, Bej AK (2004) Rapid detection of *Vibrio vulnificus* in shellfish and Gulf of Mexico water by real-time PCR. Appl Environ Microbiol 70:498–507
188. Parveen S, Murphree RL, Edmiston L, Kaspar CW, Portier KM, Tamplin ML (1997) Association of multiple-antibiotic-resistance profiles with point and nonpoint sources of *Escherichia coli* in Apalachicola Bay. Appl Environ Microbiol 63:2607–2612
189. Parveen S, Hodge NC, Stall RE, Farrah SR, Tamplin ML (2001) Phenotypic and genotypic characterization of human and nonhuman *Escherichia coli*. Water Res 35:379–386
190. Payan A, Ebdon J, Taylor H, Gantzer C, Ottoson J, Papageorgiou GT, Blanch AR, Lucena F, Jofre J, Muniesa M (2005) Method for isolation of *Bacteroides* bacteriophage host strains suitable for tracking sources of fecal pollution in water. Appl Environ Microbiol 71:5659–5662
191. Payment P, Franco E (1993) *Clostridium perfringens* and somatic coliphages as indicators of the efficiency of drinking water treatment for viruses and protozoan cysts. Appl Environ Microbiol 59:2418–2424

192. Pianetti A, Sabatini L, Bruscolini F, Chiaverini F, Cecchetti G (2004) Faecal contamination indicators, salmonella, vibrio and aeromonas in water used for the irrigation of agricultural products. Epidemiol Infect 132:231–238
193. Pina S, Puig M, Lucena F, Jofre J, Girones R (1998) Viral pollution in the environment and in shellfish: human adenovirus detection by PCR as an index of human viruses. Appl Environ Microbiol 64:3376–3382
194. Pinto R, Abad F, Gajardo R, Bosch A (1996) Detection of infectious astroviruses in water. Appl Environ Microbiol 62:1811–1813
195. Pruss A (1998) Review of epidemiological studies on health effects from exposure to recreational water. Int J Epidemiol 27:1–9
196. Puig A, Queralt N, Jofre J, Araujo R (1999) Diversity of *Bacteroides fragilis* strains in their capacity to recover phages from human and animal wastes and from fecally polluted wastewater. Appl Environ Microbiol 65:1772–1776
197. Rabinovici SJM, Bernknopf RL, Wein AM, Coursey DL, Whitman RL (2004) Economic and health risk trade-offs of swim closures at a Lake Michigan beach. Environ Sci Technol 38:2737–2745
198. Rajal VB, McSwain BS, Thompson DE, Leutenegger CM, Kildare BJ, Wuertz S (2007) Validation of hollow fiber ultrafiltration and real-time PCR using bacteriophage PP7 as surrogate for the quantification of viruses from water samples. Water Res 41:1411–1422
199. Rajal VB, McSwain BS, Thompson DE, Leutenegger CM, Wuertz S (2007) Molecular quantitative analysis of human viruses in California stormwater. Water Res 41:4287–4298
200. Ram JL, Thompson B, Turner C, Nechvatal JM, Sheehan H, Bobrin J (2007) Identification of pets and raccoons as sources of bacterial contamination of urban storm sewers using a sequence-based bacterial source tracking method. Water Res 41:3605–3614
201. Resnick IG, Levin MA (1981) Assessment of bifidobacteria as indicators of human fecal pollution. Appl Environ Microbiol 42:433–438
202. Rose JB, Mullinax RL, Singh SN, Yates MV, Gerba CP (1987) Occurrence of rotaviruses and enteroviruses in recreational waters of Oak Creek, Arizona. Water Resour 21:1375–1381
203. Saha ML, Khan MR, Ali M, Hoque S (2009) Bacterial load and chemical pollution level of the River Buriganga, Dhaka, Bangladesh. Bangladesh J Botany 38:87–91
204. Sambrook J, Russell D (2001) Molecular cloning: a labortory manual, 3rd edn. Cold Spring Harbor Laboratory, New York
205. Sandery M, Stinear T, Kaucner C (1996) Detection of pathogenic *Yersinia enterocolitica* in environmental waters by PCR. J Appl Bacteriol 80:327–332
206. Santo Domingo JW, Bambic DG, Edge TA, Wuertz S (2007) Quo vadis source tracking? Towards a strategic framework for environmental monitoring of fecal pollution. Water Res 41:3539–3552
207. Santoro AE, Boehm AB (2007) Frequent occurrence of the human-specific *Bacteroides* fecal marker at an open coast marine beach: relationship to waves, tides and traditional indicators. Environ Microbiol 9:2038–2049
208. Savichtcheva O, Okayama N, Okabe S (2007) Relationships between *Bacteroides* 16 S rRNA genetic markers and presence of bacterial enteric pathogens and conventional fecal indicators. Water Res 41:3615–3628
209. Schriewer A, Miller WA, Byrne BA, Miller MA, Oates S, Conrad PA, Hardin D, Yang HH, Chouicha N, Melli A, Jessup D, Dominik C, Wuertz S (2010) Presence of *Bacteroidales* as a predictor of pathogens in surface waters of the central California coast. Appl Environ Microbiol 76:5802–5814
210. Schulz CJ, Childers GW (2011) Fecal *Bacteroidales* diversity and decay in response to temperature and salinity. Appl Environ Microbiol 77:2563–2572
211. Scott TM, Rose JB, Jenkins TM, Farrah SR, Lukasik J (2002) Microbial source tracking: current methodology and future directions. Appl Environ Microbiol 68:5796–5803

212. Scott TM, Jenkins TM, Lukasik J, Rose JB (2005) Potential use of a host associated molecular marker in *Enterococcus faecium* as an index of human fecal pollution. Environ Sci Technol 39:283–287
213. Sedmak G, Bina D, MacDonald J (2003) Assessment of an enterovirus sewage surveillance system by comparison of clinical isolates with sewage isolates from Milwaukee, Wisconsin, collected August 1994 to December 2002. Appl Environ Microbiol 69:7181–7187
214. Sedmak G, Bina D, Macdonald J, Couillard L (2005) Nine-year study of the occurrence of culturable viruses in source water for two drinking water treatment plants and the influent and effluent of a wastewater treatment plant in Milwaukee, Wisconsin (August 1994 through July 2003). Appl Environ Microbiol 71:1042–1050
215. Semel JD, Trenholme G (1990) *Aeromonas hydrophila* water-associated traumatic wound infections: a review. J Trauma 30:324–327
216. Sercu B, Van De Werfhorst LC, Murray J, Holden PA (2009) Storm drains are sources of human fecal pollution during dry weather in three urban Southern California watersheds. Environ Sci Technol 43:293–298
217. Seurinck S, Verstraete W, Siciliano SD (2003) Use of 16 S-23S rRNA intergenic spacer region PCR and repetitive extragenic palindromic PCR analyses of *Escherichia coli* isolates to identify nonpoint fecal sources. Appl Environ Microbiol 69:4942–4950
218. Seurinck S, Defoirdt T, Verstraete W, Siciliano SD (2005) Detection and quantification of the human-specific HF183 *Bacteroides* 16 S rRNA genetic marker with real-time PCR for assessment of human faecal pollution in freshwater. Environ Microbiol 7:249–259
219. Sghir A, Gramet G, Suau A, Rochet V, Pochart P, Dore J (2000) Quantification of bacterial groups within human fecal flora by oligonucleotide probe hybridization. Appl Environ Microbiol 66:2263–2266
220. Shanks OC, Santo Domingo JW, Lamendella R, Kelty CA, Graham JE (2006) Competitive metagenomic DNA hybridization identifies host-specific microbial genetic markers in cow fecal samples. Appl Environ Microbiol 72:4054–4060
221. Shanks OC, Domingo JW, Lu J, Kelty CA, Graham JE (2007) Identification of bacterial DNA markers for the detection of human fecal pollution in water. Appl Environ Microbiol 73:2416–2422
222. Shanks OC, Atikovic E, Blackwood AD, Lu J, Noble RT, Domingo JS, Seifring S, Sivaganesan M, Haugland RA (2008) Quantitative PCR for detection and enumeration of genetic markers of bovine fecal pollution. Appl Environ Microbiol 74:745–752
223. Shanks OC, Kelty CA, Sivaganesan M, Varma M, Haugland RA (2009) Quantitative PCR for genetic markers of human fecal pollution. Appl Environ Microbiol 75:5507–5513
224. Shanks OC, White K, Kelty CA, Hayes S, Sivaganesan M, Jenkins M, Varma M, Haugland RA (2010) Performance assessment PCR-based assays targeting bacteroidales genetic markers of bovine fecal pollution. Appl Environ Microbiol 76:1359–1366
225. Shanks OC, White K, Kelty CA, Sivaganesan M, Blannon J, Meckes M, Varma M, Haugland RA (2010) Performance of PCR-based assays targeting Bacteroidales genetic markers of human fecal pollution in sewage and fecal samples. Environ Sci Technol 44:6281–6288
226. Shanks OC, Kelty CA, Archibeque S, Jenkins M, Newton RJ, McLellan SL, Huse SM, Sogin ML (2011) Community structure of cattle fecal bacteria from different animal feeding operations. Appl Environ Microbiol 77:2992–3001
227. Shibata T, Solo-Gabriele HM, Fleming LE, Elmir S (2004) Monitoring marine recreational water quality using multiple microbial indicators in an urban tropical environment. Water Res 38:3119–3131
228. Silkie SS, Nelson KL (2009) Concentrations of host-specific and generic fecal markers measured by quantitative PCR in raw sewage and fresh animal feces. Water Res 43: 4860–4871
229. Sinton LW, Hall CH, Lynch PA, Davies-Colley RJ (2002) Sunlight inactivation of fecal indicator bacteria and bacteriophages from waste stabilization pond effluent in fresh and saline waters. Appl Environ Microbiol 68:1122–1131

230. Sivaganesan M, Seifring S, Varma M, Haugland RA, Shanks OC (2008) A Bayesian method for calculating real-time quantitative PCR calibration curves using absolute plasmid DNA standards. BMC Bioinformatics 9:120
231. Sivaganesan M, Haugland RA, Chern EC, Shanks OC (2010) Improved strategies and optimization of calibration models for real-time PCR absolute quantification. Water Res 44:4726–4735
232. Soller JA, Bartrand T, Ashbolt NJ, Ravenscroft J, Wade TJ (2010) Estimating the primary etiologic agents in recreational freshwaters impacted by human sources of faecal contamination. Water Res 44:4736–4747
233. Soller JA, Schoen ME, Bartrand T, Ravenscroft JE, Ashbolt NJ (2010) Estimated human health risks from exposure to recreational waters impacted by human and non-human sources of faecal contamination. Water Res 44:4674–4691
234. Steets BM, Holden PA (2003) A mechanistic model of runoff-associated fecal coliform fate and transport through a coastal lagoon. Water Res 37:589–608
235. Stewart JR, Ellender RD, Gooch JA, Jiang S, Myoda SP, Weisberg SB (2003) Recommendations for microbial source tracking: lessons from a methods comparison study. J Water Health 1:225–231
236. Stewart JR, Gast RJ, Fujioka RS, Solo-Gabriele HM, Meschke JS, Amaral-Zettler LA, Del Castillo E, Polz MF, Collier TK, Strom MS, Sinigalliano CD, Moeller PD, Holland AF (2008) The coastal environment and human health: microbial indicators, pathogens, sentinels and reservoirs. Environ Health 7(Suppl 2):S3
237. Stoeckel DM, Harwood VJ (2007) Performance, design, and analysis in microbial source tracking studies. Appl Environ Microbiol 73:2405–2415
238. Stoeckel DM, Stelzer EA, Dick LK (2009) Evaluation of two spike-and-recovery controls for assessment of extraction efficiency in microbial source tracking studies. Water Res 43:4820–4827
239. Sullivan D, Brooks P, Tindale N, Chapman S, Ahmed W (2010) Faecal sterols analysis for the identification of human faecal pollution in a non-sewered catchment. Water Sci Technol 61:1355–1361
240. Tartera C, Lucena F, Jofre J (1989) Human origin of *Bacteroides fragilis* bacteriophages present in the environment. Appl Environ Microbiol 55:2696–2701
241. Teng LJ, Hsueh PR, Huang YH, Tsai JC (2004) Identification of *Bacteroides thetaiotaomicron* on the basis of an unexpected specific amplicon of universal 16 S ribosomal DNA PCR. J Clin Microbiol 42:1727–1730
242. Turnbaugh PJ, Hamady M, Yatsunenko T, Cantarel BL, Duncan A, Ley RE, Sogin ML, Jones WJ, Roe BA, Affourtit JP, Egholm M, Henrissat B, Heath AC, Knight R, Gordon JI (2009) A core gut microbiome in obese and lean twins. Nature 457:480–484
243. Ufnar JA, Wang SY, Christiansen JM, Yampara-Iquise H, Carson CA, Ellender RD (2006) Detection of the nifH gene of *Methanobrevibacter smithii*: a potential tool to identify sewage pollution in recreational waters. J Appl Microbiol 101:44–52
244. Ufnar JA, Ufnar DF, Wang SY, Ellender RD (2007) Development of a swine-specific fecal pollution marker based on host differences in methanogen mcrA genes. Appl Environ Microbiol 73:5209–5217
245. Ufnar JA, Wang SY, Ufnar DF, Ellender RD (2007) Methanobrevibacter ruminantium as an indicator of domesticated-ruminant fecal pollution in surface waters. Appl Environ Microbiol 73:7118–7121
246. Unno T, Jang J, Han D, Kim JH, Sadowsky MJ, Kim OS, Chun J, Hur HG (2010) Use of barcoded pyrosequencing and shared OTUs to determine sources of fecal bacteria in watersheds. Environ Sci Technol 44:7777–7782
247. USEPA (2000) Improved enumeration methods for recreational water quality indicators: Enterococci and *Escherichia coli*. EPA 821/R-97/004. US Environmental Protection Agency Office of Water, Washington, DC

248. USEPA (2002) Method 1603: *Escherichia coli* (*E. coli*) in water by membrane filtration using modified membrane-thermotolerant Escherichia coli agar (modified mTEC) EPA-821-R-02-023. US Environmental Protection Agency Office of Water, Washington, DC
249. USEPA (2002) Method 1604: total coliforms and *Escherichia coli* in water by membrane filtration using a simultaneous detection technique (MI Medium). US Environmental Protection Agency Office of Water, Washington, DC
250. USEPA (2006) Method 1600: Enterococci in Water by Membrane Filtration Using MEMBRANE-Enterococus Indoxyl-B-D-Glucoside agar (mEI) EPA-821-R-06-009. US Environmental Protection Agency Office of Water, Washington, DC
251. USEPA (2007) Critical path science plan for the development of new or revised recreational water quality criteria. 823-R-08-002. US Environmental Protection Agency Office of Water, Washington, DC
252. USEPA (2009) National water quality inventory: report to congress 2004 reporting cycle. EPA 841-R-08-001. US Environmental Protection Agency Office of Water, Washington, DC
253. USEPA (2010) Method A: Enterococci in water by TaqMan® quantitative polymerase chain reaction (qPCR) assay. US Environmental Protection Agency Office of Water, Washington, DC
254. Viau EJ, Goodwin KD, Yamahara KM, Layton BA, Sassoubre LM, Burns S, Tong H-I, Wong SHC, Boehm AB (2011) Human bacterial pathogens and fecal indicators in tropical streams discharging to Hawaiian coastal waters. Water Res 45:3279–3290
255. Volkmann H, Schwartz T, Kirchen S, Stofer C, Obst U (2007) Evaluation of inhibition and cross-reaction effects on real-time PCR applied to the total DNA of wastewater samples for the quantification of bacterial antibiotic resistance genes and taxon-specific targets. Mol Cell Probes 21:125–133
256. Wade TJ, Pai N, Eisenberg JN, Colford JM Jr (2003) Do U.S. Environmental Protection Agency water quality guidelines for recreational waters prevent gastrointestinal illness? A systematic review and meta-analysis. Environ Health Perspect 111:1102–1109
257. Wade TJ, Calderon RL, Sams E, Beach M, Brenner KP, Williams AH, Dufour AP (2006) Rapidly measured indicators of recreational water quality are predictive of swimming-associated gastrointestinal illness. Environ Health Perspect 114:24–28
258. Wade TJ, Calderon RL, Brenner KP, Sams E, Beach M, Haugland R, Wymer L, Dufour AP (2008) High sensitivity of children to swimming-associated gastrointestinal illness: results using a rapid assay of recreational water quality. Epidemiol 19:375–383
259. Wade TJ, Sams E, Brenner KP, Haugland R, Chern E, Beach M, Wymer L, Rankin CC, Love D, Li Q, Noble R, Dufour AP (2010) Rapidly measured indicators of recreational water quality and swimming-associated illness at marine beaches: a prospective cohort study. Environ Health 9:66
260. Walters SP, Gannon VPJ, Field KG (2007) Detection of *Bacteroidales* fecal indicators and the zoonotic pathogens *E. coli* O157:H7, *Salmonella*, and *Campylobacter* in river water. Environ Sci Technol 41:1856–1862
261. Walters SP, Yamahara KM, Boehm AB (2009) Persistence of nucleic acid markers of health-relevant organisms in seawater microcosms: implications for their use in assessing risk in recreational waters. Water Res 43:4929–4939
262. Walters SP, Thebo AL, Boehm AB (2011) Impact of urbanization and agriculture on the occurrence of bacterial pathogens and stx genes in coastal waterbodies of central California. Water Res 45:1752–1762
263. Westrell T, Teunis P, van den Berg H, Lodder W, Ketelaars H, Stenstrom TA, de Roda Husman AM (2006) Short- and long-term variations of norovirus concentrations in the Meuse river during a 2-year study period. Water Res 40:2613–2620
264. Wetz JJ, Blackwood AD, Fries JS, Williams ZF, Noble RT (2008) Trends in total *Vibrio* spp. and *Vibrio vulnificus* concentrations in the eutrophic Neuse River Estuary, North Carolina, during storm events. Aquat Microb Ecol 53:141–149

265. Whitman RL, Nevers MB (2003) Foreshore sand as a source of *Escherichia coli* in nearshore water of a Lake Michigan beach. Appl Environ Microbiol 69:5555–5562
266. Whitman RL, Nevers MB, Korinek GC, Byappanahalli MN (2004) Solar and temporal effects on *Escherichia coli* concentration at a Lake Michigan swimming beach. Appl Environ Microbiol 70:4276–4285
267. Whitman RL, Ge Z, Nevers MB, Boehm AB, Chern EC, Haugland RA, Lukasik AM, Molina M, Przybyla-Kelly K, Shively DA, White EM, Zepp RG, Byappanahalli MN (2010) Relationship and variation of qPCR and culturable Enterococci estimates in ambient surface waters are predictable. Environ Sci Technol 44:5049–5054
268. Wiedenmann A, Kruger P, Dietz K, Lopez-Pila JM, Szewzyk R, Botzenhart K (2006) A randomized controlled trial assessing infectious disease risks from bathing in fresh recreational waters in relation to the concentration of *Escherichia coli*, intestinal enterococci, *Clostridium perfringens*, and somatic coliphages. Environ Health Perspect 114:228–236
269. Wiggins BA (1996) Discriminant analysis of antibiotic resistance patterns in fecal streptococci, a method to differentiate human and animal sources of fecal pollution in natural waters. Appl Environ Microbiol 62:3997–4002
270. Wilkes G, Edge T, Gannon V, Jokinen C, Lyautey E, Medeiros D, Neumann N, Ruecker N, Topp E, Lapen DR (2009) Seasonal relationships among indicator bacteria, pathogenic bacteria, *Cryptosporidium* oocysts, *Giardia* cysts, and hydrological indices for surface waters within an agricultural landscape. Water Res 43:2209–2223
271. Wong M, Kumar L, Jenkins TM, Xagoraraki I, Phanikumar MS, Rose JB (2009) Evaluation of public health risks at recreational beaches in Lake Michigan via detection of enteric viruses and a human-specific bacteriological marker. Water Res 43:1137–1149
272. Wu CH, Sercu B, Van de Werfhorst LC, Wong J, DeSantis TZ, Brodie EL, Hazen TC, Holden PA, Andersen GL (2010) Characterization of coastal urban watershed bacterial communities leads to alternative community-based indicators. PLoS One 5:e11285
273. Wyn-Jones AP, Carducci A, Cook N, D'Agostino M, Divizia M, Fleischer J, Gantzer C, Gawler A, Girones R, Höller C, Husman AM, Kay D, Kozyra I, López-Pila J, Muscillo M, Nascimento MS, Papageorgiou G, Rutjes S, Sellwood J, Szewzyk R, Wyer M (2011) Surveillance of adenoviruses and noroviruses in European recreational waters. Water Res 45:1025–1038
274. Xu J, Bjursell MK, Himrod J, Deng S, Carmichael LK, Chiang HC, Hooper LV, Gordon JI (2003) A genomic view of the human-*Bacteroides thetaiotaomicron* symbiosis. Science 299:2074–2076
275. Yamahara KM, Layton BA, Santoro AE, Boehm AB (2007) Beach sands along the California coast are diffuse sources of fecal bacteria to coastal waters. Environ Sci Technol 41:4515–4521
276. Yamahara KM, Walters SP, Boehm AB (2009) Growth of enterococci in unaltered, unseeded beach sands subjected to tidal wetting. Appl Environ Microbiol 75:1517–1524
277. Yampara-Iquise H, Zheng G, Jones JE, Carson CA (2008) Use of a *Bacteroides thetaiotaomicron*-specific alpha-1-6, mannanase quantitative PCR to detect human faecal pollution in water. J Appl Microbiol 105:1686–1693
278. Young TA, Heidler J, Matos-Perez CR, Sapkota A, Toler T, Gibson KE, Schwab KJ, Halden RU (2008) Ab initio and in situ comparison of caffeine, triclosan, and triclocarban as indicators of sewage-derived microbes in surface waters. Environ Sci Technol 42:3335–3340
279. Zheng G, Yampara-Iquise H, Jones JE, Carson CA (2009) Development of *Faecalibacterium* 16 S rRNA gene marker for identification of human faeces. J Appl Microbiol 106:634–641

Chapter 10
Polio and Its Epidemiology

Lester M. Shulman

Glossary

ACPE	The Advisory Committee on Polio Eradication
AFP	Acute flaccid paralysis.
AFP surveillance	Characterization of enteroviruses in stool samples from all AFP cases especially in individuals under 15 years of age to rule-in or rule-out etiology by polioviruses.
aVDPV	A vaccine-derived poliovirus isolate whose evolutionary path is unknown or ambiguous.
bOPV	Bivalent oral polio vaccine (usually containing serotypes 1 and 3).
BSL	Biosafety standard level.
Capsid	The protein shell that surrounds a virus particle.
Capsomere	One of the individual morphological units that make up the viral capsid.
CDC	US Centers for Disease Control and Prevention
CD155 or PVr	The human encoded cell receptor for poliovirus, a member of the immunoglobulin superfamily.

This chapter, which has been modified slightly for the purposes of this volume, was originally published as part of the Encyclopedia of Sustainability Science and Technology edited by Robert A. Meyers. DOI:10.1007/978-1-4419-0851-3

L.M. Shulman (✉)
Central Virology Laboratory, Laboratory of Environmental Virology at Sheba Medical Center, Public Health Services, Israel Ministry of Health

Department of Epidemiology and Preventive Medicine, School of Public Health, Sackler Faculty of Medicine, Tel Aviv University, Tel Aviv 69978, Israel
e-mail: lester.shulman@sheba.health.gov.il; lester.shulman@post.tau.ac.il

P. Kanki and D.J. Grimes (eds.), *Infectious Diseases: Selected Entries from the Encyclopedia of Sustainability Science and Technology*, DOI 10.1007/978-1-4614-5719-0_10, © Springer Science+Business Media New York 2013

Codon	A sequence of three adjacent nucleotides on a strand of DNA or RNA that specifies which specific amino acid will be incorporated into a protein.
Codon bias	Unequal usage of synonymous codons (different codons that specify the same amino acid)
CPE	Cytopathic effect.
cVDPV	A circulating vaccine-derived poliovirus, that is, a poliovirus that has evolved from vaccine during person-to-person transmission.
eIPV	Enhanced inactivated polio vaccine.
Emergence	The appearance of a pathogen in a previously pathogen-free area.
Endemic	The constant presence of a disease to a greater or lesser extent in a particular locality.
Enteroviruses	Any of >80 different species of polioviruses, coxsackie viruses, echoviruses, and enteroviruses belonging to the genus *Enterovirus* in the family *Picornaviridae*.
Environmental surveillance (as related to polioviruses)	Investigation of sewage and recreational water for the presence of poliovirus as an indication of the presence of poliovirus-infected individuals in a community.
EPI	Expanded Program on Immunization.
Epidemic	A rapid spread of disease into a disease-free area or spread of a disease to more than the usual number of persons affected in a region with disease.
Epitopes	The component of an antigen that is recognized by and binds to an antibody.
Eradication	The complete elimination of all incidence of disease and/or the presence of the agent that causes the disease.
Evolution	Change in the genetic composition of a population or the genome of a given organism during successive generations.
GAP	Global action plan for laboratory containment.
GAP I	GAP phase I – plan for identifying all known and potential sources of poliovirus especially wild polioviruses within each country.
GAP II	GAP phase II – plan for laboratory containment of wild polioviruses.
GAP III	GAP phase III – plan to minimize post-eradication poliovirus facility-associated risks.
GAVI	Global alliance for vaccines and immunization
Genetic recombination (of polioviruses)	A situation where one portion of the genome of a poliovirus is replaced through a covalent linkage with the equivalent segment from another poliovirus or non-polio enterovirus.

Genotype	He genetic makeup of an organism as distinguished from its physical characteristics.
GMT	Geometric mean titer, usually calculated according to Karber.
GOARN	Global Outbreak Alert and Response Network.
GPEI or GEI	The Global Poliomyelitis Eradication Initiative of the WHO, adopted in 1988.
GPLN	Global Polio Laboratory Network.
Hydrophobic pocket	A hydrophobic space located under the binding site for the host encoded viral receptor that is located on the bottom of the canyon surrounding the fivefold axis of symmetry of the enteroviral capsid.
Hydrophobic pocket factors	Small hydrophobic molecules that occupy the hydrophobic pocket and that may regulate the host receptor viral capsid interaction.
Immunodeficient	Lacking one of the components of the immune system.
Immunogenicity	The relative ability of a molecule to elicit an immune response.
i.d.	Intradermal or under the skin.
i.m.	Intramuscular.
i.n.	Intranasal.
i.p.	Intraperitoneal.
i.t.	Intrathecal.
Infection	Establishment and growth of an infectious agent in the body.
IPV	Inactivated poliovirus vaccine.
IRES	Internal ribosome entry site.
ITD	Intratypic differentiation (determination if a virus isolate is vaccine, vaccine-derived, or wild).
iVDPV	A vaccine-derived poliovirus that has diverged from its respective oral poliovirus serotype during persistent infection of an immunodeficient host.
Lineages	A group of organisms that are closely related genetically.
MAPREC	Mutant analysis by PCR and restriction fragment enzyme cleavage to measure reversion of attenuation sites in vaccine strains.
MNVT	Monkey neurovirulence test, an in vivo neurovirulence test in monkeys.
mOPV	Monovalent OPV.
Neurovirulence	The ability of the poliovirus to infect and damage nerve cells causing disease of the nervous system.
Neurovirulence attenuation sites	Specific nucleotide positions along the poliovirus genome where the specific nucleotide present at that

	site will influence whether or not an individual polioviral isolate will be neurovirulent.
Neutralizing antigenic sites	Epitopes of the poliovirus that induce neutralizing antibodies.
NID	National immunization day.
NSL	Non-Sabin-like (wild) virus of vaccine-derived poliovirus (based on results of certain ITD tests).
Major disease	Poliomyelitis, AFP, or cases of infection with polio that involves invasion and permanent damage to the nervous system.
Microarray	A technology used to study many genes at once using thousands of different short molecular sequences at known position on solid support to hybridize to complementary nucleic acid sequences from different sources.
Minor disease	Nonspecific illness caused by poliovirus that may include upper respiratory tract symptoms (sore throat and fever), gastroenteritis (nausea vomiting, abdominal pain, constipation or diarrhea), and influenza-like illness.
NCCs	National Certification Committees.
NGOs	Nongovernmental organizations.
NIDs	National immunization days.
NPEV	Non-polio enteroviruses.
Nonstructural genes	Viral genes encoding proteins that are not incorporated into the structure of the capsid.
Oligonucleotide	A short sequence of nucleotides frequently synthetic.
OPV	Live attenuated oral poliovirus vaccine.
Outbreaks	(Under eradication conditions) even the presence of a single case of paralytic poliomyelitis.
Persistent poliovirus infection	An infection associated with an immunodeficient host where virus is not cleared but continues to replicate for an indefinite period of time.
Phylogenetic tree	A diagram with branches showing the inferred evolutionary relationships among various biological entities.
Picornaviridae	A viral family made up of the small (18–30 nm) ether-sensitive single stranded, positive-sense RNA viruses that lack an envelope.
Poliomyelitis	The infectious disease caused by poliovirus involving inflammation of motor neurons of the spinal cord and brainstem that leads to acute paralysis followed by atrophy of the muscles enervated by the infected motor neurons.
Poliovirus	One of three serotypes of picornaviruses that can cause acute flaccid paralysis and whose cell receptor is CD155.

Polypeptide	A small molecule constructed from linked amino acids.
Postpolio syndrome	Slow progressive muscle pain and weakness that reappears 30 or 40 years after paralysis caused by a poliovirus infection affecting muscles previously affected by polio as well as muscles that may not have been affected.
Posttranslational processing	Any modification of a protein after it has been translated.
Provoked poliomyelitis	Poliomyelitis resulting from physical trauma during infection with poliovirus.
Proofreading	An enzymatic process that checks whether a newly incorporated nucleotide in a nascent chain is the correct compliment of its corresponding nucleotide in the template.
PVR	The poliovirus receptor, CD155.
PVR Tg21 transgenic mouse	A mouse that has been genetically modified to express the human poliovirus receptor.
Quasispecies	A term used to describe a cluster, cloud, or swarm of viruses with minor differences in nucleotide sequence that arise during replication as a consequence of polymerase incorporation errors.
Rearrangement (in relation to polioviruses)	A structural alteration in the genomic sequence occurring during coinfection with two or more viruses resulting in a new genome in which parts are from different parental polio or non-polio enteroviruses
Reemergence	Emergence after an absence.
RCC	Regional Certification Committees.
RCT	Reproductive capacity temperature, the temperature at which viruses can replicate.
RNA-dependent RNA polymerase	A viral encoded polymerase that synthesizes a complimentary RNA strand from an RNA template.
RRL	Regional Reference Laboratory of the Global Polio Laboratory Network.
SAGE	Strategic Advisory Group of Experts on Immunization.
Serotype	A group of closely related virus expressing a common set of antigens.
Seroconversion	Appearance of antibodies following exposure to antigen in seronegative person, or $\geq$4-fold increase in titer of previously immune person.
Seroconversion index	The mean seroconversion rate against all three poliovirus serotypes.
SL	Sabin-like poliovirus (based on result from some ITD tests).
SIA	Supplemental immunization activities (such as NIDs, SNIDS, and mop-ups).

Silent circulation	Person-to-person transmission of virus in a community in the absence of cases of AFP.
Silent infection	Asymptomatic infection.
Silent presence	The presence of a virus in the absence of clinical cases and the absence of person-to-person transmission.
SNIDS	Sub-national Immunization Days.
Stakeholders	All governmental and nongovernmental agencies involved in the GPEI.
Structural genes (in relation to polio)	Genes encoding viral capsid proteins.
Synonymous nucleotide substitutions	Changes in the nucleotide sequence that do not result in a change in encoded amino acid.
TAG	Technical Advisory Group.
TD	Typic differentiation (determination of the serotype of a poliovirus isolate).
TOPV	Trivalent oral polio vaccine containing all three serotypes of attenuated poliovirus.
Transition	The substitution of a purine nucleotide with the other purine, or a pyrimidine nucleotide with the other pyrimidine.
Transversions	The substitution of a pyrimidine nucleotide by a purine nucleotide or vice versa.
UNICEF	United Nations Children's Fund.
Vaccine strains	Poliovirus strains approved by the WHO for production of live and inactivated polio vaccines.
VAPP	Vaccine-associated paralytic poliomyelitis.
VDPV	A vaccine-derived poliovirus that has diverged through evolution from its respective live poliovirus vaccine strain serotype by more than 1% (serotypes 1 and 3) or more than 0.6% (serotype 2) of its respective VP1 capsid protein.
Viremia	The presence of virus in the bloodstream during an infection.
VPg	Viral protein genome linked – 22 amino acid protein covalently linked to genome and complimentary negative strand.
VP1	Viral capsid protein 1.
VP2	Viral capsid protein 2.
VP3	Viral capsid protein 3.
VP4	Viral capsid protein 4.
WHA	World Health Assembly.
WHO	World Health Organization.
WPV or wild poliovirus	Any poliovirus that is not derived from attenuated oral polio vaccine strains

3'UTR	The untranslated region of the polioviral genome that is located 3′ of the open reading frame that encodes the viral polyprotein.
$3D^{pol}$	Viral encoded RNA-dependent RNA polymerase.
5'UTR	A highly structured, untranslated area of the polioviral genome located 5′ to the open reading frame that encodes the viral polyprotein. The 5'UTR is covalently linked on its 5′ base to viral protein VPg.

A Brief Definition of Polio and Its Importance

The word "polio" has been used to describe both a disease and the disease agent. Among current methods to measure the importance of or interest in a topic is to run a general web search for the term and to search the scientific literature in PubMed. A Google web search of the word "polio" in Aug 2010 yielded 31,100,000 hits, while a search in PubMed yielded 22,000 articles and 826 review articles. This review will concentrate on those aspects of the epidemiology of polio as it relates to disease eradication and the sustainability of this effort. The terms "polio" and "poliomyelitis" will be used when describing the disease and "poliovirus" and related terms such as "polio vaccine" will be used to describe the agent that causes the disease.

In order to understand the epidemiology of polio, it is important to understand the adversary. Toward this goal, this chapter starts with a detailed physical characterization of polioviruses and the pathological effects caused by poliovirus infections that are most relevant to understanding the epidemiology of polio. This is followed by a description of the global efforts to eradicate poliomyelitis and the viral agent causing the disease, and concludes with a discussion of the future directions needed to achieve and sustain eradication and prevent reemergence. Smallpox was the first human disease to be eradicated and we are currently in the endgame of eradication of polio as the second. Polio eradication is currently the largest public health program in the world and has involved both health professionals and more than ten million volunteers in all countries since the inception of the Global Poliomyelitis Eradication Initiative by the World Health Assembly in 1988 [1].

Introduction

The road toward polio eradication has been long [2] and by no means smooth. Important milestones along the march toward recognition and understanding the disease, identification of its causative agent, and toward prevention and eradication will be briefly discussed in the introduction (see also Fig. 10.1). A poliovirus isolate

Major milestones along the road towards polio eradication.

1400 BCE	First pictorial record of a person with poliomyelitis.
1789	First modern characterization of polio as a "debility of the lower extremities" by Underwood.
1813	More complete detailed description of polio by Monteggia.
1855	Description of "infantile paralysis by Heine based on systematic investigation of cases started in the 1840's.
1863	Confirmation by biopsy for motor neuron involvement in polio by Cornil.
1870	Detailed description of physiological changes in the anterior horn of the spinal cord by Charcot and Joffroy.
1889	Medin realizes that paralytic cases occur in only a small number of infected individuals during epidemics.
Late 19th Century	The emergence of outbreaks of poliomyelitis.
1905	Recognition of the infectious nature of polio by Wickman confirming Medin's earlier observations.
1908-9	The discovery of poliovirus as the causative agent of poliomyelitis by Landsteiner and Popper.
1920'2	Establishment of the first national rehabilitation center.
1929	Drinker develops the iron lung.
1931	Burnet and MacNamara report more than one non-cross reacting antigenic strain of polio; culminates in the conclusion in 1951 that there were only three serotypes.
1935-36	First clinical trials with inactivated polio vaccine by Brodie and Park and with live attenuated vaccine by Kolmer.
1937	Establishment of the first NGOs to fund support of polio victims and research: The National Foundation of Infantile Paralysis and The March of Dimes.
1939	Armstrong was the first to grow poliovirus in a non-primate (rodent) host.
1940's	Kinny introduces the concept of supportive rehabilitation.
1949	First in vitro passages in tissue cultures by Enders, Weller, and Robbins
1950's	Development and testing of attenuated vaccines by Kaprowski (1950), Sabin (1956-57) and Cox (1958).
1950's	Bottiger and Kaprowski observed that live vaccine spreads to contacts.
1951	Conclusion that there were only three serotypes of poliovirus.
1951-61	Immunizaton of > 11 million children with Sabin and with Kaprowski live oral vaccine strains.
1953-54	Salk develops and tests inactivated poliovirus vaccine licensed for use by 1955.
1954	First use of live vaccine by Kaprowski to control a large outbreak.
1954	Development of standard plaque assays for quantification of virus by Dulbecco and Voigt.
1955	The Cutter Incident in which 400,000 children were immunized with inadequately inactivated wild poliovirus.
1962	Detection of the first persistent infections with vaccine-derived polioviruses.
1968	Introduction of microcarrier cell systems for uniform large-scale vaccine production by van Wezel followed by the use of pathogen cell-free cell systems in the early 1990's.
1979	The development of primate model to test neurovirulence of poliovirus strains.
1979	Last case of wild polio in the Unites States.
1985	Pan Americans Health Organization and CDC establish the Latin American Regional Polio Network.
1986	Development of murine L20B cells transformed with, and expressing the human receptor allowing selective growth of poliovirus while non-permissive for most other human enteroviruses.

Fig. 10.1 Major milestones along the road toward polio eradication

1988	WHO establishes the Global Poliomyelitis Eradication Initiative
1989	Identfication and cloning of the poliovirus receptor by Mendelsohn, Wimmer and Racaniello.
1981	Recombinant nucleic acid techniques used by Racaniello and Baltimore to prepare an infectious clone of poliovirus.
1990's=>	Development of molecular and immunological assays to identify the serotype and determine the vaccine or non-vaccine origin of poliovirus isolates.
1990-91	Development of a transgenic mouse expressing the human encoded poliovirus receptor approved as a non-primate model for neurovirulence testing safety of all three serotypes of live OPV in 1999-2000.
1991	Last case of endogenous wild polio in the Western Hemisphere.
1991	Establishment of a global Polio Laboratory Network to monitor poliovirus infections throughout the world.
1998=>	Isolation and characterization of highly diverged vaccine derived polioviruses from environmental samples excreted by unknown individuals.
1999	Establishment of the Global Action Plan for laboratory containment of all wild polioviruses throughout the world.
1999	Last reported case of poliomyelitis anywhere in the world caused by a wild serotype 2 poliovirus.
2000-01	First prospective recognition that an outbreak was caused by a vaccine-derived poliovirus (in Haiti and the Dominical Republic).
2000	Last case of endogenous wild poliomyelitis in the Western Pacific Region.
2002	Last case of endogenous wild poliomyelitis in the Eastern European Region.
2002	Cello, Paul, and Wimmer synthesize infectious poliovirus from individual nucleotides.
2003	Failure to vaccinate in Nigeria leads to a large increase in the number of cases and exportation of wild polioviruses and vaccine derived -polioviruses to > 21 polio-free countries. This spread is being brought under control by local and regional vaccination campaigns using monovalent, divalent, and trivalent vaccines.
2005	Decision: successful eradication must include cessation of the routine use of OPV.
2010	Large successful clinical trials using fractional sub-dermal doses of IPV.

Fig. 10.1 (continued)

is classified as vaccine, vaccine-derived (VDPV), or wild-type poliovirus based on the percent nucleotide sequence homology between its capsid protein VP1 and that of the corresponding OPV vaccine serotype. An isolate with VP1 homology of 99–100% is classified as vaccine virus, 85–99% as VDPV, and $> 85\%$ as wild-type poliovirus [3]. This rule of thumb for classifying polioviruses as VDPVs has recently been modified for serotype 2 to include isolates with ≥ 6 nucleotide changes (i.e., $<1\%$) and the upper limit of 15% for VP1 divergence has been eliminated (Summary of the 16th Informal Consultation on the Global Polio Laboratory Network, Geneva, Switzerland, 2010).

The earliest record attributed to polio comes from an Egyptian Stele from 1400 BCE that depicts an Egyptian high priest or official with a walking stick and withered leg that bears a striking resemblance to a recent picture of a man with poliomyelitis (Fig. 10.2). Polio infections from this time to the nineteenth century were endemic and usually occurred in young children where most infections were probably asymptomatic. While early descriptions of "acquired clubfoot" by Hippocrates and Galen were consistent with polio, the first modern medical characterization of polio includes descriptions of "Debility of the Lower Extremities"

a b

Fig. 10.2 *Living after paralytic poliomyelitis: then and now*. Paralytic poliomyelitis occurs after a biphasic infection where viremia in a small number of systemic infections is followed by infection of the CNS. Paralysis is a direct result of destructive replication of poliovirus in motor neurons followed by atrophy of de-enervated muscles. Both pictures represent men whose skeletal muscles have been affected by infections of nerves in the anterior horn of their spinal cord. The picture on the *left* (**a**) depicts the earliest record of poliomyelitis in a man and comes from a stele from ancient Egypt created around 1500 BCE, and is strikingly similar to the image of the man in the photograph on the *right* (**b**) who has atrophy of the right foot and leg due to polio that was taken in the Far East in 2007 ((**a**) Egyptian Stele at the Ny Carlsberg Glyptotek Museum, Copenhagen, Denmark (GNU free documentation License). (**b**) Photograph #134 Centers for Disease Control and Prevention Public Image Library [CDC/NIP/Barbara Rice])

by Underwood in 1789, polio by Monteggia in 1813, "infantile paralysis" by Heine in 1840, and involvement of motor neurons in infantile paralysis by Duchenne in 1855. Involvement of motor neurons was confirmed by biopsy of the brain and spinal cord of a polio victim by Cornil in 1863 and by a detailed description of physiological changes in the anterior horn of the spinal cord by Charcot and Joffroy in 1870.

A new epidemiological aspect of polio emerged in the nineteenth century, namely, the appearance of outbreaks that increasingly affected adults as well as children [4]. Paradoxically this shift from an endemic to an outbreak pattern of disease transmission may have been facilitated by a "hygiene barrier" derived from improved community sanitation that may have resulted in a shift from fecal–oral to oral–oral transmission, an increase in naïve individuals especially among older cohorts, and primary exposure of increasingly older cohorts where disease manifestation were more severe. These epidemics became more frequent by the mid-twentieth century and involved growing numbers of people. Wickman described the acute infectious nature of polio in his analysis of a 1905 polio outbreak in New York, confirming Medin's realization in 1889 that paralytic cases were only a small part of epidemics and that even persons with mild illness could infect others. Further complications were the observation by Burnet and MacNamara in 1931 [5] that different strains of poliovirus caused disease, but infection with some strains did not protect against subsequent infection with other strains and the observation in the 1950s that poliomyelitis could be triggered by physical injury during a poliovirus infection and that there was an increased risk of paralysis in limbs that received a mechanical stress or after tonsillectomies [6]. Two important new concepts were the establishment of a national center for treatment of poliomyelitis victims at Warm Springs, Georgia, and the use of professional fund-raisers by President Roosevelt supported by others in the late 1920s. The nonpartisan National Foundation for Infantile Paralysis and the March of Dimes established in 1937 institutionalized this fundraising effort. The iron lung, developed by Drinker in 1929, and the concept of supportive rehabilitation involving the use of hot moist packs to relieve muscle spasm and physiotherapy to maintain strength of unaffected muscle fibers promoted by Kenny in the 1940s were important advances for treatment of poliomyelitis.

The study of the pathological organism that caused poliomyelitis was enabled by the discovery of a bacteria-free "filterable" etiological agent, the poliovirus, which could pass disease from one primate to another by Landsteiner and Popper in 1909. Burnet and MacNamara realized in 1931 that there was more than one type of poliovirus since exposure to some isolates did not protect against exposure to others. By 1951, the National Foundation for Infantile Paralysis concluded that there were only three serotypes of poliovirus. The study of poliovirus was aided by (a) the first passages of poliovirus in a non-primate rodent system by Armstrong in 1939, (b) passage in tissue cultures by Enders, Weller, and Robbins in 1949 [7], (c) development of plaque assays for quantification of polio by Dulbecco and Vogt in 1954 [8], (d) the use of microcarrier cell systems for vaccine production by van Wezel in 1967 [9], (e) development of monkey neurovirulence tests in 1979 [10], (f) the use of pathogen-free diploid MRC5 cells (human fetal cells derived from normal lung tissue) and permanent cell lines like Vero (a cell line prepared from the kidney of a normal adult African green monkey) for vaccine production in the early 1990s, (g) identification and cloning of the poliovirus receptor CD155 [11], (h) development of the transgenic PVr-mouse model which expresses the human poliovirus receptor as an alternative to monkeys for neurovirulence testing [12], (i) preparation of a murine cell line, L20B, expressing the human poliovirus receptor for selective growth of poliovirus [13], and (j) the development of the immunological and molecular tools (discussed in detail below) that provide the identity the serotype of the isolate, distinguish whether its origin was from a vaccine or wild strain, and provide phylogenetic information on the evolutionary relationship to other isolates.

Advances in culturing polioviruses outlined in the previous paragraph laid the foundation for developing the vaccines that have turned polio into a vaccine-preventable disease and a candidate for eradication (see below). Early experiments and clinical trials such as those in 1935–1936 with inactivated poliovirus by Brodie and Park [14] and attenuated live vaccine by Kolmer [15] were hampered by lack of awareness until 1951 that there were three serotypes. Afterward, effective inactivated vaccine was developed and tested by Salk and coworkers starting in 1953–1954 [16, 17], while Koproswski, Sabin, and Cox developed and tested attenuated oral vaccines in 1950 [18, 19], 1956–1957 [20], and 1958 [21], respectively. Between 1951 and 1962, 12.9 million children were vaccinated with Koprowski strains and 11 million with Sabin strains [19]. A number of important epidemiological observations were made during that time that continue to guide current vaccination strategies. For attenuated oral vaccines these included (a) the first demonstration by Koprowski of interference between poliovirus serotypes during coinfection [19], (b) a demonstration that maternal antibodies did not prevent an immune response in vaccinees under 6 months of age [19], (c) the observation by Koprowski and especially Bottiger that live vaccine spread to contacts [19], (d) a demonstration of persistence of antibodies at the same levels in vaccinated children for at least 3 years [19], (e) proof of concept by Koprowski that live polio vaccine could be effective in containing large outbreaks [19], and (f) documentation of high vaccine safety with both the Koprowski and Sabin OPV strains [4, 19, 22]. Safety issues relating to both the live and inactivated viral strains will be mentioned in discussions starting on pages 8150 and 8160. After extensive evaluation in hundreds of monkeys at Baylor College of Medicine, and the Division of Biological Standards at the NIH, the Sabin strains were chosen for licensure primarily on the basis of lower neurotropism, but also based on genetic stability on passage in humans and a lower ability to spread to contacts (reviewed in Sutter et al. [4] and Furesz [22]). Efforts to eradicate polio and to prevent reemergence are presented in detail in the following section. Initial paradigms attributed to the different properties of the individual vaccines have not always held true in all circumstances [23].

The Epidemiology of Polio

Epidemiological studies to discover the means of preventing a disease usually begin with the recognition of a new pattern of similar symptoms among those affected and the establishment of a case definition. Discovering means for preventing the disease may start before the disease agent is discovered and characterized, but is certainly accelerated once this characterization becomes available together with the means of quantifying intervention strategies. It is much less common to start with an agent and then search for a disease as in the case of human anelloviruses [24]. Human anelloviruses are small circular DNA viruses considered to be orphan viruses.

They were initially discovered in a patient with hepatitis, but subsequent research indicated no causal link to hepatitis and it has been very difficult to associate them with any other specific disease. However, this section of the review will start with a description of those physical aspects of poliovirus that have the most impact on epidemiology of the disease. This is because there is already a clear case definition for polio and poliomyelitis, polioviruses have been recognized as the causative agents of these diseases, numerous methods for characterizing poliovirus and preventing poliovirus infections have been developed and tested, and the disease is approaching elimination or eradication.

Structural and Functional Organization of the Poliovirus Genome

Polioviruses belong to the *Picornaviridae* virus family. The *Picornaviridae* genome consist of a single strand of positive-sense RNA approximately 7,500 nucleotides located within a protein capsid made up of 60 capsomeres that forms a virion 27–30 nm in diameter. The genome is organized from its 5′ end to its 3′end into a number of functional regions (Fig. 10.3) that include a 5′ untranslated region (5'UTR) that regulate translation and replication [25], a long open reading frame that encodes a single large polypeptide that is cleaved after translation into four structural capsid proteins and a number of nonstructural proteins including an RNA polymerase, and a short 3′ untranslated region that is attached to a poly-A tail in both viral mRNA and genomic RNA in the virion [25] (see reviews by Wimmer et al. [26], Racaniello [25], and Sutter et al. [4]). The positive-sense single strand of genomic RNA in the virion, serves directly as an mRNA template for translation to viral proteins once the virion penetrates its host cell membrane. Later it serves as a template for synthesis of a complimentary negative sense strand. The current understanding of the physical and genetic aspects of polio was greatly facilitated by the development of and the current commercial availability of methods for easily extracting viral nucleic acids from poliovirus and poliovirus-infected cells and analyzing and manipulating these sequences. Some of these studies led to the unanticipated conclusion that poliovirus capsid proteins and the sequences that encode them define polioviruses, whereas all other elements in the poliovirus genome may be substituted by genomic recombination with equivalent sequences from closely related isolates of enterovirus species C in vivo and even more distantly related rhinoviruses in the laboratory as long as functionality is maintained (reviewed by Kew et al. [3]). Finally, advances in molecular biology also enabled poliovirus to be the first virus to be synthesized from nucleotides in a test tube [27, 28].

The 5'UTR was first subdivided into a highly conserved region (nucleotides 1–650) and a hypervariable region (nucleotides 651–750) based on an analysis of 33 wild-type 3 polioviruses [29]. A series of stem-loop structures with a high degree of secondary structure were proposed to be present within the conserved region by

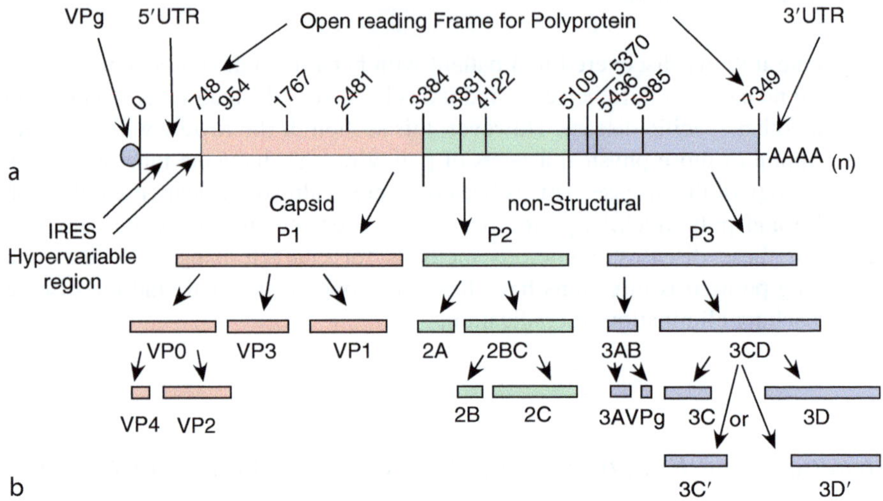

c

Serotype	Nucleotide	Gene	Amino Acid	Temperature Sensitivity
Sabin 1	A480G	5′UTR	Non-coding	Yes
	G935U	VP4	ala 65 ser	-
	U2438A	VP3	lys 225 met	-
	G2791A	VP1	ala 106 thr	-
	C2879U	VP1	leu 134 phe	-
	U6203C	3D	tyr 73 his	Yes
Sabin 2	G481A	5′UTR	Non-coding	Yes
	C2909U	VP1	thr 143 iso	-
Sabin 3	C472U	5′UTR	Non-coding	Yes
	C2034U	VP3	ser 91 phe	-
	U2493C	VP1	iso 6 thre	Yes

Fig. 10.3 Organization of the polio viral genome, posttranslational processing of the nascent poliovirus polyprotein, and the nucleotide substitutions that differentiate attenuated oral polio vaccine strains from their neurovirulent progenitors. The RNA positive-sense strand genome of Sabin 2 based on GenBank/EMBL/DDBJ entry AY184220 (**a**) is covalently linked to the viral encoded protein VPg. There is a single open reading frame flanked by a 5′ and a 3′ untranslated sequence (UTR). An internal ribosomal entry site (IRES) in the 5'UTR allows the uncapped polio genomic RNA to serve as mRNA for translation on host cell ribosomes. The open reading frame is translated into a single poliovirus polyprotein that undergoes a series of posttranslational proteolytic cleavages (**b**) while it is still being translated. Some of the intermediate products have enzymatic and/or structural functions that differ from those of the final cleavage products. Poliovirus genomic and mRNA terminates in a poly-A tail. The attenuation of neurovirulence in Sabin 2 and the other 2 serotypes, Sabin 1 (GenBank/EMBL/DDBJ entry V01150) and Sabin 3 (GenBank/EMBL/DDBJ entry X00925), of poliovirus strains used for the live polio vaccine result from the nucleotide and amino acid substitutions shown in (**c**). Reversion of these substitutions may restore a neurovirulent phenotype for the progeny of these vaccine strains. Nucleotide substitutions are indicated by the original nucleotide of the parental strain the nucleotide position, and the substituted nucleotide in the vaccine strain (*A*denine *U*racil, *G*uanine, or *C*ytosine). Amino acid substitutions are indicated by the parental amino acid, the position of the amino acid in the final cleavage product, and the amino acid in the vaccine strain (*ala*nine, *his*tidine, *iso*leucine, *leu*cine, *met*hionine, *phe*nylalanine, *ser*ine, *thr*eonine, and *tyr*osine). ((**b**) Based on: [1] Krausslich HG, et al. [37] and [2] Kitamura N, et al. [290]. (**c**) Modified from: Kew OM, et al. [3])

Pilipenko et al. [30] and Skinner et al. [31]. A single nucleotide substitution in a loop structure in stem-loop V of the 5'UTR significantly influenced the neurovirulence of poliovirus isolates from all three serotypes and affected the maximum temperature at which viral isolate replicate efficiently (see reviews by Kew et al. [3] and Sutter et al. [4] and discussions on poliovirus evolution starting on page 8137). The hypervariable region appears to be much less structured, reflecting the high degree of variation and the U nucleotide richness [29].

An Internal Ribosome Entry Site [32, 33], IRES, enables uncapped RNA from *Picornaviridae* to be translated in eukaryotic cells by host ribosomes [25]. One of the first steps in initiation of viral translation is the binding [34] of cellular RNA binding proteins PCB_1 and PCB_2 to stem-loop IV of the IRES. This enables the 40S ribosomal unit to bind to the IRES and continue the process of translation as if the RNA was a capped eukaryotic mRNA. Functional IRES elements can be interchanged among *Picornaviridae* [3]. Nucleotide differences in the conserved 5'UTR among different isolates were unevenly distributed [29] with changes tending to conserve the stem structures. In contrast, the hypervariable region did not seem to have a highly conserved secondary structure and nucleotide differences appeared to be more or less evenly spread throughout [29]. While the length of the hypervariable region was generally conserved suggesting an unknown function [29], small deletions were tolerated [35].

Picornaviridae have a genome of approximately 7,200–7,400 nt with a single open reading frame (ORF). While this ORF encodes four capsid proteins and at least seven viral proteins (Fig. 10.3), these proteins are only produced after the initial translation product, a single polypeptide, is enzymatically cleaved into smaller and smaller polyproteins during and after translation (posttranslational processing). The polypeptide is cleaved in an ordered series of steps (Fig. 10.3b), by viral encoded protease activity within the nascent polypeptide (self-cleavage) and *in trans* from viral proteases released after cleavage. Interestingly some of the intermediate cleavage products have unique activities by themselves that contribute to the replication cycle of the virus, but which differ from those of the final cleavage products (reviewed by Racaniello [36] and Krausslich et al. [37]). Properties of the polioviral capsid proteins define the epidemiology of polioviruses. The most important aspects of the structure of the four capsid proteins, their assembly into capsomeres and organization within adjacent capsomeres that relate to the epidemiology of polio, will be discussed above on page 8131. The 900–906 nucleotide sequence of the VP1 of polioviruses has become the minimum standard for determining the evolutionary relationship among polioviruses and the rate at which they evolve [4, 38, 39].

Many of the nonstructural proteins and intermediate cleavage products are multifunctional and act at a number of steps in RNA synthesis (reviewed in [4, 25]). Most of the nonstructural proteins will only be mentioned in passing since equivalent nonstructural proteins from other related picornaviruses may replace all of the nonstructural viral proteins as long as functional sites including cleavage recognition sites are maintained (reviewed in [3]). The resultant chimeric recombinants behave as polioviruses. One nonstructural protein, the RNA polymerase, will be discussed in some detail (see page 8135) because of its profound effect on polio epidemiology regardless of its source.

The secondary structure in the 3'UTR that may play a role in translation and replication of picornaviral genomic RNA has been reviewed [25]. A nucleotide difference between Sabin serotype 1 and its wild parent influences the temperature at which serotype 1 can replicate [40]. A poly-A tail is present on both genomic and mRNA that stimulates the cap-independent, internal ribosome entry site (IRES)-driven translation of poliovirus RNA in a mammalian cell-free system by tenfold [41].

Both genomic and minus strand RNA are linked to the small viral encoded protein, VPg, (Fig. 10.3a) through pUpU bound to tyrosine, the third amino acid from NH terminal end of VPg, by a phosphodiester bond [42]. VPg is also present in infected cells in an unmodified form and bound to pUpU through the same 0^4-phosphotyrosine bond found in the covalently linked forms [42, 43]. The uridylylation of VPg takes place on the opposite side of the polymerase that binds RNA. A host encoded unlinking enzyme that cleaves the 0^4-phosphotyrosine bond between VPg and RNA has been described [44] although its role in replication has not been established. The poliovirus encoded VPg can be replaced by VPg from echoviruses [43].

A mature infectious poliovirus consists of a single sense strand of polyadenylated RNA covalently linked to a viral encoded protein, VPg, surrounded by an icosahedral protein coat, the capsid, made up of 60 capsomeres that each contain a single copy of each of the four viral capsid proteins. Adjacent capsomeres are organized around both fivefold and threefold axes of symmetry and the surface around these axes is organized into a series of regular protrusions and depressions (Fig. 10.4). The capsid structure is metastable [45] rather than rigid and internal parts of capsid may even be transiently expressed on the surface ([46], review in [25]) exposing additional epitopes such as PALTAVE inVP1 [47].

Capsid proteins are the first viral encoded proteins to appear on the nascent poliovirus polyprotein and are cleaved from the nascent polyprotein into an intermediate polyprotein, P1, by $2A^{pro}$ while the full-length polyprotein is still being synthesized. P1 is processed into final cleavage products VP1 and VP3 and an intermediate cleavage product VP0. VP0 is only cleaved into VP2 and VP4 during the final stages of maturation of the virion. The protein chains of VP1, VP2, and VP3 are arranged in wedge-like structures with extruding loops that interact to form the major (NAgIa, NAgIIa, and NAgIIIa) and minor (NAgIb, NAgIIb, and NAgIIIb) neutralizing antigenic epitopes [48]. Amino acid differences within neutralizing antigenic sites divide polioviruses into three serotypes with limited cross-reactivity [49]. The amino acids of the neutralizing antigenic sites have been mapped onto the three-dimensional structures of the viral capsid of type 2 poliovirus as colored spheres, Fig. 10.4. Those that are unique to the neutralizing antigenic epitopes are colored yellow. Some amino acid residues in and adjacent to these neutralizing antigenic sites (red spheres in Fig. 10.4) are also involved in receptor binding and this may have restricted the number of serotypes [50] and influenced

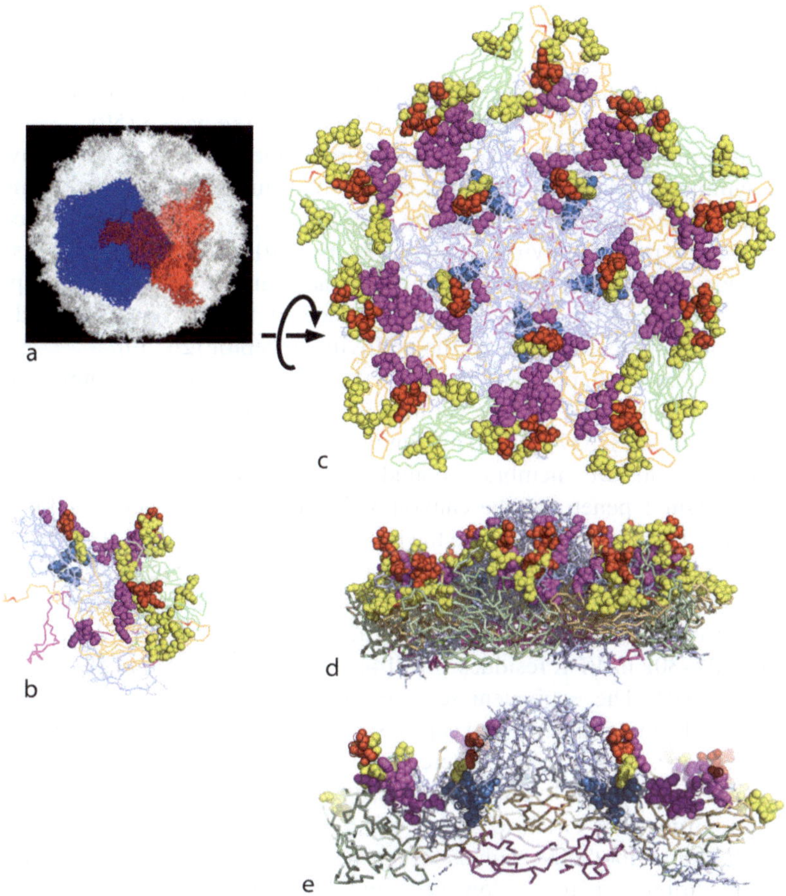

Fig. 10.4 The hydrophobic pocket and amino acid residues in the neutralizing antigenic epitopes and receptor binding sites of the Sabin 2 polio vaccine strain. The three-dimensional structures represent capsomeres 1–5 from human serotype 2 poliovirus, Genbank/EMBL/DDBJ entry 1eah. The backbones of the amino acid chains of the capsid proteins are represented by *light blue*, *pale green*, *light orange*, and *magenta colored ribbons* for VP1, VP2, VP3, and VP4, respectively. Amino acid residues at the surface of the hydrophobic pocket are represented by *blue spheres*. Amino acid residues within the epitopes recognized by neutralizing antibodies are represented by *yellow spheres*, those involved in receptor recognition and binding are represented by *magenta spheres*, and amino acid residues shared by both antigenic sites and receptor binding sites are represented by *red spheres*. The figure was prepared using the MacPYMOL program (DeLano Scientific LLC, www.pymol.org). Figure (**a**) is a representation of the entire capsid of poliovirus showing the positions of the threefold (in *red*) and fivefold (in *blue*) symmetrical organization of the capsomeres. Each poliovirus capsomere (**b**) contains a single copy of each of the four viral capsid proteins. Five capsomeres are assembled around a fivefold axis of symmetry shown in (**c**) and by *blue* in (**a**). They also assemble around a threefold axis of symmetry shown in *red* in (**a**). Figure (**c**) represents an external view of the five capsomeres at the fivefold axis of symmetry. Figure (**d**) is the side view of the same five capsomeres formed by rotating the figure in (**c**) in the direction of the *circular arrow*, so that the lower structures in (**c**) are nearest the viewer and the internal surfaces of the capsid proteins are facing downward. Figure (**e**) is a transverse section of the figure in (**d**) at the position of the *straight arrow* in (**c**) to more clearly illustrate the topography of the surface of the virion. An animated "Interactive 3D Complement" (I3DC) for the structures in this figure appears in *Proteopedia* at http://proteopedia.org/w/Polio_Epidemiology

evolution in these epitopes in the absence of immunoselection especially during the emergence of vaccine-derived polioviruses (VDPVs) (see page 8159).

The three-dimensional view of the structure of the capsomeres at the fivefold axis of symmetry reveals an elevated central plateau with a hole in the middle surrounded by a depression called the canyon [51, 52]. The fivefold axis of symmetry for type 2 poliovirus is shown in Fig. 10.4. A number of conserved amino acids and amino acids within and adjacent to the serotype-specific neutralizing epitopes are located on the surface of the canyon walls and have been implicated in interaction with the poliovirus receptor [26, 50].

The human encoded, poliovirus receptor, CD155, belongs to the immunoglobulin super gene family and has one variable and two constant immunoglobulin-like domains (residues 28–337) [11]. This human encoded gene has alternative splice sites that result in two membrane-bound and two secreted isoforms [53]. The variable domain 1 penetrates the canyon and binds to amino acid residues from all three external capsid proteins and the principal binding sites are at the bottom of the canyon above the hydrophobic pocket (blue spheres in Fig. 10.4) and on the outer side of the canyon rim [50, 54]. The residues of type 1 poliovirus involving receptor virion binding include residues 102–108, 166–169, 213–214, 222–236, 293–297, 301–302 in VP1, residues 140–144, 170–172 in VP2, and 58–62, 93, and 182–186 in VP3. The equivalent residues for serotype 2 poliovirus have been mapped onto the three-dimensional capsid structure as red (shared with neutralizing antigenic epitopes) and magenta spheres for those associated only with the receptor binding sites (Fig. 10.4). Cryo-electron microscope studies have shown the binding of the poliovirus to the virion to be a two-step process [54]. The initial binding of the receptor to amino acid residues along the canyon wall results in little or no change in virion structure. However, this binding rapidly sets into motion conformational changes leading to the 135 S or A particle state that initiates uncoating and the start of the infections cycle [45, 54].

The human poliovirus receptor has been cloned and used to establish a murine cell line, L20B, where expression of the poliovirus receptor allows infection and growth of polio from clinical and other samples but not most other human non-polio enteroviruses [13, 55, 56]. Transgenic mice, PVR Tg-21 mice that express the human poliovirus receptor, not only support poliovirus infection and present with neurological symptoms, but allow determination of the relative neurovirulence of the isolates [12, 57–59].

A hydrophobic pocket (blue spheres in Fig. 10.4) located below the canyon floor is normally occupied by pocket factors such as sphingosine-like molecules including palmitic and myristic acids and hydrophobic compounds, that stabilize the capsid, enable receptor docking and whose removal is a necessary prerequisite for uncoating [25, 45, 54, 60].

Small molecules such as pleconaryl and isoflavenes can bind in this hydrophobic pocket and exert antiviral effects by affecting the binding of the receptor or enhancing the stability of the virion and preventing uncoating [25]. Because of the metastable nature of the capsid, mutations distant from the receptor and drug binding sites can compensate mutations in the respective binding sites [45, 61].

Molecular analysis studies of isolates shed during persistent infections of immunodeficient patients [62, 63] and from phylogenetically related aVDPVs from environmental samples help pinpoint amino acid substitutions in capsid proteins that determine antigenicity, receptor recognition, attenuation of neurovirulence, and properties of the hydrophobic pocket. Other changes, some of which are at interfaces between the threefold or fivefold interfaces of capsomeres may also affect these properties indirectly.

In order for single stranded, positive-sense genomic RNA to be incorporated into progeny of the infecting virus, a complimentary negative RNA strand must first be synthesized using the original single stranded positive-sense RNA genome as template, and this complimentary negative strand must then be used as template for synthesis of new positive-strand RNAs. While some positive-sense copies are incorporated into progeny virions as genomic RNA, other newly synthesized positive-strand RNAs serve as templates to repeat and amplify RNA replication and/or for translation to produce more viral proteins. Eukaryotic cells that serve as the host for poliovirus replication lack a polymerase that can synthesize complimentary RNA from an RNA template. Therefore the virus must encode its own polymerase. Since the single stranded positive-sense RNA genome of the infecting virion is also an mRNA that is immediately translated, the virion does not have to incorporate the polymerase into the virion itself to start replication. The translation product of the $3D^{pol}$ gene (Fig. 10.3) is the required RNA-primed RNA polymerase. Both the intermediate cleavage products that contain the $3D^{pro}$ and the final cleavage product are multifunctional and the crude replication complex also contains other viral proteins and protein cleavage intermediates such as 2BC, 2C, and 3AB as well as host proteins (reviewed in [4, 25]). The binding site for RNA template and primer are on one face of $3D^{pol}$ and a binding site for the uridylylation of VPg, a prerequisite for covalent linking of VPg to viral RNA, is on the opposite face [25].

One important contrast between genomic DNA replication in eukaryotic host cells and genomic RNA replication in *Picornaviridae* relates to the fidelity of replication. Specifically, there is an elaborate proofreading mechanism combined with pathways for correcting misincorporations during replication of Eukaryotic DNA that is lacking in the RNA-primed RNA polymerase complex for viral replication [64, 65]. This leads to such a high evolutionary rate for polioviruses that it borders on error catastrophe [66, 67] (discussed further below).

Poliovirus Infections in Cells, Individuals, and Populations

This section will deal with the epidemiological aspects of poliovirus infections at three levels, infections in single cells, infections in a single individual, and infections in populations of individuals. The normal infectious cycle of a poliovirus starts with recognition and attachment to poliovirus-specific cell receptors on susceptible cells of human or closely related primate origin. It continues with penetration and uncoating, translation of viral RNA, posttranslational processing of viral polyproteins, replication of viral genomes, assembly and maturation of progeny viruses culminating in the release of infectious polioviruses. During this process, the virus employs and modifies host cell functions to optimize viral yield. The observation that viral RNA and cDNA is infectious when transfected into permissive host cells has allowed recovery of virus from extracted genomic RNA, cloned cDNA or RNA translated from cloned DNA [68–70], genomic RNA immobilized on FTA paper (WHO 16th Informal Consultation Of The Global Polio Laboratory Network, September 2010, Geneva, Switzerland), and from polioviral RNA synthesized in a test tube from individual nucleotides [27, 71]. The infectious cycle has been reviewed extensively. The reader is referred to the following reviews for further reading [4, 25, 72].

Poliovirus Infections at the Level of the Host Cell

All polioviruses recognize a single host cell receptor [50], CD155, also known as the poliovirus receptor (PVR). Identification and cloning of the poliovirus receptor CD155 [11] allowed the creation of cell lines [13] and animal models [58, 59] for the study of polioviral infections in non-primate hosts. The interaction of virion and receptor is complex [25]. The capsid structure is dynamic allowing the transient presence of internal portions and epitopes of capsid proteins on the outer surface of the virion [25, 46] including the N-terminus of VP1 even before uncoating. The shape of the receptor and its position relative to the host cell membrane and the canyon on the virion into which it fits bring the fivefold axis of capsomeres in close proximity to the cell membrane [54].

Conformational changes, induced shortly after the virion–receptor interaction, are required to initiate the uncoating process (reviewed in Racaniello [25]). The capsid begins to disassociate during a transition to the A particle. The A particle contains the viral RNA but has lost its VP4 capsid proteins. The N-terminal of the VP1 externalizes and may insert into the plasma membrane. Viral RNA is believed to enter the cell at or near the fivefold axis through a continuous channel formed in part by VP1 that continues through the cell membrane [54]. VP4 plays a part in formation of the pore. The pore for poliovirus entry is probably not formed within endosomes [25]. Small molecules that sit in the hydrophobic pocket may influence these conformational changes without affecting receptor binding [61, 73–75].

The only viral proteins in the virion are the four capsid proteins and the VPg covalently linked to the genomic RNA. The internal ribosomal entry site (IRES) on uncoated polioviral RNA enables translation of the viral polyprotein on host cell ribosomes (reviewed in [3, 25]). VPg appears to be cleaved from this RNA and subsequently synthesized viral RNA that will be used as mRNA [44]. Nuclear trafficking of cellular proteins is downregulated shortly after infection resulting in accumulation of host nuclear proteins in the cytoplasm that could function alone or in combination with viral encoded proteins in viral RNA translation, synthesis, and packaging [73]. Downregulation may be due in part to specific degradation of two host transporters, Nup 153 and p62. A full-length polyprotein is not observed in spite of being encoded by the single long open reading frame since posttranslational cleavage of the polyprotein is initiated as soon as the portion encoding the $2A^{pro}$ has been translated. Many of the nonstructural proteins and intermediate cleavage products are multifunctional and act directly or indirectly at a number of steps in the RNA synthesis pathway (reviewed in [4, 25]). One example is the aforementioned viral encoded protease, $2A^{pro}$, that also shuts off host protein synthesis by cleaving eIF4G. eIF4G is required for translation of capped eukaryotic mRNAs, while the C-terminal of the cleavage product enhances IRES activity [25]. $2A^{pro}$ is also important for negative strand but not positive-strand RNA synthesis [76]. Another example is the intermediate cleavage product, $3CD^{pro}$, that also participates in the posttranslational processing of the polyprotein.

The last protein of the polyprotein to be translated is the 3D polymerase. RNA-primed RNA synthesis is initiated once the 3D has been released from the polyprotein reviewed in [25]. VPg-pUpU or VPg itself could act as a precursor for RNA synthesis by hybridizing to template RNA [44, 77]. The binding site for template and primer are on one side and that for VPg is on the other side. A replicate intermediate is formed and consists of a positive-sense RNA with 6–8 nascent negative strand RNAs. The negative sense strand serves in turn as template for synthesis of a 30-fold excess of new sense strand RNAs. Full-length dsRNAs can be isolated from infected cells. Altogether the genomic RNA is amplified up to 50,000-fold. VPg is bound to both genomic RNA and negative sense RNA.

Poliovirus and other picornaviruses employ a quasispecies reproductive strategy [64, 78] where the lack of proofreading rapidly results in a mixture of progeny with modified genomes containing randomly positioned single nucleotide substitutions. Genomic recombination is a second method of evolution where a single event results in substitutions of many nucleotides from a different poliovirus or closely related non-polio enterovirus for the equivalent sequence in the original poliovirus. The majority of single nucleotide substitutions are deleterious or neutral; however some may confer a reproductive advantage for progeny for growth in the current or future host and/or for host-to-host transmission. Evolutionary changes become "fixed" by selective outgrowth of individual members of the quasispecies that pass through bottlenecks within and between hosts [79]. Two evolutionary pathways, the very high number of progeny (>10,000 per infected cell) and outgrowth by chance selection and/or a selective advantage, result in one of the highest observed rates of molecular evolution [39].

RNA is synthesized from four nucleotides, two pyrimidines (uracil and cytosine) and two purines (adenine and guanine). The most common route for polioviral evolution is by nucleotide misincorporation (single nucleotide substitution) in the absence of both proofreading and post-incorporation excision–repair pathways. The nucleotide position that is substituted is probably random but may be influenced to some extent by secondary structure and the adjacent nucleotides. Quasispeciation arises from the fact that the remaining progeny retain the original nucleotide at the position of each unique substitution in an individual progeny virus, while within the cloud of progeny each isolate may have a unique substitution at a different position in the genome.

Among the isolates that make up the quasispecies, substitutions should be found at each position in the genome at an equal frequency, at least in theory. However, substitutions are much more frequently observed in some positions than in others. Two related factors contribute significantly to the nonuniform distribution (see page 8140) of observed substitutions along the genome. The first is that almost all observations have been made with RNA extracted from the quasispecies that arose during replication of a viable virus directly in the primary host or after amplification of one or more isolates from the quasispecies in vivo in a second host or ex vivo in tissue culture. The second is that substitutions in some positions produce nonviable or less fit offspring that are eliminated during this amplification process.

Sequence-specific variability, based on the individual nucleotide base and its nearest neighbors [67], and inherent characteristics of the polymerase are other factors that contribute to the nonuniform distribution. If misincorporations were unbiased, transversions (the substitution of a pyrimidine by a purine or vice versa) would be expected to occur at twice the rate of transition (the substitution of a purine with a purine or a pyrimidine with a pyrimidine). However empiric observations have revealed a polymerase-based bias of approximately ten to one in favor of transitions [39]. To currently include sequence data from the genomes of nonviable progeny requires either amplification of individual genomes by a process that does not require an active poliovirus infection but that includes high fidelity with proofreading and excision-repair (reverse transcribing the genomic RNA and cloning the cDNA of all viable and nonviable poliovirus progeny into plasmids that can be amplified in bacterial strains with high fidelity, proofreading, and error correction) or by direct sequencing of individual gnomes without amplification (chip/array sequencing technology) [66, 80]. Neither approach is currently very easy to apply since both would require individually processing large numbers of genomes equivalents, although Crotty et al. were able to calculate a rate of 2.1×10^{-2} substitutions per site by direct measurement of mutations in the VP1 of 55 cloned genomes after a single cycle of in vitro virus growth. Massive parallel next-generation sequencing may offer the best approach for analyzing viable and nonviable members of a quasispecies [289].

Wild poliovirus genomes frequently recombine (recombination) with polioviruses and closely related non-polio enterovirus genomes [81, 82]. This recombination can only occur during concurrent infection of a single cell by both parental isolates. Intratypic recombination may occur even within capsid proteins [83–85].

The noncapsid regions of polioviruses are most similar in sequence to other members of the enterovirus C genotype that includes Coxsackie A virus (CAV) serotypes 1, 11, 13, 15, 17, 18, 19, 20, 21, 22, and 24, and these sequences are readily shuffled among polioviruses and the other members of this group [86, 87]. In fact polioviruses show evidence of having evolved from C-cluster Coxsackie A viruses and may reemerge from them after eradication [87]. Interspecific recombination contributes to the phenotypic biodiversity of polioviruses and may favor the emergence of circulating vaccine-derived polioviruses, cVDPVs [88].

Recombination is not site specific, does not require extensive homology between genomes at the crossover site, and most likely occurs by an exchange of templates by the synthesis of complimentary RNA by the RNA-primed RNA polymerase rather than by breakage rejoining [89]. Intratypic (same serotype) and intertypic (different serotype) recombination in in vitro occurred at 1.3×10^{-3} and 7.6×10^{-6}, respectively [89], while recombinations between polio and NPEVs occurred at a frequency of 10^{-6} [87]. Administration of trivalent oral vaccine anywhere and in areas where wild polioviruses and genotype C viruses co-circulate provides the conditions for concurrent infections and polio vaccine-polio vaccine, polio vaccine-wild polio, and polio vaccine-NPEV, as well as endemic wild polio-NPEV recombinations. For examples of such recombinations see molecular analyses of isolates from the cVDPV outbreaks in Haiti and Dominican Republic [90] and Indonesia [91] and in individual cases [92].

Molecular epidemiology is the study of disease and factors controlling the presence or absence of a disease or pathogen using molecular data (DNA, RNA, or protein sequences). The next portion of this section will concentrate on those aspects of molecular epidemiology that impact on the epidemiology of polio.

Polioviruses and other enteroviruses are among the organisms with the highest rate of misincorporation ([67, 78], and reviewed [39]). Misincorporation comes at a high cost, namely, only approximately 10% of the >10,000 progeny from a single infected cell are viable [67]. This high frequency of misincorporation helps to explain the high ratio of physical to infectious particles [93]. Studies with ribavirin [66, 94], an antiviral drug acting as a nucleoside analogue, have shown that the misincorporation rate of polioviruses is close to the catastrophe error rate, that is, the transition point where a modest increase in misincorporation results in a drastic decrease in viability. In these experiments a 9.7-fold increase in mutagenesis resulted in a 99.3% loss in viral genome infectivity after a single round of replication, while a less than twofold increase in the natural mutational frequency resulted in a 50% loss of viability.

Fitness is based on the overall performance during viral replication [67], a complex process, involving recognition of and binding to the host cell, uncoating/entry, initiation of protein synthesis before RNA replication takes place, regulation of replication and translation once RNA replication is initiated, culminating with assembly, maturation and externalization of mature virus and survivability until subsequent infection of the next host cell or organism. Changes can affect more than one of these processes. For example, an increase in the mutational rate in infections in the presence of the nucleoside analog ribavirin not only led to an increase in nonviable genomes, but also caused a reduction in the total number of viral genomes produced [66]. One of the advantages of the quasispecies nature of poliovirus offspring is that isolates with a selective advantage to new growth conditions may already exist in the population [66]. Studies on mixed infections in PVR Tg21-transgenic mice suggest that random selection may play a role in the selection of which genomic variants within a mixed infection in the gut infect the CNS since virus isolated from the CNS was not always the most neurovirulent [95]. Other experiments showed that increased fidelity of the polymerase reduced viral fitness in the PVR Tg21-transgenic mice [96] or in tissue culture [97]. An alternate explanation for selection was proposed by Andino and colleagues [98]. They provide evidence that the quasispecies is not just a collection of individual variants, but a group of interactive variants and that fitness and selection may occur at the level of the population rather than at the level of individual genomes. In their study, an increase in polymerase fidelity affected viral adaptation and pathogenesis in addition to genome variability. Data supporting the suggestion that minor components can alter the phenotype of quasispecies comes from retroviral infections [99] and studies with VSV [100].

A number of studies have shown that single nucleotide misincorporations by the polio RNA-primed RNA polymerase accumulate at a more or less constant rate that can be used as a "molecular clock" to estimate evolutionary time between isolates and to determine whether sequence differences between two polioviruses isolated within a given time interval are consistent with a shared, direct evolutionary pathway between them [38, 39, 101, 102]. In general, the rate of accumulation and fixation of single nucleotide substitutions appears to be similar for all isolates regardless of kind (all three serotypes of wild, vaccine, or vaccine-derived polioviruses), type of polymerase (original intact polymerase or chimeric or complete recombinant from the same serotype, a different serotype or even a group C non-polio enterovirus), or type of infection (transient in immune competent individuals, persistent in immunodeficient patients, or even in the very elderly where waning immunity may play a role in selection) [4, 39, 79, 90, 101, 103–110]. Moreover, the rate of third codon position synonymous substations appeared to be fairly constant throughout the period of virus excretion in a persistently infected individual [107]. Using molecular observations from full-length genome sequences from viruses isolated during a 10 year long outbreak established from a single imported founder virus, Jorba et al. [39] calibrated five clocks based on five different classes of nucleotide substitutions. The constants for total substitutions (K_t), synonymous third position substitutions in coding regions (K_s), synonymous transitions (A_s), synonymous transversions (B_s), and non-synonymous substitutions (K_a) were $1.03 \pm 0.10 \times 10^{-2}$, $1.00 \pm 0.08 \times 10^{-2}$, 0.96 ± 0.09 _ 10^{-2}, 0.10 ± 0.03 _ 10^{-2}, and $0.03 \pm 0.01 \times 10^{-2}$ substitutions/site/year, respectively. The rates were similar whether calculated using linear regression, a maximum likelihood/single-rate dated tip method, and Bayesian inference. The first two constants were mostly controlled by the third. As for saturation, third position synonymous transitions become evident by 10 years and complete saturate within 65 years while saturation of synonymous transversions was predicted to be minimal at 20 years and incomplete even at 100 years. This wide variation in calculated time constants depending on the type of substitutions together with differences in the estimated time until all possible changes become saturated, provides a flexibility that allows one or more clock to be applied to characterize the range from evolution in outbreaks between very closely related isolates with short intervals between isolations, to comparison between much more distantly related polioviruses or related enteroviruses. It is interesting to note that the molecular clocks are fairly constant given that intratypic and intrageneous recombination can result in complete or partial substitution of the polymerase whose intrinsic properties presumably govern the rate of misincorporation. Mutations may increase non-synonymous mutation rates [111] while others decrease them [96]. Multiple recombination events [83, 85] must be ruled out or taken into account when calculating time clocks based on the number of nucleotide differences.

Different factors that affect fitness and determine the viability of individual viral offspring result in differences in observed substitution rates and patterns in the different functional elements of the genome shown in Fig. 10.3. Namely, using the rate of substitutions in the VP1 capsid protein as reference, the rates of substitutions are approximately half in the conserved region of the 5'UTR, approximately threefold higher in the hypervariable region of the 5'UTR, and equivalent or somewhat lower in the remainder of the ORF [4, 39, 101, 109]. The data for nucleotide substitutions in the nonstructural P2 and especially the P3 regions of the ORF and the 3'UTR are less accurate and less informative due to frequent recombinations among polioviruses and between polioviruses and non-polio enteroviruses within these regions.

The genetic code introduces a bias in the position of observed substitutions. Substitutions in the third position of a codon are least likely to result in an amino acid change and these synonymous substitutions are by far the most abundantly observed in wild poliovirus, polio vaccine and VDPV infections [4, 79]. Non-synonymous substitutions that occur in the initial stages of vaccine infections restore replicative fitness and in many cases neurovirulence [3, 112] while those in persistent infections may influence receptor–virus interaction (see page 8133, 8159).

Three-dimensional requirements also bias the observed distribution of substitutions. Maintenance of stem of the stem-loop structure in the conserved region of the 5'UTR especially within the IRES appears to be one of the major constraints on viability. For example, complimentary paired double substitutions that maintained stem structure were frequently found in the loop V of evolutionarily related environmental isolates [109] whereas a single nucleotide substitution in a loop of Loop V is a dominant determinant of attenuation of neurovirulence and growth at elevated temperatures. Other examples of three-dimensional effects are mutations that occur at distances from functional sites that influence the viral response to antiviral drugs in the hydrophobic pocket [61] and mutations that occur at the interfaces between proteins and at the N-terminals of VP1 and VP4 that may affect structural stability and the receptor-induced transitions [45].

Due to the complex nature of polioviral replication and multi-functionality of viral enzymes and viral three-dimensional structures, selective pressures that operate on one structure or function may affect another seemingly unrelated property. One of these apparent paradoxes is the fact that some RNA viruses including poliovirus may diverge antigenically in the absence of immune selection [113]. One of the features that distinguishes polio vaccine evolution during persistent infections in total B-cell deficient immunodeficient patients from evolution during person-to-person transmission in immune competent but naïve individuals is that isolates from the former but not the latter have high numbers of amino acid substitutions in and around neutralizing antigenic sites [3, 4, 62, 114]. Antibody titers tend to wane in the elderly. The finding of amino acid substitutions at or near neutralizing antigenic sites during infection of the elderly with type 1 monovalent mOPV [103] may suggest that waning immunity may create a situation resembling the early stages in establishment of persistence in immunodeficient patients. Since some of the amino acids and structural organizations are shared by neutralizing antigenic sites and receptor binding sites, the high mutation rate in neutralizing antigenic sites is more likely the result from selective pressures governing receptor–virus interaction during establishment and maintenance of persistence than on non-humoral mediated immune selection or selection by variations in anti-polio antibodies in the IVIg regimens these immunodeficient patients receive to compensate for their B-cell deficiencies. This sharing of functions has also been suggested to be one of the reasons why there are only three serotypes of poliovirus [26, 50]. It is commonly accepted that the evolution of a fourth serotype would require receptor switching of a non-polio enterovirus to the use of the PVR, CD155. A somewhat paradoxical alternative for emergence of a fourth serotype may be through antigenic evolution during persistent infections in immunodeficient individuals as a result of selective pressures relating to receptor binding. Consistent with this is the observation that cohorts of immunized individuals who had high titers against vaccine strains had significantly reduced geometric mean titers against highly diverged neurovirulent vaccine-derived viruses that were isolated from environmental samples [109, 115] and individual titers against some of these isolates were <1:8 in 7% of the adults [109, 116].

Any discussion on molecular evolution of polioviruses and their effects on polio epidemiology would be incomplete without a discussion on vaccine-derived polioviruses, VDPVs (see page 8157 and the section on future directions). Evolution of live attenuated polio vaccine occurs by the same processes as in wild polioviruses, namely, by accumulation of single nucleotide substitutions and through genomic recombination. Evolution of VDPVs occurs during person-to-person circulation in cohorts of naïve or under-immunized individuals especially after interruption of vaccination, or during persistent infections of immunodeficient individuals [3, 80, 106, 114, 117, 118]. The letters "c" or "i" for viruses that evolved during person-to-person circulation or during persistent infections of immunodeficient individuals, respectively, are appended before "VDPV" when the evolutionary pathway is known. The prefix "a" is added instead when the pathway is ambiguous or unknown.

One of the goals of the Global Eradication Initiative (see page 8150) is to reach a stage where all wild poliovirus transmission is terminated and vaccination can be discontinued. "Emergence of VDPVs," "failure to vaccinate," and "vaccine failure" discussed below have been the three major reasons for the delay in achieving eradication of poliomyelitis. Providing that enough money and effort can be mobilized, current vaccines and vaccine strategies are probably sufficient to enable immediate solutions for "vaccine failure" and "failure to vaccinate." Promising alternatives applying experience with adjuvants and better applicators have revived the possibility of using techniques explored in the 1950s such as fractional subdermal doses of inactivated virus, IPV [16, 119–121].

Vaccine-derived viruses consistently emerge as a consequence of the inherent genetic instability of poliovirus [122]. Moreover, many of the first sites that mutate restore replicative fitness, reverse attenuation of neurovirulence, and restrictions on growth at elevated temperatures. cVDPVs behave like wild polio [106, 112]. These cVDPVs clearly present a serious health threat [114]. The minimal amino acid changes in neutralizing antigenic sites that occur during person-to-person transmission of cVDPVs [3, 4, 123] allow rapid control through OPV immunization campaigns [106, 114, 124]. In contrast, iVDPV infections have not always been curable [63, 125], and the numbers and identities of anonymous persistent secretors are unknown [109]. The problem of reemergence of poliomyelitis through cVDPVs and especially iVDPVs requires a coordinated global program to discontinue the use of OPV with substitution of alternative vaccination strategies to prevent the appearance of large cohorts of unimmunized individuals during the period when OPV or cVDPVs may still circulate and iVDPV and aVDPV infections persist [116, 122, 126–128]. In fact, eradication should be redefined to include the elimination of both wild and vaccine-derived viruses [122].

OPV strains, like their wild counterparts, readily recombine the noncapsid encoding portions of their genomes with other polioviruses and related non-polio enteroviruses at very early stages in emergence. Evidence for this comes from analysis of poliovirus RNA from vaccine-associated paralytic poliomyelitis cases, VAPP, where, for example, >50% of polioviruses isolates had recombinant genomes [81, 129–132]. Supporting this is evidence from environmental surveillance where vaccine viruses with minimal divergence (0.5–1%) in their VP1 sequences had already recombined with polio and non-polio enteroviral genomes in regions encoding nonstructural proteins [133]. The ability to simultaneously reverse multiple mutations by recombination could foil efforts to develop improved oral vaccines. Introducing mutations that decrease the chance for reversal by single nucleotide replacement such as incorporation of polymerases with improved fidelity or a total redesign of the genome of each serotype based on rare codon usage [3] could be bypassed by recombination.

Finally, the general consensus is that selective pressure or a higher mutation rate due to local sequence or secondary structure leads to a higher frequency of mutations at certain "hot spots" [93]. However after reviewing the pattern of changes in substitution frequencies throughout the genome it may be more accurate to think of the real frequency of substitutions as that observed in the so-called hyper variable region of the 5'UTR which may be under minimal selective pressure, and consider all other regions as "cold or colder spots" with lower observed rates of substitution derived from negative selection driven by the requirement for viability.

Virion assembly, maturation, and release in picornaviral infections have been reviewed [25]. The ratio of viral particles to infection particles ranges between 30:1 and 1,000:1. Many of the viral particles are noninfectious due to lethal mutations in their genomic RNA and/or incomplete maturation.

Poliovirus Infections at the Level of the Individual Host

The incidence of poliovirus infections is significantly higher in summer and autumn in temperate zones, becoming less seasonal as the environment becomes more tropical (reviewed in [3]). Improved sanitation and vaccination have reduced natural endemic infections in the very young and together with incomplete vaccine coverage has led to an increasing number of infections in older individuals.

There are two major routes of host-to-host transmission. The most common and most efficient is fecal–oral, followed by oral–oral transmission as Dowdle et al. described for the fate of poliovirus in the environment and their review of the infectious dose for transmission in humans [134]. It has been postulated that there has been a shift from the former to the latter route, as the level of community hygiene improved [56]. The infective dose after ingestion of Sabin vaccine strains is approximately 100-fold higher than that for wild poliovirus, 1000 $CCID_{50}$ compared to 10 $CCID_{50}$, respectively [3, 134, 135]. Nerve damage in the lower spinal cord results in paralysis of the lower limbs (spinal poliomyelitis), whereas damage in the upper spinal cord and medulla may result in bulbar poliomyelitis and paralysis of breathing [72]. The percent of infections ending in paralytic poliomyelitis is further reduced in highly immunized populations. This ratio of asymptomatic cases to paralytic cases has implications for surveillance strategies (see page 8154) based on investigation of all AFP cases. Between these two extremes falls the "minor disease" [72, 136], approximately 5% of infections with wild polio that result in abortive poliomyelitis with fever, fatigue headache, sore throat, and/or vomiting, and another 1–2% result in non-paralytic poliomyelitis with aseptic meningitis, pain, and muscle spasms. The incubation period is between 7 and 14 days but ranges between 2 and 35 days [72]. Virus can be recovered from the throat, blood, and feces by 3–5 days. It was initially thought that viremia was infrequent, but this was based on observations in patients with paralytic poliomyelitis who most likely already had high circulating titers of neutralizing antibodies [72, 137]. However when observations were made early after exposure, for example, in contacts of cases, a high frequency of viremia was demonstrated, implying that the viremia might play a vital role in the development of paralytic poliomyelitis [136, 138]. This was strengthened by concurrent experiments that demonstrated a protective effect against CNS lesions by antiserum in experimentally infected primates. The genetic basis for neurovirulence of poliovirus isolates is addressed below on page 8146.

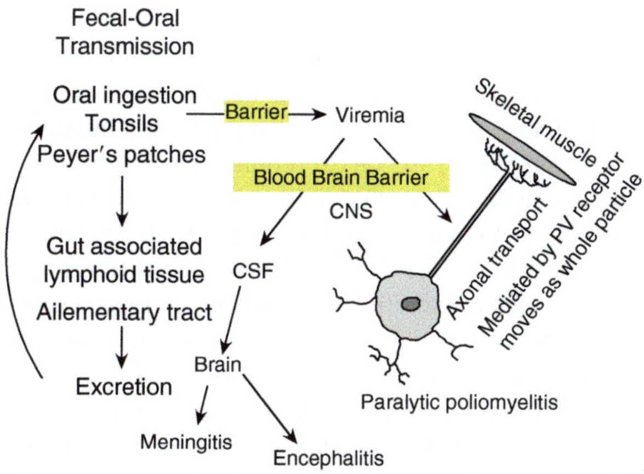

Fig. 10.5 Poliovirus infections. Poliovirus is transmitted from host-to-host by a fecal–oral and to a lesser extent oral–oral routes of transmission. Virus first infects cells in the tonsils, Peyer's patches, and gut-associated lymphoid tissues and viral progeny are excreted in feces. This phase is followed by a systemic infection during which there is viremia for a short period of time. In some individuals virus crosses the blood-brain barrier by entering the CSF, by axonal transport along nerve cells, and possibly from infected white blood cells that enter the brain. These individuals may develop meningitis, encephalitis, or paralytic poliomyelitis. Destructive viral replication in nerves of the anterior horn of the spinal cord may lead to irreversible atrophy of de-enervated muscles while bulbar paralysis occurs when cranial nerves are infected. Most (>90%) infections of naïve individuals even with the most neurovirulent strains are asymptomatic, 5% result in meningitis, encephalitis, and/or transient paralysis. Only 0.1–1% of the infections will result in permanent paralytic poliomyelitis or death

Paralytic poliomyelitis, encephalitis, and aseptic meningitis occur after a biphasic infection where viremia in some systemic infections is followed by infection of the CNS [136, 138] (Fig. 10.5). Studies of virus in the CNS and stools in VAPP patients suggested that the virus that invades the CNS was randomly selected [95]. Acute flaccid paralysis (AFP) is a direct result of destructive replication in motor neurons followed by atrophy of de-enervated muscles. Skeletal muscles are affected when nerves in the anterior horn of the spinal cord are infected and bulbar paralysis occurs when cranial nerves are infected [4, 139]. The maximum effect on muscles occurs within a few days after the start of symptoms. Muscle recovery can occur when infection only results in temporary loss of nerve function. Residual paralysis may last from months to the life of the infected individual [72].

Poliovirus infections are not the only cause of AFP. Non-polio AFP occurs with an incidence of 1 per 100,000 children (see page 8153 for the implication this has for surveillance). Guidelines that help epidemiological investigators distinguish AFP caused by polio from AFP caused by other causes are reviewed in Sutter et al. [4]. Final diagnosis requires laboratory confirmation of a poliovirus infection.

Poliovirus infections start as a local infection of cells in the tonsil, intestinal M cells, Peyer's patch of the ileum, and the mesenteric lymph nodes [3, 72].

This replication in the gut results in the excretion of poliovirus during defecation by all individuals with asymptomatic as well as symptomatic infections and is the basis for fecal–oral transmission. It also provides the rational for supplementary environmental sewage surveillance (see page 8154) for poliovirus infections. A review of publications between 1935 and 1995 on excretion of polioviruses by Alexander and associates [140] indicated that in most infections of naïve children, wild polioviruses were excreted for 3–4 weeks with a mean rate of 45% at 28 days, and 25% of the cases were still excreting during the sixth week. In contrast, fewer than 20% excreted vaccine strains after 5 weeks. Excretion of polio ranged from a few days to several months [141]. The highest probability of detecting poliovirus positive stool samples was reported to be at 14 days after the onset of paralysis [140] and is the basis of stool sample collection for diagnosis of polio AFP surveillance (see page 8153). The disappearance of poliovirus from sewage samples and from stool samples of immunized children within 6–8 weeks after an immunization campaign [142] or after transition to exclusive immunization with IPV [143] provides additional confirmation for the short duration of excretion. Persistent poliovirus infections are the exception and will be discussed in more detail below. Interestingly, more than one evolutionarily linked lineage of the same serotype may co-circulate in the gut of such persistently infected individuals [79, 104, 107, 109].

Excretion and the duration of excretion are dependent on host factors and on vaccination history of the infected individual. Immunization history may start with passive immunization from maternal antibodies. However, maternal antibodies have an estimated half-life of approximately 1 month [144]. Based on a comparison between titers in cord blood and at 6 weeks, the half-lives for maternal neutralizing antibodies against type 1, 2, and 3 polio were 30.1 days, 29.2 days, 34.6 days, respectively [119]. Immunization history obviously also includes polio vaccinations and natural exposure to endemically circulating wild poliovirus and waning immunity in aging cohorts.

Skeletal muscle injury, including injury caused from intramuscular injections, increases the likelihood of poliomyelitis in children infected with wild or vaccine poliovirus. Mouse model studies have suggested that in this provocative poliomyelitis, the muscle injury facilitates viral entry to nerve axons and subsequent damage to the motor neurons in the spinal cord [145].

Some individuals who had poliomyelitis develop new muscle pains, hypoventilation, new or increased weakness or fatigue and paralysis decades later after a period of relative stability. This reappearance of polio-related symptoms is referred to as postpolio syndrome. There is a large body of literature relating to postpolio syndrome that will be left up to the reader to pursue. Suggested starting points include the websites of the Post-Polio Health International (www.post-polio.org), the Mayo Clinic (www.MayoClinic.com), a 1992 paper on the "Epidemilogy of the post-polio syndrome" by Ramlow et al. [146], and a 2010 review on the pathophysiology and management of postpolio syndrome by Gonzalez et al. [147]. There is still a debate whether persistent poliovirus or mutated poliovirus contribute to the development of postpolio syndrome [147]. The risk factors include the extent of permanent residual impairment after recovery from the poliovirus infection, an increased recovery after AFP possibly related to the extra stress on compensatory neural pathways and overuse of weakened muscles, the age of onset of the initial illness, and physical activity performed to the point of exhaustion.

Natural infections with poliovirus stimulate both humoral and cell-mediated immunity (see [149, 150] and reviews [4, 148]). Neutralizing antibodies appear in exposed individuals around the time that paralytic symptoms become evident in the few individuals who develop symptomatic infections [72]. Neutralizing IgG and IgM antibodies are also induced in response to immunization with inactivated polio vaccine. The neutralizing antibodies induced after exposure to live or inactivated poliovirus prevent disease by blocking virus spread to motor neurons of the central nervous system [3]. Once seroconversion occurs after vaccination, individuals are protected from disease for life, although circulating antibody titers may wane late in life and may drop below protective levels against one or more serotype in some individuals.

The epitopes on vaccine-derived and wild poliovirus strains that induce neutralizing antibodies may differ from those on vaccine strains. Neutralizing antibody titers $\geq$1:8 against each of the three Sabin OPV serotypes are considered protective; however higher titers may be needed to compensate for the relatively lower antigenicity of wild and vaccine-derived strains [151]. For example, the highest serum neutralizing antibody titers were recorded from individuals immunized exclusively with OPV or IPV when the live challenge virus was the same as that used in vaccination, slightly lower for the respective heterologous strain, and significantly lower for wild and vaccine-derived strains. Serum from some individuals who had titers of $>$1:50 against Sabin vaccine strains had titers of $<$1:8 against some wild or vaccine-derived of at least one serotype, suggesting that titers of 1:64, 1:32, and 1:16 against Sabin serotypes 1, 2, and 3, respectively, might be more appropriate to ensure minimal protective coverage [151].

Primary infection in the intestinal tract by wild poliovirus or live attenuated polio vaccine induces secretory IgA antibodies in addition to IgM and IgG antibodies. One of the rationales for the use of live attenuated polio vaccine was that while disease would be prevented by humoral antibody production stimulated by either OPV or IPV, the extent of infection or reinfection and shedding would also be reduced by induction of secretory IgA antibodies by active infection of intestinal cells with live vaccine in mimicry of the natural route of infection [148, 152–155]. IgA induced in the gut plays an important role in terminating primary infection in the intestinal tract and the tonsils [3, 152] affecting both fecal–oral and oral–oral transmission. In practice IPV also induce some intestinal immunity although less than OPV and the duration of excretion in individuals immunized with IPV appears to be longer [23, 56, 156–158].

There is some indication that the duration and possibly memory of intestinal protection is relatively short and the time for clearance of virus relatively longer than the 3–6 or 7–14 days incubation period of the minor and paralytic diseases [152, 156, 159]. Complete blockage of replication in the intestines may occur in only 25–40% of fully immunized children [158, 159]. The rapid decline in intestinal immunity means that polio can establish transient infections even in persons with adequate humoral immunity and circulate silently in that community. Lower efficiency of oral vaccines under certain conditions further complicates efforts to break chains of poliovirus transmission. Passive immunization with maternal antibodies, which has a short half-life, also affects oral vaccine efficacy (see page 8143).

It is not clear what role cell-mediated immunity may play in the control of polioviral infections. Cell-mediated immune responses were observed early after wild poliovirus infections by the macrophage migration inhibition (MIF) technique but were not observed a later time [160], whereas intradermal administration of subfractional doses caused induration and erythema of 3 mm diameter or above, in 14 of 18 vaccinees that indicated a cell-mediated immune response [161]. In addition, at least in a mouse model, all three serotypes stimulated cross-reactive and serotype-specific T helper cell responses detected by both in vitro proliferation and interleukin (IL)-2/IL-4 production [162].

How can poliovirus infections be prevented? The main tools in the global eradication of poliomyelitis have been the introduction of universal vaccination (vaccine) and improvements in hygiene. The primary goal of routine immunization is to protect the individual [163]. The secondary goal is to immunize a high enough proportion of the population so that the entire population will become protected. As eradication approaches completion, it is becoming more and more apparent that additional approaches will need to be employed in parallel with and perhaps instead of vaccination to extinguish the last pockets of endemic person-to-person transmission and persistent infections. All of these approaches will be necessary to prevent and control reemergence of polio after eradication. For more information, the readers are referred to excellent reviews on inactivated poliovirus vaccine by Plotkin and Vidor [164] and live oral poliovirus vaccines by Sutter et al. [4]. Sources for early history can be found in *A History of Poliomyelitis* by Paul [165] and *Polio Vaccine: The First 50 Years and Beyond* edited by E. Griffiths et al. [166].

The road to the development of effective vaccines against poliovirus was long and paralleled the growing understanding and ability to manipulate viral infections in the laboratory. Most of the important early milestones were listed in the last paragraph of the introduction and in Fig. 10.1. Mass vaccination trials and studies involving millions of vaccinees played an early and important part in acceptance of universal polio vaccination as a means for fighting poliomyelitis [19]. It must be stressed that problems and other difficulties during this progression stimulated numerous basic and epidemiological research studies that have resulted in improvements culminating in the current safe high-potency oral and inactivated vaccines that have reduced the number of annual paralytic poliomyelitis cases from >350,000 per year in 1988 to approximately 1,500 in the last few years. Difficulties in reducing this further are discussed below on page 8166. Criteria for quality control for production of polio vaccines introduced by the WHO in 1962 have been updated in relation to newly acquired knowledge about the epidemiology of poliovirus and polio vaccine. One of the major risks associated with the use of live vaccine is that progenies of the vaccine readily accumulate mutations some of which may reverse attenuation. The highest risk for vaccine-associated paralytic paralysis, VAPP, comes from Sabin 3, the vaccine serotype that also has the highest variability across production lots [3]. However the risk of OPV-associated polio is less than 0.3 per million doses [22] with the risk highest in naive children receiving their first dose [81]. The risk (see page 8156) of not using oral vaccine for global eradication compared to its use at the current stage in the Global Poliovirus Eradication Initiative remains overwhelmingly in favor of its use [1].

Three incidents nearly derailed early efforts to develop and employ effective vaccines. The most glaring of these, primarily from the point of views of negative publicity for use of polio vaccines, was the "Cutter Incident" in 1955 where wild poliovirus was inadequately inactivated probably because of failure to remove clumps that may have sequestered and protected infective vaccine virus, and a nonlinear tailing-off of inactivation at low titers [22, 167, 168]. Altogether more than 400,000 children were inoculated with an inadequately inactivated vaccine batch produced by the cutter vaccine production facility which resulted in 94 cases of poliomyelitis among primary vaccinees, 126 cases in family contacts, and another 40 cases among community contacts and 10 deaths. The publicity caused great concern throughout the world until the cause was discovered and corrective measures applied. The second, apparent failure of early vaccines to protect against subsequent infection and paralytic disease due to an initial lack of awareness in the 1950s that there were three non-cross-reacting serotypes of polio has already been mentioned. The third problem, the contamination of live polio vaccine with SV40 virus, a simian virus, continues to raise concerns about long-term effects from human zoonotic infection with this virus that was shown to cause cancer in mice [22, 169–171]. The SV40 was inadvertently introduced through the use of SV40-infected simian cell cultures in some early vaccine production batches. So far there is little evidence for any contribution to the incidence of tumors in the humans who received SV40.

The many vaccination formulation and vaccination schedules that have been employed during the effort to eradicate polio and the rational for their use have been reviewed in depth [4, 164]. Changes in schedules and formulations mean that in any one region different cohorts in the total population will have received different vaccine formulations and immunization schedules. This complicates determining duration of protection and interpretation of events. The evolution of vaccination policy in Israel [172–174], a graph showing the history of poliomyelitis in Israel (Fig. 10.6), and the two disagreeing discussions that were published within the same report on the underlying causes that enabled the last outbreak in Israel in 1988 [175] are a good example of this difficulty.

Isolation of poliovirus with attenuated neurovirulence was a prerequisite for the development of oral polio vaccines (OPV; see reviews [3, 4, 166]). Vaccine candidates were either derivatives of neurovirulent or even highly neurovirulent (e.g., Sabin 3) isolates selected for attenuation after passage in primates, primate cell cultures, and/or non-primate cell cultures or starting from isolates with low neurovirulence (e.g., Sabin 2). Neurovirulence refers to the ability of an isolate to cause an infection adversely affecting functions of the CNS, keeping in mind that for any neurovirulent isolate, only 5% of infections cause transitory adverse CNS effects and less than 1% cause permanent paralytic poliomyelitis. The total number

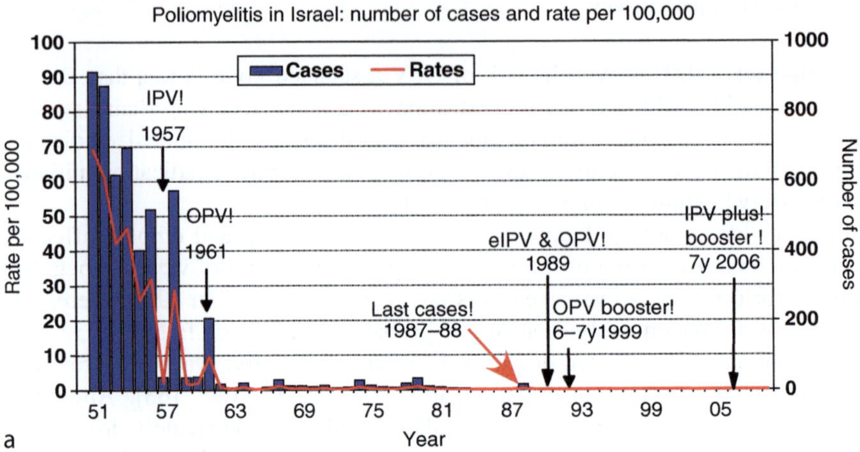

a

Polio vaccine immunization schedules in Israel 1957–2010

1957–1960 exclusive use of IPV (mainly local production)
- varied number of doss (2 to 4)
- varied quantity of antigen per dose
- initially administered in campaigns then by routine vaccination schedule!
- estimated coverage 75% to 90%

1961–1963 exclusive use of OPV
- coverage 95%
- 1961 mOPV1
- 1962 add mOPV2 and mOPV3
- 1963 tOPV 3 dose schedule (2, 6 and12 months)

1964–1978 4 doses by (2, 4, 6, 24 mo) coverage 81% to 91%

1979–1981 same as in1964–78 plus a supplimental mOPV1 once a year for ages 0–2 yrs

1982–1987 three programs
- a. 4 OPV doses in 12/14 health sub-districts
- b. 4 OPV doses as in a. plus one dose mOPV1 in selected groups at risk
- c. exclusive eIPV 3 doses in 2 of 14 healthsub-districts
  at least one dose in combination with DTP.

1989–2005 combined eIPV/OPV
- 2 doses of eIPV (2 and 4 mo)
- 2 OPV doses (4 and 6 mo)
- simultaneous IPVplus OPV at 12–16 mo)
- 1990 IPV booster added at 6–7 yr

2006–2010 exclusive use of eIPV
- 4 doses of eIPV (2, 4, 6, 12 mo) plus booster 7 yrs)

1957–2010 Immigration of familes with children who were vaccinated by different vaccination schedules used at their countries of origin.

b

Fig. 10.6 *Prevention of poliomyelitis through universal vaccination and evolving vaccination strategies as illustrated by the history of cases and vaccination schedules in Israel between 1957 and 2010*. Figure (**a**) represents the annual number of cases (*blue bars*) of laboratory confirmed poliomyelitis cases and the rate per 100,000 children (*red line*) between 1951 and 2010. The *red arrow* indicates the last cases of poliomyelitis that occurred during an outbreak in 1987–1988. Israel has been poliomyelitis-free since 1989. *Black arrows* indicate major changes in vaccination policy. Previous attack rates of 14.2 and 146.9 per 100,000 in 1949 and 1950, respectively, signaled the transition from endemic to epidemic epidemiology of poliomyelitis in Israel. A full list of vaccination schedules is indicated in (**b**) (Data presented in (**a**) was supplied with permission by the Israel Center for Disease Control. The vaccination schedules were taken from Swartz TA. The Epidemiology of Polio in Israel A Historical Perspective. Tel Aviv: Dyonon Pub. Ltd.; 2008 [172])

of nucleotide differences between vaccine strains and their respective parental strains was found to be 57 nucleotides and 21 amino acids for serotype 1 [176–178], 2 nucleotide differences and 1 amino acid difference for serotype 2 [179, 180], and 10 nucleotide differences and 3 amino acid differences for serotype 3 [181, 182]. Sequence analysis coupled with genetic manipulation has allowed investigators to pinpoint which of these nucleotide differences between vaccine candidates and vaccine strains account for the loss of neurovirulence (Fig. 10.3) [93, 183]. "Quantitative determination of the contributions of each substitution is complicated by several factors: (a) The role of minor determinants of attenuation is difficult to measure, (b) some substitutions have pleiotropic effects on phenotype, (c) some Sabin strain phenotypes require a combination of substitutions, (d) second-site mutations can suppress the attenuated phenotype in various ways, and (e) the outcome of experimental neurovirulence tests may vary with the choice of experimental animals (monkeys versus transgenic mice) or the route of injection (intraspinal versus intracerebral)" [3]. The propensity of vaccine to evolve and revert to neurovirulent phenotype is discussed throughout the current review.

The safety and effectiveness of live attenuated polio vaccine strains in preventing poliomyelitis was very clearly demonstrated in large clinical studies involving millions of children in the 1950s [19]. A number of factors including vaccination schedules, the presence of maternal antibodies, hygiene, and nutritional status of the individual influence the efficiency of induction of seroconversion by OPV strains. Early studies showed that viral interference between strains in the trivalent vaccine and from concurrent infections with non-polio enteroviruses also influences vaccination outcome [19]. Multiple doses of OPV are recommended to ensure seroconversion rather than to boost waning immunity [184]. The number of OPV doses that is needed to reach 90–95% seroconversion rates in naïve children is not the same for all populations. For example, three doses will seroconvert 90–95% of naïve children in developing countries, whereas in certain regions within developing countries such as India, the same three doses will only seroconvert a maximum of 60% of vaccinees [184]. Supplemental immunization activities (SIAs) employed sometimes more than once a year are needed to ensure adequate primary vaccination coverage and to fight endemic circulation of wild poliovirus or reintroductions of wild poliovirus. In SIAs, all children in national or subnational regions are immunized in national immunization day (NID) and/or subnational immunization day (SNID) campaigns, respectively, with a dose of OPV irrespective of vaccine history. The costs of the additional doses needed to raise seroconversion rates to above 90%, significantly raise the cost for effective immunization with OPV and require the coordination and use of many paid and voluntary staff. In fact, in the end it may actually be easier to immunize three times with IPV (even at current costs) than with the additional number of doses of OPV especially when access to populations is difficult and environmental conditions challenge maintenance of viability of the live vaccine. This counters both the lower cost and difficulty of administration rationales for using OPV instead of IPV. Mass immunization campaigns have rapidly boosted herd immunity [3].

The take of OPV is negatively influenced by the presence of maternal antibodies. Nonetheless, when infants are fed OPV at birth, 30–60% excrete virus, 20–40% of infants seroconvert, and the subsequent take of OPV is better when a birth dose is given (reviewed in [184, 185]).

Vaccination formulation must also take into account differences in the efficacy of induction of intestinal immunity by the different vaccine serotypes [155]. For example, type two was 100% effective with two doses, whereas types 1 and 3 needed more than three doses. Since the elimination of wild type 2 in 1999 [186], and the significant decrease in the number of endemic regions where wild type 1 and 3 co-circulate, SIAs have increasingly turned to the use of monovalent and divalent OPV. Monovalent OPV vaccines improve seroconversion rates compared with tOPV [187]. However routine immunization still requires the use of tOPV to prevent the accumulation of large cohorts of individuals who are naïve to type 2 poliovirus and who could serve as a reservoir for transmission of neurovirulent type 2 VDPV as has occurred in Nigeria [106]. New guidelines for the use of mOPV1, mOPV2, and dOVP1+3 have been recently issued [188].

The use of inactivated poliovirus is an alternative approach to vaccination against polio (reviewed in [164]). Salk developed an inactivated polio vaccine, IPV, using neurovirulent strains of the three serotypes of poliovirus. IPV was successfully tested by a placebo-controlled trial in over 400,000 children and in unblinded observations on another 1,000,000 children before certification for use in the mid-1950s [16, 17, 189]. A relatively higher difficulty in production, greater production costs, higher difficulty in administration, and the initial belief that only live vaccine would efficiently evoke intestinal immunity led to the choice of OPV for most routine national vaccination programs [122]. Countries are currently switching to vaccination with IPV in combination with OPV or more often in place of OPV because of its relative safety record (no VAPP cases), improvements in manufacture that have increased effectiveness and reduced the cost difference between a dose of OPV and IPV, and the paradoxical success of OPV in reducing poliomyelitis as an epidemic disease in most countries [164]. Additional motivation has come from the increasing awareness that fully neurovirulent vaccine-derived polioviruses behave like wild polioviruses [133, 190, 191] that must be eliminated and prevented from emerging in order to attain final success for poliomyelitis eradication.

Early studies on genetic and antigenic variation such as a study of Sauket strains, the type 3 used in production of IPV [192] were instrumental in the establishment of rigorous standards for seed stocks for vaccine production. The original IPV formulation has since been improved. This enhanced IPV, eIPV, has a higher immunogenicity and protective efficacy than IPV [157]. A number of factors contributed to this improvement. These included new production protocols, new tissue culture techniques including a microcarrier-based technology, and a more optimal balanced formulation of the three serotypes. It can be administered alone or can be combined with other vaccines such as DTP. The immunogenicity of eIPV was at least as good as that of OPV and there was good long-term immunity [157]. Subdermal administration of fractional doses of IPV was one of the approaches tried in the early 1950s

[16]. Subsequently seroconversion rates from fractional doses were shown to be adequate but somewhat lower than for full dose intramuscular injections fractional doses [119, 193]. In another trial, similar seroconversion rates were observed but there were lower median titers in those receiving fractional doses [120]. Fractional doses effectively boost titers in previously immunized individuals [194]; however there is no long-term information on the rate of waning immunity in individuals treated with these fractional doses. Large non-inferiority studies testing subdermal administration of fractional doses of IPV using needle-free devices such as recently by Mohammed et al. [120] and Resik et al. [119] offer one quite promising practical solution for realizing cessation of use of live OPV with affordable alternative vaccines as recommended by the Advisory Committee on Polio Eradication in 2004 [195].

Antiviral drugs offer a promising complimentary or alternative approach to the use of vaccine to control poliovirus infections especially for persistent infections in immunodeficient individuals, during the final stages of eradication, and for post-eradication reemergence [122]. Presumably theses drugs may also work to control severe infections by non-polio enteroviruses or have been chosen because they have been shown to do so. Drugs with unique virus-specific targets such as capsid proteins, the hydrophobic pocket, the RNA-primed RNA polymerase, protease inhibitors, protein 3A inhibitors, nucleoside analogs, proteinase 2c inhibitors, and compounds with unknown mechanisms of action have been reviewed [196]. There is still a long way to go to find truly effective universal anti-polio or anti-enteroviral drugs, thus only a few examples will be provided.

Pocket factor drugs such as WIN 51711 [74], isoflavenes [61], pleconaryl [197], and capsid inhibitor V-037 [198] prevent viral entry by interfering with receptor binding or by preventing conformational changes needed for viral capsid uncoating. One of the difficulties in developing pocket factor drugs comes from the quasispecies nature of enteroviral infections, where mutants may rapidly emerge [61, 63] or there may be viral isolates already present in the quasispecies that have mutations in the capsid that may either render the isolate resistant or even dependent on the drug for growth. Furthermore these resistance mutations may not even have to be at the drug binding site (see, e.g., [61]).

Ribavirin is a drug that normally interferes with mRNA capping. While enterovirus mRNA is uncapped, the polio polymerase can incorporate ribavirin into both negative and positive-strand progeny RNA molecules increasing mutagenesis above the catastrophe limit causing a decrease in the reproductive capacity of the viruses [66, 94] (discussed above on page 8138).

Passive immunization has also been tested as a means of preventing polio. Administration of immunoglobulin shortly after exposure to polio may reduce the incidence or severity of paralytic disease although its general use is not practical due to the short time during which it is effective [199]. Intravenous preparations of immunoglobulin prepared from human populations exposed to enteroviral infections have however helped to decrease chronic meningoencephalitis infections by these enteroviruses in agammaglobulinemic patients [200]. Regular intravenous treatment may help prevent poliomyelitis in immunodeficient individuals but may not prevent virus replication and shedding [62]. Passive immunization with immunoglobulin or human milk rich in anti-polio IgA together with another antiviral pleconaryl may have helped to resolve at least one persistent poliovirus infection [125]. However, efforts to cure another persistent poliovirus infection with human milk and ribavirin, or other antiviral treatments, did not prove successful [63] and this individual has continued to excrete highly diverged vaccine-derived poliovirus for more than 20 years [62]. Anecdotally, shedding of intestinal mucosa associated with a superinfection with Shigella sonnei may have helped to cure another persistent excretor [62].

Poliovirus Infections in Populations

Poliovirus infections in populations have been the subject of many reviews over the years. The older reviews are still of interest not only because of the information they review but because they also provide a picture of policies and knowledge available at the time. The following paragraphs will concentrate on those aspects of poliovirus infections in populations that impact the most on the endgame strategy of poliomyelitis eradication. The discussion will start with a brief overview of the changing nature of the epidemiology of poliovirus infections. This will be followed by a description of the Global Polio Eradication Initiative and will end with a discussion of the three main problems that have led to a delay in its realization, namely, "failure to vaccinate," "vaccine failure," and "vaccine-derived polioviruses." When reading this section which will highlight some of the current problems and their solutions, the reader must keep in mind the overwhelming success of the Global Polio Eradication Initiation to date: a major reduction in the number of endemic countries where polio is still transmitted from 126 to 4; a decrease in the number of annual cases by >99% that prevented life-long paralysis in more than five million children between 1988 and 2005; eradication of one wild poliovirus serotype in 1999; and elimination of the majority of wild lineages throughout the world.

The nature of poliovirus infections in populations has gone through a number of phases. Before the appearance of outbreaks of poliomyelitis starting in the nineteenth century, poliovirus circulated endemically. Infections occurred in the very young, and conferred lifelong immunity against reinfections with the same serotype. The epidemics became more frequent, grew in size, and infections included older children and adults who were not naturally immunized. Vaccination has drastically reduced the number of people who have been exposed to natural infection with wild polioviral strains. There also appears to be a shift in person-to-person transmission routes. Oral–oral transmission has increased in importance while fecal–oral transmission decreased as a result of improved hygiene [56]. Control of poliomyelitis requires breaking all chains of wild poliovirus transmission by immunizing all children with three doses of polio vaccine or at least enough children so that herd immunity protects the remaining population. The percent of the population that needs to be immunized for establishment of herd immunity against wild poliovirus is >85% for developed countries and >90–95% in tropical developing countries. In 2010, there were still three groups of countries: four "endemic countries," Afghanistan, India, Nigeria, and Pakistan where the transmission of wild viruses has not yet been completely halted, "polio-free countries" where vaccination has broken all endemic chains of wild poliovirus circulation and there have been no cases other than VAPP within the last 3 or more years, and "importation countries" that were formerly polio-free, but where there are poliomyelitis cases caused by wild poliovirus imported from one of the "endemic countries" and where there may be local transmission of the imported wild poliovirus. Between 2002 and 2006, there were 26 importation counties, 7 with viruses imported from India and 19 with viruses imported from Nigeria [4]. There were 21 importation countries in 2009 and 13 in 2010. These included ongoing outbreaks in Tajikistan and the Russian Federation and apparently expanding outbreaks in Angola and the DRC. The former represents the first cases from wild poliovirus in the European region since regional eradication was declared in 2006 and the latter could potentially spread to polio-free countries in Africa and other regions. The list of importation countries is updated on a weekly and monthly basis and can be accessed at the web page of the Global Polio Eradication Initiative, www.polioeradication.org. Transmission routes are dependent on population movements. One or more outbreak founders may be introduced by infected persons coming from an endemic or importation country or by returning travelers from such countries. One event with very high risk for spreading poliovirus from one country to another is the Hajj in Saudi Arabia. To counter this threat, it is now mandatory for foreign pilgrims coming on Hajj and Umrah to be vaccinated for communicable diseases including polio, especially pilgrims with young children arriving from countries with polio cases. The children must have received a dose of OPV 6 weeks before their arrival and another upon arrival.

Smallpox was eliminated as a circulating human pathogen in 1977 after an 11-year extensive vaccination and surveillance program [201]. Only two sources of smallpox virus have been reserved for research purposes, one in the United States and one in Russia. Final eradication will be achieved when these last two remaining, contained sources of smallpox virus are finally destroyed. Proposing a similar approach in 1988, the World Health Assembly set a goal of eradicating poliomyelitis by the year 2000 (resolution WHA41.28). A group of experts at the global, regional, and country level set the criteria and conduct the process of certification of eradication [202]. These experts must be independent from the vaccine administration and polio laboratories. Global polio eradication, first targeted for completion by 2000, was limited to the eradication of all wild polioviruses with the caveat that "the occurrence of clinical cases of poliomyelitis caused by other enteroviruses, including attenuated polio vaccine viruses, does not invalidate the achievement of wild poliovirus eradication" (Report of the first meeting of the Global Commission for the Certification of the Eradication of Poliomyelitis. Geneva: World Health Organization; 1995. WHO document WHO/

EPI/GEN/95.6). OPV vaccination would cease within a few years after eradication of poliomyelitis from wild polio and industrialized nations would save not only the large sums of money needed for the maintenance and rehabilitation of individuals with paralytic poliomyelitis but the costs of vaccine and vaccine administration as well [203–206].

In order to easily eradicate an infectious agent, (a) the agent should replicate in a single host with no intermediate vector, alternative reservoir species, or carrier state, (b) vaccines and/or anti-infectious agents must be available to break chains of transmission whether directly between susceptible organisms or after exposure to the infectious agent in the environment, (c) if transmission involves environmental exposure, there must be a finite and relatively short survival time for the infectious agents in the environment, (d) all or the majority of infections should be clinically apparent with unique symptoms, and (e) there must be an easy and cost-effective surveillance system for detection of the infectious agent, identifying infected individuals, and for determining the efficacy of treatments in individuals and populations [3, 207]. Deviations from some of these requirements for eradication have made eradication of polio more difficult from the start [56]. In particular, most (>95%) poliovirus infections are clinically asymptomatic while symptoms associated with the few infections that result in poliomyelitis are not unique to poliovirus infections and although effective vaccines were available, there is no easy and cost-effective method to determine the effectiveness of vaccination in vaccinated individuals. This makes surveillance and the identification and isolation of infected individuals much more difficult. When the GPEI resolution was passed in 1988 and even as late as 1996, it was stated [208] that there was no indication of chronic excretors; however, persistent infections (see page 8161) do exist and have emerged as one of the difficulties in achieving eradication. Differences between smallpox and polio have made it relatively more difficult to achieve polio eradication. For example, live polio vaccine is made from three temperature-sensitive serotypes of poliovirus because there were three non-cross-reacting poliovirus serotypes, whereas the vaccine for smallpox was a single more stable unrelated bovine virus. This has complicated vaccine formulation and administration. In addition since the vaccine contains live poliovirus there are also a number of more severe safety issues concerning pre- and post-eradication vaccine production compared with the smallpox vaccine.

Specific eradication strategies for polio included (a) high routine immunization with at least three doses of vaccine for all children and an additional birth dose in countries where polio has remained endemic (b) SIAs, either national or subnational immunization campaigns targeting children under 5 years of age (c) surveillance, primarily investigation of all cases of AFP with an increase in the number of supplementary surveillance programs such as sewage surveillance and enterovirus surveillance as the endgame of eradication approaches, and (d) house-to-house mopping-up immunization campaigns to block final chains of wild polio transmission [1, 163, 203]. The WHO requires genomic sequencing of all isolates of potential interest to the Polio Eradication Initiative. An isolate is of interest when results from either standard immunological and/or molecular tests conducted by

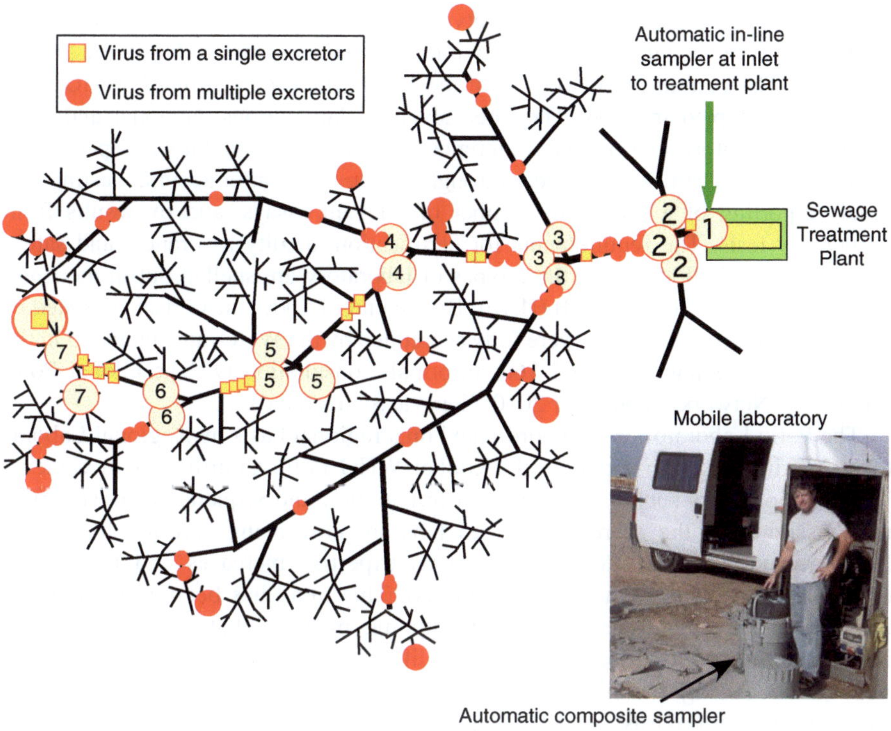

Fig. 10.7 *Population-based environmental surveillance for poliovirus.* The figure is a schematic representation of a network of sewage drainage pipes leading to a sewage treatment plant starting from individual homes, schools, or places of work or entertainment (*thinnest lines*) and converging into larger and larger *trunk lines* (*thicker lines*) until entering the treatment plant. There is an in-line automatic sampler at the inlet to the treatment plant (*green arrow*). Portable automatic samplers like those illustrated in the photograph (*black arrow*) can be lowered into sites at branch points to determine which of the branches contain virus detected by downstream sampling sites. *Yellow squares* represent virus excreted by a single infected individual living at the periphery of the system (*large yellow square*). *Large red circles* represent the situation where more than a single individual is infected and the viruses that they excrete are represented by the smaller *red circles*. The incrementally increasing *black numbers in the circles* represent upstream the order in which the portable samplers can be placed at major branch points to approach and determine the location of the single excretor (Adapted from Hovi T et al. [143] and Shulman LM et al. [291])

accredited laboratories using standard methods (see next section) indicate that the isolate has behaved differently than standard Sabin strains of the corresponding serotype.

A Global Polio Laboratory Network, GPLN, was established to monitor poliovirus infections throughout the world using standardized methods, cell lines, reagents, and reporting methods [209–213]. These standard methods (WHO/EPI.CDS/POLIO/90.1) have been reviewed and revised and improved as knowledge about the epidemiology of polio expands and as new analytical methods become available (Fig. 10.7). This includes even the flow charts or “algorithms” for culturing viruses, identifying polioviruses, and characterizing the serotype (Typic Differentiation) and determining wild, vaccine, or vaccine-derived virus within specific time frames (Intratypic Differentiation). The current fourth version of the *Polio Laboratory Manual* (WHO/IVB/04.10) was adopted in 2004. The three levels of laboratories, National and Subnational Laboratories, Regional Reference Laboratories, and Global Specialized Laboratories, are certified each year through on-site visits, after accurate testing and timely reporting of a minimum number of relevant assays, and by results from proficiency tests. The requirements for certification, quality assurance, and safety and the responsibilities of the three types of laboratory are spelled out in the *Polio Laboratory Manual* (WHO/IVB/04.10). By the end of 2009, the GPLN consisted of 146 laboratories of which 139 were fully accredited by the WHO, and another 5 in the process of accreditation (WHO 16th Informal Consultation Of The Global Polio Laboratory Network, September 2010, Geneva, Switzerland).

The Polio Laboratories are coordinated on a regional basis by Regional Laboratory Coordinators who report to the Global WHO Polio Coordinator in Geneva. Identification tasks such as intratypic differentiation and sequence analysis originally assigned to the more specialized labs are now being certified for use in National and Regional Reference Labs as expertise increases and methods – especially molecular methods are simplified. This trend has been accelerated by the increasing difficulty and costs of shipping material that may contain live infectious wild polioviruses between the different levels of laboratories and the need for decreasing the time between isolation and final notification of characterization of the poliovirus isolates. Rapid identification is especially critical for eradication efforts in endemic regions and for identifying introductions to polio-free regions from these endemic regions.

The laboratories work in close coordination with epidemiologists and medical staff in the investigation of all cases of AFP and/or supplementary enterovirus surveillance and with municipal employees and epidemiologists where supplemental environmental surveillance is utilized to screen for poliovirus presence and circulation. In late 2010, a commercially available method for preparing noninfective viral RNA suitable for molecular analysis based on automatic nucleic acid extraction, immobilization, and storage on Flinders Technology Associates (FTA) filter papers was being evaluated to increase safety and drastically reduce costs of shipping material between laboratories (Summary and Recommendations of the 16th Informal Consultation Of The Global Polio Laboratory Network, 2010, Geneva). Using this technology, viable virus could be reconstituted from the RNA after it is transfected into eukaryotic cells.

Intratypic differentiation (determination of the serotype of an isolate) was based on results from one immunological ELISA test [214] and one molecular-based test, either probe hybridization [215], diagnostic RT-PCR [216], RT-PCR and RFLP [217], RealTime-RT-PCR [218], or micro-array-based systems [80]. At the 16th Informal Consultation Of The Global Polio Laboratory Network, 2010, Geneva, the recommendation for ITD testing from ITD-accredited laboratories was modified to one of the following three options: (a) two RealTime RT-PCR procedures, one for ITD and one for detecting vaccine-derived poliovirus and sending all non-Sabin-like viruses or Sabin-like viruses with non-Sabin-like VP1 to higher level labs for full-length sequencing and molecular analysis of VP1, (b) one validated ITD method (ITD, or molecular) and shipment of all viruses to higher level labs for full-length sequencing and molecular analysis of VP1 or (c) on-site full-length VP1 sequencing of all isolates or referral of all virus isolates to higher level labs for full-length sequencing and molecular analysis of VP1. Results are confirmed by sequence analysis of the entire VP1 capsid gene. Molecular analysis of the sequences of the genomic RNA encoding the VP1 capsid gene is in fact the definitive method to determine whether an isolate is a vaccine strain, a VDPV, or wild isolate. Sequence analysis of the VP1, all four capsid genes (e.g., P1), and even the entire genome, infers evolutionary relatedness to other isolates in endemic or external reservoirs. The methodology and results from such analyses that help trace the origin of viruses founding outbreaks have been clearly presented in reviews by Kew et al. [101] and Sutter et al. [4]. Sequence data is kept in databases maintained by the specialized laboratories of the Polio Laboratory Network, such as the CDC in Atlanta, GA, Pasteur Institute in Paris, and the HTL in Finland. Phylogenetic comparisons of sequences from new isolates, routinely provided by the CDC, indicate evolutionary relationships to previously isolated polioviruses from the same region and trace importations to or from external reservoirs. Important epidemiological information can be obtained from this phylogenetic analysis. For example, a significant gap between a new sequence and all other known sequences indicates a gap in surveillance while an importation implies that there are cases or silent circulation of related viruses in the region containing the reservoir that it is most closely related to. Knowledge obtained about time clocks [39] for nucleotide substitutions (see page 8139) allow investigators to infer whether nucleotide differences between two isolates are consistent with local transmission or represent separate introductions (e.g., see Manor et al. [102]).

The currently recommended standard method for poliomyelitis surveillance is based on the isolation and molecular and serological analysis of all viruses from all cases of acute flaccid paralysis, AFP, to rule-in or rule-out polioviral etiology [219]. The definition of an outbreak varies depending on whether endogenous poliovirus transmission has remained unbroken or the area has been found to be polio-free. In the latter, given the goal of eradication, a single AFP case can be considered to be an outbreak. The previous section describes what is needed for timely high quality testing of all poliovirus isolates from cases and from other sources such as environmental surveillance. Much time, effort, and money has been spent on maintaining and improving lab quality assurance and performance of laboratories in the Global Laboratory Network. However two factors outside of the control of the laboratories strongly influence the final result. The first is sample collection and the second is the conditions under which the sample is stored and shipped to the first processing lab. The most appropriate sample with the highest probability of detecting poliovirus is a 5-g stool sample. For AFP cases, two stool samples (not rectal swabs) should be collected 1–2 days apart within 14 days of onset of paralysis. This is based on a review of studies measuring the timing of viral excretion (discussed on page 8143) in infected individuals [140] and the fact that detectable viral excretion is sometimes intermittent. Standardized tissue culture conditions using limited passages of poliovirus sensitive L20B, HEp2C, and RD cells provided by Specialized Laboratories of the Global Polio Laboratory Network are used according to standard operating procedures to isolate polioviruses from clinical samples [219]. An amended algorithm for isolating polioviruses designed to reduce the workload and the time between receipt of sample and identification of viruses of interest has been successfully evaluated in a number of National Poliovirus Laboratories (WHO 16th Informal Consultation Of The Global Polio Laboratory Network September 2010, Geneva, Switzerland). Standard typing and intratypic differentiation assays are based on results from serological assays and molecular assays as described above with final characterization based on the full-length sequence of VP1. Additional regions such as the 5'UTR, the entire P1 region encoding all four capsid proteins, the 3D polymerase, or the entire genome may be sequenced for higher resolution and to determine whether and to what extent genomic recombination has occurred.

Molecular data from any polioviral isolates recovered from the stool samples provides information about the serotype of the isolate or isolates, and differentiates between VAPP, persistent VDPV, and circulating VDPV or wild polioviruses. The different time clocks for single nucleotide substitutions [39] and unique recombination patterns are important tools for these analyses. Timely AFP surveillance also provides the necessary critical information about the temporal and geographic distribution of the isolates for efficient and economical infection and outbreak response. The rational for AFP surveillance is based on the observation that AFP from all non-polioviral causes occurs with an incidence of 1 per 100,000 in children up to the age of 15. When all AFP cases are investigated and the AFP incidence is within the range for non-polio causes, the absence of poliovirus in the stool samples from any AFP case is considered to indicate absence of circulating poliovirus in the region under surveillance. A surveillance area is considered to be wild poliovirus-free when adequate AFP surveillance levels for greater than 3 years indicate absence of wild poliovirus, and entire WHO-designated regions are considered to be free from endogenous wild polioviruses when all countries within that region are wild poliovirus-free.

Wild poliovirus positive AFP cases in previously polio-free areas or WHO regions can occur. Molecular analysis then reveals the most likely external reservoir from which the virus was transmitted [101]. Two recent examples of country-to-country transmission (see page 8156) which have seriously impacted the eradication initiative are the spread of wild polioviruses to >21 polio-free countries [220] as consequences of the temporary cessation of vaccination in Nigeria starting in 2003 and the spread of wild polio into the European region in 2010 [221] enabled by low vaccine coverage in Tajikistan. Sequence analysis of poliovirus isolates from cases in Mumbai confirmed cessation of local chains of transmission and the reintroduction of viral lineages still circulating in the north of India [222].

Most countries employ AFP surveillance. However not all are able to reach the required incidence of AFP investigations. Some of these countries supplement AFP surveillance with enteroviral surveillance and/or environmental surveillance (Fig. 10.7). Enteroviral surveillance is the systematic identification of the enterovirus genotypes in all clinical infections in general or more specifically from all cases with associated meningitis and encephalitis, symptoms that may appear more frequently than AFP in patients with poliovirus infections (approximately 5% of poliovirus infections compared to 0.5–1% for AFP). In some countries enteroviral surveillance and/or environmental surveillance are used exclusively.

A number of different sampling techniques have been used to obtain environmental samples including grab sampling during peak sewage usage, trapping with silicates or gauze, and automatic composite sampling of sewage aliquots at given time intervals over a 24-h period (Guidelines for environmental surveillance of poliovirus circulation, WHO/V&B/03.03 [223]). All sample storage and shipment must be at low temperatures (4–8°C) to maintain viability since currently certified tests require an amplification step in tissue culture. The FTA technology trial referred to above may eliminate the need for maintaining low temperatures.

The usefulness of L20B cells to isolate polio isolates in the presence of high titers of non-polio enteroviruses [222] has already been mentioned. It is still important to characterize the viruses that grow since L20B cells can also support growth of some other human and bovine enteroviruses, as well as less well-characterized viruses [56]. Additional steps involving molecular screening and growth at elevated temperatures has enabled investigators to more easily identify and characterize wild and vaccine-derived polio in the presence of high titers of vaccine viruses [102, 215, 216, 218, 224, 225]. Selective growth of non-vaccine poliovirus at elevated temperatures [226] is based on a relative small decrease of titer for these viruses compared to a much higher reduction in yield for vaccine strains at elevated temperatures. The main molecular determinant responsible for this difference is a single nucleotide change in loop V of the 5'UTR which can revert or be modified by other changes, hence some polioviruses of interest may escape detection and some minimally diverged vaccine virus may be included among the selected isolates. Confidence that polioviruses isolated by environmental surveillance reflect circulating viruses comes from the high sequence homology between environmental samples and isolates from cases [38, 143, 227]. One of the major contributions of environmental surveillance reviewed by Hovi et al. [143] is that it can be used to establish the presence and/or circulation of wild or vaccine-derived polioviruses before the appearance of AFP cases [102, 222, 227–229]. Environmental surveillance has also revealed the presence of presumptive primary vaccinees excreting OPV in Switzerland where vaccination is by exclusive use of IPV [56].

Different methods for analyzing environmental samples are also currently employed in different laboratories [223] since unlike AFP surveillance [230] there is as of yet no single standard method. The probability of detecting poliovirus in environmental samples [229] depends on the duration and amount of poliovirus excreted by one or more infected individuals (see page 8143), the effect of physical and mechanical factors on the dilution and survival of poliovirus in the sewage system (reviewed by Dowdle [134]), and the frequency of collection and laboratory processing of the environmental samples [223]. A model based on these factors [229] predicted that environmental surveillance could outperform AFP surveillance for small outbreaks as well as detect circulation before the appearance of cases. The location of the sampling site relative to the excretor and the number of excretors (Fig. 10.7) also determines the probability of detection [143]. In general polioviruses can be quantitatively recovered from the environment [231]. Decreasing this distance between the excretor and the sample collection site is more effective in increasing the probability of detection and less labor intensive than increasing the sampling frequency [109]. Environmental surveillance is resource and labor intensive and may require large capacity high-speed centrifuges that are not commonly present in most National Poliovirus Laboratories [143]. It requires judicious choice of potential target populations, a competent laboratory, a plan for routine surveillance and reporting, and the cooperation of municipal authorities. The WHO has recommended principles for selecting sites, sampling strategies, and interpretation of results [223]. Lengthy periods of poliovirus-free monitoring are needed to confirm that poliovirus transmission has stopped since even the most comprehensive surveillance covers only subgroups of the entire population of potential excretors [232].

Many poliovirus positive environmental samples contain one or at best a few polioviruses of interest indicating that the surveillance is operating at the lower limits of detection. Thus while negative findings cannot rule-out the presence of polioviruses at levels below detection, they gain significance when they are part of a long sequence of negative results from frequent routine surveillance at the given site. A positive finding of a wild poliovirus or a VDPV can trigger a response ranging from public announcements to remind individuals scheduled for routine vaccinations to be vaccinated in time in areas with high vaccine coverage to scheduling NIDS or SNIDS in regions where immunization coverage is below that required for establishing herd immunity. Sequence analysis can differentiate between multiple importations and local circulation when more than one poliovirus is isolated within a short interval of time [102].

Detection of "orphan polioviruses" or virus that are not closely linked to previously sequenced isolates indicates gaps or suboptimal surveillance. The length of the gap is roughly proportional to 1% single nucleotide divergence per year [39]. Orphan viruses [91, 105, 191] were responsible for most cVDPV outbreaks. This is in contrast to the situation in Nigeria where intensive AFP surveillance triggered by the circulating wild polioviruses also revealed the initial stages in circulation of multiple lineages of predominantly type 2 cVDPVs [106]. The presence of type 2 cVDPVs, iVDPVs, and aVDPVs is of concern since despite elimination of transmission of wild type 2 in 1999, neurovirulent serotype 2 poliovirus is still among us [124, 128].

A number of countries that switched from OPV or combinations of OPV and IPV to exclusive use of IPV conducted environmental surveillance for OPV after the transition (reviewed in [143]). The OPV rapidly disappeared from the environmental samples but imported OPV-like isolates have been isolated from time to time presumably imported from OPV-using countries [233].

One of the important milestones toward achieving eradication is the containment of all potential sources of the pathogen. In 1999, a process for containing all laboratory stocks of wild poliovirus was initiated by the World Health Assembly entitled *Global action plan for the laboratory containment of wild polioviruses* or GAP 1 (WHO/V&B/99.32). A revised plan, GAP II, included two phases: (1) the identification in all facilities of all known poliovirus stocks and any material that could potentially contain live wild poliovirus, for example, any stool specimens that were collected at times when poliovirus was endemic, and (2) the containment of these stocks by destroying them, rendering them noninfectious, or transferring them to a minimal number of laboratories certified by the WHO as having appropriate BSL3/polio biosafety facilities and justification to work with wild viruses. A draft of the next version, GAP III, extends containment of wild polio to now also include containment of OPV/Sabin strains, and concentrates on minimizing risks associated from facilities that work with polioviruses and vaccine production and storage facilities after eradication of wild poliovirus transmission and cessation of OPV vaccination. Pathways of exposure from these facilities and assessment of the risks from a literature review have been calculated [134]. After risk analysis, a goal was set to reduce the number of such facilities globally to <20 essential facilities that meet required safeguards.

The original target date for polio eradication was not met. By 2001, the WHO Global Commission for the Certification of the Eradication of Poliomyelitis extended eradication to include elimination of circulating VDPVs (Certification of the Global Eradication of Poliomyelitis Report of the sixth meeting of the Global Commission for the Certification of the Eradication of Poliomyelitis. Washington D.C., 28–29 March 2001 WHO/V&B/01.15). The current target date for interruption of all wild poliovirus has been moved to 2013 (Global Polio Eradication Initiative – Programme of Work 2009 and financial resource requirements 2009–2013. WHO/POLIO/09.02). This section will conclude with a discussion of the three major problems that have accounted for the delay in completing the GPEI, "failure to vaccinate," "vaccine failure," and the emergence of "vaccine-derived viruses."

Among the reports available online at the WHO website for polio eradication, www.polioeradication.org, is a report on the annual percentage of children in each country who received a minimum of three doses of polio vaccine annually since 1980. This report provides a complete picture of current polio immunization status. However the variability in coverage between countries and the annual fluctuation within countries illustrates the problem of failure to vaccinate and is also an indication of problems in sustaining the high coverage necessary for successful eradication.

Wild and vaccine-derived poliovirus can penetrate and circulate within areas where vaccination coverage is low or where vaccination has been discontinued [106, 117, 234]. When this occurs this is an example of "failure to vaccinate." The temporary cessation of vaccination in Nigeria in 2003 [235, 236] is usually presented as the classic example for the consequences of a failure to vaccinate. The situation was complicated by suboptimal coverage within Nigeria when vaccination resumed within 12 months and the suboptimal coverage in other countries that had person-to-person contacts with infected Nigerians. Thirteen countries with 52% coverage have had multiple introductions of wild poliovirus from Nigeria, while another eight countries with higher 83% coverage did not have repeated outbreaks [1]. Use of type 1 mOPV was successful in reducing the number of cases from wild type 1 [190, 234] but the decreased use of tOPV lead to a significant increase in cases due to wild type 3. Subsequent use of mOPV1, mOPV3, and bOPV has effectively reduced the number of cases due to wild types 1 and 3 [237]. Unfortunately suboptimal immunization with any vaccine that contained type 2, presented fertile conditions for the emergence of multiple lineages of type 2 CVDPVs some of which have continued to circulate well into 2010 [106] (see discussion on page 8157, 8161).

A 2010 outbreak that started from a wild type 1 virus imported into Tajikistan from India spread into the Russian Federation. This was the first outbreak due to wild poliovirus in the WHO European region since it was declared polio-free in 2002 [221]. Again failure to vaccinate with coverage sufficient to maintain herd immunity was the main factor that facilitated the outbreak. As of June 2010, wild poliovirus cases from this outbreak accounted for more than 70% of all wild polio cases reported in 2010. Four NIDS with mOPV1 were conducted since the start of the outbreak.

When poliomyelitis occurs in vaccinated individuals it is categorized as "vaccine failure." Current reports on India (20th Meeting of the India Expert Advisory Group for Polio Eradication Delhi, India, 24–25 June 2009 www.polioeraication.org) indicate that most of India is poliomyelitis-free with the exception of the north where both wild type 1 and type 3 still circulate. Type 1 and 3 mOPV have helped to reduce the number of circulating lineages and to constrict the areas within which the wild polioviruses are still circulating and causing cases. However, lack of cases in the south does not mean absence of wild poliovirus as environmental surveillance has revealed silent wild polio in Mumbai. The reservoir is not only a problem for India. Populations with suboptimal coverage in other countries are also at risk, as shown by the 2010 outbreak in Tajikistan and the Russian Federation that was traced back to northern India [221].

Vaccine failure in children in India refers to the finding that antibody response or seroconversion in children required more than the recommended three doses of tOPV [238, 239]. In Uttar Pradesh and Bihar in north India, local conditions exist where even administration of five doses of OPV does not induce the expected seroconversion rates. Approximately 15 doses were required to reach population immunity [240]. The fact that the age when the disease is acquired had not shifted upward was taken as an additional indication of vaccine failure [1]. Various trials of efficacy of mOPVs and bOPV and fractional IPV are underway to evaluate their short-term effectiveness in halting endemic transmission and their long-term performance in maintaining protective titers. In India mucosal immunity in response to vaccination with OPV varied depending on location, serotype, and vaccine formulation [241].

The high number of additional doses needed to achieve herd immunity in some regions such as in northern India combined with the additional cost of OPV in annual and sub-annual vaccination campaigns must be taken into account when comparing the cost effectiveness of OPV and IPV in inducing effective herd immunity.

A less serious type of vaccine failure is based on observations that do not completely confirm the paradigm that OPV prevents replication during subsequent exposures to poliovirus. Israeli children who had concluded a primary immunization schedule consisting of three doses of OPV and three doses of IPV by 18 months of age had seroconverted for all three serotypes with geometric mean titers >1,000 [159]. One month after the last vaccination they were challenged with an additional dose of OPV. Up to 60% of children excreted at least one OPV serotype between 1 and 3 weeks, the upper range of similar studies reviewed in that report. There was no evidence of transmission to siblings or mothers of these children, most likely because of good hygiene [159]. These rates were comparable to other similar studies [159]. Hygiene and high coverage probably also contributed to the fact that there was also no evidence for OPV circulation in IPV-vaccinated populations in the United States living adjacent to OPV-vaccinated populations in Mexico [242].

A number of comprehensive reviews on vaccine-derived polioviruses have been published [3, 62, 79, 105, 114, 243–246]. As described above (page 15), vaccine-derived polioviruses evolve either during person-to-person transmission (cVDPVs) or during relatively rare [247] persistent infections in immunodeficient patients (iVDPVs) (see reviews [3, 4, 105, 114]). To date (2010) there have been 12 cVDPV outbreaks [106]. Using the definition of outbreak in the context of eradication (i.e., even a single case), the number of outbreak may be even higher. For example, in 2009, 21 cases due to cVDPVs were found in four countries in addition to 153 in Nigeria and a case in Guinea traced back to Nigeria [248], and the cases in Nigeria represent emergence of multiple independent lineages [106].

Most outbreaks caused by cVDPVs have been caused by a single lineage that spread rapidly through a susceptible cohort within the general population. The outbreaks were only discovered after silent circulation of the VDPVs for more than 1 year or more indicating gaps in surveillance. By the time such outbreaks became evident, the genomes of the isolates had usually recombined with those of the progeny of other vaccine serotypes or closely related non-polio enteroviruses. When OPV is introduced or reintroduced into a population with cohorts of naïve individuals as in Nigeria, adequate surveillance revealed that in early stages of reemergence more than one independent lineage may emerge [106]. There is also a potential for recombination of cVDPVs with wild-type viruses in areas where both co-circulate as shown from retrospective molecular analysis of isolates during endemic circulation of wild polioviruses [82]. Luckily from the point of view of eradication, cVDPV outbreaks resemble outbreaks of poliomyelitis from wild polioviruses introduced into polio-free areas and their chains of transmission can be broken by similar vaccination responses [106, 234].

In certain circumstances poliovirus can establish persistent infection in immunodeficient individuals. The types of immune deficiencies of known chronic excretors have been reviewed [3, 4, 62, 105]. The genomes of the Sabin strains are unstable [114] and reversion of nucleotide changes that attenuated neurovirulence appear even among the progeny virus excreted by primary OPV vaccinees. These reversions are believed to improve the replicative fitness of the isolates [123] and are responsible for the rarity of vaccine-associated paralytic poliomyelitis cases (VAPP; 1 per 500,000–1,000,000 vaccinations of naïve infants, and 7,000 times higher for some immunodeficient individuals) and cVDPV outbreaks [3, 105]. Thus it is not surprising that reversion of attenuation also occurs at an early stage in chronic excretors [62]. During 4 months of observation of long-term excretion in a healthy child [62] type 1 virus diverged by 1.1% and evolved toward full reversion to wild-type phenotypic properties similar to the Mahoney parent of the Sabin 1 strain. It is less obvious why these isolates so quickly predominate in the quasispecies of persistently infected immunodeficient individuals. The process by which persistence is established and maintained may present selection through bottlenecks within a single individual that is similar to the bottleneck by which only a single progeny or a subpopulation of the quasispecies is passed onto the next host in person-to-person transmission among immune competent hosts. Selection by passage through bottlenecks was also suggested to explain evolution of wild poliovirus during long-term expression [249]. Examination of the genomes of iVDPV isolates differentiates them from cVDPVs in that significantly fewer heterotypic recombinations occur [4, 110] and intrageneous recombination appears to be largely absent [4, 105]. Interestingly, more than one highly divergent lineage may be recovered from a single stool sample from persistently infected individuals [3, 110]. This suggests that persistence and evolution occur in separate sites although intratypic recombination indicates that some mixed infections in a single cell must occur. Only a single serotype was detected in most (30 of 33) long-term excretors identified between 1962 and 2006 [4] and this pattern has continued to date. Amino acid changes in capsid proteins may allow polo to establish persistence in cells of the CNS [250].

Molecular analysis of phylogenetically related highly diverged (>10%) aVDPVs isolated from sewage in Finland, Slovakia, and Israel reveals a pattern of amino acid substitutions in or near neutralizing antigenic epitopes and lack of intrageneous recombination that resembles the pattern of changes in iVDPVs and is qualitatively different from evolutionary changes in cVDPVs [109, 251]. This pattern and the extended periods of time over which phylogenetically related polioviruses have been isolated from the same sewage systems and surveillance sites within those systems strongly suggests that replication of the related viruses has taken place in one immunodeficient host, or a very limited number of individuals in contact with such a host. Routine monthly sewage surveillance of catchment areas representing 35–40% of the population in Israel intermittently and repeatedly revealed the presence of highly diverged type 2 VDPVs 2 between 1998 and 2010. Phylogenetic analysis indicated that the isolates came from two different and unrelated persistent infections. Isolates form one foci have been isolated intermittently for 12 years and the second for 4 years. In addition there was a single, respectively, and a single isolation of a highly diverged type 1 VDPV. The situation in Finland is particularly interesting and unusual, since evidence suggests that the infected individual is simultaneously and persistently infected with three highly diverged VDPV serotypes [251].

Most mutations in iVDPVs and aVDPVs are synonymous and are observed in third position codons. These synonymous substitutions occur at similar rates to those for poliovirus during person-to-person transmission [39, 101, 107, 109]. Non-synonymous amino acid substitutions affect antigenicity, neurovirulence, receptor binding motifs, hydrophobic pockets, and drug sensitivity.

The prevalence of aVDPV excretors is unknown, but additional countries with excretors of aVDPVs are being reported as environmental surveillance is introduced into more and more regions [143, 246, 252–256]. Hovi et al. [143] have proposed expanding the suggestion that the GPLN include regular monitoring for cVDPVs [257] through increasing the number of laboratories that employ routine environmental surveillance. It is important to determine the exact nature of the immune status of these types of persistent excretors since it may be different than that for identified persistently infected individuals. Unfortunately the individuals infected with these aVDPS remain unidentified, and will most likely remain so for a long time [143]. The most frustrating attempt to locate such an excretor occurred in Slovakia where moving sampling sites progressively upstream successively restricted the excretor to a population of 500 individuals before detection ceased [143].

There is no consensus on the extent that persistent VDPVs may affect the realization of eradication [133, 163]. Determining the number of unidentified persistent infections is becoming more urgent as eradication of wild polio approaches (WHO 16th Informal Consultation Of The Global Polio Laboratory Network. 22–23 September 2010, Geneva, Switzerland, WHO/HQ. Summary of Discussions and Recommendations). Some researchers believe that VDPVs may pose an insurmountable problem [114, 127] while others feel that the problem is less severe [3, 4, 126]. Most of the identified persistent excretors had primary B-cell-related immunodeficiencies [4, 105]. While molecular epidemiological analysis has indicated that highly diverged neurovirulent anonymous VDPVs isolated from sewage in Finland, Slovakia, and Israel [109, 233, 258] resemble the molecular epidemiology of poliovirus isolates excreted over time by identified excretors of iVDPVs, the exact nature of the immune status that has presumably enabled infection to persist in these aVDPV excretors remains unknown. The time course of excretions, the rate of nucleotide substitutions in virus isolated from identified persistent excretors, and genomic recombination patterns have been consistent with the establishment of persistence and evolution of the virus in these individuals rather than transmission of an iVDPV. Highly diverged iVDPVs (as opposed to cVDPVs and less diverged iVDPVs) have not been found during routine AFP surveillance of cases of immune competent individuals [3]. One clear indication that iVDPV-like aVDPVs are transmissible comes from a study of silent transmission in infected children in an under-vaccinated community in Minnesota [259] where 8 of 23 infants had evidence of type 1 poliovirus of VDPV infection. While this absence of documented transmissibility of very highly diverged VDPVs is encouraging, it may only be circumstantial, since most of the highly diverged neurovirulent aVDPVs have been found in the environment of countries with high vaccine coverage and good hygiene barriers that have also prevented circulation and appearance of wild poliovirus cases even after neurovirulent wild polio was introduced from external reservoirs [102].

The amino acid substitutions in neutralizing antigenic epitopes/receptor binding residues in iVDPVs and iVDPV-like aVDPVs may have helped specialize these virus isolates for microenvironments within the gut during persistent infections. These same changes might affect/reduce transmission via the oral–oral route in communities where there is high vaccine coverage and good hygiene. If true, this might significantly reduce the threat to eradication, despite the highly neurovirulent nature of these isolates in animal model systems and the decreased geometric mean neutralizing antibody titers against some of these excreted iVDPV and aVDPV isolates in the general public in the highly immunized communities where these isolates are found [109, 233]. It must also be taken into account that identified and unknown excretors are free to travel to communities with poor vaccine coverage and substandard hygiene where the fecal–oral transmission route is more important and continuous person-to-person transmission easier to maintain.

Future Directions: The Endgame Stage of Eradication and Sustainability of Postpolio Eradication

This section will start with an overview of current accomplishments to provide a suitable background for the discussion of future directions and sustainability. One of the best online sources for keeping up to date on eradication can be found at www.polioeradication.org.

There has been a >99% overall reduction in the number of cases since adoption of the Global Poliovirus Eradication Initiative in 1988 [248]. The Region of the Americas (AMR) was certified to be free from all three serotypes of indigenous wild polioviruses in 1994 [260, 261] and the last case anywhere in the world from wild type 2 polio occurred in India in 1999 [186, 262]. Subsequently the Western Pacific Region (WPR) in 2000 [263] and the European Region (EUR) in 2002 [264] have also been certified to be free from indigenous wild polioviruses. These successes have been due to the dynamic nature of the eradication program where vaccination strategies have been adapted in response to specific problems and to changing conditions emerging as the endgame approached [1]. Four countries remain where indigenous poliovirus has continued, Afghanistan and Pakistan in the WHO Eastern Mediterranean Region (EMR), Nigeria in the WHO African Region (AFR), and India in the WHO South-East Asia Region (SEAR).

The accumulated costs for the vaccination program have exceeded 4.5 billion US dollars. National governments (list by alphabetical order: Australia, Austria, Belgium, Canada, Denmark, Finland, Germany, Ireland, Italy, Japan, Luxembourg, the Netherlands, Norway, the United Kingdom, and the United States) have provided a significant portion of the necessary funding. NGOs (WHO, UNICEF, Rotary International, the Bill and Melinda Gates Foundation, and the International

Red Cross and Red Crescent societies), the World Bank, and corporate partners (Aventis Pasteur, De Beers) have also made significant contributions toward purchase of vaccines and for applied and basic polio research. In addition to paid professional staff, more than ten million volunteers have assisted in the global vaccination program. Their knowledge of local practices and beliefs has provided a significant asset to the GPEI [1].

One of the goals of eradication was to reach a stage where vaccination could be discontinued, as was the case for smallpox vaccinations after eradication of smallpox [265]. The estimated annual savings would be enormous and could be used to fund other global health initiatives. Similarly the organizational capabilities experience expertise and facilities of member Laboratories in the Global Polio Laboratory Network would also be employed to solve other health-related problems. The three main problems that have delayed eradication originally scheduled for 2000 have been discussed. Among these problems, chronic excretion of vaccine-derived viruses probably remains the most serious obstacle since the number of excretors remains unknown and there are no universally recognized methods of curing chronic excretion in those chronic excretors who have been identified.

Between January 2009 and June 2010, the Global Polio Laboratory Network analyzed 258,000 fecal specimens from 130,000 AFP cases for the presence of poliovirus, and between January 2009 and September 2010 it detected introductions of wild poliovirus into 23 previously "polio-free countries," countries where indigenous polio transmission had been interrupted. Nineteen were in the African region and included seven countries (Burkina Faso, Benin, Chad, Côte d'Ivoire, Mauritania, Niger, and Togo) where the outbreak isolates were related to previous importations into those countries as was the case for Sudan in the Eastern Mediterranean Region. One of the more serious setbacks for eradication was the introduction of wild poliovirus into Tajikistan from Uttar Pradesh in India marking the first outbreak in the European region since it was certified poliovirus-free in 2002 [221]. The large outbreak (>450 confirmed cases) ensued spread into the Russian Federation and resulted in an immediate tenfold increase in the amount of samples that needed to be processed by the Polio Laboratories in the region. In all of these importation countries and/or regions, large-scale coordinated SIAs were conducted. The spread of wild poliovirus to poliovirus-free countries from Nigeria and India via Tajikistan illustrate the need to maintain high population immunity until all transmission of wild virus has ceased. Similarly, the emergence of multiple lineages of neurovirulent VDPVs in Nigeria and the increasing frequency of isolation of aVDPVs as more countries adopt environmental surveillance reinforce the need to discontinue use of OPV globally in a coordinated effort or staged manner. See Ehrenfeld et al. [266] for a review and discussion of key issues that have affected and will affect the GPEI, including: safety for volunteers in areas of strife, the low efficiency of OPV to induce herd immunity in certain settings, the requirement for maintaining high coverage even after eradication, the inherent mutability of OPV, problems for establishing the safety and efficacy and costs of new vaccines, new vaccine formulations, and scaling up alternatives to OPV.

The saying "*May you live in interesting times*," often attributed to an ancient Chinese proverb or curse, appears appropriate for describing the current status in the quest to eradicate wild polioviruses. Eradication, which is tantalizingly close, will require substantial changes in vaccination policy and practice [117]. It must also include appropriate emergency response measures to control reemergence. "The ideal vaccine choice for the stockpile should be effective in any outbreak scenario, protect all vaccinees with one dose, spread to and protect the unvaccinated population, and have no detrimental effect" [267]. Long-term effects should be considered. While mOPV might be the most effective in rapidly controlling an outbreak and spreading and protecting unvaccinated individuals [267], plans that preferably do not require use of OPV adjacent to areas with high concentrations of unvaccinated individuals would be better in the long run [117, 268]. The reader is referred to the website of poliovirus eradication (www.polioeradication.org) for the latest information on past, current, and future policies.

Three problems have delayed the realization of eradication as has been discussed. Currently available vaccines can overcome "failure to vaccinate," provided that enough doses of vaccines are made available, that there is the political will to use them, and that natural or man-made disasters do not prevent reaching the children for vaccination. Preliminary results from newly approved monovalent and bivalent oral polio vaccines and clinical studies using fractional doses of IPV indicate that there may already be a solution for "vaccine failure" which is exemplified by the poor seroconversion rates for OPV in northern India [240]. The third major problem, "vaccine-derived polioviruses" is more complex, since VDPVs can evolve by person-to-person transmission (cVDPVs) or during persistent infections (iVDPVs). The spread of cVDPVs can be interrupted using the same methods as used to stop transmission of wild poliovirus (paradoxically including use of OPV in vaccination campaigns), since cVDPVs behave like wild virus [106, 234]. Moreover, while it is easy to say that current GPEI plans to coordinate global cessation of the use of OPV will prevent VAPP [206] and emergence of new cVDPVs, at this juncture the actual process is quite complicated and associated with a number of risks. The main problem that will need to be solved is the shedding of highly neurovirulent VDPVs into the environment for prolonged periods by identified and unidentified, persistently infected individuals. There is currently no universal solution to this problem [63, 125]. As long as shedding persists (perhaps as long as some of these individuals remain alive), containment as envisioned in GAP III will be incomplete and high vaccination coverage will need to be maintained. A related but more difficult problem that will need to be solved is (a) to determine the prevalence of unidentified, presumably persistently infected individuals who are responsible for shedding the highly diverged aVDPVs that have been isolated from environmental surveillance, and (b) to identify the presumably persistently infected individuals to determine the physiological conditions that enabled persistence and to try and clear the persistent infection with current or future antiviral treatments. As the endpoint of eradication of wild poliovirus is approached, the number of cases of poliomyelitis will decrease while the number of silent infections may increase as a result of high vaccine coverage. Under these conditions supplemental surveillance programs such as enterovirus surveillance and environmental surveillance will become an even more important tool for providing geographical information for designing NIDS, SNIDS, and final mopping-up campaigns for eradication and for monitoring for post-eradication reemergence.

"Although OPV has been the mainstay of the eradication program, its continued use is ironically incompatible with the eradication of paralytic disease (since) vaccine-derived viruses consistently emerge as a consequence of the inherent genetic instability of poliovirus [122]." "Eradication of vaccine" suggested in 1997 [204, 205] has become recommended policy on condition that provisions of GAP III for safety in vaccine production and polio laboratories are met [264, 269]. A model describing the impact of cVDPVs on eradication indicated that the probability of eradicating polio with continuous use of OPV was not very likely [270]. Alternative vaccination should be continued during and especially after the transition to maintain high coverage and to avoid the buildup of large susceptible populations during the time when there is the highest risk for reemergence of OPV strains [117, 134, 270–272]. Low population immunity remains the main known risk factor for the emergence and spread of cVDPVs [234, 243]. Since most of the cVDPVs in outbreaks circulated silently for months or years (VP1 divergence $>2\%$) before detection in AFP cases, it is imperative that surveillance be improved and expanded to high-risk regions to detect silent circulation of VDPVs as early as possible.

Endgame vaccination strategies have been reviewed [3, 122, 163, 266, 268] and include (1) indefinite use of OPV, (2) cessation of all polio immunization (3) transition to use of IPV, by synchronous coordinated cessation of all use of OPV with (a) limited use of IPV or (b) replacement of all OPV with IPV, (4) country-by-country cessation of OPV use with options (3a) or (3b), (5) sequential removal of Sabin strains from OPV, as eradication proceeds, (6) development of new vaccines, and (7) indefinite use of IPV or new vaccines. The synchronous cessation of OPV has several problems particularly if inexpensive alternatives are not in place when it occurs, since this vacuum may result in large populations of naïve individuals, in whom, polio could reemerge, after periods of silent circulation, with high force and rapid spread. Such a scenario also does not address the potential risks of unidentified chronic excretors. A gradual shift to IPV may avoid some of the programmatic disadvantages that coordinating a synchronous shift would have on vaccination programs and vaccination production facilities. It also potentially provides a longer window for industry to increase production, integrate information from current fractional vaccine dosage and alternative routes of administration trials, and overcome problems of biocontainment and antigenicity associated with optional use of killed OPV as a substitute for the wild strains used in IPV production.

The WHO and UNICEF regularly consult informally with vaccine manufacturers to discuss the implications and practicality of vaccine policy decisions (summaries are available from the Internet using variations of a search for "WHO/UNICEF Informal Consultation with IPV and OPV Manufacturers"). For example, the 3rd WHO/UNICEF Informal Consultation with IPV and OPV

Manufacturers (2003) included a discussion of post-eradication needs and biocontainment requirements and the 5th (2006) included updated information on progress of the GPEI and OPV cessation strategies.

The financial requirements for the transition period are complicated and have been set forth by the WHO (WHO Global Polio Eradication Initiative – Programme of Work 2009 and financial resource requirements 2009–2013 WHO/POLIO/09.02). The bottom line is that alternatives to OPV must be affordable [234]. Three recent reports deal in depth with the economics and practicality of universal replacement of OPV with IPV: (a) Global Post-eradication IPV Supply and Demand Assessment: Integrated Findings, March 2009, and (b) The supply landscape and economics of IPV-containing combination vaccines: Key findings, May 2010, both commissioned by the Bill & Melinda Gates Foundation and prepared by Oliver Wyman, Inc., and (c) Improving the affordability of inactivated poliovirus vaccines (IPV) for use in low- and middle-income countries – An economic analysis of strategies to reduce the cost of routine IPV immunization, April 20, 2010, prepared for PATH by Hickling, Jones, and Nundy. The second report [273] presents a thorough review of the current options and risks for new vaccines and vaccine formulations for achieving and maintaining eradication. New generations of inactivated polio vaccines may need to be developed for post-eradication use [266, 274] and they may have to be used indefinitely.

A number of decisions must be made now, some based on incomplete knowledge, because of the long lead time needed between planning facilities and final production of regulatory agency-approved products. For example, while fractional doses significantly reduce costs, they are less effective than full doses and there is little data on kinetics of waning, while questions still exist concerning sufficient antigenicity of Sabin IPV. Additional complications involve testing and regulatory approval of new products or formulations (see discussion on regulations and standardization of IPV and IPV combination vaccines in Baca-Estrada and Griffiths [275] and the views of vaccine producers [276, 277]). The good news is that when "new" polio vaccine, type 1 mOPV, was needed, it was produced by two companies and licensed in three countries in a relatively short time, 6 months [135, 278, 279]. (Quotation marks were used around the word new since in actuality millions of monovalent doses of each serotype had been used before introduction of tOPV [280] when old licenses had been left to expire). Licensing was also aided by the fact that monovalent batches were produced and safety tested before being combined to produce tOPV and only qualified tOPV producers were approached to provide mOPVs [279]. Ironically if mOPVs are more effective than respective serotypes in tOPV because of increased titers and longer replication times, the increased number of nucleotide substitutions may increase the potential for cVDPV outbreaks by the serotype used [103] or conversely from the remaining serotypes (or serotype if bOPV is used). Supporting this is the emergence of significantly higher numbers of type 1 viruses with increased antigenic divergence from Sabin 1 after a birth dose of mOPV1 and a second exposure to Sabin 1 [111]. Most (71%) were isolated from stools from infants who did not seroconvert after the birth dose [111]. Rapid licensing of bOPV on January 10, 2010, followed release of efficacy results on June 2009 (issue 6 PolioPipeline, summer 2010). The bad news is that combination vaccines containing IPV cannot be frozen raising questions about long-term stability and appropriate reference standards [275].

There have been a number of attempts to rationally redesign the sequence of vaccine seed strains to make them more stable and safer to use in vaccine production facilities in the post-eradication period [3]. One drawback is that there is no empirical data on how these new viruses will behave in the field especially in relation to genome stability and the ability to recombine with heterotypic or intragenic enteroviruses. Changing codons to equivalent but rarely used synonymous codons based on studies of codon use bias or increasing the frequency of CpG and UpA dinucleotides are methods to change the substitution rate [97, 281, 282]. Others modifications have led to polioviruses that can grow in nonhuman cell lines for production but have very low ability to infect human cells.

Widespread vaccination will continue at least during the 3-year period between the last case due to wild poliovirus and certification that wild poliovirus transmission has been interrupted globally. However, global vaccination should be continued for much longer since by one model [283], after 3 years there would only be a 95% certainty that all silent circulation had in fact ceased and the probability after 5 years ranged between 0.1% and 1%, while a more recent model has predicted a very high probability of reemergence within 10 years after eradication by VDPVs or accidental release of virus from vaccine production facilities, a polio laboratory, or bioterror [272]. Consequently vaccination will need to continue for at least 10 years after eradication. A special issue of the journal Risk Analysis (Volume 26, Issue 6, 2006) has been devoted to risks associated with polioviruses before, during, and after eradication of wild poliovirus. Finally vaccination with IPV may be continued indefinitely at least in countries where aVDPVs continue to be isolated from the environment with attendant risk from a polio vaccine production facility operating in a polio-free era (see discussion above on GAP III). To reiterate, current contingency plans for use of OPV in response to reemergence need to be revised based on the data on circulation of live vaccine strains after temporary and/or partial cessation of vaccination [117].

The final and one of the most important problems that must be successfully dealt with is to answer the question: "How can current achievements and eradication be sustained once the endgame has been concluded?" Ideally a major public health undertaking such as eradication requires a cost-benefit analysis, sufficient funds at the beginning, the means for achievement at hand, and the political and social will to carry the process through to the end. Delays and problems with fund-raising, especially when they occur during the endgame, may derail the entire effort [1]. Some problems with sustainability are associated with management and not scientific problems [1, 284]. However, new unanticipated scientific problems may appear which further delay polio eradication. After all, awareness of the potential problems from cVDPVs in communities with low vaccine coverage and chronic excretors of VDPV primarily appeared during the endgame of eradication when the global burden of cases had decreased by >99% and only after appropriate analytic tools to easily document and confirm VDPV had became widely available [3]. An example is the revelation of the 10-year circulation of cVDPVs in Egypt starting in 1983 [285] by retrospective phylogenetic examination of VP1 sequences.

The means to prevent disease and contain the spread of virus transmission when it emerges are already in hand. Safer and more cost-effective measures are in the pipeline that include schedule reduction and fractional doses, adjuvant use, optimizing of processing, Sabin or modified Sabin IPV, and noninfectious IPV [119, 273]. Sustainability for achieving eradication will depend on vaccine policy decisions made today, on the length of time it takes to eliminate all wild poliovirus transmissions, on political will and advocacy, on the motivation of volunteers and the level of local community involvement, program-related fatigue, and on the absence of complications [286] from bio-error, bioterror, or mother nature [291]. However the major determining factor will probably be the availability of financial resources [135]. Limited resources mean competition between routine immunization and eradication efforts during endgame. A predicted 1.3 billion USD funding gap in June 2010 is already forcing a reprioritization of planned activities (Global Polio Eradication Initiative Monthly Situation Report June 2010 www.polioeracdication.org.) "Even if there are no competing health needs, it is unlikely that immunization could be maintained indefinitely against a non-existent disease at a level that is sufficient to prevent vaccine-derived viruses evolving to cause epidemics"[163]. Programmatic setbacks such as those associated with failure to vaccinate (Nigeria and Tajikistan), vaccine failure (northern India), the frequency of repeated vaccination campaigns, or post-eradication reemergence must not be allowed to derail the current momentum and lead to program-related fatigue [1]. Detailed planning must be made for any post-eradication outbreaks (see Jenkins and Modlin [267] and Tebbins et al. [268]) and provisions to implement them including stockpiling must be in place. Finally it is well worth reading "The pathogenesis of poliomyelitis: what we don't know" by Nathanson [287] and "Gaps in scientific knowledge for the post-eradication world" by Minor [288].

Note

Polio is the second human pathogen for which there is an ongoing global program for eradication that has reached the endgame. The first, smallpox, successfully completed the endgame and is now in the stage of post-eradication sustainability. Bioterror is the main threat to sustainability of smallpox eradication. This chapter will describe some of the difficulties with completing the endgame of polio eradication and then in sustaining postpolio eradication. More than the usual number of items are included in the glossary to make it easier for the reader to follow the progress of eradication as it unfolds in the large number of official documents that deal with a global eradication program and which contain the usual copious number of professional acronyms.

Bibliography

1. Lahariya C (2007) Global eradication of polio: the case for "finishing the job". Bull World Health Organ 85(6):487–492
2. Daniel TM, Robbins FC (1997) A history of poliomyelitis. In: Daniel TM, Robbins FC (eds) Polio. University of Rochester Press, Rochester, pp 5–22
3. Kew OM, Sutter RW, de Gourville EM, Dowdle WR, Pallansch MA (2005) Vaccine-derived polioviruses and the endgame strategy for global polio eradication. Annu Rev Microbiol 59:587–635
4. Sutter RW, Kew OM, Cochi SL (2008) Poliovirus vaccine – live. In: Plotkin SA, Orenstein WA, Offit PA (eds) Vaccines, 5th edn. W.B. Saunders/Elsevier, Philadelphia, pp 631–685
5. Burnet FM, MacNamara J (1931) Immunological differences between strains of poliomyelitic virus. Br J Exp Pathol 12:57–61
6. Wyatt HV (1985) Provocation of poliomyelitis by multiple injections. Trans R Soc Trop Med Hyg 79(3):355–358
7. Enders J, Weller T, Robbins FC (1949) Cultivation of the lansing strain of poliomyelitis virus in cultures of various human embryonic tissues. Science 109(2822):85–87
8. Dulbecco R, Vogt M (1954) Plaque formation and isolation of pure lines with poliomyelitis viruses. J Exp Med 99(2):167–182
9. van Wezel AL (1967) Growth of cell-strains and primary cells on micro-carriers in homogeneous culture. Nature 216(5110):64–65
10. Cockburn WC (1988) The work of the WHO consultative group on poliomyelitis vaccines. Bull World Health Organ 66(2):143–154
11. Mendelsohn CL, Wimmer E, Racaniello VR (1989) Cellular receptor for poliovirus: molecular cloning, nucleotide sequence, and expression of a new member of the immunoglobulin superfamily. Cell 56(5):855–865
12. Dragunsky E, Nomura T, Karpinski K, Furesz J, Wood DJ, Pervikov Y et al (2003) Transgenic mice as an alternative to monkeys for neurovirulence testing of live oral poliovirus vaccine: validation by a WHO collaborative study. Bull World Health Organ 81(4):251–260
13. Wood DJ, Hull B (1999) L20B cells simplify culture of polioviruses from clinical samples. J Med Virol 58(2):188–192
14. Brodie M (1935) Active immunization against poliomyelitis. Am J Public Health Nations Health 25(1):54–67
15. Kolmer JA (1936) Vaccination against acute anterior poliomyelitis. Am J Public Health Nations Health 26(2):126–135
16. Salk JE (1953) Studies in human subjects on active immunization against poliomyelitis. I. A preliminary report of experiments in progress. J Am Med Assoc 151(13):1081–1098
17. Salk JE, Bazeley PL, Bennett BL, Krech U, Lewis LJ, Ward EN et al (1954) Studies in human subjects on active immunization against poliomyelitis. II. A practical means for inducing and maintaining antibody formation. Am J Public Health Nations Health 44(8):994–1009
18. Koprowski H, Jervis G, Norton T (1952) Immune responses in human volunteers upon oral administration of a rodent-adapted strain of poliomyelitis virus. Am J Epidemiol 55(1):108–126
19. Koprowski H (2006) First decade (1950–1960) of studies and trials with the polio vaccine. Biologicals 34(2):81–86
20. Sabin AB (1957) Properties and behavior of orally administered attenuated poliovirus vaccine. J Am Med Assoc 164(11):1216–1223
21. Cox HR, Cabasso VJ, Markham FS, Moses MJ, Moyer AW, Roca-Garcia M et al (1959) Immunological response to trivalent oral poliomyelitis vaccine. Br Med J 2(5152):591–597
22. Furesz J (2006) Developments in the production and quality control of poliovirus vaccines – historical perspectives. Biologicals 34(2):87–90
23. John TJ (2001) Anamalous observations on IPV and OPV vaccination. In: Brown F (ed) Progress in polio eradication: vaccine strategies for the end game. Karger, Basel, pp 197–208

24. Davidson I, Shulman LM (2008) Unraveling the puzzle of human anellovirus infections by comparison with avian infections with the chicken anemia virus. Virus Res 137(1):1–15
25. Racaniello VR et al (2007) Picornaviridae: the viruses and their replication. In: Chief E-I, Knipe DM, Howley PM, Editors A, Griffin DE, Lamb RA (eds) Fields virology, 5th edn. Wolters Klewer/Lippincott Williams & Wilkins, Philadelphia, pp 795–838
26. Wimmer E, Hellen CU, Cao X (1993) Genetics of poliovirus. Annu Rev Genet 27:353–436
27. Wimmer E (2006) The test-tube synthesis of a chemical called poliovirus. The simple synthesis of a virus has far-reaching societal implications. EMBO Rep 7:S3–S9
28. Molla A, Paul AV, Wimmer E (1991) Cell-free, de novo synthesis of poliovirus. Science 254(5038):1647–1651
29. Poyry T, Kinnunen L, Hovi T (1992) Genetic variation in vivo and proposed functional domains of the 5′ noncoding region of poliovirus RNA. J Virol 66(9):5313–5319
30. Pilipenko EV, Blinov VM, Romanova LI, Sinyakov AN, Maslova SV, Agol VI (1989) Conserved structural domains in the 5′-untranslated region of picornaviral genomes: an analysis of the segment controlling translation and neurovirulence. Virology 168(2):201–209
31. Skinner MA, Racaniello VR, Dunn G, Cooper J, Minor PD, Almond JW (1989) New model for the secondary structure of the 5′ non-coding RNA of poliovirus is supported by biochemical and genetic data that also show that RNA secondary structure is important in neurovirulence. J Mol Biol 207(2):379–392
32. Pelletier J, Sonenberg N (1988) Internal initiation of translation of eukaryotic mRNA directed by a sequence derived from poliovirus RNA. Nature 334(6180):320–325
33. Jang SK, Krausslich HG, Nicklin MJ, Duke GM, Palmenberg AC, Wimmer E (1988) A segment of the 5′ nontranslated region of encephalomyocarditis virus RNA directs internal entry of ribosomes during in vitro translation. J Virol 62(8):2636–2643
34. Gamarnik AV, Andino R (1997) Two functional complexes formed by KH domain containing proteins with the 5′ noncoding region of poliovirus RNA. RNA 3(8):882–892
35. Liu HM, Zheng DP, Zhang LB, Oberste MS, Kew OM, Pallansch MA (2003) Serial recombination during circulation of type 1 wild-vaccine recombinant polioviruses in China. J Virol 77(20):10994–11005
36. Racaniello VR (2001) Picornaviridae: the viruses and their replication. In: Fields BN, Knipe N, Howley P (eds) Virology, 4th edn. Lippincott Williams & Wilkins, Philadelphia, pp 685–722
37. Krausslich HG, Nicklin MJ, Lee CK, Wimmer E (1988) Polyprotein processing in picornavirus replication. Biochimie 70(1):119–130
38. Shulman LM, Handsher R, Yang CF, Yang SJ, Manor J, Vonsover A et al (2000) Resolution of the pathways of poliovirus type 1 transmission during an outbreak. J Clin Microbiol 38(3):945–952
39. Jorba J, Campagnoli R, De L, Kew O (2008) Calibration of multiple poliovirus molecular clocks covering an extended evolutionary range. J Virol 82(9):4429–4440
40. Bouchard MJ, Lam DH, Racaniello VR (1995) Determinants of attenuation and temperature sensitivity in the type 1 poliovirus Sabin vaccine. J Virol 69(8):4972–4978
41. Bergamini G, Preiss T, Hentze MW (2000) Picornavirus IRESes and the poly(A) tail jointly promote cap-independent translation in a mammalian cell-free system. RNA 6(12):1781–1790
42. Crawford NM, Baltimore D (1983) Genome-linked protein VPg of poliovirus is present as free VPg and VPg-pUpU in poliovirus-infected cells. Proc Natl Acad Sci USA 80(24):7452–7455
43. Kuhn RJ, Tada H, Ypma-Wong MF, Dunn JJ, Semler BL, Wimmer E (1988) Construction of a "mutagenesis cartridge" for poliovirus genome-linked viral protein: isolation and characterization of viable and nonviable mutants. Proc Natl Acad Sci USA 85(2):519–523
44. Ambros V, Pettersson RF, Baltimore D (1978) An enzymatic activity in uninfected cells that cleaves the linkage between poliovirion RNA and the 5′ terminal protein. Cell 15(4):1439–1446

45. Wien MW, Chow M, Hogle JM (1996) Poliovirus: new insights from an old paradigm. Structure 4(7):763–767
46. Li Q, Yafal AG, Lee YM, Hogle J, Chow M (1994) Poliovirus neutralization by antibodies to internal epitopes of VP4 and VP1 results from reversible exposure of these sequences at physiological temperature. J Virol 68(6):3965–3970
47. Harkonen T, Lankinen H, Davydova B, Hovi T, Roivainen M (2002) Enterovirus infection can induce immune responses that cross-react with beta-cell autoantigen tyrosine phosphatase IA-2/IAR. J Med Virol 66(3):340–350
48. Page GS, Mosser AG, Hogle JM, Filman DJ, Rueckert RR, Chow M (1988) Three-dimensional structure of poliovirus serotype 1 neutralizing determinants. J Virol 62(5):1781–1794
49. Bodian D, Morgan IM, Howe HA (1949) Differentiation of types of poliomyelitis viruses; the grouping of 14 strains into three basic immunological types. Am J Hyg 49(2):234–245
50. Harber J, Bernhardt G, Lu HH, Sgro JY, Wimmer E (1995) Canyon rim residues, including antigenic determinants, modulate serotype-specific binding of polioviruses to mutants of the poliovirus receptor. Virology 214(2):559–570
51. Hogle JM, Chow M, Filman DJ (1985) Three-dimensional structure of poliovirus at 2.9 A resolution. Science 229(4720):1358–1365
52. Filman DJ, Syed R, Chow M, Macadam AJ, Minor PD, Hogle JM (1989) Structural factors that control conformational transitions and serotype specificity in type 3 poliovirus. EMBO J 8(5):1567–1579
53. Koike S, Ise I, Nomoto A (1991) Functional domains of the poliovirus receptor. Proc Natl Acad Sci USA 88(10):4104–4108
54. Belnap DM, McDermott BM Jr, Filman DJ, Cheng N, Trus BL, Zuccola HJ et al (2000) Three-dimensional structure of poliovirus receptor bound to poliovirus. Proc Natl Acad Sci USA 97(1):73–78
55. WHO t (1998) Scheme adopted for use for L20B cells. Polio LaB network quarterly update IV(4):1–2
56. Zurbriggen S, Tobler K, Abril C, Diedrich S, Ackermann M, Pallansch MA et al (2008) Isolation of sabin-like polioviruses from wastewater in a country using inactivated polio vaccine. Appl Environ Microbiol 74(18):5608–5614
57. Abe S, Ota Y, Koike S, Kurata T, Horie H, Nomura T et al (1995) Neurovirulence test for oral live poliovaccines using poliovirus-sensitive transgenic mice. Virology 206(2):1075–1083
58. Horie H, Koike S, Kurata T, Sato-Yoshida Y, Ise I, Ota Y et al (1994) Transgenic mice carrying the human poliovirus receptor: new animal models for study of poliovirus neurovirulence. J Virol 68(2):681–688
59. Ren RB, Costantini F, Gorgacz EJ, Lee JJ, Racaniello VR (1990) Transgenic mice expressing a human poliovirus receptor: a new model for poliomyelitis. Cell 63(2):353–362
60. Smyth M, Pettitt T, Symonds A, Martin J (2003) Identification of the pocket factors in a picornavirus. Arch Virol 148(6):1225–1233
61. Salvati AL, De Dominicis A, Tait S, Canitano A, Lahm A, Fiore L (2004) Mechanism of action at the molecular level of the antiviral drug 3(2H)-isoflavene against type 2 poliovirus. Antimicrob Agents Chemother 48(6):2233–2243
62. Martin J (2006) Vaccine-derived poliovirus from long term excretors and the end game of polio eradication. Biologicals 34(2):117–122
63. MacLennan C, Dunn G, Huissoon AP, Kumararatne DS, Martin J, O'Leary P et al (2004) Failure to clear persistent vaccine-derived neurovirulent poliovirus infection in an immunodeficient man. Lancet 363(9420):1509–1513
64. Domingo E, Martinez-Salas E, Sobrino F, de la Torre JC, Portela A, Ortin J et al (1985) The quasispecies (extremely heterogeneous) nature of viral RNA genome populations: biological relevance–a review. Gene 40(1):1–8
65. Kinnunen L, Huovilainen A, Poyry T, Hovi T (1990) Rapid molecular evolution of wild type 3 poliovirus during infection in individual hosts. J Gen Virol 71(Pt 2):317–324

66. Crotty S, Cameron CE, Andino R (2001) RNA virus error catastrophe: direct molecular test by using ribavirin. Proc Natl Acad Sci USA 98(12):6895–6900
67. Eigen M (2002) Error catastrophe and antiviral strategy. Proc Natl Acad Sci USA 99(21):13374–13376
68. Racaniello VR, Baltimore D (1981) Cloned poliovirus complementary DNA is infectious in mammalian cells. Science 214(4523):916–919
69. Pollard SR, Dunn G, Cammack N, Minor PD, Almond JW (1989) Nucleotide sequence of a neurovirulent variant of the type 2 oral poliovirus vaccine. J Virol 63(11):4949–4951
70. van der Werf S, Bradley J, Wimmer E, Studier FW, Dunn JJ (1986) Synthesis of infectious poliovirus RNA by purified T7 RNA polymerase. Proc Natl Acad Sci USA 83(8):2330–2334
71. Cello J, Paul AV, Wimmer E (2002) Chemical synthesis of poliovirus cDNA: generation of infectious virus in the absence of natural template. Science 297(5583):1016–1018
72. Melnick JL (1996) Current status of poliovirus infections. Clin Microbiol Rev 9(3):293–300
73. Gustin KE, Sarnow P (2002) Inhibition of nuclear import and alteration of nuclear pore complex composition by rhinovirus. J Virol 76(17):8787–8796
74. Fox MP, Otto MJ, McKinlay MA (1986) Prevention of rhinovirus and poliovirus uncoating by WIN 51711, a new antiviral drug. Antimicrob Agents Chemother 30(1):110–116
75. Zeichhardt H, Otto MJ, McKinlay MA, Willingmann P, Habermehl KO (1987) Inhibition of poliovirus uncoating by disoxaril (WIN 51711). Virology 160(1):281–285
76. Jurgens CK, Barton DJ, Sharma N, Morasco BJ, Ogram SA, Flanegan JB (2006) 2Apro is a multifunctional protein that regulates the stability, translation and replication of poliovirus RNA. Virology 345(2):346–357
77. Takegami T, Kuhn RJ, Anderson CW, Wimmer E (1983) Membrane-dependent uridylylation of the genome-linked protein VPg of poliovirus. Proc Natl Acad Sci USA 80(24):7447–7451
78. Eigen M (1993) Viral quasispecies. Sci Am 269(1):42–49
79. Gavrilin GV, Cherkasova EA, Lipskaya GY, Kew OM, Agol VI (2000) Evolution of circulating wild poliovirus and of vaccine-derived poliovirus in an immunodeficient patient: a unifying model. J Virol 74(16):7381–7390
80. Cherkasova E, Laassri M, Chizhikov V, Korotkova E, Dragunsky E, Agol VI et al (2003) Microarray analysis of evolution of RNA viruses: evidence of circulation of virulent highly divergent vaccine-derived polioviruses. Proc Natl Acad Sci USA 100(16):9398–9403
81. Guillot S, Caro V, Cuervo N, Korotkova E, Combiescu M, Persu A et al (2000) Natural genetic exchanges between vaccine and wild poliovirus strains in humans. J Virol 74(18):8434–8443
82. Dahourou G, Guillot S, Le Gall O, Crainic R (2002) Genetic recombination in wild-type poliovirus. J Gen Virol 83(Pt 12):3103–3110
83. Zhang Y, Wang H, Zhu S, Li Y, Song L, Liu Y et al (2010) Characterization of a rare natural intertypic type 2/type 3 penta-recombinant vaccine-derived poliovirus isolated from a child with acute flaccid paralysis. J Gen Virol 91(Pt 2):421–429
84. Tao Z, Wang H, Xu A, Zhang Y, Song L, Zhu S et al (2010) Isolation of a recombinant type 3/type 2 poliovirus with a chimeric capsid VP1 from sewage in Shandong, China. Virus Res 150(1–2):56–60
85. Blomqvist S, Savolainen-Kopra C, Paananen A, El Bassioni L, El Maamoon Nasr EM, Firstova L et al (2010) Recurrent isolation of poliovirus 3 strains with chimeric capsid protein Vp1 suggests a recombination hot-spot site in Vp1. Virus Res 151(2):246–251
86. Brown B, Oberste MS, Maher K, Pallansch MA (2003) Complete genomic sequencing shows that polioviruses and members of human enterovirus species C are closely related in the noncapsid coding region. J Virol 77(16):8973–8984
87. Jiang P, Faase JA, Toyoda H, Paul A, Wimmer E, Gorbalenya AE (2007) Evidence for emergence of diverse polioviruses from C-cluster coxsackie A viruses and implications for global poliovirus eradication. Proc Natl Acad Sci USA 104(22):9457–9462

88. Riquet FB, Blanchard C, Jegouic S, Balanant J, Guillot S, Vibet MA et al (2008) Impact of exogenous sequences on the characteristics of an epidemic type 2 recombinant vaccine-derived poliovirus. J Virol 82(17):8927–8932
89. Kirkegaard K, Baltimore D (1986) The mechanism of RNA recombination in poliovirus. Cell 47(3):433–443
90. Kew O, Morris-Glasgow V, Landaverde M, Burns C, Shaw J, Garib Z et al (2002) Outbreak of poliomyelitis in Hispaniola associated with circulating type 1 vaccine-derived poliovirus. Science 296(5566):356–359
91. Estivariz CF, Watkins MA, Handoko D, Rusipah R, Deshpande J, Rana BJ et al (2008) A large vaccine-derived poliovirus outbreak on Madura Island–Indonesia, 2005. J Infect Dis 197(3):347–354
92. Cherkasova EA, Korotkova EA, Yakovenko ML, Ivanova OE, Eremeeva TP, Chumakov KM et al (2002) Long-term circulation of vaccine-derived poliovirus that causes paralytic disease. J Virol 76(13):6791–6799
93. Chumakov KM, Norwood LP, Parker ML, Dragunsky EM, Ran YX, Levenbook IS (1992) RNA sequence variants in live poliovirus vaccine and their relation to neurovirulence. J Virol 66(2):966–970
94. Reyes GR (2001) Ribavirin: recent insights into antiviral mechanisms of action. Curr Opin Drug Discov Devel 4(5):651–656
95. Georgescu MM, Balanant J, Ozden S, Crainic R (1997) Random selection: a model for poliovirus infection of the central nervous system. J Gen Virol 78(Pt 8):1819–1828
96. Pfeiffer JK, Kirkegaard K (2005) Increased fidelity reduces poliovirus fitness and virulence under selective pressure in mice. PLoS Pathog 1(2):e11
97. Burns CC, Campagnoli R, Shaw J, Vincent A, Jorba J, Kew O (2009) Genetic inactivation of poliovirus infectivity by increasing the frequencies of CpG and UpA dinucleotides within and across synonymous capsid region codons. J Virol 83(19):9957–9969
98. Vignuzzi M, Stone JK, Arnold JJ, Cameron CE, Andino R (2006) Quasispecies diversity determines pathogenesis through cooperative interactions in a viral population. Nature 439(7074):344–348
99. Tao B, Fultz PN (1995) Molecular and biological analyses of quasispecies during evolution of a virulent simian immunodeficiency virus, SIVsmmPBj14. J Virol 69(4):2031–2037
100. de la Torre JC, Holland JJ (1990) RNA virus quasispecies populations can suppress vastly superior mutant progeny. J Virol 64(12):6278–6281
101. Kew OM, Mulders MN, Lipskaya GY, de Silva E, Pallansch MA (1995) Molecular epidemiology of polioviruses. Sem Virol 6:401–405
102. Manor Y, Blomqvist S, Sofer D, Alfandari J, Halmut T, Abramovitz B et al (2007) Advanced environmental surveillance and molecular analyses indicate separate importations rather than endemic circulation of wild type 1 poliovirus in Gaza district in 2002. Appl Environ Microbiol 73(18):5954–5958
103. Boot HJ, Sonsma J, van Nunen F, Abbink F, Kimman TG, Buisman AM (2007) Determinants of monovalent oral polio vaccine mutagenesis in vaccinated elderly people. Vaccine 25(24):4706–4714
104. Kew OM, Sutter RW, Nottay BK, McDonough MJ, Prevots DR, Quick L et al (1998) Prolonged replication of a type 1 vaccine-derived poliovirus in an immunodeficient patient. J Clin Microbiol 36(10):2893–2899
105. Kew OM, Wright PF, Agol VI, Delpeyroux F, Shimizu H, Nathanson N et al (2004) Circulating vaccine-derived polioviruses: current state of knowledge. Bull World Health Organ 82(1):16–23
106. Jenkins HE, Aylward RB, Gasasira A, Donnelly CA, Mwanza M, Corander J et al (2010) Implications of a circulating vaccine-derived poliovirus in Nigeria. N Engl J Med 362(25):2360–2369

107. Martin J, Dunn G, Hull R, Patel V, Minor PD (2000) Evolution of the Sabin strain of type 3 poliovirus in an immunodeficient patient during the entire 637-day period of virus excretion. J Virol 74(7):3001–3010
108. Martin J, Odoom K, Tuite G, Dunn G, Hopewell N, Cooper G et al (2004) Long-term excretion of vaccine-derived poliovirus by a healthy child. J Virol 78(24):1310–13847
109. Shulman LM, Manor Y, Sofer D, Handsher R, Swartz T, Delpeyroux F et al (2006) Neurovirulent vaccine-derived polioviruses in sewage from highly immune populations. PLoS One 1:e69
110. Yang CF, Chen HY, Jorba J, Sun HC, Yang SJ, Lee HC et al (2005) Intratypic recombination among lineages of type 1 vaccine-derived poliovirus emerging during chronic infection of an immunodeficient patient. J Virol 79(20):12623–12634
111. van der Sanden S, Pallansch MA, van de Kassteele J, El-Sayed N, Sutter RW, Koopmans M et al (2009) Shedding of vaccine viruses with increased antigenic and genetic divergence after vaccination of newborns with monovalent type 1 oral poliovirus vaccine. J Virol 83(17):8693–8704
112. Yakovenko ML, Cherkasova EA, Rezapkin GV, Ivanova OE, Ivanov AP, Eremeeva TP et al (2006) Antigenic evolution of vaccine-derived polioviruses: changes in individual epitopes and relative stability of the overall immunological properties. J Virol 80(6):2641–2653
113. Domingo E, Diez J, Martinez MA, Hernandez J, Holguin A, Borrego B et al (1993) New observations on antigenic diversification of RNA viruses. Antigenic variation is not dependent on immune selection. J Gen Virol 74(Pt 10):2039–2045
114. Agol VI (2006) Vaccine-derived polioviruses. Biologicals 34(2):103–108
115. Blomqvist S, Savolainen C, Laine P, Hirttio P, Lamminsalo E, Penttila E et al (2004) Characterization of a highly evolved vaccine-derived poliovirus type 3 isolated from sewage in Estonia. J Virol 78(9):4876–4883
116. Shulman LM, Manor Y, Sofer D (2010) Poliovirus vaccine and vaccine-derived polioviruses. N Engl J Med 363(19):1870
117. Korotkova EA, Park R, Cherkasova EA, Lipskaya GY, Chumakov KM, Feldman EV et al (2003) Retrospective analysis of a local cessation of vaccination against poliomyelitis: a possible scenario for the future. J Virol 77(23):12460–12465
118. Gumede N, Venter M, Lentsoane O, Muyembe-Tamfum J, Yogolelo R, Puren A et al (2010) Identification of vaccine-derived polioviruses (VDPVS) in the DRC from 2005 to 2010. Commun Dis Sur Bull 8(3):43–45
119. Resik S, Tejeda A, Lago PM, Diaz M, Carmenates A, Sarmiento L et al (2010) Randomized controlled clinical trial of fractional doses of inactivated poliovirus vaccine administered intradermally by needle-free device in Cuba. J Infect Dis 201(9):1344–1352
120. Mohammed AJ, AlAwaidy S, Bawikar S, Kurup PJ, Elamir E, Shaban MM et al (2010) Fractional doses of inactivated poliovirus vaccine in Oman. N Engl J Med 362(25):2351–2359
121. Nirmal S, Cherian T, Samuel BU, Rajasingh J, Raghupathy P, John TJ (1998) Immune response of infants to fractional doses of intradermally administered inactivated poliovirus vaccine. Vaccine 16(9–10):928–931
122. Ehrenfeld E, Glass RI, Agol VI, Chumakov K, Dowdle W, John TJ et al (2008) Immunisation against poliomyelitis: moving forward. Lancet 371(9621):1385–1387
123. Yakovenko ML, Korotkova EA, Ivanova OE, Eremeeva TP, Samoilovich E, Uhova I et al (2009) Evolution of the Sabin vaccine into pathogenic derivatives without appreciable changes in antigenic properties: need for improvement of current poliovirus surveillance. J Virol 83(7):3402–3406
124. Modlin JF (2010) The bumpy road to polio eradication. N Engl J Med 362(25):2346–2349
125. Buttinelli G, Donati V, Fiore S, Marturano J, Plebani A, Balestri P et al (2003) Nucleotide variation in Sabin type 2 poliovirus from an immunodeficient patient with poliomyelitis. J Gen Virol 84(Pt 5):1215–1221
126. Arita I, Nakane M, Fenner F (2006) Public health. Is polio eradication realistic? Science 312(5775):852–854

127. Chumakov K, Ehrenfeld E, Wimmer E, Agol VI (2007) Vaccination against polio should not be stopped. Nat Rev Microbiol 5(12):952–958
128. Shulman LM, Manor Y, Sofer D, Mendelson E (2009) Type 2 polio still in our midst. Science 324(5925):334
129. Furione M, Guillot S, Otelea D, Balanant J, Candrea A, Crainic R (1993) Polioviruses with natural recombinant genomes isolated from vaccine-associated paralytic poliomyelitis. Virology 196(1):199–208
130. Lipskaya GY, Muzychenko AR, Kutitova OK, Maslova SV, Equestre M, Drozdov SG et al (1991) Frequent isolation of intertypic poliovirus recombinants with serotype 2 specificity from vaccine-associated polio cases. J Med Virol 35(4):290–296
131. Georgescu MM, Delpeyroux F, Tardy-Panit M, Balanant J, Combiescu M, Combiescu AA et al (1994) High diversity of poliovirus strains isolated from the central nervous system from patients with vaccine-associated paralytic poliomyelitis. J Virol 68(12):8089–8101
132. Minor PD, John A, Ferguson M, Icenogle JP (1986) Antigenic and molecular evolution of the vaccine strain of type 3 poliovirus during the period of excretion by a primary vaccinee. J Gen Virol 67(Pt 4):693–706
133. Shulman LM, Manor Y, Sofer D, Swartz T, Mendelson E (2006) Oral poliovaccine: will it help eradicate polio or cause the next epidemic? Isr Med Assoc J 8(5):312–315
134. Dowdle W, van der Avoort H, de Gourville E, Delpeyroux F, Desphande J, Hovi T et al (2006) Containment of polioviruses after eradication and OPV cessation: characterizing risks to improve management. Risk Anal 26(6):1449–1469
135. Wood DJ (2006) Polio vaccine: the first 50 years and beyond. Summary of the meeting and next steps. Biologicals 34(2):171–174
136. Horstmann DM, Mc CR, Mascola AD (1954) Viremia in human poliomyelitis. J Exp Med 99(4):355–369
137. Steigman AJ, Sabin AB (1949) Antibody response of patients with poliomyelitis to virus recovered from their own alimentary tract. J Exp Med 90(4):349–372
138. Bodian D, Paffenbarger RS Jr (1954) Poliomyelitis infection in households; frequency of viremia and specific antibody response. Am J Hyg 60(1):83–98
139. Racaniello VR, Ren R (1996) Poliovirus biology and pathogenesis. Curr Top Microbiol Immunol 206:305–325
140. Alexander JP Jr, Gary HE Jr, Pallansch MA (1997) Duration of poliovirus excretion and its implications for acute flaccid paralysis surveillance: a review of the literature. J Infect Dis 175(Suppl 1):S176–S182
141. Minor PD (1992) The molecular biology of poliovaccines. J Gen Virol 73(Pt 12):3065–3077
142. Mas Lago P, Caceres VM, Galindo MA, Gary HE Jr, Valcarcel M, Barrios J et al (2001) Persistence of vaccine-derived poliovirus following a mass vaccination campaign in Cuba: implications for stopping polio vaccination after global eradication. Int J Epidemiol 30(5):1029–1034
143. Hovi T, Shulman LM, van der Avoort H, Deshpande J, Roivainen M, de Gourville EM (2011) Role of environmental poliovirus surveilllance in global polio eradication and beyond, a review. Epidemiol Infect 18:1–13
144. Cohen-Abbo A, Culley BS, Reed GW, Sannella EC, Mace RL, Robertson SE et al (1995) Seroresponse to trivalent oral poliovirus vaccine as a function of dosage interval. Pediatr Infect Dis J 14(2):100–106
145. Gromeier M, Wimmer E (1998) Mechanism of injury-provoked poliomyelitis. J Virol 72(6):5056–5060
146. Ramlow J, Alexander M, LaPorte R, Kaufmann C, Kuller L (1992) Epidemiology of the post-polio syndrome. Am J Epidemiol 136(7):769–786
147. Gonzalez H, Olsson T, Borg K (2010) Management of postpolio syndrome. Lancet Neurol 9(6):634–642
148. Ogra PL (1995) Comparative evaluation of immunization with live attenuated and inactivated poliovirus vaccines. Ann N Y Acad Sci 754:97–107

149. Ogra PL, Karzon DT (1969) Distribution of poliovirus antibody in serum, nasopharynx and alimentary tract following segmental immunization of lower alimentary tract with poliovaccine. J Immunol 102(6):1423–1430
150. Ogra PL, Karzon DT (1969) Poliovirus antibody response in serum and nasal secretions following intranasal inoculation with inactivated poliovaccine. J Immunol 102(1):15–23
151. Sofer D, Handsher R, Abramovitz B, Shilon K, Manor Y, Halmut T, et al (2008) Determining vaccination efficacy: is the current minimum anti-polio neutralization antibody titer of >1:8 against Sabin strains high enough? Meeting of the three division of the international union of microbiological societies 5–15 Aug 2008, Istanbul
152. Valtanen S, Roivainen M, Piirainen L, Stenvik M, Hovi T (2000) Poliovirus-specific intestinal antibody responses coincide with decline of poliovirus excretion. J Infect Dis 182(1):1–5
153. Nishio O, Sumi J, Sakae K, Ishihara Y, Isomura S, Inouye S (1990) Fecal IgA antibody responses after oral poliovirus vaccination in infants and elder children. Microbiol Immunol 34(8):683–689
154. Ogra PL, Fishaut M, Gallagher MR (1980) Viral vaccination via the mucosal routes. Rev Infect Dis 2(3):352–369
155. Samoilovich E, Roivainen M, Titov LP, Hovi T (2003) Serotype-specific mucosal immune response and subsequent poliovirus replication in vaccinated children. J Med Virol 71(2): 274–280
156. Onorato IM, Modlin JF, McBean AM, Thoms ML, Losonsky GA, Bernier RH (1991) Mucosal immunity induced by enhance-potency inactivated and oral polio vaccines. J Infect Dis 163(1):1–6
157. Vidor E, Caudrelier P, Plotkin S (1994) The place of DTP/eIPV vaccine in routine paediatric vaccination. Rev Med Virol 4(4):261–277
158. Laassri M, Lottenbach K, Belshe R, Wolff M, Rennels M, Plotkin S et al (2005) Effect of different vaccination schedules on excretion of oral poliovirus vaccine strains. J Infect Dis 192(12):2092–2098
159. Swartz TA, Green MS, Handscher R, Sofer D, Cohen-Dar M, Shohat T et al (2008) Intestinal immunity following a combined enhanced inactivated polio vaccine/oral polio vaccine programme in Israel. Vaccine 26(8):1083–1090
160. Lasch EE, Livni E, Englander T, El-Massri M, Marcus O, Joshua H (1978) The cell mediated immune response in acute poliomyelitis and its use in early diagnosis. Dev Biol Stand 41:179–182
161. Samuel BU, Cherian T, Sridharan G, Mukundan P, John TJ (1991) Immune response to intradermally injected inactivated poliovirus vaccine. Lancet 338(8763):343–344
162. Katrak K, Mahon BP, Minor PD, Mills KH (1991) Cellular and humoral immune responses to poliovirus in mice: a role for helper T cells in heterotypic immunity to poliovirus. J Gen Virol 72(Pt 5):1093–1098
163. Minor PD (2004) Polio eradication, cessation of vaccination and re-emergence of disease. Nat Rev Microbiol 2(6):473–482
164. Plotkin SA, Vidor E (2008) Poliovirus vaccine - inactivated. In: Plotkin SA, Orenstein WA, Offit PA (eds) Vaccines, 5th edn. W.B. Saunders/Elsevier, Philadelphia, pp 605–629
165. Paul J (1971) A history of poliomyelitis. Yale University Press, New Haven
166. Griffiths E, Wood D, Barreto L (2006) Polio vaccine: the first 50 years and beyond. Biologicals 34(2):73–74
167. Nathanson N, Langmuir AD (1963) The cutter incident. Poliomyelitis following formaldehyde- inactivated poliovirus vaccination in the United States during the Spring of 1955. I. Background. Am J Hyg 78:16–28
168. Nathanson N, Langmuir AD (1963) The cutter incident. Poliomyelitis following formaldehyde- inactivated poliovirus vaccination in the United States during the Spring of 1955. II. Relationship of poliomyelitis to cutter vaccine. Am J Hyg 78:29–60
169. Eddy BE, Borman GS, Berkeley WH, Young RD (1961) Tumors induced in hamsters by injection of rhesus monkey kidney cell extracts. Proc Soc Exp Biol Med 107:191–197

170. Shah K, Nathanson N (1976) Human exposure to SV40: review and comment. Am J Epidemiol 103(1):1–12
171. Mortimer EA Jr, Lepow ML, Gold E, Robbins FC, Burton GJ, Fraumeni JF Jr (1981) Long-term follow-up of persons inadvertently inoculated with SV40 as neonates. N Engl J Med 305(25):1517–1518
172. Swartz TA (2008) The epidemiology of polio in Israel an historical perspective. Dyonon, Tel Aviv
173. Tulchinsky T, Abed Y, Handsher R, Toubassi N, Acker C, Melnick J (1994) Successful control of poliomyelitis by a combined OPV/IPV polio vaccine program in the West Bank and Gaza, 1978–93. Isr J Med Sci 30(5–6):489–494
174. Tulchinsky TH, Goldblum N (2001) Polio immunization. N Engl J Med 344(1):61–62
175. Slater PE, Orenstein WA, Morag A, Avni A, Handsher R, Green MS et al (1990) Poliomyelitis outbreak in Israel in 1988: a report with two commentaries. Lancet 335(8699):1192–1195
176. Kawamura N, Kohara M, Abe S, Komatsu T, Tago K, Arita M et al (1989) Determinants in the 5′ noncoding region of poliovirus Sabin 1 RNA that influence the attenuation phenotype. J Virol 63(3):1302–1309
177. Nomoto A, Omata T, Toyoda H, Kuge S, Horie H, Kataoka Y et al (1982) Complete nucleotide sequence of the attenuated poliovirus Sabin 1 strain genome. Proc Natl Acad Sci USA 79(19):5793–5797
178. Christodoulou C, Colbere-Garapin F, Macadam A, Taffs LF, Marsden S, Minor P et al (1990) Mapping of mutations associated with neurovirulence in monkeys infected with Sabin 1 poliovirus revertants selected at high temperature. J Virol 64(10):4922–4929
179. Macadam AJ, Pollard SR, Ferguson G, Skuce R, Wood D, Almond JW et al (1993) Genetic basis of attenuation of the Sabin type 2 vaccine strain of poliovirus in primates. Virology 192(1):18–26
180. Ren RB, Moss EG, Racaniello VR (1991) Identification of two determinants that attenuate vaccine-related type 2 poliovirus. J Virol 65(3):1377–1382
181. Macadam AJ, Arnold C, Howlett J, John A, Marsden S, Taffs F et al (1989) Reversion of the attenuated and temperature-sensitive phenotypes of the Sabin type 3 strain of poliovirus in vaccinees. Virology 172(2):408–414
182. Westrop GD, Wareham KA, Evans DM, Dunn G, Minor PD, Magrath DI et al (1989) Genetic basis of attenuation of the Sabin type 3 oral poliovirus vaccine. J Virol 63(3):1338–1344
183. Minor PD, Macadam AJ, Stone DM, Almond JW (1993) Genetic basis of attenuation of the Sabin oral poliovirus vaccines. Biologicals 21(4):357–363
184. Okonko IO, Babalola ET, Adedeji AO, Onoja BA, Ogun AA, Nkang AO et al (2008) The role of vaccine derived polioviruses in the global eradication of polio-the Nigeria experience as a case study. Biotechnol Mol Biol Rev 3(6):135–147
185. Sutter RW, Cochi SL, Melnick JL (1999) Live attenuated poliovirus vaccines. In: Plotkin S, Orenstein WA (eds) Vaccines, 3rd edn. W.B. Saunders, Philadelphia, pp 364–408
186. Mmwr T (2001) Apparent global interruption of wild poliovirus type 2 trasmission. MMWR Morb Mortal Wkly Rep 50(12):222–224
187. Grassly NC, Wenger J, Durrani S, Bahl S, Deshpande JM, Sutter RW et al (2007) Protective efficacy of a monovalent oral type 1 poliovirus vaccine: a case-control study. Lancet 369(9570):1356–1362
188. WKly Epidemiol Rec (2009) Advisory committee on poliomyelitis eradication: recommendations on the use of bivalent oral poliovirus vaccine types 1 and 3. Wkly Epidemiol Rec 84(29):289–290
189. Salk JE, Krech U, Youngner JS, Bennett BL, Lewis LJ, Bazeley PL (1954) Formaldehyde treatment and safety testing of experimental poliomyelitis vaccines. Am J Public Health Nations Health 44(5):563–570
190. Jenkins HE, Aylward RB, Gasasira A, Donnelly CA, Abanida EA, Koleosho-Adelekan T et al (2008) Effectiveness of immunization against paralytic poliomyelitis in Nigeria. N Engl J Med 359(16):1666–1674

191. Wright PF, Modlin JF (2008) The demise and rebirth of polio–a modern phoenix? J Infect Dis 197(3):335–336
192. Minor PD, Schild GC, Ferguson M, Mackay A, Magrath DI, John A et al (1982) Genetic and antigenic variation in type 3 polioviruses: characterization of strains by monoclonal antibodies and T1 oligonucleotide mapping. J Gen Virol 61(Pt 2):167–176
193. Simoes EA, Padmini B, Steinhoff MC, Jadhav M, John TJ (1985) Antibody response of infants to two doses of inactivated poliovirus vaccine of enhanced potency. Am J Dis Child 139(10):977–980
194. Samuel BU, Cherian T, Rajasingh J, Raghupathy P, John TJ (1992) Immune response of infants to inactivated poliovirus vaccine injected intradermally. Vaccine 10(2):135
195. Wkly Epidemiol Record (2004) Conclusions and recommendations of the Ad Hoc Advisory Committee on Poliomyelitis Eradication, Geneva, 21–22 September 2004. Wkly Epidemiol Rec 79(41):401–407
196. De Palma AM, Purstinger G, Wimmer E, Patick AK, Andries K, Rombaut B et al (2008) Potential use of antiviral agents in polio eradication. Emerg Infect Dis 14(4):545–551
197. Pevear DC, Tull TM, Seipel ME, Groarke JM (1999) Activity of pleconaril against enteroviruses. Antimicrob Agents Chemother 43(9):2109–2115
198. Oberste MS, Moore D, Anderson B, Pallansch MA, Pevear DC, Collett MS (2009) In vitro antiviral activity of V-073 against polioviruses. Antimicrob Agents Chemother 53(10):4501–4503
199. Levy AH (1962) The uses of gamma globulins in the prophylaxis of infection. J Chronic Dis 15:589–598
200. McKinney RE Jr, Katz SL, Wilfert CM (1987) Chronic enteroviral meningoencephalitis in agammaglobulinemic patients. Rev Infect Dis 9(2):334–356
201. Breman JG, Arita I (1980) The confirmation and maintenance of smallpox eradication. N Engl J Med 303(22):1263–1273
202. Smith J, Leke R, Adams A, Tangermann RH (2004) Certification of polio eradication: process and lèssons learned. Bull World Health Organ 82(1):24–30
203. Hull HF, Ward NA, Hull BP, Milstien JB, de Quadros C (1994) Paralytic poliomyelitis: seasoned strategies, disappearing disease. Lancet 343(8909):1331–1337
204. Dove AW, Racaniello VR (1997) The polio eradication effort: should vaccine eradication be next? Science 277(5327):779–780
205. Hull HF, Aylward RB (1997) Ending polio immunization. Science 277(5327):780
206. Wood DJ, Sutter RW, Dowdle WR (2000) Stopping poliovirus vaccination after eradication: issues and challenges. Bull World Health Organ 78(3):347–357
207. Wright PF, Kim-Farley RJ, de Quadros CA, Robertson SE, Scott RM, Ward NA et al (1991) Strategies for the global eradication of poliomyelitis by the year 2000. N Engl J Med 325(25):1774–1779
208. WHO (1996) Field guide for supplementary activities aimed at achieving polio eradication, 1996 Revision: WHO/EPI/GEN/95.01 Rev.1
209. Hull BP, Dowdle WR (1997) Poliovirus surveillance: building the global polio laboratory network. J Infect Dis 175(Suppl 1):S113–S116
210. Wkly Epidemiol Record (2002) Expanding contributions of the global laboratory network for poliomyelitis eradication, 2000-2001. Wkly Epidemiol Rec 77(17):133–137
211. Wkly Epidemiol Record (2003) Laboratory surveillance for wild and vaccine-derived polioviruses, January 2002-June 2003. Wkly Epidemiol Rec 78(39):341–346
212. Wkly Epidemiol Record (2004) Laboratory surveillance for wild and vaccine-derived polioviruses, January 2003-June 2004. Wkly Epidemiol Rec 79(44):393–398
213. de Gourville E, Duintjer Tebbens RJ, Sangrujee N, Pallansch MA, Thompson KM (2006) Global surveillance and the value of information: the case of the global polio laboratory network. Risk Anal 26(6):1557–1569

214. van der Avoort HG, Hull BP, Hovi T, Pallansch MA, Kew OM, Crainic R et al (1995) Comparative study of five methods for intratypic differentiation of polioviruses. J Clin Microbiol 33(10):2562–2566
215. De L, Nottay B, Yang CF, Holloway BP, Pallansch M, Kew O (1995) Identification of vaccine-related polioviruses by hybridization with specific RNA probes. J Clin Microbiol 33(3):562–571
216. Kilpatrick DR, Nottay B, Yang CF, Yang SJ, Mulders MN, Holloway BP et al (1996) Group-specific identification of polioviruses by PCR using primers containing mixed-base or deoxyinosine residue at positions of codon degeneracy. J Clin Microbiol 34(12):2990–2996
217. Balanant J, Guillot S, Candrea A, Delpeyroux F, Crainic R (1991) The natural genomic variability of poliovirus analyzed by a restriction fragment length polymorphism assay. Virology 184(2):645–654
218. Kilpatrick DR, Yang CF, Ching K, Vincent A, Iber J, Campagnoli R et al (2009) Rapid group-, serotype-, and vaccine strain-specific identification of poliovirus isolates by real-time reverse transcription-PCR using degenerate primers and probes containing deoxyinosine residues. J Clin Microbiol 47(6):1939–1941
219. WHO (2004) Polio laboratory manual 4th edn, 2004. WHO/IVB/04.10 (database on the internet)
220. Wkly Epidemiol Record (2006) Resurgence of wild poliovirus type 1 transmission and effect of importation into polio-free countries, 2002–2005. Wkly Epidemiol Rec 81(7):63–68
221. Wkly Epidemiol Record (2010) Poliomyelitis in Tajikistan - first importation since Europe certified polio-free. Wkly Epidemiol Rec 85(18):157–158
222. Deshpande JM, Shetty SJ, Siddiqui ZA (2003) Environmental surveillance system to track wild poliovirus transmission. Appl Environ Microbiol 69(5):2919–2927
223. WHO (2003) Guidelines for environmental surveillance of poliovirus circulation (database on the internet). WHO, Dept of Vaccines and Biologicals; http://www.who.int/vaccines-documents/DoxGen/H5-Surv.htm. Available from: http://www.who.int/vaccines-documents/DoxGen/H5-Surv.htm
224. Manor Y, Handsher R, Halmut T, Neuman M, Abramovitz B, Mates A et al (1999) A double-selective tissue culture system for isolation of wild-type poliovirus from sewage applied in a long-term environmental surveillance. Appl Environ Microbiol 65(4):1794–1797
225. Shulman LM, Manor Y, Handsher R, Delpeyroux F, McDonough MJ, Halmut T et al (2000) Molecular and antigenic characterization of a highly evolved derivative of the type 2 oral poliovaccine strain isolated from sewage in Israel. J Clin Microbiol 38(10):3729–3734
226. Nakano JH, Hatch MH, Thieme ML, Nottay B (1978) Parameters for differentiating vaccine-derived and wild poliovirus strains. Prog Med Virol 24:178–206
227. Vinje J, Gregoricus N, Martin J, Gary HE Jr, Caceres VM, Venczel L et al (2004) Isolation and characterization of circulating type 1 vaccine-derived poliovirus from sewage and stream waters in Hispaniola. J Infect Dis 189(7):1168–1175
228. Manor Y, Handsher R, Halmut T, Neuman M, Bobrov A, Rudich H et al (1999) Detection of poliovirus circulation by environmental surveillance in the absence of clinical cases in Israel and the Palestinian authority. J Clin Microbiol 37(6):1670–1675
229. Ranta J, Hovi T, Arjas E (2001) Poliovirus surveillance by examining sewage water specimens: studies on detection probability using simulation models. Risk Anal 21(6): 1087–1096
230. WHO (2004) Immunization, vaccines and biologicals. Polio laboratory manual, 4th edn. World Health Organization, Geneva, WHO/IVB/04.10
231. Hovi T, Stenvik M, Partanen H, Kangas A (2001) Poliovirus surveillance by examining sewage specimens. Quantitative recovery of virus after introduction into sewerage at remote upstream location. Epidemiol Infect 127(1):101–106
232. Hovi T (2006) Surveillance for polioviruses. Biologicals 34(2):123–126

233. Roivainen M, Blomqvist S, Al-Hello H, Paananen A, Delpeyreux F, Kuusi M et al (2010) Highly divergent neurovirulent vaccine-derived polioviruses of all three serotypes are recurrently detected in Finnish sewage. Euro Surveill 15(19):pii/19566
234. Wkly Epidemiol Record (2008) Conclusions and recommendations of the Advisory Committee on Poliomyelitis Eradication, Geneva, 27–28 November 2007. Wkly Epidemiol Rec 83(3):25–35
235. Kapp C (2003) Surge in polio spreads alarm in northern Nigeria. Rumors about vaccine safety in Muslim-run states threaten WHO's eradication programme. Lancet 362(9396):1631–1632
236. Samba E, Nkrumah F, Leke R (2004) Getting polio eradication back on track in Nigeria. N Engl J Med 350(7):645–646
237. Wkly Epidemiol Record (2009) Advisory Committee on Poliomyelitis Eradication: recommendations on the use of bivalent oral poliovirus vaccine types 1 and 3. Wkly Epidemiol Rec 84(29):289–290
238. John TJ (1972) Problems with oral poliovaccine in India. Indian Pediatr 9(5):252–256
239. John TJ (1976) Antibody response of infants in tropics to five doses of oral polio vaccine. Br Med J 1(6013):812
240. Grassly NC, Fraser C, Wenger J, Deshpande JM, Sutter RW, Heymann DL et al (2006) New strategies for the elimination of polio from India. Science 314(5802):1150–1153
241. Grassly NC, Jafari H, Bahl S, Durrani S, Wenger J, Sutter RW et al (2009) Mucosal immunity after vaccination with monovalent and trivalent oral poliovirus vaccine in India. J Infect Dis 200(5):794–801
242. Gary HE Jr, Smith B, Jenks J, Ruiz J, Sessions W, Vinje J et al (2008) Failure to detect infection by oral polio vaccine virus following natural exposure among inactivated polio vaccine recipients. Epidemiol Infect 136(2):180–183
243. Wringe A, Fine PE, Sutter RW, Kew OM (2008) Estimating the extent of vaccine-derived poliovirus infection. PLoS One 3(10):e3433
244. Mmwr T (2007) Update on vaccine-derived polioviruses–worldwide, January 2006-August 2007. MMWR Morb Mortal Wkly Rep 56(38):996–1001
245. MMWR (2009) Update on vaccine-derived polioviruses – worldwide, January 2008-June 2009. MMWR Morb Mortal Wkly Rep 58(36):1002–1006
246. Dowdle W, Kew O (2006) Vaccine-derived polioviruses: is it time to stop using the word "rare?". J Infect Dis 194(5):539–541
247. Halsey NA, Pinto J, Espinosa-Rosales F, Faure-Fontenla MA, da Silva E, Khan AJ et al (2004) Search for poliovirus carriers among people with primary immune deficiency diseases in the United States, Mexico, Brazil, and the United Kingdom. Bull World Health Organ 82(1):3–8
248. Wkly Epidemiol Record (2010) Progress toward interrupting wild poliovirus transmission worldwide. 2009. Wkly Epidemiol Rec 85(18):178–184
249. Hovi T, Lindholm N, Savolainen C, Stenvik M, Burns C (2004) Evolution of wild-type 1 poliovirus in two healthy siblings excreting the virus over a period of 6 months. J Gen Virol 85(Pt 2):369–377
250. Pelletier I, Duncan G, Pavio N, Colbere-Garapin F (1998) Molecular mechanisms of poliovirus persistence: key role of capsid determinants during the establishment phase. Cell Mol Life Sci 54(12):1385–1402
251. Roivainen M, Blomqvist S, Al-Hello H, Paananen A, Delpeyroux F, Kuusi M et al (2010) Highly divergent neurovirulent vaccine-derived polioviruses of all three serotypes are recurrently detected in Finnish sewage. Euro Surveill 15(19):pii/19566
252. ECDC (2009) Risk assessment from the ECDC on the finding of vaccine-derived polio virus in Finland February 17, 2009. \Documents and Settings\daah\Local Settings\Temporary Internet Files\OLK35A\Risk assessment polio 2009_02_17-JG (2).doc
253. Pavlov DN (2006) Poliovirus vaccine strains in sewage and river water in South Africa. Can J Microbiol 52(8):717–723

254. Yoshida H, Horie H, Matsuura K, Miyamura T (2000) Characterisation of vaccine-derived polioviruses isolated from sewage and river water in Japan. Lancet 356(9240):1461–1463
255. Paximadi E, Karakasiliotis I, Papaventsis D, Papageorgiou G, Markoulatos P (2008) Recombinant Sabin environmental isolates in Greece and Cyprus. J Appl Microbiol 104(4):1153–1162
256. CDC (2009) Update on vaccine-derived polioviruses – worldwide, January 2008-June 2009. MMWR Morb Mortal Wkly Rep 58(36):1002–1006
257. Arya SC, Agarwal N (2007) Global polio laboratory network: future pursuit and commitments. J Clin Virol 38(4):362–363
258. Maderova E, Slacikova M, Cernakova B, Sobotova Z, Nadova K (2005) First isolation of vaccine-derived poliovirus in Slovakia. Euro Surveill 10(8):E050818 3
259. Alexander JP, Ehresmann K, Seward J, Wax G, Harriman K, Fuller S et al (2009) Transmission of imported vaccine-derived poliovirus in an undervaccinated community in Minnesota. J Infect Dis 199(3):391–397
260. MMWR (1994) Certification of poliomyelitis eradication – the Americas, 1994. MMWR Morb Mortal Wkly Rep 43(39):720–722
261. Strebel PM, Sutter RW, Cochi SL, Biellik RJ, Brink EW, Kew OM et al (1992) Epidemiology of poliomyelitis in the United States one decade after the last reported case of indigenous wild virus-associated disease. Clin Infect Dis 14(2):568–579
262. MMWR (2001) Erratum: apparent global interruption of wild poliovirus type 2 transmission. MMWR 50(12):249
263. MMWR (2001) Certification of poliomyelitis eradication–Western Pacific Region, October 2000. MMWR Morb Mortal Wkly Rep 50(1):1–3
264. Wkly Epidemiol Record (2005) Conclusions and recommendations of the Advisory Committee on Poliomyelitis Eradication, Geneva, 11–12 Oct 2005. Wkly Epidemiol Rec 80(47):410–416
265. Heymann DL, Sutter RW, Aylward RB (2006) A vision of a world without polio: the OPV cessation strategy. Biologicals 34(2):75–79
266. Ehrenfeld E, Modlin J, Chumakov K (2009) Future of polio vaccines. Expert Rev Vaccines 8(7):899–905
267. Jenkins PC, Modlin JF (2006) Decision analysis in planning for a polio outbreak in the United States. Pediatrics 118(2):611–618
268. Tebbens RJ, Pallansch MA, Alexander JP, Thompson KM (2010) Optimal vaccine stockpile design for an eradicated disease: application to polio. Vaccine 28(26):4312–4327
269. WHO (2005) WHO framework for national policy makers in OPV-using countries – Cessation of routine oral polio vaccine (OPV) use after global polio eradication 05.02.: WHO/POLIO/05.02
270. Wagner BG, Earn DJ (2008) Circulating vaccine derived polio viruses and their impact on global polio eradication. Bull Math Biol 70(1):253–280
271. Fine PE, Sutter RW, Orenstein WA (2001) Stopping a polio outbreak in the post-eradication era. Dev Biol (Basel) 105:129–147
272. Tebbens RJ, Pallansch MA, Kew OM, Caceres VM, Jafari H, Cochi SL et al (2006) Risks of paralytic disease due to wild or vaccine-derived poliovirus after eradication. Risk Anal 26(6):1471–1505
273. Oliver Wyman Inc (2010) The supply landscape and economics of IPV-containing combination vaccines: Key findings. May 2010: Commissioned by the Bill & Melinda Gates Foundation
274. Chumakov K, Ehrenfeld E (2008) New generation of inactivated poliovirus vaccines for universal immunization after eradication of poliomyelitis. Clin Infect Dis 47(12):1587–1592
275. Baca-Estrada M, Griffiths E (2006) Regulation and standardization of IPV and IPV combination vaccines. Biologicals 34(2):159–161
276. Duchene M (2006) Production, testing and perspectives of IPV and IPV combination vaccines: GSK biologicals' view. Biologicals 34(2):163–166
277. Graf H (2006) Manufacturing and supply of monovalent oral polio vaccines. Biologicals 34(2):141–144

278. El-Sayed N, El-Gamal Y, Abbassy AA, Seoud I, Salama M, Kandeel A et al (2008) Monovalent type 1 oral poliovirus vaccine in newborns. N Engl J Med 359(16):1655–1665
279. Farag MM (2006) Licensing of monovalent OPV1 vaccine. Biologicals 34(2):145–149
280. Caceres VM, Sutter RW (2001) Sabin monovalent oral polio vaccines: review of past experiences and their potential use after polio eradication. Clin Infect Dis 33(4):531–541
281. Plotkin JB, Dushoff J (2003) Codon bias and frequency-dependent selection on the hemagglutinin epitopes of influenza A virus. Proc Natl Acad Sci USA 100(12):7152–7157
282. Macadam AJ, Ferguson G, Stone DM, Meredith J, Knowlson S, Auda G et al (2006) Rational design of genetically stable, live-attenuated poliovirus vaccines of all three serotypes: relevance to poliomyelitis eradication. J Virol 80(17):8653–8663
283. Eichner M, Dietz K (1996) Eradication of poliomyelitis: when can one be sure that polio virus transmission has been terminated? Am J Epidemiol 143(8):816–822
284. Vashishtha VM (2004) But do we have other options? Indian J Pediatr 71(2):183–184
285. Yang CF, Naguib T, Yang SJ, Nasr E, Jorba J, Ahmed N et al (2003) Circulation of endemic type 2 vaccine-derived poliovirus in Egypt from 1983 to 1993. J Virol 77(15):8366–8377
286. Shulman LM, Manor J, Sofer D, Mendelsohn E (2009) Environmental surveillance for polioviruses in Israel: Bio-error, Bio-terror or just mother nature. In: Marks RS, Lobel L, Amadou SA (eds) Advanced detection of viral pathognes. Neobionics, Omer, Israel, pp 111–121
287. Nathanson N (2008) The pathogenesis of poliomyelitis: what we don't know. Adv Virus Res 71:1–50
288. Minor P (2006) Gaps in scientific knowledge for the post eradication world. Biologicals 34(2):167–170
289. Willerth SM, Pedro HA, Pachter L, Humeau LM, Arkin AP, Schaffer DV (2010) Development of a low bias method for characterizing viral populations using next generation sequencing technology. PLoS One 5(10):e13564
290. Kitamura N, Semler BL, Rothberg PG, Larsen GR, Adler CJ, Dorner AJ et al (1981) Primary structure, gene organization and polypeptide expression of poliovirus RNA. Nature 291(5816):547–553
291. Shulman LM, Manor Y, Sofer D, Mendelson E (2012) Environmental surveillance for poliovirus in Israel. In: Marks RS (eds) Bio-error, Bio-terror, or just mother nature. In viral detection biosensors, Pan Stanford Publishing, Singapore (In press)

Chapter 11
Tropical Health and Sustainability

J. Kevin Baird

Glossary

ACT	Artemisinin-combined therapies, a large class of paired drugs that represent the front-line treatments for malaria globally.
BCMS	Board for the Coordination of Malaria Studies, a US government entity established during World War II and later dissolved to coordinate management of the development of new antimalarial therapies.
BMGF	Bill & Melinda Gates Foundation.
CFR	Case fatality rate, the percentage of patients with a given clinical condition not surviving.
DDT	Pesticide ((1,1,1-trichloro-2,2-di(4-chlorophenyl)ethane)), once often used in IRS applications.
G6PD	Glucose-6-phosphate dehydrogenase, an inherited deficiency of which causes patients to be vulnerable to mild to severe drug-induced acute intravascular hemolysis.
GFATM	Global Fund for AIDS, Tuberculosis and Malaria, a consortium of International donors committing resources to those health issues.
GMEC	Global Malaria Eradication Campaign of the 1950s and 1960s.

This chapter, which has been modified slightly for the purposes of this volume, was originally published as part of the Encyclopedia of Sustainability Science and Technology edited by Robert A. Meyers. DOI:10.1007/978-1-4419-0851-3

J.K. Baird (✉)
Eijkman–Oxford Clinical Research Unit, Jalan Diponegoro, Jakarta 10430, Indonesia

Centre for Tropical Medicine, Nuffield Department of Medicine, University of Oxford, Oxford, UK
e-mail: kevin.baird@ndm.ox.ac.uk; jkevinbaird@yahoo.com

P. Kanki and D.J. Grimes (eds.), *Infectious Diseases: Selected Entries from the Encyclopedia of Sustainability Science and Technology*,
DOI 10.1007/978-1-4614-5719-0_11,

GMP	Good manufacturing practice, a high standard of manufacturing certified by experts.
IRS	Indoor residual spraying, in malaria control the practice of spraying the interior walls of homes with insecticide in order to attack mosquitoes that feed on humans.
ITN	Insecticide-treated net, or a net that covers a bed at night to protect sleeping people from mosquito bites and malaria infection.
Malaria	Infection by protozoan parasites (*Plasmodium* species) carried by mosquitoes, often serious and fatal.
MMV	Medicines for Malaria Venture, a public-private partnership committed to developing and licensing new antimalarials therapies.
North/south	Generally characterizes global geographic divide between so-called developed and developing nations.
R&D	Research and development, deliberate, systematic application of S&T in striving toward specific understanding or objectives.
RDT	Rapid diagnostic test, an immunochromatographic, point-of-care kit used at village level to diagnose pathogens (malaria) at low cost.
S&T	Science and technology, techniques for expanding understanding of the physical universe.
Sustainability	Technologies or systems that operate effectively in the absence of long-term external financial or technical assistance.
Tropical health	Health issues specific to the tropical zones, especially endemic infectious diseases and underdeveloped healthcare delivery.
WHO	World Health Organization.
WRAIR	The Walter Reed Army Institute of Research in Washington, DC, where development of antimalarials by the US government has been supported.

Definition of the Subject and Its Importance

Tropical health may be considered relevant to the range of human maladies that occur either predominantly in the tropics or become exacerbated by poorly resourced healthcare delivery systems typical of many nations in the tropics. More often, it is both of those factors, as with malaria. Sustainability implies the introduction of novel interventions aimed at mitigating the burdens and risks of such maladies by improving or replacing existing instruments that may be inadequate or suboptimal. Sustainability further implies a design of those instruments amenable to the capacities of those left to use them, following the inevitable retreat of sponsors and their fiscal and technical resources. Further, the instrument must prove effective in mitigating the problem at which it aims. The research and development (R&D) of sustainable technologies against tropical infections represents a relentlessly urgent task, often linked to preventable deaths in very substantial numbers.

The limited ability of underdeveloped healthcare systems to implement many or most new health technologies greatly compounds the difficulty of achieving sustainability. New science and technology (S&T) must be adapted to settings often completely alien to its developers, both in terms of ability to satisfactorily apply it and its suitability in achieving desired impacts. The subject of tropical health and sustainability is thus concerned with the gap between new technological tools and the ability of tropical healthcare systems to leverage them in improving human health. Further, the intended tools may be poorly conceived as a consequence of fundamental misunderstanding of the problem and, even with appropriate implementation, contribute little to real progress against it.

These distinct gaps, such as the poor grasp of the health problem on one side and of new technologies on the other, may be historically viewed along a North/South geographic divide. The nations that industrialized in the nineteenth century and led the scientific and technological revolution of the twentieth century were predominantly above the Tropic of Cancer. Below that divide, most nations remained largely agrarian and relatively less developed as far as S&T were concerned. Today, many nations of that South are often characterized as developing or resource limited and nations of the North, of course also continuously developing, are characterized as developed or resource rich. The North still commands global scientific and technological superiority, at least within the context of how the North views that critical field of human endeavor. Sustainable S&T in tropical health thus typically refer to tools developed in the North but put to practical and effective use in the South.

The importance of tropical health and sustainability lies in both the Northern and Southern perspectives of what this subject may mean in practice. In the North it implies applying a formidable suite of S&T capacities in improving Southern health. The Southern perspective may focus upon broad improvement of healthcare delivery by any means and not necessarily by the application of what may be poorly understood or even unknown Northern technologies. This perspective may be misconstrued in the North as the larger part of the problem, that is, a lack of awareness of the possibilities of Northern technology. This in turn leads the North to assume leadership roles in applying their R&D capacities to Southern health problems. Poor grasp of those health challenges may instead be the larger part of S&T sustainability failures in addressing them. This entry explores that problematic paradigm and alternatives to it.

Introduction

The frontiers of science and technology (S&T) represent increasingly expensive investments in a future driven by expanding complexity and sophistication of research and development (R&D) systems. Nations endowed with the fiscal and technical wherewithal for such investments have historically dominated important scientific and technological progress. Moreover, these advances have often been driven by conspicuous national self-interest, for example, geopolitical competition with the Soviet Union catalyzed the US National Aeronautics & Space Administration into a technology behemoth. The US Department of Defense may be described in the same terms. Quite distinct from military superiority, nations also recognize the extraordinary economic potential of S&T, and it is increasingly linked to the political and economic security of nations. Contemporary reviews of national S&T agendas express this explicitly [1, 2]. The advancement of other nations in S&T may be construed as a threat by the empowering of potential economic competitors or even political or military rivals. One nation or group of nations may be inherently reluctant to promote the advancement of the S&T capacities of other nations.

Northern nations certainly acknowledge the vital importance of improving the global human condition and they do not lack in humanitarian compassion and generosity. Vast resources and technical know-how have flowed from North to South in the hope of alleviating endemic human suffering in much of the tropics. According to the Organization for Economic Development and Assistance, $104 billion was donated to developing countries by developed countries in 2007 (http://www.oecd.org/dataoecd/47/25/41724314.pdf). About $6 billion of that figure went to the health sector (in 2006), and recipient governments expended three times that amount from their own coffers for the same purpose [3]. However, very few of these resources are applied to advancing Southern or Northern health S&T/R&D per se – these funds almost invariably go to the application of proven rather than experimental technological solutions to human health problems. The North aims to apply proven technologies against Southern health problems through the broad aid schema. These funds and their application in the business of improving Southern health may be thought of as downstream from the S&T/R&D capacities aimed at improving the efficacy of these interventions.

In the S&T context of this encyclopedia, focus may be given to those R&D capacities and their effectiveness in real support of the downstream application of interventions against human health problems. This entry further focuses on a single important health issue, malaria, as an example of how Northern S&T works on a serious Southern health problem. The relatively vast sums available for malaria R&D (a recent development) likely create dynamics distinct from neglected tropical diseases like leishmaniasis, trypanosomiasis, filariasis, and many others. However, "neglected" may remain an appropriate term for malaria despite the funds being expended on it. An important question is the efficacy or impact of those funds in creating technologies that contribute to solving the problem.

Malaria causes very high burdens of morbidity and mortality, in addition to significantly stunting economic development [4]. At least hundreds of thousands perish each year and hundreds of millions suffer debilitating illness [5, 6]. In the tropical South, such burdens have remained entrenched despite the emergence of modern medications against the parasite and insecticides against the mosquito vectors over 90 years ago [7]. Malaria is preventable, treatable, and its cycle of transmission may be permanently broken with available S&T. This is known with certainty. Why then does malaria remain such an enormous human problem? This question strikes at core

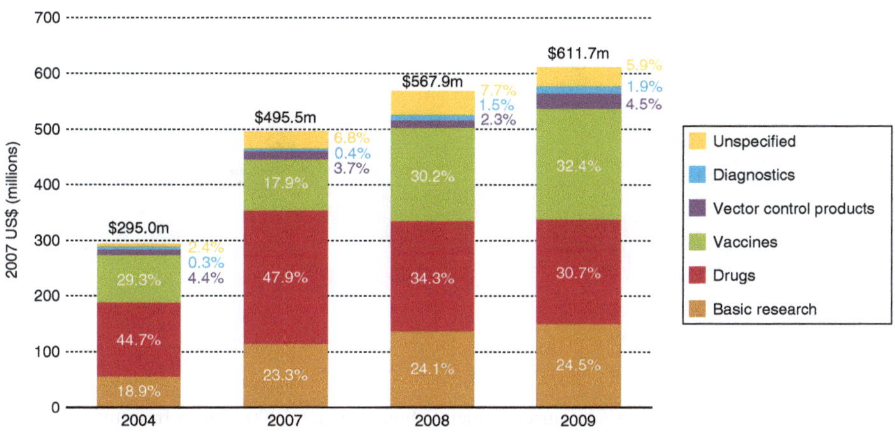

Fig. 11.1 Global malaria R&D spending (Taken from reference [8] with permission of the Malaria Vaccine Initiative)

key issues in tropical health sustainability in the context of North-South R&D collaboration.

In 2011, a consortium of organizations funded by the Bill & Melinda Gates Foundation (BMGF) published an enormously informative and useful summary of global malaria R&D activities and financing (Fig. 11.1), predominantly for the period 2004–2009 [8]. The perspective taken in this entry reaches further back, to the Northern commitment to and later abandonment of malaria as an S&T endeavor between 1900 and 1965. The modern era of malaria R&D began around 1980 when Northern laboratories began applying fast-breaking biotechnological methodologies in earnest. These broader views provide insights on North-South S&T dynamics and a means of assessing the strengths and weaknesses historically embedded within them.

Forging Tools

The analogy of endemic malaria as a craftsman's problem to be worked by applying the range of tools in a toolbox provides a useful conceptualization of human management of that problem. The craftsman's finished project may be thought of as a void once occupied with the human misery caused by malaria. The analogy finds further usefulness in recognizing the craftsman as representing the governments and institutions in endemic zones doing the physical work of crafting that project. They pick up the tools and use them. In the first half of the twentieth century, the Northern craftsman designed and forged the tools in his toolbox. As may be seen by the retreat of malaria between 1900 and 1965 [9] (Fig. 11.2), this approach produced effective outcomes. Malaria disappeared from the North

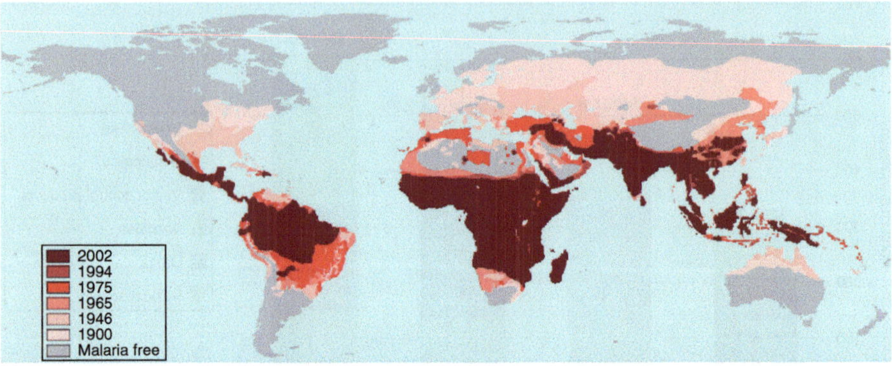

Fig. 11.2 The retreat of malaria to current endemic bounds since 1900 (Taken from reference [9] with permission of the authors and publishers)

and began to be thought of as an exclusively Southern disease. The retreat of malaria involved complex dynamics [7], especially economic, but its initiation only after science generated sufficient understanding and tools makes compelling argument for the vital efficacy of S&T in achieving such gains.

A glance at where malaria remains entrenched today reveals where that early Northern toolbox has proved inadequate to the task of elimination. Experts will see determinants of that failure according to their respective area of expertise. The biologist will point to the tropical climate as especially conducive to the transmission of malaria parasites by their anopheline mosquito definitive hosts. The medic will highlight inadequate healthcare systems. The economist will blame the relatively impoverished condition of the nations composing the malaria belt. Sociologists and politicians will note armed conflict and poor or unstable systems of governance dotted across that belt. And each expert would be partly correct. All of these issues contribute to the contemporary intransigence of endemic malaria in the tropics and even some temperate zones, despite the long availability of technologies adequate to the problem elsewhere. Northern R&D strives to outfit the Southern craftsman with an improved toolbox that may overcome these factors.

The Northern R&D community engaging malaria sees itself as the designer and forger of those tools. Historic Northern dominance in S&T and inarguably superior standing in R&D capacities relative to those of the South, logically lead to the assumption of this role. Northern laboratories engaging the malaria problem express this role explicitly. As one example among a wide range of Northern R&D institutions, the following comes from the US National Institutes of Health, National Institute for Allergy & Infectious Diseases (NIAID) 2008 publication, "NIAID Research Agenda for Malaria," and the section headed, "Malaria Prevention & Control Strategies" [10]: *"To reverse the trend toward rising malaria prevalence and its return to countries where it had been eliminated, new drugs, diagnostics, and vector management tools, as well as effective vaccines, are urgently needed. There also is an acute need to understand the factors that favor the effective use of new interventions and of prevention and control interventions already available. To develop both new products and a more complete understanding of factors that assure their effective introduction and use, public-private partnerships are essential.*" The message also comes across in the 2011 "Staying the Course?" monograph [8]: "*... investment in malaria R&D would create new tools that would more rapidly decrease the malaria burden....*"

The presumed superior usefulness of those Northern tools to the task at hand, as opposed to those that could possibly be developed by the Southern craftsman himself, may be rationally challenged. However, Southern R&D capacity today cannot match that in the North and long-term investment in Southern S&T as an alternative approach has been, and continues to be, trumped by the often expressed sentiment, most recently in the malaria R&D monograph [8], that: "*Providing slow and inadequate funding over the next five to six years will not only delay the time when donors can begin to reduce their malaria R&D funding but – more importantly – will unnecessarily delay saving millions of lives in the developing world.*" The unspoken corollary is the humanitarian urgency in fighting malaria renders the long and difficult work of building Southern R&D capacity strategically irrelevant. Northern R&D strategizes a solution for global malaria upon what is effectively faith in its ability to deliver definitive solutions in the short term.

The contemporary position of an organization called the Multilateral Initiative on Malaria (MIM) supports this assessment. MIM was formed for the specific purpose of creating malaria R&D capacities in Africa. The malaria R&D monograph [8] devotes a single sentence to the organization and offers no insight on its funding levels, despite the rich details on funding embedded throughout the book. MIM's funding page on its website (www.mimalaria.org) was blank in July 2011. It may be surmised from other numbers provided in the monograph that the MIM budget was perhaps about $1.4 million dollars in 2009 (it is also unclear where that funding originated) [8]. It may thus be seen that of the $612 million dollars allocated to malaria R&D in 2009, 0.2% of that was committed to building Southern malaria R&D capacity, and only in Africa. At least within the Northern R&D agenda expressed in that monograph, standing Northern R&D capacity and the urgency of lives lost seems to preclude meaningful investment in long-term transfer of R&D capacities southward.

Providing Tools

If Northern R&D is to be the designated driver of tool design and delivery to the Southern craftsman, an assessment of its performance in this role may be instructive and useful. Performance evaluations look backward rather than forward, even though ultimately aimed at better performance in the future. Assessments of possible or hoped-for performance add little to the utility of such evaluations. Useful evaluation of R&D performance thus demands suspension of its primary

Table 11.1 The malaria control toolbox 2011

Tools	State of the art	Era developed	Sector developed	Utility
Diagnosis of infection	Immunochromatographic cassettes	1990s	Private	Limited
Diagnosis of G6PD deficiency	NADP + reduction assays	1960s	Private	Poor
Therapy against acute attack	ACTs	1990s	Private	Good
Therapy against relapse	Primaquine	1940s	US Army	Poor
Therapy against transmission	Primaquine	1940s	US Army	Poor
Medical insecticides	Pyrethroids, carbamates, organophosphates	1970s	Private	Poor
Insecticide treated nets	Long-lasting pyrethroid impregnated nets	1990s	Private	Limited

focus, the future. How has the Northern R&D posture of the past delivered upon a future represented by today? What lessons may be gathered from that experience and applied in a strategic sense to better performance in reaching a planned and sustainable future?

The primary tools of malaria control today are rapid diagnostic tests (RDTs), artemisinin-combined therapies (ACTs), insecticide-treated bed nets (ITNs), and, increasingly, indoor residual insecticide spraying (IRS) [11]. None of these tools emerged from mainstream contemporary Northern malaria R&D imperatives or programs (Table 11.1). RDTs and ITNs were almost entirely commercially developed, and IRS represents an almost abandoned and recently rediscovered tool developed almost a century ago. The vital artemisinin component of ACTs came from the Chinese Army, and the concept and promise of ACTs emerged in the late 1980s and early 1990s from individuals (particularly Prof. Nicholas White at Mahidol University in Thailand) and private industry (Novartis began codeveloping an ACT with public Chinese partners in 1990) rather than any Northern R&D agenda impelling them in that direction [12]. Up to the present day, with the exception of drugs, the other three vital tools in the standing malaria control toolbox represent the two most severely underfunded avenues of the malaria R&D agenda: diagnostics (1% of funding) and mosquito vector-based interventions (4% of funding) [8].

The proposed strategy to deal with this underfunding is to hold all other areas (basic, drugs, vaccines) stable and to reserve the hoped-for 2% increases in annual funding to those neglected avenues [8]. In short, the expressed strategy is to give those two areas, currently funded at $30 million, an incremental (and hoped-for) increase of about $12 million per year. If this report represents Northern S&T consensus (it does not purport to do so), nothing in basic, drugs, or vaccines research may be sacrificed beyond the savings of achieved targets (and those go to offer relief to donors). The Northern toolmaker's investment preferences seem to reflect optimism for delivery of more practical and useful tools than diagnosis and warding off mosquitoes. Drugs and vaccines (and the basic research that ultimately drives those two avenues) command the Northern R&D agenda at 89% of expenditures between 2004 and 2009 [8]. Has the toolmaker gotten this balance right?

Another legitimate question may be, "Who is this toolmaker?" There is no formal body that gathers and determines the direction and magnitude of malaria R&D funding decisions. The malaria R&D monograph [8] is simply a summary of where investments have been made and by whom. The data are presented for the expressed purpose of guiding those bringing fiscal resources to the malaria R&D table. Thus, the people embedded in funding organizations and having the responsibility of distributing those funds constitute the toolmaker and his collective resources – money, ideas, expertise, a plan, and the will to achieve it.

In 2009, $612 million was expended on malaria R&D. The $99 million from this figure expended by biotechnological firms may be discounted because the funds do not set anyone else's research agenda – they overwhelmingly fund their own R&D imperatives and operations [8]. Two thirds of the remaining $513 million came from the BMGF ($184 million), the US NIH ($116 million), and the US Department of Defense ($38 million). Three European donors followed with a collective $77 million (European Commission, Wellcome Trust, UK Medical Research Council). The balance of $98 million came from a variety of other donors. Decision makers in R&D funding among these organizations, collectively and proportionally, embody the major global malaria toolmaker.

Diagnostics Are Critical

Malaria in the real world is poorly understood by most in the North. The terminology "real world" captures the sum of complex medical, scientific, ecological, social, economic, and political determinants of endemic transmission and disease. It also expresses real (evidence-based) versus imagined (supposition-based) understanding of the character of entrenched endemic malaria. Strategists look at malaria in terms of the burdens of morbidity and mortality imposed by each of the five species of plasmodia routinely infecting humans. This entry considers only two of those species, *Plasmodium falciparum* (cause of falciparum malaria) and *Plasmodium vivax* (cause of vivax malaria) because these likely represent the vast majority of the global malaria burden.

Estimating Burden of Morbidity

Few in the North appreciate the great distance between the real malaria case numbers and those reported by Ministries of Health (MoH) and globally summarized by the World Health Organization (WHO) each year [13]. The crucial importance of that distance is even less appreciated as a key determinant of S&T

sustainability failures against endemic malaria. It reflects not just incomplete understanding of the magnitude of those burdens [6, 14], but also what ultimately drives the unseen and unknown burdens in place and time, that is, in the real world. This vulnerability is driven by unsustainable S&T for the diagnosis of malaria in endemic zones.

Data from Indonesia, to be considered typical among endemic nations, illustrate unsustainable diagnostic S&T and the consequences of the failure to mobilize a nearly adequate R&D response to it. In 2008, WHO reported between 1.4 and 1.8 million annual malaria cases in Indonesia during the years 2000–2006. In 2009, they modified estimates for the same years to between 1.3 and 3.0 million. These were based on confirmed cases reported to them by the Indonesian authorities ranging from 0.22 to 0.44 million per year [13, 15]. The differences in those numbers reflect suppositions regarding the efficiency of Indonesian diagnosis and reporting systems. Studies by the Malaria Atlas Project (MAP) based upon prevalence surveys and Bayesian statistical modeling, estimated 12 million clinical attacks by falciparum malaria alone [16]. Capture of laboratory-confirmed and reported cases of malaria reflected by the numbers provided by MoH implies a roughly 10% case diagnosis rate by the WHO estimates of total cases. That rate drops to less than 1% using the MAP case estimate which excludes vivax malaria. By either estimate, despite the order of magnitude difference between them, the majority of malaria cases in Indonesia go undiagnosed and unreported [17].

The Ministry of Health of Indonesia, like almost all others, reserves therapy with ACTs to cases having a confirmed diagnosis, and all others receive ineffective chloroquine or sulfadoxine/pyrimethamine because there are few practical options [18]. They do so under the rules of the Global Fund for AIDS, Tuberculosis, and Malaria (GFATM) – the donor paying the bill for most ACTs in Indonesia. GFATM reasonably does not wish to have precious ACTs wasted for any given febrile illness. It may be appreciated that this enormous problem is not a challenge of better drugs, but better diagnosis. Failure to deliver diagnostic services and reporting directly causes most malaria in endemic zones to go undiagnosed, unreported, and not treated.

Estimating Burden of Mortality

Reports and estimates of malaria mortality rates in Indonesia also range freely across several decimal places. The Government of Indonesia's Central Bureau of Statistics conducted national Household Health Surveys in 1995 and 2001, and estimated 30,000–38,000 deaths due to malaria had occurred in each of those years [17]. In 2001, the WHO reported 68 deaths caused by malaria in Indonesia [15]. This reported number was 494 in 2006, and WHO estimated the number to be closer to 3,000 deaths [15]. A recent study in India challenged the government's estimates of malaria mortality at 15,000 deaths in 2008 [19]. Investigators conducted verbal autopsies (interviewing surviving family members of people lost to premature death on the clinical nature of the fatal illness), using a rigorous sampling methodology to represent the broader population. They analyzed 122,000 deaths during 2001–2003, and their findings led to an estimate of 205,000 deaths annually from malaria, and death by malaria carried a 1.8% risk by age 70. They found that 86% of deaths caused by malaria did not occur in any healthcare facility, and thus went unreported. Verbal autopsy methodologies of course come with uncertainties, but it may be that fewer than 1 in 10–100 or more deaths caused by malaria are reported from endemic zones. Unsustainable S&T for diagnosis and reporting explain these dangerous gaps in understanding mortality burdens and their demographic/geographic distributions.

The species of plasmodia responsible for death cannot be ascertained in verbal autopsy, but it is now known that in India specifically, *P. vivax* at least occasionally and perhaps routinely causes death in malaria patients [20–23]. The same has been found in other regions [24–31]. Long known as "benign tertian malaria," such reports of fatal vivax malaria have been met with appropriate, albeit stubborn skepticism in malariology circles. The conventional view has been that only the "malignant tertian malaria" of *P. falciparum* routinely causes the range of life-threatening syndromes linked to malaria.

Benign Tertian Malaria Fallacy

Careful study of the genesis of the term "benign tertian malaria" in the late 1800s and the clinical experience with malaria therapy of neurosyphilis in the 1920s and 1930s provides important insights on the contemporary view of vivax malaria as an insignificant contributor to global malaria mortality. The notion of benign and malignant malarias had been well known long before Laveran's discovery of the plasmodia in 1880 [32]. After broad acceptance of species identity between *P. falciparum* and *P. vivax* at around 1890, a few elite clinical study centers in Europe and America sought to reconcile taxonomic identity to those clinical classifications of malaria. As pointed out by Kitchen in 1949 [33], taxonomic identity should have been the frame of reference rather than ambiguous and confused clinical classifications. As early as 1901, a pathologist in New York City describing autopsy findings of a fatal case of vivax malaria challenged the certainty of *P. vivax* as a benign infection [34]. He wrote, "*While the statement of the Italian authorities has long held true that no autopsy in a case of malaria with infection by the large tertian parasite, as the infection is never fatal, the present case requires modification of that view. . ..*"

The advent of malaria for therapy of neurosyphilis in the 1920s added substantial depth to the understanding of clinical course and taxonomic identity. Despite contemporary views by some that vivax malaria in those patients did not cause severe or threatening disease [35], death caused by vivax malaria carried consistent and apparently strain-specific risks ranging from 3% to 15% case fatality rates (CFR). At treatment centers across the USA, the CFR was extraordinarily consistent at 3–6% after therapy had been optimized [36–41]. Prior to both learning to screen out poor candidates for treatment and gaining expertise in successful clinical management, the CFR had been between 15% and 30%. At least one series of fatal treatments had been followed by autopsy studies that confirmed malaria as the cause of death (rather than any underlying disease, including neurosyphilis) in at least half of the patients [40]. Most physicians reporting these experiences expressed the conviction that malaria had killed most of their nonsurviving patients. Nicol [42] wrote in 1932, "*One must acknowledge that, though in a few cases death is caused by intercurrent disease, malaria, if not directly responsible, must be regarded as a contributing factor; in most cases the malaria itself is the cause.*" It should be recalled that in that era unsuccessful therapy of neurosyphilis would almost always end in death for the patient. This fact, and the only 35% efficacy of malaria therapy against the disease under the best of circumstances [42], explains the aggressive and extreme treatment. More severe and punishing repeated malaria paroxysms (episodic extreme chills and spiking fevers) improved the odds of successful therapy. Nicol [42] closed the above-cited quotation with, "*Risks, however, must be taken in treating a disease which, if untreated, proves fatal.*"

Clinics in the UK usually used the Madagascar strain of *P. vivax* and this strain for treatment of neurosyphilis typically killed 10–14% of patients [43]. Clinics in the Netherlands used a local strain of *P. vivax* that carried a 0% CFR, but had very low therapeutic efficacy against neurosyphilis. A switch to the Madagascar strain improved efficacy but came with an 8% CFR [44]. One clinic in the UK evaluated *P. falciparum* for therapy and reported a 4% CFR [45]. It may thus be appreciated that most strains of vivax malaria, permitted to follow a severe course of infection, may threaten life in at least 5% of patients. Contemporary studies in hospitals report CFR among patients classified as having severe illness at between 5% and 10%, a rate roughly equal to that among patients with severe falciparum malaria in the same hospitals.

The chapter on clinical vivax malaria penned by S.F. Kitchen [33] in the classic 1949 text on malaria does not deal directly with the neurosyphilis mortality data. The author describes some controversy on the issue of severe and fatal vivax malaria but dismisses it as implausible largely on the basis of this parasite rarely causing hyperparasitemia – a well-known risk factor for death in falciparum malaria and the basis of its "malignant tertian malaria" moniker. Relatively low parasitemia is routinely documented in contemporary reports of severe and fatal disease in vivax malaria, and this may reflect an unknown pathogenesis distinct from falciparum malaria. The seemingly nonthreatening low parasitemias of vivax malaria may have deceived earlier workers. Indeed, Kitchen [33] wrote, "*...present knowledge concerning the provocative potentialities of low-grade parasitemias does not permit denial of the possibility that the presence of the Plasmodium [vivax], even in very small numbers, may serve as an incitant of entirely foreign conditions....*" Today it is known, for example, that *P. vivax* is physiologically capable of accessing the bone marrow [46] and is sometimes found in that tissue [47].

Although further studies are required to gauge the burden of severe morbidity and mortality of vivax relative to falciparum malaria in endemic zones, the infection is certainly not benign. Regardless of relative burden, the persistent view of vivax malaria as a benign entity may be considered incompatible with available evidence and probably dangerous to patients and populations living with this threat.

Vivax Malaria and Global Malaria Burden

The long-engrained notion that only *P. falciparum* kills profoundly shapes how the global malaria problem is perceived as a whole and, ultimately, strategized to deal with. A statement very similar to this may be found in popular, peer-reviewed, and authoritative technical media [48–50], "80% of cases and 90% of the world's deaths caused by malaria occur in Africa." Vivax malaria is relatively rare in sub-Saharan Africa as a consequence of the almost universal inherited absence of a molecule called Duffy factor on the surface of red blood cells which *P. vivax* requires for invasion.Thus, the dominance of *P. falciparum* on that continent, and the relative dominance of *P. vivax* elsewhere (along with the presumption of its benign nature) largely accounts for that popular supposition. As the severe morbidity in vivax malaria and very high estimated all malaria mortality data from India and elsewhere suggest [19–31], the assertion is scientifically unsound and perhaps likely to be a gross distortion of malaria as it occurs in the real world.

The statement likely overstates the relative burden of falciparum malaria and simply dismisses the contribution of vivax malaria to the global burden of morbidity and mortality. Hay and colleagues [16] estimated African clinical attacks of falciparum malaria accounted for 60% of the global burden of that species. Almost all of the remaining estimated attacks by falciparum malaria occurred in Central, South, and Southeast Asia, where the substantial burden of vivax malaria, yet to be reliably estimated, also occurs [51]. The number of clinical attacks caused by *P. vivax* is estimated between 100 and 400 million [9]. As has been explained, the mortality burden imposed by this parasite (and very likely by *P. falciparum* as well) may be described as unknown. Adding the lower end of the vivax malaria morbidity range to the global falciparum clinical attack numbers for 2007 of Hay et al. [16] suggests at least 50% of the global burden of clinical attacks by any species occurs in Central, South, and Southeast Asia.

If careful study confirms heavy burdens of morbidity and mortality caused by endemic vivax malaria in Asia and elsewhere, the response to this threat would largely be limited to chemotherapies no longer likely to work (chloroquine), another that never worked effectively in the first place (primaquine), or others untried against this species. This hypothetical scenario contains an uncontestable core fact: the burden of morbidity or mortality of vivax malaria anywhere it occurs is not known. And yet, Northern science up to 2011 remains almost wholly focused on falciparum malaria, especially in an African context. Research on *P. vivax* between 2007 and 2009 accounted for only 3.1% of R&D funding [8]. There may be serious consequences to this possible miscalculation driven by unsustainable S&T for the diagnosis of malaria in endemic zones.

Proven resistance to the first-line treatment of acute vivax malaria, chloroquine, now occurs across the Indonesian archipelago, at sites through the Mekong region, and cases have been reported from the Indian subcontinent [52]. Another drug, primaquine, must also be administered to prevent further attacks caused by dormant forms in the liver. The efficacy of that drug has not been unambiguously measured since the 1950s. The hemolytic toxicity of primaquine in glucose-6-phosphate dehydrogenase (G6PD) deficient patients – so prevalent in endemic zones (5–25%) – coupled with the failure to field sustainable diagnostic S&T for that disorder, likely drives therapeutic effectiveness of primaquine close to null in endemic zones [53, 54]. There are no standardized means of assessing the efficacy of these drugs against this parasite, and no mechanisms of activity against the parasite or of resistance to the drugs are known. Worse still, the standing drug development paradigm, designed for and largely serving the fielding of drugs against the acute attack of falciparum malaria, is very poorly suited for fielding new therapies for radical cure of vivax malaria [55].

RDTs Have Limited Utility

Several dozen commercially available immuno-chromatographic RDTs for malaria compete for the large volume of global sales. Most of these kits perform very well in specific settings, but in others they completely fail to meet operational requirements. Ideally, the kit is used for the diagnosis of malaria in acutely ill patients in endemic zones lacking access to a definitive microscopic diagnosis of malaria by Giemsa-stained blood film. When falciparum malaria is the offending agent and parasitemia is higher than 200 parasites/μL, sensitivity and specificity each come in at well above 90% in most kits. Below that threshold, sensitivity falls precipitously. In vivax malaria, the corresponding threshold is about 500 parasites/μL for most kits [56]. The threshold of parasitemia causing illness in nonimmune patients falls far below both of those diagnostic thresholds for RDTs [57]. Also, cross-sectional surveys in endemic zones typically show median parasitemia levels at or below those thresholds, especially for vivax malaria [58, 59]. Thus, RDTs are not suitable for diagnosis in nonimmune patients or for detection of parasitemia among asymptomatic residents of endemic zones.

Asymptomatic carriers of malaria represent the rule rather than the exception in almost all endemic zones. These infections almost universally present very low levels of parasitemia far beyond the diagnostic reach of RDTs [59]. Cross-sectional surveys in Cambodia employing ultrasensitive polymerase chain reaction (PCR) diagnostics revealed that only about half of those positives were also positive by expert microscopy, and far fewer by RDT. Thus, elimination strategies must consider mass administration of drugs in populations having very low levels of endemic malaria, as a consequence of diagnostics, inadequate to the task of finding and treating the few infected people [60]. Current RDT technology leaves wide diagnostic gaps in endemic zones.

Diagnostics R&D Performance Evaluation

The failure to develop sustainable diagnostic and reporting S&T for endemic zones may have distorted the view of global malaria with possibly serious consequences. By the nearly complete neglect of unrecognized heavily burdened regions [61] and a heavily burdening species [62, 63], one may have been lured, by diagnostic blindness, into a position of profound strategic weakness. Assuming that vivax malaria indeed kills with some regularity in India and elsewhere, it may be reasonable to assert that one does not have even the barest grasp of malaria in the real world, that is, it is not known how many acute attacks kill how many people, where, or by which parasite, much less who is most vulnerable and why. Poor diagnostics is the reason and the Northern R&D toolmaker appears unable to prioritize that task (1% of R&D funding [8]).

Insecticides Are Critical

Since the discovery in 1880 by Alphonse Laveran that plasmodia cause malaria, and of their transmission by anopheline mosquitoes by Ronald Ross in 1898, malariologists have fiercely argued strategies of control focused upon treatment versus interventions aimed at the anopheline vectors. In the early 1900s, Robert Koch championed the former, and Ross the latter [64]. Both applied their concepts in the field (Koch with quinine in New Guinea, and Ross with crude anti-mosquito measures in West Africa) with almost no success. However, key successes in the era that followed, along with the discovery of dichlorodiphenyltrichloroethane (DDT), ensured the dominance of mosquito vector control in the global malaria eradication effort undertaken during the 1950s and 1960s and, later, misplaced disillusionment with that strategy.

Abandonment of a Proven Tool

Between 1904 and 1914, William Crawford Gorgas effectively attacked the mosquito vectors of the yellow fever and malaria in Panama that had decisively dashed the French efforts of the 1880s to build a canal [65]. Watson in British Malaya (today Malaysia) tamed malaria on economically important plantations by deliberate and highly specific environmental modifications aimed at the behavior of the local anopheline vector. His approach, called species sanitation, was successfully applied by Dutch and Indonesian malariologists in the Netherlands East Indies of the 1920s and 1930s [66]. In the same period, William Soper successfully eliminated from Brazil the imported and terribly efficient African mosquito vector of malaria, *Anopheles gambiae* [67]. Mueller discovered DDT in Germany at about the same time, and the US military brought it to bear as a defensive weapon against malaria and other insect vector-borne diseases during and immediately after the Second World War.

The technological stage was thus set in the late 1940s for the conception and strategic design of an ambitious effort to eradicate malaria through the 1950s and 1960s. The USA pushed and largely financed the WHO-led Global Malaria Eradication Campaign (GMEC), which was essentially an attack upon anophelines using principally DDT. The role of another product liberated from wartime Germany, the extraordinarily safe and effective treatment for acute malaria called chloroquine [68], was secondary or a mop-up in that strategy.

DDT exploited a behavior of anophelines – most species go indoors at night, feed upon sleeping people and, heavily laden with a blood meal, rest upon interior walls. DDT applied to those walls provided an opportunity for lethal contact with insecticide even many months after its application. Indoor residual insecticide spraying, or IRS, proved remarkably effective in malaria control in almost every setting. Global malaria numbers in the era of its systematic application plummeted, except in most of Africa where IRS was not applied in earnest for want of the logistical infrastructure to do so. In India, for example, over 800,000 deaths due to malaria each year fell to, reportedly, none [69]. Malaria transmission ceased in most of the Caribbean and was sharply curtailed all across Central and South America. Endemic malaria disappeared from Southern Europe, North Africa, Japan, Taiwan, the Korean peninsula, and most of Java [70].

Northern agriculture effectively derailed the success of DDT by hijacking the medically important product. The aerial spraying of DDT across many millions of square miles of croplands resulted in vast quantities of this very stable compound being introduced into the environment at disastrous cost. In 1971 in the USA alone, the year before such practice was legally banned, 14 million pounds of DDT was used in the agricultural sector [71]. The practice harmed beneficial insect populations and accumulated dangerously in higher predators, even those above the Arctic Circle. Theft of a medically important instrument by Northern agriculture spilled southward. The onset of resistance to DDT by anopheline mosquitoes may have been driven by agricultural applications. In the late 1960s, as many of these problems were coming fully to light, the GMEC formally collapsed [72].

Adopting Impractical Tools

Northern agricultural insecticide industries turned to their S&T engines to solve the problem of the loss of DDT with viable alternatives. They required insecticides only stable long enough for the quick work of killing crop pests. Three classes of insecticides emerged to dominate that market: organophosphates, carbamates, and pyrethroids. These short-lived compounds, their costs, along with relatively high toxicity or noxious character make them poorly suited for IRS. Nonetheless, in the decades that followed, and up to the present day, Southern IRS against malaria and other vector-borne diseases of the tropics has depended upon these insecticides developed and licensed for Northern agricultural practices [73]. The impracticality of the new insecticides contributed to the broad contraction or collapse of IRS programs in the endemic tropics [74]. Those products stand as clear examples of the sometime very poor fit of Northern S&T imperatives and products to Southern health problems.

Cost-effective, safe, and practical long-lasting insecticides sprayed inside homes were nowhere in that Northern R&D agenda for the simple reason that Northern homes were not being invaded by insects carrying life-threatening infections.

Revitalizing IRS

Roberts and colleagues [75] documented quantitative insights on the health costs incurred by Northern aversion to DDT especially, and IRS in general. A science- and public health-led campaign narrowly averted a 2001 effort to ban DDT for any use anywhere in the world [76]. The aversion to DDT, borne of its reckless agricultural application, irrationally inhibited endorsement or support to Southern IRS operations for decades. In 2007, WHO reversed its long-standing recommendation against IRS in areas of stable transmission and today recommends DDT for such operations [77]. In that reversal, WHO expressed, "*DDT is still needed and used for vector control simply because there is no alternative of both equivalent efficacy and operational feasibility....*" The reason for that simple truth is clear: Northern insecticides R&D abandoned the task of medical insecticides 60 years ago and there has been no Southern R&D capacity to pick it up.

Northern sociopolitical aversion to IRS has at long last diminished to greater support and emphasis on this vital tool in the control toolbox. In late 2009, for example, USAID advertised a US$130 million grant for support of IRS operations in Tanzania. However, unleashing such resources to apply poorly suited short-lived insecticides may greatly dampen efficiency and impact. A recent study described a microencapsulated organophosphate insecticide (chlorpyrifos methyl, Dow Agrosciences) with >90% activity persisting at least 9 months when used in experimental IRS [78]. This work, and the growing realization of the threat of resistance to both pyrethroids and DDT [79, 80], emphasizes the need for much more R&D on insecticides.

Reliance upon Northern agricultural S&T for medically useful insecticides should be acknowledged as at least inefficient and perhaps impractical. The discovery and development of medical insecticides should not be simply incidental to agricultural pesticides. The malaria R&D monograph [8], however, points hopefully to Northern industrial pesticide firms as a means of supplementing the pitifully small amounts of funding committed to insecticides.

Insecticides R&D Performance Evaluation

Northern R&D on discovery of medically important insecticides for use in IRS or ITNs has been nascent since the discovery of DDT in the 1930s. This failure, viewed in light of the enormous impact of DDT against malaria in the middle of the twentieth century, may be considered the most conspicuous of all Northern R&D failures on global malaria. The stubborn and irrational Northern aversion to insecticide tools for malaria control may at last be waning, but R&D investment in earnest exploration of improving these tools remain grossly inadequate. Hopefully, however, the BMGF expressed interest in spatial repellency of some compounds by making a substantial investment in a proof-of-concept trial in Indonesia in 2009 (JKB is an investigator in that effort), and their Grand Challenges funding scheme invested in mosquito attractant/repellant molecule discovery. Agricultural insecticide developers apply algorithms of chemical lethality and exclude compounds on the basis of repellant characteristics that would diminish opportunities for lethal contact with an insecticide product. If spatial repellency proves effective in risk reduction in endemic zones, it would reopen the universe of chemical exploration possibilities with development algorithms wholly distinct from insecticides. Perhaps not coincident with its operational efficacy in the past century, DDT happens to be one of the most effective compounds known in eliciting avoidance behavior in anopheline mosquitoes [81].

Drug Delivery

Humanity always needs new and better anti-infective drugs – diverse microbial populations mount their formidable Darwinian defenses that make old drugs useless, and new infections emerge. The commercial pharmaceutical industries of the North generate streams of new therapies in a fiercely competitive market environment that sharply hones the science and technologies (and regulatory prowess) of surviving companies. The prospect of profit drives and sustains this superb and very expensive engine of biotechnological innovation. However, at least several major multinational pharmaceutical firms have stepped into the malaria R&D arena with fairly substantial investments. During 2007–2009, they brought 14% of the $1.68 billion R&D investment (including work on vaccines), but kept this funding almost entirely internal [8]. Industry also frequently participates in antimalarial drug development through donor-funded product development partnerships (PDPs) like the Medicines for Malaria Venture (MMV) funded by BMGF, Wellcome Trust, WHO, and others. PDPs spent $114 million donor funds on malaria R&D in 2009. Private enterprises (both major multinationals and small to medium businesses) and academic or other not-for-profit institutions divided that pie in roughly equal halves [8].

State of the Art

Understanding these investments and the dynamics driving them requires a brief look at the class of therapies called the ACTs. The artemisinin family of drugs (artemisinin, artesunate, artemether, arte-ether, and dihydro-artemisinin), all rapidly excreted, combines with another typically slowly excreted antimalarial drug (mefloquine, piperaquine, lumefantrine, pyronaridine, or others) to both improve efficacy and protect the artemisinins against onset of parasite resistance to them [12]. ACTs today are manufactured commercially at many factories scattered across Southeast and South Asia, the Middle East, and Africa. These companies produce dozens of ACTs in as many formulations and means of packaging. The artemisinins have no patent protections, nor do the majority of drugs partnered with them. These products also share other important characteristics: almost none have been produced by the standards of Good Manufacturing Practice (GMP) as applied and certified by regulators, and few have co-formulated the components, preferring instead to package them together as separate pills. Few of these manufacturers possess the resources or influence to press for the aggressive arrest and prosecution of counterfeiters of their products. Collectively, these important flaws permit patients to take medication inadequate for cure and thereby expose each therapeutic agent to real risk of onset of resistance.

The GFATM will not fund ACT products of unproven quality. Therapies must be approved by WHO, and such approval includes certified compliance to GMP [82]. In effect, the availability of ACTs to Southern governments hinges upon access to GMP product. One strategy at work with GFATM funding policy is to push inferior drugs out of the marketplace with the superior drugs simply by making them at least as affordable. Thus, GMP product emerges as a key determinant of success in this strategy. In the 13 years of its existence, MMV has fielded two such ACTs (Coartem™, Novartis; Coarsucam™, Sanofi Aventis). Two others approach registration in 2011 (Euartesim™, Sigma Tau; and Pyramax®, Shin Poong).

Even as GMP co-formulated ACTs now approach closer to broader reality, resistance to ACTs has emerged in the Mekong region [83] where commercially driven monotherapy with artesunate and counterfeiting has been rampant [84]. Expansion of this problem into the rest of South and Southeast Asia, and ultimately Africa (as occurred with previous antimalarials), would constitute a global health catastrophe because currently there are no therapeutic replacements [85]. This problem could possibly have been averted by early use of GMP, co-formulated ACTs, and aggressive, commercially driven protection against counterfeiting. The neglect of malaria chemotherapy as a whole by Northern pharmaceutical S&T, and also by malaria institutional R&D stalwarts during the second half of the 1900s, today has the South precariously near persistent, untreatable, and unbeatable endemic malaria. How did such neglect arise? Addressing this question requires examination of the genesis and content of the chemotherapeutics toolbox of 1965.

Chemotherapeutics Accomplishment: 1900–1965

Assessment of current strategy for delivery of drugs, as summarized above, requires consideration of where Northern S&T left off when it effectively abandoned the problem of malaria chemotherapy in 1965 as a mission accomplished. Examining malaria R&D capacities and imperatives in the preceding half century provides essential understanding of that turning point.

Scientific chemotherapy of malaria emerged in the organic chemistry crucible of the German aniline dye industry in the closing years of the 1800s. Screening for activity against parasites in bird models of malaria, the German laboratories of I.G. Farben during the 1920s and 1930s discovered many active classes and compounds [86]. The two most important were the 4- and 8-aminoquinolines, and focus upon them during the frenetic war-spurred search for better therapies in the 1940s delivered what were to be considered the ultimate antimalarial drugs.

Chemotherapy of malaria in 1940–1941 remained overwhelmingly dominated by quinine. The few synthetic antimalarial drugs in the market fared poorly due to unpleasant or even dangerous side effects. The German chemotherapists had superior drugs coming and plotted to unhinge the Dutch monopoly on chemotherapy of malaria manifest by their control of 95% of the world trade in quinine [87]. The following year, the Nazis occupied Holland and the Imperial Japanese armed forces occupied Java and Sumatra, where virtually all Dutch quinine originated. Axis powers denied the Allied forces access to the only drug then widely used to treat malaria.

The Allies understood the coming warfare would be conducted where malaria posed a serious threat. Endemic malaria loomed ominously in the Western Pacific especially, and their worst fears were soon realized: the first offensive ground combat involving American troops in that war was at Guadalcanal in the Solomon Islands in mid-1942; those troops suffered malaria attack rates of 1,700/1,000 man-years [88]. When the US Army Americal Division was evacuated from Guadalcanal to nonmalarious Fiji, their chemoprophylaxis was withdrawn and they suffered attack rates by *P. vivax* of 3,700/1,000 man-years [89].

The US Government mobilized to solve their serious malaria problem. The malariologists enlisted in the effort, and their sponsors, viewed success in their endeavors as directly impacting the likelihood and speed of military victory, especially in the Pacific [90]. The Board for the Coordination of Malaria Studies (BCMS), managed from within the US National Science Foundation, oversaw the evaluation of over 14,000 potential drugs from discovery to clinical trials in dozens of laboratories and clinics around the USA, Australia, and Britain. Their top priority was *P. vivax* because it accounted for five of every six cases among Allied troops. The Allies quickly mobilized production of a synthetic antimalarial (lifting it from the German S&T machine that generated it) called atabrine (also called mepacrine or quinacrine) and put it to work for both prophylaxis and treatment [91]. This hasty solution, however, had serious drawbacks. The skin of troops adhering to atabrine prophylaxis turned a distinct yellow hue and compliance had to be strictly enforced.

More importantly, atabrine had an unexpected and dangerous interaction with the only drug then available against relapse, pamaquine.

Pamaquine was the first synthetic antimalarial licensed and marketed for therapy against malaria (as Plasmochin). This 8-aminoquinoline compound exerted excellent activity against the then poorly understood phenomenon of relapse in vivax malaria. Its activity against the sexual stages responsible for infecting mosquitoes was also known. Following its distribution in the 1920s, however, it earned the reputation of being sometimes an exceptionally dangerous drug [92]. As is well known today, the 8-aminoquinolines as a class exhibit potentially lethal toxicity in patients with an inherited deficiency of glucose-6-phosphate dehydrogenase (G6PD). Unwitting administration to many such patients caused fatalities [92] and the drug never saw widespread use. The Allies, however, facing their serious relapse problems in the Pacific theater of World War II, mobilized pamaquine to that zone and serious toxicity problems followed [93, 94]. The BCMS issued a classified report in early 1943 demonstrating that when atabrine (but not quinine) was administered with pamaquine, plasma levels of the latter elevated tenfold. An already dangerous drug to some became a dangerous drug to all, and in mid-1943 the US Surgeon General ordered it withdrawn as therapy against relapse (although it could still be used, and was, in lower doses against transmission) [95]. The Allies were effectively forced to surrender to their serious relapse problem.

The BCMS resolved to address this problem. They screened several hundred candidate 8-aminoquinolines for toxicity in rats and monkeys and sent 24 drugs to clinical trials in prisoner volunteers in the USA. That screening had critically important limitations, but the BCMS in its haste to deliver a replacement therapy accepted those acknowledged shortcomings. The screening for antimalarial activity of those compounds in birds, for example, was understood to be nearly irrelevant and the developers largely ignored those data [90]. Efficacy could only be ascertained against clinical relapse in humans. Knowing the 8-aminoquinolines to be uniquely effective against clinical relapse as a class, and relatively toxic, the screening became a search for the least toxic 8-aminoquinoline without a means of screening for efficacy before evaluation in clinical trials. By almost any contemporary standard of science, this search for a replacement therapy for pamaquine may be characterized as incomplete and barely adequate. The product thus discovered reflected these qualities.

The superior candidate drug to emerge from this program was primaquine. It exhibited a slightly higher therapeutic index than pamaquine, and its anticipated toxicity (acute intravascular hemolysis) in subjects with G6PD deficiency, discovery of which occurred in that clinical development program, was confirmed [96]. As the Korean War unfolded during the early 1950s primaquine saw wide application with good therapeutic results. Chloroquine had been registered as therapy of choice against acute attacks of vivax malaria in 1946 [97]. These two companion drugs thus became therapy of choice for radical cure (meaning complete cure) against vivax malaria, and it remains today as the only such therapy with that licensed therapeutic indication.

As the American war in Vietnam loomed in the early 1960s, further clinical work on chloroquine and primaquine provided products useful to that military effort. The Americans developed methods of treatment and prophylaxis using these drugs that would not threaten the safety of its many G6PD deficient troops (the abnormality did not excuse them from eligibility for conscription and the US Army chose to not screen against it). They discovered that the most common African variant of G6PD deficiency (A-) could safely tolerate weekly doses of the drug over 8 weeks, as opposed to the relatively threatening daily dosing of 14 days. Remarkably, the efficacy of primaquine against relapse appeared to rest wholly upon total dose with almost no regard to schedule of that dosing; the same total dose of primaquine against *Plasmodium cynomolgi* in rhesus macaques – a superb model of vivax relapse – showed equally good efficacy when administered as a single dose, or multiple doses over weeks [98]. US military medicine thus delivered the "C-P" pill for protection of their troops against both acute attacks (with a weekly dose of 300 mg chloroquine) and relapse (with a weekly dose of 45 mg primaquine in the same tablet). As an added therapeutic bonus, it was shown that single 45 mg doses of primaquine in the company of chloroquine effectively killed the infectivity of the sexual stages responsible for malaria transmission [99, 100].

Thus, in the mid-1960s, corresponding with the nadir of global malaria in the wake of the GMEC, Northern S&T effectively abandoned malaria chemotherapy as a mission accomplished. That mission may be appreciated as having been driven by the threat of malaria to troops engaged in warfare in the tropics. The technically flawed and partial explorations that had been conducted nonetheless provided products suited to those Northern requirements and further or better explorations seemed pointless. The South would live with those chemotherapeutic solutions for the coming decades, and still does.

Chemotherapeutics Limbo: 1965–1995

The malaria chemotherapeutics status quo in 1965 and the reluctance of S&T in the North over the ensuing 30 years to effectively engage the problem leaves us where we stand today – poorly equipped to deal with the malaria problem of the South in a physical and strategic sense. An examination of the management of the highest chemotherapeutic priority – treating the acute attack – during this era, and our response to it, provides useful historic insights.

As has been described, the Northern R&D engine delivered chloroquine and primaquine around 1950 as the primary workhorse drugs for global malaria. Resistance to chloroquine by falciparum malaria emerged in the late 1950s and early 1960s. By the 1980s the problem had consolidated globally and new therapies trickled out of the few Northern laboratories engaged in chemotherapeutic discovery and development. Alternatively, older and less ideal drugs developed during the effort of the 1940s and 1950s were dusted off and put to work, for example, sufladoxine-pyrimethamine, proguanil, and others. However, resistance to those therapies emerged with astonishing speed and thoroughness [101]. Quinine combined with antibiotics like tetracycline or doxycycline then, in that crisis setting, found application against chloroquine-resistant falciparum malaria. Oral quinine, however, may be one of the most impractical therapies: it requires three doses daily for 7 days and causes a wide range of very unpleasant side effects; a syndrome called cinchonism that often deranges compliance to therapy.

Very few genuinely new therapies emerged during that era, most of them emanating from the US Army program at the Walter Reed Army Institute of Research (WRAIR) [102]. That program developed and fielded mefloquine (Lariam™) by the mid-1980s and it found broad application for chemoprophylaxis. However, patent protection, perceptions of its poor tolerability in many people, along with the emergence of resistance to it in areas of Southeast Asia, limited its broader application in almost all endemic zones. More recently, however, it has been partnered with an artemisinin, especially in the Mekong region, but not from GMP manufacturers. This exposes the drug to risk of poor manufacture and counterfeiting, and the packaging as separate pills risks poor compliance and onset of resistance to both artemisinin and mefloquine components. One co-formulated artesunate-mefloquine therapy has been registered in Brazil and other Latin American countries through the Drugs for Neglected Diseases Initiative. Nonetheless, mefloquine may be considered a minor contribution to the toolbox against global malaria.

The other genuinely new therapy to emerge at the tail end of the same era (and with roots in WRAIR), atovaquone-proguanil (Malarone™) followed a similar path to global malaria irrelevance. Glaxo-Wellcome (later to become GlaxoSmithKline (GSK)) took up the cause of developing this drug for the prevention and treatment of malaria during the early 1990s. Their rapid success in registering it for these indications reflects both corporate determination and skill, and superior chemotherapeutic properties: very safe and well tolerated, highly effective against acute attacks of all the malarias (including multidrug resistant falciparum malaria), co-formulated, and available as a GMP product. The single known serious problem with this drug at the time, its relatively high cost, seemed to exclude it from the global malaria toolbox. GSK, however, generously aimed to see this extraordinary therapy put to work. In cooperation with various government and nongovernment partners, they created the Malarone Donation Program with the expressed aim of providing a million treatments with this drug annually at no cost to the end users [103]. However, flaws in this strategy would undo that program.

Evidence emerged that a single point mutation in *P. falciparum* caused it to be completely and irreversibly resistant to atovoquone + proguanil [104]. Malariologists worried that its broad application would quickly result in its loss, and perhaps other important drugs via cross-resistance [105]. The donation program coped with this potential danger by setting down very strict criteria for its clinical use. Consequently, at pilot sites in Kenya, only 0.68% of malaria patients received the treatment and that minority were among the least likely to suffer a fatal outcome [106]. Hence, mortality rates among the vastly excluded majority were not affected by the intervention. African authors of that "lessons learned" report [106] explained that new drugs were not the imperative; poor access to any drug drove transmission and high mortality. The complex social and political process that drove the donation program to pilot implementation in Africa excluded African voices wholly at first and later invited them but suppressed their dissenting views [107]. The Northern voice and conviction that this new tool required implementation in the South, along with their fiscal authority and influence, dominated the strategic consensus. Despite valiant and conspicuously generous efforts by GSK to contribute to the global malaria problem with this drug, the drug plays a very minor role in malaria treatment and control in endemic zones.

Northern S&T between 1965 and 1995 effectively failed to yield a solution for the highest chemotherapeutic imperative in endemic zones: treatment of the acute attack. At Southern clinics around the globe in the 1990s and well into the 2000s patients were still being treated with drugs invented half a century earlier. These drugs were less safe and effective than chloroquine in the heyday of its efficacy, and a cost in human lives was rapidly accumulating [108]. This is when the contemporary efforts to field ACTs as front-line therapies, already described, at last took root and continue today. Treating the acute attack, one of several chemotherapeutic challenges with malaria, may be far less an S&T challenge than finding it, diagnosing it, and reporting it. Assuming that the diagnostics S&T challenge is met, have the chemotherapeutic tools that are needed to capitalize on the accurate identification of the infected person been assembled? Northern R&D obsession with the acute attack unfortunately comes with neglect of the other possibly useful drugs for managing endemic malaria at that critical time/place interface when the parasites meet the range of chemotherapeutic instruments aimed at them.

Chemotherapeutics Neglect

The chemotherapy of malaria consists of five distinct compartments, treatment of acute attack (using blood schizontocidal agents) being one. The other compartments include: drugs aimed against the dormant stages in liver, called hypnozoites, responsible for relapse (using hypnozoitocides); drugs aimed against active stages in the liver, called tissue schizonts, that lead to acute attacks (using tissue schizontocides); drugs affecting the sexual stages in blood, called gametocytes, responsible for infection of biting mosquitoes and transmission of malaria (using gametocytocides); and drugs that prevent the development of parasites in the mosquito gut (called sporontocides). Figure 11.3 illustrates these compartments.

The blood schizontocidal antimalarials constitute the vast majority of antimalarial drugs in use today. These drugs, with few exceptions and when taken as directed, exert therapeutic activity only in that compartment. A few exhibit minor or limited activity elsewhere, for example, atovaquone may act as a tissue schizontocide against falciparum but not vivax malaria; chloroquine kills the gametoctyes of

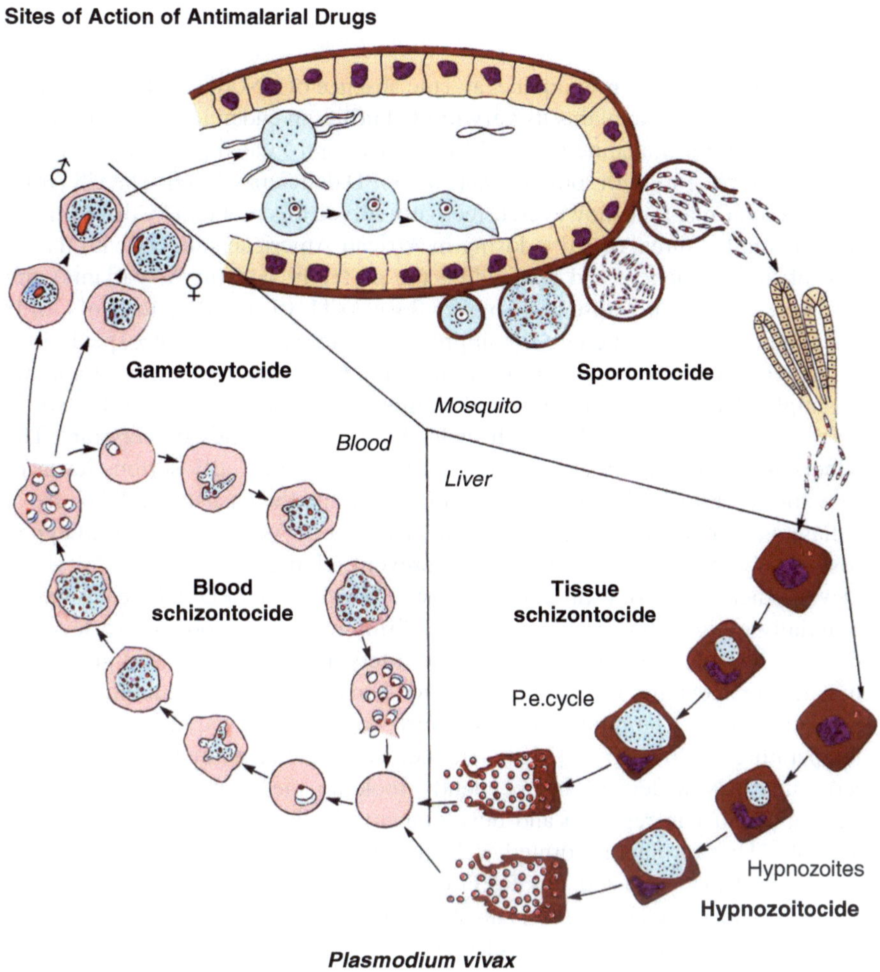

Fig. 11.3 Life cycle of *P. vivax* and chemotherapeutic compartments. The life cycle of *P. falciparum* would be essentially similar to this with the lone exception of the hypnozoite shunt in the cycle, which that parasite lacks (Taken from reference [53] with the permission of the American Society of Microbiology. Figure by Prof. Wallace Peters with artwork by Ms. Andrea Darlow)

vivax but not falciparum malaria; and the artemisinins will kill gametocytes of falciparum malaria but not if already mature. One drug in clinical use today remarkably exhibits therapeutic activity in all of these compartments. That drug is primaquine, and despite its extraordinary dexterity in killing a wide range of morphologically and physiologically distinct forms of parasites (including across species), it is a tool of very limited utility in the global malaria toolbox. This is a consequence of both failing to tame its sometimes dangerous side effects and to develop practical means of gauging its therapeutic efficacies.

G6PD Deficiency

The hemolytic toxicity of primaquine, and the absence of a Northern R&D response to the problem, largely explains its very limited utility in endemic zones. Sustained daily dosing with primaquine (15 or 30 mg daily for 14 days) poses a potentially lethal threat to patients with some variants of G6PD deficiency [109]. The US Army developers of primaquine, however, effectively solved this problem by discovery of both the nonlethal threat of daily dosing in African-American soldiers with the A- variant of the deficiency, and by fielding the perfectly safe (in A- variants) and efficacious single weekly dose of 45 mg for 8 weeks [110]. The diagnosis of G6PD deficiency requires a cold chain for shipping and storage of reagents, specialized skills and equipment: micropipettors, water bath, test tubes, and ultraviolet lamp, for example. This poses little technical challenge to care providers in the North, but in the endemic zones of the South such laboratory capacities are very rare. The providers who most often see malaria must chance potentially serious harm or choose not to prescribe the drug. G6PD deficiency and the unsustainable nature of that diagnosis in endemic zones (as it has been for 50 years) render primaquine almost wholly ineffective as a therapy to prevent relapse.

Development of a point-of-care diagnostic device for G6PD deficiency would substantially mitigate this problem. The US Army recently supported the development of such a tool in order to improve the safety and effectiveness of an experimental replacement for primaquine, another hemolytic 8-aminoquinoline called Tafenoquine [111]. The product of that effort, however, can only be used reliably at temperatures below 25°C [112]. Another commercial company in the USA (AccessBio©, New Jersey) independently took up the point-of-care challenge and has a product in research and development that is simple, inexpensive, and heat-stable. That kit has performed well in initial field evaluations in Cambodia [113]. AccessBio© initiated this R&D effort with no support or guidance from mainstream malaria R&D funders, and WHO provided the funding for that evaluation (JKB is an investigator on that effort).

Primaquine suited Northern settings and requirements, and adapting it to safe use in Southern settings has never been an R&D imperative. Among the many dozens of G6PD deficiency variants, the primaquine sensitivity phenotype has been characterized in only three geographic settings (African A-, Mediterranean B-, and, only recently, Mahidol from Thailand). These variants represent mild, severe, and moderate primaquine sensitivity phenotypes, respectively. Although residual G6PD enzyme activity has been characterized in most variants, its use as a surrogate for the primaquine sensitivity phenotype is a supposition not supported with clinical evidence. WHO has long recommended a single 45 mg dose without screening for G6PD deficiency, but largely upon the (inadequate) safety evidence from the USA that involved healthy African-American volunteers having the mild A- variant of this disorder [94]. However, in a very limited series of studies in the USA and Italy, a single 45 mg dose caused acute intravascular hemolysis of 25% of red blood cells in otherwise healthy volunteers having the Mediterranean B- variant (severe primaquine sensitivity) [114, 115]. Safe dosing in one variant may not hold for others.

Gametocytocide

Poor understanding of primaquine sensitivity phenotypes creates very serious practical problems in malaria control. This may be most appreciated by the application of a single 45 mg dose of primaquine as a gametocytocide against falciparum malaria [116]. A 25% hemolysis in an acutely ill and already anemic patient may contribute significantly to onset of life-threatening complications. Remarkably, the safety of primaquine gametocytocidal therapy in patients with malaria, much less G6PD deficiency variants of any type, has not been systematically carried out. Limited recent studies from Africa suggest a significant threat to malaria patient safety [117, 118]. Countries aiming to apply the 45 mg dose of primaquine as a supplement to their control or elimination strategies must first demonstrate safety among the variants of G6PD deficiency occurring in their areas of operations. In 2011, the WHO undertook sponsorship of such studies in Cambodia using its own relatively meager R&D funding resources (JKB is an investigator on that effort).

The safety and efficacy of primaquine as a gametocytocide in falciparum malaria were evaluated with chloroquine therapy for the acute attack over 50 years ago [99, 100]. Chloroquine is no longer used in this capacity, and no clinical evaluation of the safety and efficacy of primaquine as a gametocytocide when administered with any ACT has occurred. The artemisinins exert activity against immature gametocytes and could possibly yield a lower and less threatening optimal dose of primaquine. Northern R&D, however, expresses no agenda for clinical research on primaquine for any therapeutic indication.

The most conspicuous neglect of the gametocytocidal compartment has been the failure to conduct a systematic chemical survey of compounds for this activity. Northern S&T obsession with the acute attack and neglect of other avenues to address the problem of endemic malaria may be most evident in this line of investigation. An examination of the fairly extensive body of work on gametocytocidal properties of chemotherapeutic agents reveals that almost all such evaluations have been conducted on drugs with proven therapeutic activity against the asexual blood stages causing acute malaria [119]. Killing gametocytes has apparently been thought of strictly as a therapeutic bonus to treatment of the acute attack. The possibility of a safe, effective, and perhaps already widely available drug at low cost providing a means to effectively arrest transmission has not been explored. Northern S&T could have, with relative technical ease, conducted such explorations over 35 years ago. A 2011 call for proposals for such work from BMGF represents a welcome sign of dawning realization of the importance and enormous promise of this approach.

Hypnozoitocide

The primary therapeutic application of primaquine, as a hypnozoitocide, offers further evidence of Northern R&D inability to respond effectively to Southern problems with critical tools in its chemotherapeutics toolbox. Although primaquine has been the only therapeutic agent against relapse for the past 60 years, almost no research of its mechanisms of activity against the hypnozoites has been carried out. There is no understanding of its therapeutic target, much less which of its many metabolites may be most active. Resistance to primaquine by hypnozoites almost certainly occurs, but there is no validated means of confirming that phenotype and complete ignorance of the corresponding genotypes. Simply observing responses to therapy does not adequately address that problem because parasitemia occurring weeks to months after treatment may represent either therapeutic failure (relapse despite therapy) or simply a new infection. No means of distinguishing those events by molecular genotyping has been established, largely due to the biological complexity imposed by the hypnozoite [120]. Unambiguous estimates of primaquine efficacy thus require the absence of risk of reinfection. Such studies have not been done since the development of primaquine in experimentally challenged American prisoners in the 1940s and 1950s [121]. Consequently, there is almost no grasp of the therapeutic efficacy of the only drug against relapse in vivax malaria anywhere it occurs. Primaquine may already have been lost to resistance and there is no validated means of evaluating that possibility.

Chemotherapeutics Performance Evaluation

Over the past six decades, through periods of ebbing and flowing fiscal support, Northern R&D has focused available energies and resources, and still does, on the acute attack of falciparum malaria. The almost complete neglect all other forms of chemotherapy currently in the toolbox of the Southern craftsman working the problem now imperils successful elimination goals. And we were warned as L.T. Coggeshall [122] wrote in 1952, "*The greatest misconceptions about the treatment of malaria, especially in the past, has arisen from the fact that too many considered it a single disease. Malaria is not a disease – it is a variety of diseases.*" There are therapies for one of the malarias and, effectively, almost none for the others. One key element of this neglect has been the failure to adapt *P. vivax* to continuous in vitro cultivation in Northern laboratories as occurred with *P. falciparum* in the 1970s [123]. Setting aside the improbability of adequate R&D investment in achieving that feat for *P. vivax*, this failure does not excuse Northern R&D from dealing with this parasite. While it is true the problem limits Northern laboratories from effectively engaging the problem in the North, transfer of the required R&D capacities southward stands as an obvious but inadequately funded solution. In this and other issues of chemotherapeutics neglect outlined here, it remains unclear whether the 3.1% share of malaria R&D spending on *P. vivax* [8], and perhaps even less on gametocytocidal therapies, is the cause or effect of their neglect. Either way, "inadequate" effectively describes Northern R&D performance in these vital pieces of the malaria control toolbox. In a chemotherapeutics sense, the Southern craftsman trying to disassemble endemic malaria may be thought of as equipped with a fixed-gauge wrench for nuts of many gauges.

The Vaccine Development Gambit

In 2011, a phase III clinical trial of a malaria vaccine called RTS, S in 15,640 children aged up to 17 months is underway at 11 study sites in Africa [124]. This bold and expensive effort reflects the enormous technical and fiscal challenges in malaria vaccine development. It also expresses the Northern determination to produce this particular tool despite the extraordinary gamble of resources and will.

The leading edge biotechnology-driven effort to develop vaccines against malaria has been an epic, contentious, and, thus far anticlimactic journey of 30 years [125]. The modern effort to engineer a protein subunit vaccine began in earnest and with a great deal of optimism in the early 1980s with emergent recombinant DNA technology swinging open a very promising door. Having genetically engineered bacteria to produce malaria proteins in kilogram quantities seemed a means of conquering malaria with relative ease. Instead, through a long series of preclinical and clinical R&D efforts around the globe, *P. falciparum* has consistently shrugged off the eloquently derived antibodies and T-cells arrayed against it [125–127].

The steepness of the R&D challenge in vaccination stems from the failure to adequately understand the enormously complex immune response provoked by malaria. That response varies terrifically across variables like age, history of exposure, intensity of exposure, and host and parasite genetics [128]. The dawn of the scientific understanding of immunity to malaria began in 1899 when Robert Koch visited Dutch colleagues at what would become the Eijkman Institute in Batavia on the island of Java in the East Indies. Koch noticed that densities and frequencies of parasitemia declined sharply with age at heavily endemic Ambarawa in central Java, whereas at much less endemic Sukabumi in western Java those values remained relatively level across ages. Koch surmised that heavy exposure to infection, over a period of many years, imbued protection against high levels of parasitemia and attendant disease (see citations in [128]). The same pattern has been consistently observed over the decades and naturally acquired immunity has been considered the cumulative product of chronic and heavy exposure to infection. Immunologists envision the human immune system "learning" the antigenic repertoire of the parasite by repeated exposures and eventually, at around adolescence, mastering it [129].

Likewise, vaccinologists viewed their challenge as one of "teaching" that immune response by exposure to the proper parasite antigen or set of antigens. Though very simply expressed, this characterizes the malaria vaccine development strategy. Accepting the supposition that very young children in heavily endemic Africa constitute the majority of deaths due to malaria, the expressed R&D priority of vaccinating them is sensible. This demographic group is indeed vulnerable, exceedingly so compared to their parents. Adult Africans in heavily endemic areas, though almost constantly infected, very rarely suffer even minor illness (unless pregnant, especially for the first time). The susceptibility of their children to anemia, cerebral malaria, pulmonary complications, and other life-threatening syndromes stands in stark contrast [130]. They seem to have not "learned" to tame this infection and the appropriately "instructive" vaccine would mitigate their losses to malaria. This simplistic concept suffers logical pitfalls, however, evident with closer examination of the dynamics at work between those very young hosts and the parasites invading them.

A single episode of malaria in a nonimmune traveler, left untreated for too long, very often threatens rapid devolution into a life-threatening event. African children, on the other hand, experience about six to eight new infections each year in many endemic zones. Over 95% of them survive this rigorous challenge and go on to become malaria-hardy adults. This survival, surely impacted by access to therapy, nutrition, genetics, and probably many other factors, must also reflect successful immunological management of the always-present threat of sliding into lethal complications [128]. The loss of some of these children to malaria begins to less resemble a failure to "learn" a specific immune response from its parasite instructor, but more like an intrinsic failure to mount the appropriate immune response that permitted survival among almost all others in their cohort. And yet, vaccine R&D remains focused on parasite antigens rather than the intrinsic basis for success versus failure in what amounts to a 95% efficacious vaccination by the parasite itself. There is almost no understanding of what constitutes an appropriate versus inappropriate immune response to natural infection and the determinants of that life versus death turning of events. This gap in understanding extends to responses to experimental vaccines.

The consequences of such poor understanding may be found embedded in the process of malaria vaccine development and it imposes extraordinary financial risk. Unlike drug development, where compounds may be rejected or pushed forward on the basis of unambiguous laboratory determinations of activity against the parasite, malaria vaccine developers possess no such tool of discernment. No single immune response or set of responses has been shown to reliably predict protection in the infected human at any age or history of exposure, much less specifically infants and small children chronically exposed to infection. Vaccine candidates advance (or do not) through development algorithms on the basis of rational suppositions about protection in humans. Immunogenicity in terms of specific humoral or cellular targets, for example, may be the basis of development decisions even though those responses may have no evidence-validated role or compelling and consistent correlation with achieved protection in humans. Phase I and II trials in healthy nonimmune European or American adults for vaccines aimed at African infants and effectively already immune children compounds this problem. These generally acknowledged pitfalls [8] effectively defer credible vaccine development decisions to phase II and III trials. The vaccine RTS,S, now in phase III trials, was invented in 1987, tested in humans in 1992, evaluated in Africans in 1995, and evaluated in small numbers of African children between 2004 and 2007. The inadequate understanding of immunity to malaria as it occurs in the real world imposes a vaccine development paradigm resembling a roulette wheel with each turn of it coming with decades of work and many millions of dollars. Surely only repeated success could sustain such a development paradigm, and yet it is one having 30 years of failure. What sustains it?

R&D Investment

Today, malaria vaccine proponents glare over their fiscal parapets at their antimalarial drug development competitors vying for the same large share of malaria R&D resources. Each points to the failings and shortcomings of the other in making their cases for funding before donors and sponsors. The vaccines camp usually, albeit narrowly, wins the resources contest (Fig. 11.1). The reasons for the consistency of this large investment, despite decades of frustration, perhaps reflect how we perceive the malaria problem and, arguably, national self-interest in S&T capacity, agility, and economic spin-off.

Malaria vaccine development has and continues to probe the very far leading edges of contemporary and potentially industrial biotechnological immunology [131]. Beginning in earnest in the 1980s, the best and brightest were generously resourced and engaged in a globally important scientific endeavor. Northern laboratories blossomed and hummed with activity [127]. The US and European biotechnology industries enjoyed many technological and economic spin-offs from the effort. One of the malaria vaccine biotechnology pioneers, Wayne T. Hockmeyer, retired from the US Army in the late 1980s and went directly on to found Molecular Vaccines Inc. (later renamed MedImmune), a biotechnology concern purchased by AstraZeneca in 2007 for $15.2 billion. The vaccines industry saw and participated in proofs-of-concept of novel approaches to vaccine development like DNA vaccines and arrays of sophisticated adjuvants. In short, over 30 years of malaria vaccine development produced many scientific, technological, and economic dividends without delivering, yet, a licensed vaccine.

This is the nature of science, where exploration leads to unexpected destinations. Simply having a technical problem and the wherewithal to address it is key to S&T success even without directly solving the primary problem. The nature of funding in malaria vaccine development illustrates the linkage between S&T capacity and national S&T self-interest. The laboratories that executed malaria vaccine development work were supported by the nations where the laboratories were located, with few minor exceptions. American-funded efforts went primarily to American laboratories, as was true of European efforts, and even one bold effort in Colombia [132] was funded by that government. What is likely at work is the irresistible drive to reap the peripheral benefits of S&T endeavors within borders. No nation in the endemic zones, save the valiant Colombian example, undertook vaccine discovery and development. That may be due to a lack of fiscal resources, standing technical capacity, and an available technologically endowed workforce. But those determinants themselves provide grim reflection of the unwillingness of Northern nations, collectively, to make substantial direct investment in S&T foundations in the South.

It may be argued that the North harvests S&T/R&D dividends derived from a Southern problem – and reasonably counter-argued that the investments are their own and, because of standing capacities, the less costly and speediest avenue to solving the problem. The Northern R&D posture would be supportable if it delivered the most useful tools, but it has not and does not seem strategically positioned to do so. Northern R&D engaging the malaria problem yields tremendous and important advances in S&T, but it historically exhibits very limited abilities in delivery of practical and useful tools that significantly impact malaria as a global health problem.

Sustainable S&T in Tropical Health

Northern R&D remains the technological fulcrum for the public health lever against Southern infections, and this may be an important contributor to stymied and perhaps unsustainable progress. Gordon Harrison wrote in his 1978 history of malaria [64] of the failure of the global eradication effort in the prior decade, "*Failure so universal, so apparently ineluctable, must be trying to tell us something. The lesson could be of course that we have proved incompetent warriors. It could also be that we have misconstrued the problem.*" The failure of Northern R&D agendas to field relevant tools against endemic Southern malaria must also be trying to tell us something. Bench malaria science of Northern S&T is a powerful engine of scientific discovery and human progress engaged in the fight against malaria, and the aim here is not to demean those earnest endeavors by compassionate and dedicated people. The aim is to spark reconsideration of reliance solely upon that engine for practical solutions.

Southern scientists may be criticized for having abandoned the field to Northern scientists, but this is unjust and self-serving of the critic. As the Malarone Donation Program already described illustrates, Northern bias, imperatives, and fiscal influence may displace Southern voice and initiative out of the strategic equation. This is where the Northern perception of a lack of Southern awareness of useful technologies plays in the unhelpful assumption of Northern leadership on decisions regarding tools intended for Southern use. In the Malarone example, Northern science perceived a powerful new drug available at no cost, whereas African stakeholders perceived a futile intervention not addressing the real problem. Southern scientists put up futile active and passive resistance to the program because the proposed intervention deflected alternatives and the prospect of progress. The report of Shretta et al. [110] documents this history and should be studied by Northern scientists engaging the malaria problem.

What the Southern scientists pointed to was not the drug, but the availability of diagnosis and treatment where most deadly malaria occurred [109, 110]. It may be that Northern S&T perceives such a problem as one it is not particularly suited to engage. In other words, the problem seems practical rather than technological, at least in how the North perceives its technology imperatives. The Northern perception may be that problems that do not engage the suite of Northern technology assets need not involve them. That may be, but their technological imperatives and financial resources tend to suffocate alternative approaches by Southern scientists co-opted into a Northern R&D agenda.

Malaria vaccines represent the primary Northern S&T answer, or at least hope, to solve lethal malaria at isolated sites all across the global endemic belt. Vaccination in lieu of very hard to sustain diagnosis and treatment services seems the ideal solution. Vaccines have pushed the burden of morbidity and mortality due to infections to the lowest point in all of human history, and such a track record is hard to argue against. However, one may forcefully argue that hope is a poor offensive strategy against entrenched malaria. Determination, will, resources, and faith in the success of vaccines may be defied by the unfolding of an unknowable future. Preparing for possible failure of vaccines neither endorses nor ensures it, and responsibly envisioning a future without vaccines prudently commands attention to alternative technological approaches. Putting the development of those approaches into Southern hands may be the surest means of success with most appropriate technologies providing sustainable solutions.

Southern Scientific and Technological Development

This entry has emphasized the problems with Southern reliance upon Northern technological innovation to diminish preventable endemic infectious disease burdens. A South empowered with indigenous S&T capacities would likely generate more effective instruments that more quickly lift those burdens. Such capacity would likewise, as it did in the North, spin off substantial economic benefits that indirectly drive endemic diseases down and health up. Achieving this would require deliberate action by those nations of the North capable of delivering such capacities, and the foresight to see very significant short-term losses recovered many times over in the longer term.

It may be argued, despite the MIM example already discussed, that the North already invests in the development of S&T capacities of the South. Each year many Southern scientists receive academic and technological training in the North with generous support from Northern sponsors in the government, academia, and private sectors. This approach, however, may be unlikely to yield real capacities in Southern S&T. Many of these students acquire skill sets that cannot be applied in their home nations. There are few institutions with the technological capacities of the Northern universities they attend. These students typically either remain in the North as valued technologically skilled immigrants, or they go home and assume positions that do not hone and deepen their technological abilities. Many of the latter subsequently relocate to developed nations selfishly promoting immigration policies designed specifically to lure them away [133]. The promise of advanced S&T training for Southern scientists in the North rings hollow without Southern R&D institutions where those skills may be put to effective work.

The key may be establishing S&T "nurseries" in the South having specific technological suites and focused objectives. In the example of malaria, few could argue against the potential direct and indirect benefits of an institution committed wholly to the development of medically useful insecticides or repellants, or another dedicated to a thorough chemical exploration of potential therapeutic agents aimed solely at sexual blood stages and arresting malaria transmission. The production of GMP malaria vaccines would nurture a commercial industrial biotechnological base capable of expansion to other vaccine-preventable illnesses and pharmaceuticals. S&T sustainability successes standing today in the South offer examples.

The Eijkman Institute for Molecular Biology in Jakarta, Indonesia is named after the Dutch scientist who discovered vitamin B as the cause and cure of beriberi with work executed largely in that city between 1886 and 1925. Christiaan Eijkman was awarded the Nobel Prize in 1929 for the discovery of vitamins [134]. During this heyday, the institute employed both Dutch and Indonesian scientists functioning as full peers heading their own avenues of biomedical research of tropical infections. The war in the Pacific saw the Dutch scientists arrested and imprisoned. The Indonesian scientists carried on with the work of the institute largely without direct interference from the Japanese forces. Tragically, however, an occupier suspicious of intellectual elites executed many Indonesian doctors and medical scientists [135]. The director of the Eijkman Institute in 1944, Achmad Mochtar, offered a false confession of the murder of 900 Indonesian slave laborers by sabotage of a vaccine administered to them. He did this in exchange for the liberty of almost the entire Indonesian technical staff of the Eijkman Institute being held in prison and brutally tortured. In July 1945, the occupiers executed Mochtar. This event, and the bloody war for independence from the Netherlands that immediately followed, effectively destroyed the institute and it formally closed in 1965 [136].

In 1990, the dynamic then Minister of Science and Technology (later to become the third President of Indonesia), B.J. Habibie, devoted considerable government resources and his formidable force of personality to reopen the institute as a nursery for biotechnological innovation. He segregated it from the ministerial hierarchies in order to protect it from the political vagaries of the Indonesian federal system. The laboratory became a semiautonomous institutional appendage to his ministry. Habibie, himself a product of the European Airbus Industrie S&T engine, sought an Indonesian scientist immersed, tried, and tested in the culture of Northern S&T. He found and recruited Prof. Sangkot Marzuki, then a tenured and highly productive cellular and molecular biologist at Monash University in Australia. The fledgling laboratory in Jakarta received generous support from the government until the economic collapse and political turmoil of 1998. The laboratory nonetheless survived by its own scientific and technological wits, becoming a largely self-sustaining research institution through the conduct of work paid for by competitive, merit-driven grants. Their focus on molecular biology endows them today with sophisticated gene sequencing and analysis capacities. Those are being leveraged to attract funding for specific work on discrete pathogens endemic to Indonesia, in addition to a variety of human genetic, wildlife conservation, and even criminal law enforcement molecular biological problems.

The primary point with this example is that young Indonesian scientists, whether trained at home or abroad, have a stable institution where they may sharpen their skills with ever-steeper technological challenges under the mentorship of both senior Indonesian scientists and their foreign partners. The scientists in the institute are not government employees and their salaries are thus neither guaranteed nor constrained. They are able to commit fully to the S&T mission, with the certain understanding that secure employment demands scientific productivity and overcoming stiff competition from other labs engaged in similar work around the world. More importantly, consistent success provides a career of advancement in abilities, responsibilities, and compensation reflective of such success.

Almost all scientists today manage their professional lives in this way, no matter where they happen to live. Doing science at the leading edge is a hard-earned privilege reserved for those able to demonstrate the ability to contribute to the furthering of that edge. Instilling this understanding, and providing places where sustained success is possible and deliberately nurtured, substantially furthers the S&T capacity of the Republic of Indonesia as a whole. It is not the S&T per se that makes this possible; it is the institute, its foundations, and its culture of uncompromising scientific excellence. B.J. Habibie and the leadership of the Eijkman Institute understood this and achieved it.

That institute is certainly not the only example of significant progress in the deliberate long-term commitment to the systematic nurturing of Southern S&T capacities. The Pasteur Institutes around the world provide excellent nurseries for local S&T, as do the laboratories funded by the Wellcome Trust in Kenya, Tanzania, Thailand, and Vietnam. Indeed, the former Pasteur Institute (founded in 111) at Bandung, West Java, created most of the vaccines used in Southeast Asia prior to the Second World War, and the foundations of that investment remain today in the same physical location occupied by the government-owned industry that produces most of Indonesia's vaccines. Not coincidentally, Indonesia's vibrant and lucrative pharmaceutical manufacturing industries surround that facility. Bringing S&T capacities southward obviously has and can provide sustainable S&T and R&D enterprises and their biotechnological industrial spin-offs.

The North must overcome its impatience and evidently misplaced self-confidence in its ability to effectively engage Southern health problems. S&T is the key to successfully resolving those burdens, but probably not as practiced in the North. Supplanting Southern S&T into the leadership role of R&D aimed at its infectious disease burdens should be deliberately planned, financed, and executed.

Future Directions

The certain failures and likely strategic missteps by Northern S&T put forth in this entry aim at dispassionate and rational reflection on the means of addressing Southern health problems in the S&T/R&D arena. Accepting Northern R&D agendas as poorly suited to the task, despite the humanitarian urgency, brings appropriate and useful focus to mitigating deficiencies in Southern R&D capacities. Resisting the fiscal and temporal expediencies that push R&D capacities and agendas northward may prove the surest means of sustainable and substantial gains in tropical health. Placing durable

R&D capacities in the South as a primary rather than tangential benefit of fruitful North-South S&T collaboration generally characterizes a direction aiming at that envisioned future.

Southern scientists have much to learn from their Northern counterparts, especially those in positions of institutional leadership of R&D endeavors. The intensely competitive leading edge of S&T requires institutional foundations that create and reward healthy scientific competitiveness. Northern R&D laboratories that fail to compete quickly vanish, and the cruel Darwinian-like forces that shape effective Northern R&D must also be brought to bear on Southern R&D enterprises – their vitality and sustainability requires it. Transfer of those R&D survival skills is perhaps where Northern S&T may be put to most effective use in the South. And the task of transfer should not be viewed as passive, as by example, but direct and deliberate participation in the building of those skills and capacities in Southern R&D enterprises. Acknowledging limited Southern S&T capacities as the core impediment to improving Southern health, rather than possible or hoped for tools emanating from Northern S&T, naturally leads in that direction.

Such a course requires strategic thinking, long-term commitment, flexibility, and inspired international cooperation and trust from both North and South partners undertaking it. Northern investment ought to aim at institutional R&D competitiveness rather than the R&D agenda itself. Logically, however, this inevitably leads to establishing successful R&D endeavors of a specific goal-oriented nature. Scientific productivity would be engrained in any rational North-South institutional R&D capacity-building cooperative endeavor. In short, success in the core institutional capacity-building objective necessarily carries with it R&D success and rewards for both partners. An effective win-win formula ensures their mutual R&D survival, and therefore commitment to the endeavor. Perhaps most importantly, the Northern partner would not be bringing to the partnership a predetermined R&D agenda. That would be decided between the partners and the sponsor of their endeavor.

That third party in this envisioned collaboration, the sponsor, represents the primary challenge in realizing it. At least in the realm of malaria R&D today, that sponsor is most likely to be an American funding agency with what it views as clear product- or knowledge-development paths and technical milestones to accomplish in achieving success, that is, fielding a product that it decided would be useful. The Northern obsession with the development of products of its conception for Southern health progress should be acknowledged as a significant impediment and tempered if not abandoned.

The assumption of Southern R&D leadership on Southern health issues would require many years of determined will and real reform on both sides of the Tropic of Cancer to achieve and sustain. It may require pragmatic lessening of the faith Northern R&D exhibits in its ability to deliver solutions in the short term. Southern awareness of the intensely competitive nature of biomedical R&D should rise to a level consistent with firm commitment to that very difficult course. Southern R&D enterprises must not depend wholly upon funding from noncompetitive sources, and gauge their own relevance upon the objective evidence of competitive survival.

Southern public and private donors also need to participate and engage R&D endeavors managed by Southern institutions. Northern dominance of the R&D agenda in part stems from the reluctance of Southern donors to make these investments and thus bring that voice and influence to discussion and consensus. Their absence becomes ever more conspicuous with the booming economies of East, South East, and South Asia in comparison to the now troubled Northern economies. Establishing stable and sustainable R&D enterprises in the South may imbue such donors with the confidence and even self-interest to make such investments. Northern investments in Southern R&D should at least be matched by Southern governments and donors positioned to contribute. The South desperately requires humanitarian visionaries and champions, as in the models of Sir Henry Wellcome, Bill and Melinda Gates, Warren Buffett, and others in the North, to generously and intelligently unleash available capital upon very serious tropical health challenges.

North-South S&T/R&D cooperation on tropical health, after all, aims at lifting the mingled burdens of endemic diseases and poverty in the South. Northeast Asia largely achieved this feat over the past 50 years and today is the mercantile engine of the global economy. Brazil in South America has made large strides in health and trade and with it drives a substantial portion of the North American economy. Healthy people liberated from the pursuit of survival assume the pursuit of happiness that drives modern economies. They become educated, economically creative, and productive, selling to each other the accoutrements of easier and more fulfilling lives. Acknowledging North-South S&T/R&D parity as a key element in such progress moves far broader development agendas rapidly forward. Success in these endeavors may ultimately remove the conceptual utility of dividing the world into these two disparate spheres.

Acknowledgments JKB is supported by the Wellcome Trust grant #B9RJIXO. Prof. Jeremy Farrar, Dr. Simon Hay, and Dr. Trevor Jones provided helpful reviews of the manuscript.

Bibliography

Primary Literature

1. Galama T, Hosek J (2008) U.S. competitiveness in science and technology. RAND Corporation, Santa Monica
2. Committee on Ensuring the Best Presidential and Federal Advisory Committee Science and Technology Appointments, National Academy of Sciences, National Academy of Engineering, Institute of Medicine (2005) Science and technology in the national interest: ensuring the best presidential and federal advisory committee science and technology appointments. The National Academies Press, Washington, DC, p 208
3. Lu C, Schneider MT, Gubbins P, Leach-Kemon K, Jamison D, Murray CJL (2010) Public financing of health in developing countries: a cross-national systematic analysis. Lancet 375:1375–1387

4. Sachs J, Malaney P (2002) The economic and social burden of malaria. Nature 415:680–685
5. WHO (2010) World malaria report 2010. WHO, Geneva
6. Hay SI, Okiro EA, Gething PW et al (2010) Estimating the global clinical burden of *Plasmodium falciparum* malaria in 2007. PLoS Med 7:e1000290
7. Gething PW, Smith DL, Patil AP, Tatem AJ, Snow RW, Hay SI (2010) Climate change and the global malaria recession. Nature 465:342–345
8. PATH (2011) Staying the course? Malaria research and development in a time of economic uncertainty. PATH, Seattle
9. Hay SI, Guerra CA, Tatem AJ, Noor AM, Snow RW (2004) The global distribution and population at risk of malaria: past, present, and future. Lancet Inf Dis 4:327–336
10. NIH (2008) NIAID agenda for malaria, Bethesda, Maryland
11. WHO (2008) Global Malaria Action Plan, Geneva
12. Nosten F, Luxemburger C, ter Kuile FO, Woodrow C, Eh JP, Chonsuphajaisddhi T, White NJ (1994) Treatment of multi-drug resistant *Plasmodium falciparum* with a 3-day artesunate-mefloquine combination. J Infect Dis 170:971–977
13. WHO (2008) World malaria report. WHO, Geneva
14. Snow RW, Guerra CA, Noor AM, Myint HY, Hay SI (2005) The global distribution of clinical episodes of *Plasmodium falciparum* malaria. Nature 434:214–217
15. WHO (2009) World malaria report. WHO, Geneva
16. Hay SI, Okira EA, Gething PW, Patil AP, Tatem AJ, Guerra CA, Snow RW (2010) Estimating the global clinical burden of *Plasmodium falciparum* malaria in 2007. PLoS Med 6:e1000290
17. Elyazar I, Hay SI, Baird JK (2011) Malaria distribution, prevalence, drug resistance and control in Indonesia. Adv Parasitol 74:41–175
18. Harijanto PN (2010) Malaria treatment using artemisinin in Indonesia. Acta Med Indones 42:51–56
19. Dhingra N, Jha P, Sharma VP et al (2010) Adult and child malaria mortality in India. Lancet 376:1768–1774
20. Kochar DK, Saxena V, Singh N et al (2005) *Plasmodium vivax* malaria. Emerg Inf Dis 11:132–134
21. Kochar DK, Tanwar GS, Khatri PC et al (2010) Clinical features of children hospitalized with malaria - a study from Bikaner, Northwest India. Am J Trop Med Hyg 83:981–989
22. Srivastava S, Ahmad S, Shirazi N, Verma SK, Puri P (2011) Retrospective analysis of vivax malaria patients presenting to tertiary care referral centre of Uttarakhand. Acta Trop 117:82–85
23. Sharma A, Khanduri U (2009) How benign is benign tertian malaria? J Vector Borne Dis 46:141–144
24. Barcus MJ, Basri H, Picarima H et al (2007) Demographic risk factors for severe and fatal vivax and falciparum malaria among hospital admissions in northeastern Indonesian Papua. Am J Trop Med Hyg 77:984–991
25. Genton B, D'Acremont V, Rare L et al (2008) *Plasmodium vivax* and mixed infections are associated with severe malaria in children: a prospective cohort study from Papua New Guinea. PLoS Med 5:e127
26. Tjitra E, Anstey NM, Sugiarto P et al (2008) Multidrug-resistant *Plasmodium vivax* associated with severe and fatal malaria: a prospective study in Papua, Indonesia. PLoS Med 5:e128
27. Andrade BB, Reis-Filho A, Souza-Neto SM et al (2010) Severe *Plasmodium vivax* malaria exhibits marked inflammatory imbalance. Malar J 9:13
28. Beg MA, Khan R, Baig SM et al (2002) Cerebral involvement in benign tertian malaria. Am J Trop Med Hyg 67:230–232
29. Lomar AV, Vidal JE, Lomar FP et al (2005) Acute respiratory distress syndrome due to vivax malaria: case report and literature review. Braz J Infect Dis 9:425–430
30. Lawn SD, Krishna S, Jarvis JN, Joet T, Macallan DC (2003) Case reports: pernicious complications of benign vivax malaria. Trans R Soc Trop Med Hyg 97:551–553

31. Spudick JM, Garcia LS, Graham DM, Haake DA (2005) Diagnostic and therapeutic pitfalls associated with primaquine-tolerant *Plasmodium vivax*. J Clin Microbiol 43:978–981
32. Sternberg GM (1884) Malaria and malarial diseases. Wood, New York, p 329
33. Kitchen SF (1949) Vivax malaria. In: Boyd MF (ed) Malariology, vol II. W.B. Saunders, Philadelphia
34. Ewing J (1902) Contribution to the pathological anatomy of malarial fever. J Exp Med 6:119–180
35. Bassat Q, Alonso P (2011) Defying malaria: fathoming severe *Plasmodium vivax* disease. Nat Med 17:48–49
36. Eldridge WW et al (1925) Treatment of paresis: results of inoculation with the organism of benign tertian malaria. JAMA 84:1097–1101
37. Ferraro A et al (1927) The malaria treatment of general paresis. J Nerv Ment Dis 65:225–239
38. Freeman W, Eldridge WW, Hall RW (1934) Malaria treatment of dementia paralytica: results in 205 cases after five to eleven years. South Med J 27:122–126
39. O'Leary PA, Welsh AL (1933) Treatment of neurosyphilis with malaria: observations on nine hundred and eighty-four cases in the last nine years. JAMA 101:498–501
40. Fong TCC (1937) A study of the mortality rate and complications following therapeutic malaria. South Med J 30:1084–1088
41. Krauss W (1932) Analysis of reports of 8,354 cases of IMPF-malaria. South Med J 27:537–541
42. Nicol WD (1932) A review of seven year's malarial therapy in general paralysis. J Ment Sci 78:843–866
43. James SP, Nicol WD, Shute PG (1936) Clinical and parasitological observations on induced malaria. Proc R Soc Med 29:27–42
44. Verhave JP (2010) personal communication
45. James SP, Nicol WD, Shute PG (1932) A study of induced malignant tertian malaria. Proc R Soc Med 25:37–52
46. Handayani S, Chiu DT, Tjitra E et al (2009) High deformability of *Plasmodium vivax*-infected red blood cells under microfluidic conditions. J Infect Dis 199:445–450
47. O'Donnell J, Goldman JM, Wagner K et al (1998) Donor-derived *Plasmodium vivax* infection following volunteer unrelated bone marrow transplantation. Bone Marrow Transplant 21:313–314
48. Finkel M (2007) Bedlam in the blood: malaria. National Geographic, July 2007, pp 32–67
49. UN Millennium Project (2005) Coming to grips with malaria in the new millennium. Earthscan, London, p 129
50. US CDC (2011) Grand rounds: the opportunity for and challenges to malaria eradication. MMWR 60:476–480
51. Guerra CA, Howes RE, Patil AP et al (2010) The international limits and population at risk of *Plasmodium vivax* malaria in 2009. PLoS Negl Trop Dis 8:e774
52. Baird JK (2004) Resistance to chloroquine by *Plasmodium vivax*. Antimicrob Agents Chemother 48:4075–4083
53. Baird JK (2009) Resistance to therapies for *Plasmodium vivax*. Clin Microbiol Rev 22:508–534
54. Baird JK (2005) Effectiveness of antimalarial drugs. N Engl J Med 352:1565–1577
55. Baird JK (2011) Resistance to chloroquine unhinges vivax malaria therapeutics. Antimicrob Agents Chemother 55:1827–1830
56. WHO (2009) Rapid diagnostic test performance: results of WHO product testing of malaria RDTs: round 1 (2008). WHO, Geneva
57. Epstein J, Rao S, Williams F et al (2007) Safety and clinical outcome of experimental challenge of human volunteers with *Plasmodium falciparum*-infected mosquitoes: an update. J Infect Dis 196:145–154
58. Syafruddin D, Krisin, Asih P et al (2009) Seasonal prevalence of malaria in West Sumba district, Indonesia. Malar J 8:8

59. Harris I, Sharrock WW, Bain LM et al (2010) A large proportion of asymptomatic Plasmodium infections with low and sub-microscopic parasite densities in the low transmission setting of Temotu Province, Solomon Islands: challenges for malaria diagnostics in an elimination setting. Malar J 9:254
60. WHO (2011) Consideration of mass drug administration for the containment of artemisinin resistant malaria in the Greater Mekong Sub-Region. Report of a consensus meeting, 27–28 Sept 2010, Geneva
61. Snow RW, Guerra CA, Mutheu JJ, Hay SI (2008) International funding for malaria control in relation to populations at risk of stable *Plasmodium falciparum* transmission. PLoS Med 5:e142
62. Price RN, Tjitra E, Guerra CA et al (2007) Vivax malaria: neglected and not benign. Am J Trop Med Hyg 77(suppl 6):79–87
63. Baird JK (2007) Neglect of vivax malaria. Parasitol Today 23:533–539
64. Harrison G (1978) Mosquitoes, malaria and man: a history of the hostilities since 1880. John Murray, London, p 314
65. McCullough D (1978) The path between the seas: the creation of the Panama Canal 1870–1914. Simon & Schuster, New York, p 698
66. Takken W, Snellen WB, Verhave JP, Knols BGJ, Atmoswedjono S (1990) Envirnmental measures for malaria control in Indonesia – an historical review on species sanitation, Wageningen Agricultural University Papers 90-7, p 167
67. Killeen GF (2003) Following Soper's footsteps: northeast Brazil 63 years after eradication of *Anopheles gambiae*. Lancet Infect Dis 3:663–666
68. Coatney GR (1963) Pitfalls in a discovery: the chronicle of chloroquine. Am J Trop Med Hyg 12:121–128
69. Sharma VP (1996) Re-emergence of malaria in India. Indian J Med Res 103:26–45
70. Baird JK (2000) Resurgent malaria at the millennium: control strategies in crisis. Drugs 59:719–743
71. Gianessi LP, Puffer CA (1992) Reregistration of minor pesticides: some observations and implications. In: USDA (ed) Inputs situation and outlook report. USDA, Washington, DC, pp 52–60
72. Packard RM (1998) 'No other logical choice': global malaria eradication and the politics of international health in the post-war era. Parassitologica 40:217–229
73. WHO (2006) Pesticides and their application for the control of vectors and pests of public health importance, 6th edn. Department of Control of Neglected Tropical Diseases, WHO Pesticide evaluation scheme (WHOPES), Geneva, p 125
74. Roberts DR, Manguin S, Mouchet J (2000) DDT house spraying and re-emerging malaria. Lancet 356:330–332
75. Roberts DR, Laughlin LL, Hsheih P, Legters LJ (1997) DDT, global strategies, and a malaria control crisis in South America. Emerg Inf Dis 3:295–302
76. Attaran A, Maharaj R (2000) Ethical debate: doctoring malaria badly; the global campaign to ban DDT. BMJ 321:1403–1405
77. WHO (2007) Position statement: the use of DDT in malaria vector control. Global Malaria Program, Geneva
78. N'Guessan R, Boko P, Odjo A, Chabi J, Akogbeto M, Rowland M (2010) Control of pyrethroid and DDT-resistant *Anopheles gambiae* by application of indoor residual spraying or mosquito nets treated with a long-lasting organophosphate insecticide, chlorpyriphos-methyl. Malar J 9:44
79. Protopopoff N, Verhaeghen K, Van Borel W et al (2008) A significant increase in kdr in *Anopheles gambiae* is associated with an intensive vector control intervention in Burundi highlands. Trop Med Int Health 13:1479–1487
80. Czeher C, Labbo R, Arzika I, Duchemin JB (2008) Evidence of increasing Leu-Phe knock-down resistance mutation in *Anopheles gambiae* from Niger following a nationwide long-lasting insecticide-treated nets implementation. Malar J 7:189

81. Grieco JP, Achee NL, Chareonviriyaphap T et al (2007) A new classification system for the actions of IRS chemicals traditionally used for malaria control. PLoS One 2(8):156–167
82. The Global Fund to Fight AIDS, Tuberculosis and Malaria. Global Fund Policy on Quality Assurance for Pharmaceutical Products: procurement of single and limited source pharmaceuticals. Tenth Board Meeting, Geneva, 21–22 Apr 2005, GF-B10-9, Annex 4, p 47
83. Carrara VI, Zwang J, Ashley EA et al (2009) Changes in the treatment responses to artesunate-mefloquine on the northwestern border of Thailand during 13 years of continuous deployment. PLoS One 4:e4551
84. Dundorp AM, Newton PN, Mayxay M et al (2004) Fake antimalarials in Southeast Asia are a major impediment to malaria control: multinational cross-sectional survey on the prevalence of fake antimalarials. Trop Med Int Health 9:1241–1246
85. White NJ (2010) Artemisinin resistance – the clock is ticking. Lancet 376:2051–2052
86. Field JW (1938) Notes on the chemotherapy of malaria. Bulletin from the Institute for Medical Research, Federated Malay States 2:63–81
87. Fiammetta R (2003) The miraculous fever tree: malaria and the quest for a cure that changed the world. Harper Collins, New York
88. Joy RJT (1999) Malaria in American troops in the South and Southwest Pacific in World War II. Med Hist 43:192–207
89. Downs WG, Harper PA, Lisansky ET (1947) Malaria and other insect-borne diseases in the South Pacific Campaign, 1942–1945. II. Epidemiology of insect-borne diseases in Army troops. Am J Trop Med 27:69–89
90. Shannon JA (1946) Chemotherapy of malaria. Bull NY Acad Med 22:345–357
91. Slater LB (2009) War and disease: biomedical research on malaria in the 20th century (Critical issues in health and medicine). Rutgers University Press, New Brunswick, p 272
92. Brosius OT (1927) Section II. Plasmochin in malaria. In: Sixteenth Annual Report. Medical Department, United Fruit Company, Boston, pp 26–81
93. Board for the Coordination of Malaria Studies (1943) Results of study of plasmochin toxicity: cases occurring in large scale use of the drug for suppression and treatment of malaria. Malaria Report No.290. Archive of the National Academy of Sciences, Washington, DC
94. Board for the Coordination of Malaria Studies (1943) Study of the curative action of plasmochin in vivax malaria: studies in Army and Navy installations. Malaria Report No. 358. Archive of the National Academy of Sciences, Washington, DC
95. Office of the Surgeon General (1943) The drug treatment of malaria, suppressive and clinical. Circular letter No. 153. JAMA 123:205–208
96. Carson PE, Flanagan CL, Ickes CE, Alving AS (1956) Enzymatic deficiency in primaquine-sensitive erythrocytes. Science 14:103–139
97. Most H et al (1946) Chloroquine for treatment of aute attacks of vivax malaria. JAMA 131:963–967
98. Schmidt LH et al (1977) Radical cure of infections with *Plasmodium cynomolgi*: a function of total 8-aminoquinoline dose. Am J Trop Med Hyg 26:1116–1128
99. Burgess RW, Bray RS (1961) The effect of a single dose of primaquine on gametocytes, gametogony and sporogony of *Laverania falciparum*. Bull World Health Organ 24:451–456
100. Rieckmann KH, McNamara JV, Frischer H, Stockert TA, Carson PE, Powell RD (1968) Gametocytocidal and sporontocidal effects of primaquine and of sulfadiazine with pyrimethamine in a chloroquine-resistant strain of *Plasmodium falciparum*. Bull World Health Organ 38:625–632
101. Peters W (1987) Chemotherapy and drug resistance in malaria, 2nd edn. Academic, London, p 1100
102. Ockenhouse CF, Magill A, Smith D, Milhous W (2005) History of US military contributions to the study of malaria. Mil Med 170(Suppl 1):12–16
103. Oyediran AB, Heisler M (1999) Malarone donation program. J Travel Med 6(Suppl 1):28–30

104. Korsinczky M, Chen N, Kotecka B, Saul A, Rieckmann K, Chen Q (2000) Mutations in *Plasmodium falciparum* cytochrome b that are associated with atovaquone resistance are located at a putative drug-binding site. Antimicrob Agents Chemother 44:2100–2108
105. Ringwald P, Basco LK (1998) Malarone-donation programme in Africa. Lancet 351:673–674
106. Oyediran AB, Ddumba EM, Ochola SA, Lucas AO, Koporc K, Dowdle WR (2002) A private-public partnership for malaria control: lessons learned from the Malarone Donation Programme. Bull World Health Organ 80:817–821
107. Shretta R, Walt G, Brugha R, Snow RW (2001) A political analysis of corporate drug donation: the example of Malarone in Kenya. Health Policy Plan 16:161–170
108. Attaran A, Barnes KI, Curtis C et al (2004) WHO, the Global Fund, and medical malpractice in malaria treatment. Lancet 363:237–240
109. Cappelini MD, Fiorelli G (2008) Glucose-6-phosphate dehydrogenase deficiency. Lancet 371:64–74
110. Alving AS, Johnson CF, Tarlov AR, Brewer GJ, Kellermeyer RW, Carso PE (1960) Mitigation of the hemolytic effect of primaquine and enhancement of its action against exoerythrocytic forms of the Chesson strain of *Plasmodium vivax* by intermittent regimens of drug administration. Bull World Health Organ 22:621–631
111. Crockett M, Kain KC (2007) Tafenoquine: a promising new antimalarial agent. Expert Opin Investig Drugs 16:705–715
112. Tinley KE, Loughlin AM, Jepson A, Barnett ED (2010) Evaluation of a qualitative enzyme chromatographic test for glucose-6-phosphate dehydrogenase deficiency. Am J Trop Med Hyg 82:210–214
113. Khim S, Nguon C, Guillard B, Socheat D, Chy S, Sum S, Nhem S, Bouchier C, Tichit M, Christophel E, Taylor R, Baird JK, Menard D (2011) Performance of the CareStart G6PD deficiency screening test. PLoS One (in press)
114. Pannacciulli I, Tizianello A, Ajmar F, Salvidio E (1965) The course of experimentally induced hemolytic anemia in a primaquine-sensitive Caucasian: a case study. Blood 25:92–95
115. George GN, Sears DA, McCurdy PR, Conrad ME (1967) Primaquine sensitivity in Caucasians: hemolytic reactions induced by primaquine in G6PD deficient subjects. J Lab Clin Med 70:80–93
116. Baird JK, Surjadjaja S (2011) Consideration of ethics in primaquine therapy against malaria transmission. Trends Parasitol 27:11–16
117. Shekalaghe S, Drakely C, Gosling R et al (2007) Primaquine clears submicroscopic *Plasmodium falciparum* gametocytes that persist after treatment with sulphadoxine-pyrimethamine and artesunate. PLoS One 2:e1023
118. Shekalaghe SA, ter Braak R, Daou M et al (2010) In Tanzania, hemolysis after a single dose of primaquine coadministered with an artemisinin is not restricted to glucose-6-phosphate dehydrogenase deficient (G6PD A-) individuals. Antimicrob Agents Chemother 54:1762–1768
119. Butcher GA (1997) Antimalarial drugs and the mosquito transmission of Plasmodium. Int J Parasitol 27:975–987
120. Imwong M, Snounou G, Purkittayakamee S et al (2007) Relapses of *Plasmodium vivax* infection usually result from activation of heterologous hypnozoites. J Infect Dis 195:927–933
121. Baird JK, Hoffman SL (2004) Primaquine therapy for malaria. Clin Infect Dis 39:1659–1667
122. Coggeshall LT (1952) The treatment of malaria. Am J Trop Med Hyg 1:124–131
123. Trager W, Jensen JB (1976) Human parasites in continuous culture. Science 193:673–675
124. Malaria Vaccine Initiative (2011) Fact sheet: the RTS, S malaria vaccine candidate, Mar 2011, Washington, DC
125. Sherman IW (2009) The elusive malaria vaccine: miracle or mirage? American Society for Microbiology, Washington, DC, p 391
126. Committee on US Military Malaria Vaccine Research (2006) Battling malaria: strengthening the U.S. military vaccine program. National Academies Press, Washington, DC, p 144

127. Desowitz RS (1993) The malaria capers: tales of parasites and people. W.W. Norton, New York, p 288
128. Doolan DL, Dobano C, Baird JK (2009) Acquired immunity to malaria. Clin Microbiol Rev 22:13–36
129. Hviid L (2005) Naturally acquired immunity to *Plasmodium falciparum* in Africa. Acta Trop 95:270–275
130. Marsh K, Forster D, Waruiru C et al (1995) Indicators of life-threatening malaria in African children. N Engl J Med 72:334–336
131. Crompton PD, Pierce SK, Miller LH (2010) Advances and challenges in malaria vaccine development. J Clin Invest 120:4168–4178
132. Graves PM, Gelband H (2009) Vaccines for preventing malaria (SPf66). Cochrane Libr 2:1–30
133. Carrington WJ, Deragiache E (1999) How extensive is the brain drain? Finance Dev 36:46–49
134. Carpenter KJ (2000) Beriberi, white rice and vitamin B: a disease, a cause and a cure. University of California Press, Berkeley, p 296
135. Post P, Frederick WH, Heidebrink I, Sato S (eds) (2009) The encyclopedia of Indonesia in World War II. Brill, Leiden, Netherlands, p 710
136. Stone R (2010) Righting a 65-year-old wrong. Science 329:30–31

Books and Reviews

Feacham GA, Phillips AA, Targett GA (eds) (2009) Shrinking the malaria map: a prospectus on malaria elimination. University of California San Francisco, Global Health Group, San Francisco, p 187

Lobo L (2010) Malaria in the social context: a study in western India. Routledge, New Delhi, p 213

Packard RM (2007) The making of a tropical disease: a short history of malaria. The Johns Hopkins University Press, Baltimore, p 320

Shah S (2011) The fever: how malaria has ruled humankind for 500,000 years. Picador, New York, p 320

Sherman IW (2010) Magic bullets to conquer malaria: from quinine to qinghaosu. American Society for Microbiology Press, Washington, DC, p 298

Snowden F (2006) The conquest of malaria: Italy, 1900–1962. Yale University Press, New Haven, p 304

Winther PC (2005) Anglo-European science and the rhetoric of empire: malaria, opium and British rule in India 1756–1895. Lexington Books, New York, p 450

Yip K-C (ed) (2009) Disease, colonialism and the state: malaria in modern East Asian history. Hong Kong University Press, Hong Kong, p 161

Chapter 12
Tuberculosis, Epidemiology of

Giovanni Sotgiu, Matteo Zignol, and Mario C. Raviglione

Glossary

Acquired immunodeficiency syndrome (AIDS)	Clinical syndrome caused by the Human Immunodeficiency Virus (HIV). Its pathogenesis is related to a qualitative and quantitative impairment of the immune system, particularly a reduction of the CD4+cell count (surrogate marker of the disease). After an average of 10 years if untreated, HIV + individuals can develop opportunistic diseases (i.e., infections and neoplasias rarely detected in immunocompetent subjects). The natural history of the disease can be dramatically modified with administration of combination therapy composed of at least three antiretroviral (ARV) drugs.
Human immunodeficiency virus (HIV)	Virus that causes Acquired Immunodeficiency Syndrome (AIDS). It belongs to *Retroviridae* family and was discovered in 1983 by Luc Montagnier and Robert Gallo. It is transmitted mainly through

This chapter, which has been modified slightly for the purposes of this volume, was originally published as part of the Encyclopedia of Sustainability Science and Technology edited by Robert A. Meyers. DOI:10.1007/978-1-4419-0851-3

G. Sotgiu
Epidemiology and Biostatistics Unit, Department of Biomedical Sciences,
University of Sassari, Sassari, Italy
e-mail: gsotgiu@uniss.it

M. Zignol • M.C. Raviglione (✉)
STOP TB Department (STB), HIV/AIDS, TB & Malaria Cluster (HTM), World Health Organization, Geneva, Switzerland
e-mail: zignolm@who.int; raviglionem@who.int

P. Kanki and D.J. Grimes (eds.), *Infectious Diseases: Selected Entries from the Encyclopedia of Sustainability Science and Technology*,
DOI 10.1007/978-1-4614-5719-0_12,

	sexual intercourse, exchange of contaminated syringes among intravenous drug users, and contaminated blood transfusion. HIV-1 is the type most frequently detected worldwide.
Incidence	Rate describing the number of new cases of disease occurring within a unit of time in a defined cohort at risk of disease (expressed in cases per 100,000 population per year).
Latent tuberculosis infection (LTBI)	Infection caused by *Mycobacterium tuberculosis* transmitted mainly through the air. Clinical and/or radiological signs of latent tuberculosis infection cannot be detected in the majority of the cases. The infection can be presumptively diagnosed by a positive tuberculin skin testing and/or a positive interferon-γ release assay (IGRA), being able to identify a persistent adaptive immunological reactivity against mycobacterial antigens. In most individuals mycobacteria can be eliminated through chemoprophylaxis. It is estimated that one third of the human population is infected by *Mycobacterium tuberculosis* worldwide.
Mortality	Rate describing the number of deaths from a disease occurring within a unit of time in a defined cohort at risk of death (expressed in deaths per 100,000 population per year).
Mycobacterium tuberculosis	Bacterium that causes tuberculosis, discovered by Robert Koch in 1882. It is genetically closely related to other mycobacteria with which it forms a complex (*Mycobacterium africanum*, frequently detected in Western Africa, and *Mycobacterium bovis*, frequently detected in the past in cows and transmitted to human beings through unpasteurized milk).
Prevalence	Number of cases of disease in a defined population at a specific point in time; it is mainly presented as a relative frequency (i.e., proportion usually expressed per 100,000 population).
Tuberculosis	Infectious disease caused by *Mycobacterium tuberculosis*. It usually involves the lungs (pulmonary tuberculosis) but can also affect other organs (i.e., kidneys, central nervous system, lymph nodes, bones, etc.: extrapulmonary tuberculosis). Pulmonary tuberculosis, which is the most frequent clinical form, can be classified as smear-positive or smear-negative according to the result of the sputum bacteriological examination. The former is a major public

health problem being highly contagious. Only a few individuals develop tuberculosis after a mycobacterial infection, and most of them soon after infection: it is estimated that the lifetime risk is 5–10% in HIV-negatives and 5–15% yearly in HIV-positives.

Definition of the Subject and Its Importance

Tuberculosis is a disease caused by bacilli belonging to the *Mycobacterium tuberculosis* complex, which includes the species *Mycobacterium tuberculosis* (most frequently detected in human beings), *Mycobacterium africanum*, and *Mycobacterium bovis*. It is primarily an airborne disease. Mycobacteria enter into the human airways through the inhalation of droplet nuclei, i.e., particles containing mycobacteria aerosolized by coughing, sneezing, talking, or singing. Other rare, recognized ways of transmission are the ingestion of cow milk contaminated by *Mycobacterium bovis* (frequent route of transmission in the past), cutaneous inoculation in laboratory workers and pathologists, and sexual intercourse [1].

Tuberculosis is a major public health issue worldwide. Together with HIV/AIDS, it is one of the main infectious killers of individuals in their productive years in low-income countries [2]. However, tuberculosis is a highly curable disease if diagnosed and treated with combination chemotherapy. As a result of the global implementation of the World Health Organization's (WHO) STOP TB Strategy, during the period 1995–2009, 41 million tuberculosis patients were successfully treated and 6 million deaths averted compared to what would have happened if current standards were not implemented [3, 4].

Introduction

The epidemiology of tuberculosis studies the dynamics and interactions between *Mycobacterium tuberculosis* and human beings in a specific environment. Furthermore, it qualitatively and quantitatively evaluates all the covariates (for instance, the administration of antituberculosis drugs) that could interfere with the outcomes of the natural history of this interaction (i.e., exposure, infection, disease, and death) [5].

Basic and Descriptive Epidemiology of Tuberculosis

The global burden of tuberculosis is measured using epidemiological indicators (incidence, prevalence, mortality, and case fatality). Estimates of these indicators are computed yearly by the World Health Organization which collates information obtained from surveillance systems (notification system and mortality registries), special epidemiological studies (surveys of tuberculosis prevalence and in-depth analyses of surveillance data), and experts' opinion [6].

Epidemiological Indicators

Tuberculosis incidence: number of new cases of tuberculosis occurring within a specific time period (usually 1 year) in a defined cohort. It is usually presented as a rate per 100,000 inhabitants and describes the probability of developing tuberculosis in a specific time period [6, 7].

Tuberculosis prevalence: number of cases of tuberculosis in a defined population at a specific point in time. It is presented as an absolute or a relative (usually per 100,000 inhabitants) frequency and can be considered the product of the incidence of tuberculosis by the duration of the disease. This indicator assesses the global, national, regional, and local burden of tuberculosis.

Tuberculosis mortality: number of deaths from tuberculosis occurring within a specific time period (usually 1 year) in a cohort of individuals with tuberculosis disease. It is often presented as a rate per 100,000 people and describes the probability of dying from tuberculosis in a specific time period.

Tuberculosis case fatality: number of deaths from tuberculosis occurring within a specific time period (usually 1 year) in a defined cohort of individuals with tuberculosis. It is presented as a percentage.

Incidence of sputum smear-positive pulmonary tuberculosis cases: number of new sputum smear-positive cases of pulmonary tuberculosis occurring within a specific time period (usually 1 year) in a defined cohort. It is usually presented as a rate per 100,000 population. It identifies the most contagious tuberculosis cohort, that is, the most important source of infection.

Incidence of sputum culture-positive pulmonary tuberculosis cases: number of new sputum culture-positive cases of pulmonary tuberculosis occurring within a specific time period (usually 1 year) in a defined cohort. It is usually presented as a rate per 100,000 population.

Etiologic Epidemiology of Tuberculosis

Natural History

The dynamics of the probabilistic model of tuberculosis are complex and the associated covariates, as well as their quantitative effects, are not always known. The potential interaction between a human being and the strains of *Mycobacterium tuberculosis* could result in the following non-deterministic outcomes: subclinical infection, pulmonary and/or extrapulmonary disease, and death. The incidence of the above mentioned outcomes is regulated by the occurrence and the additive and/or the synergistic combination of multiple risk factors. Several environmental, bacterial, and human risk factors have been identified. Some of them could play an important role since the first crucial event, that is, the exposure to a source of mycobacteria [1, 5, 8].

Exposure

A critical exposure to transmission of mycobacteria from a contagious patient to a susceptible individual could be a close physical contact or permanence in a small room with limited ventilation [5].

The most important variable significantly increasing the likelihood of exposure to a source of *Mycobacterium tuberculosis* is the prevalence of contagious patients in a specific setting. This is estimated as the product of the incidence of infectious individuals by the duration of their infectiousness. The number of incident cases greatly varies between settings and depends on numerous factors.

Duration of infectiousness is strictly related to the capacity of the health system for early detection and adequate treatment of an index case [5, 8]. After a microbiological diagnosis, which should include drug-sensitivity testing, proper anti-tuberculosis treatment should be started without delay [3]. Correct drug combinations and dosages should be used to avoid sub-optimal anti-bacterial activity [5, 9, 10]. Strict clinical follow-up of patients in nosocomial or community settings should be undertaken to ensure adherence to anti-tuberculosis medications for the full duration of treatment. Compliance could be compromised by the long duration of therapy and/or the emergence of adverse events. The capacity of a patient to adhere to antituberculosis therapy depends on factors related to the health system, socioeconomic conditions of the individual, type of therapy, and the disease and patient's characteristics. The concept of early case detection and immediate treatment is supported by the evidence that on average 30–40% of the close contacts of a sputum smear-positive pulmonary case are estimated to be infected at the time of diagnosis. The public health consequences of a diagnostic and/or therapeutic delay are dramatic [5].

It is clear that quality and quantity of human relations modify the risk of critical exposures and, consequently, the risk of acquiring the infection [5]. In particular, population density is a critical element: the higher the population density, the greater the risk of interactions. High population density can be experienced in urban areas or in household crowding associated with poor housing [5, 8]. Climatic condition is an additional environmental variable that plays a relevant role in the likelihood of exposure. In cold climates, people tend to spend more time in indoor activities, increasing the risk of exposure because of the higher chances of close contact. On the other side, in warm climates people tend to spend more time outdoors, diminishing frequency and duration of case–contact interactions per unit of space. At the same time natural indoor ventilation and the effect of the solar ultraviolet rays that kill mycobacteria contribute to control the spreading of infection [5].

Other relevant non-environmental modifiers of the risk of exposure are sex and age. In several low-income countries and social groups, males and females have different opportunities of social contact and therefore a different risk of exposure. For instance, in some countries females are excluded from public activities and spend most of their time indoors. The median age of tuberculosis cases, and consequently of contagious sources of infection, is considerably different between low- and high-income countries, with elderly people representing the majority of the autochthonous patients in industrialized areas and young adults counting for the largest proportion of patients in low-income countries [5, 11–13].

Infection

The risk of infection by *Mycobacterium tuberculosis* is directly associated to the probability of exposure to infectious particles produced by a tuberculosis patient through coughing, sneezing, talking, or singing [1, 5, 8, 14]. Such risk is related to the concentration of contagious droplet nuclei containing mycobacteria and to the time of exposure [1, 5].

Infectious particles must be suspended in the air in order to be inhaled by a contact of a tuberculosis patient. The speed of falling to the ground is directly correlated to the square of their diameter. However, the tendency of liquid particles to evaporate and, consequently, to reduce their diameter, increases the speed of their descent. The effect of evaporation on very large liquid particles is less important. Furthermore, humidity and temperature could influence the evaporation: high humidity hinders the evaporation of infectious particles and, indirectly, increases the speed of dropping to the ground. Overall, the effect of the diameter of the particle is more influential than that of humidity [5].

Droplet nuclei should be 1–5 μm large to be inhaled and retained in the pulmonary alveoli. It has been demonstrated that a diameter >5 μm increases the probability of being entrapped in the upper airways through the continuous movement of both the vibrissae and the muco-ciliary system. On the other hand, contagious droplets whose diameter is <1 μm can arrive to the alveolar spaces, but the probability of retention in the peripheral pulmonary alveoli is very low [1, 5].

A crucial variable that can dramatically change the air density of contagious droplet nuclei in a specific indoor environment (for instance, in a nosocomial setting) is the natural, mixed-mode, or mechanical ventilation [15]. Several authors described transmission of mycobacterial strains in health-care settings with flawed or missing ventilation systems both in low- and high-income countries. The typology of the ventilation system should be chosen taking into account several variables: the building structure, outdoor climatic conditions and air quality, purchase and maintenance costs, and national and local regulations. However, modifiers should not alter the main principle of a ventilation system that must promote an airflow direction from the infectious source to the air exhaust area [5, 15]. Ventilation rate can be evaluated measuring the volume of the space (i.e., air changes per hour – ACH) or the number of individuals in a space (i.e., liters/second/

person): every individual should have a definite supply of fresh air to dilute the concentration of mycobacteria. The World Health Organization recommends more than 11 air changes per hour for an isolation room of 24 m^3 (i.e., at least 80 l/s/person). Natural ventilation may be improved by increasing the size of windows and positioning them on opposite walls. Well-designed fans can improve the airflow direction when natural ventilation rates are not adequate (mixed-mode ventilation). However, continuous maintenance of those fans is essential. In some circumstances it may be necessary to use upper room or shielded ultraviolet germicidal irradiation (UVGI) devices. In settings with frequent climatic changes or high risk of transmission of multi-drug resistant mycobacterial strains, UVGI devices could be helpful in addition to a mechanical ventilation system.

Several measures should be implemented in household settings to reduce the risk of transmission: improvement of natural ventilation, education on cough etiquette, and respiratory hygiene. Furthermore, a tuberculosis patient should sleep isolated in a well-ventilated room or, if his/her clinical conditions are favorable, should spend as much time as possible outdoors.

Personal protective equipment may considerably reduce the possibility of infection, especially when the ventilation system is weak. In this regard only high-efficiency particulate air-filter respirators filtering out droplet nuclei sized 1–5 μm can be protective. Particulate respirators meeting or exceeding the N95 standards set by the United States Centers for Disease Control and Prevention/National Institute for Occupational Safety and Health (CDC/NIOSH) or the CE certified FFP2 standards are recommended for health-care workers (particularly during the management of multi-drug resistant tuberculosis cases and aerosol-generating procedures) and visitors. Health education on the importance of respirators in communities at risk should be carried out in order to avoid stigma, and continuous training of health-care providers on the correct use (fit testing) and on indications of respirators should be organized. Surgical masks can be worn by contacts of a patient but do not protect from exposure [15]. The probability of being infected can be decreased if the source of droplet nuclei is isolated or if both mouth and nose are covered.

The most efficient measure to stop transmission is to rapidly detect and adequately treat a tuberculosis case [3, 5, 8, 15]. Patients are normally not infectious after 2–3 weeks of effective treatment. However, not all tuberculosis patients are equally infectious and there is a correlation between the number of mycobacteria in a milliliter (ml) of sputum and the efficiency of transmission. At least 5,000 mycobacteria in 1 ml of sputum are necessary to classify a sputum smear examination as positive.

Several studies demonstrated that sputum smear-positive patients are significantly more contagious than those who are sputum smear-negative and culture-positive. Furthermore, the latter are not much more infectious than tuberculosis patients who have a negative smear and culture [5, 8].

Although sputum smear-positive patients are the most contagious population, those who are sputum smear-negative need equal attention, as 17% of mycobacterial transmissions, in some industrialized settings, are due to this group of patients [5].

According to estimates computed in the pre-antibiotic era and used to model the tuberculosis epidemic, an undiagnosed sputum smear-positive patient can infect 10–12 contacts annually. Since after 2 years, the probability of death or spontaneous conversion to sputum smear-negativity is high, the average number of contacts infected by a single smear-positive case was estimated to be 20–24 [5, 8, 16]. The delay of diagnosis and/or of treatment of just 2–3 months could result in infection of several contacts [5].

Numerous factors determine patient's delay to diagnosis: these include alcohol or substance abuse, low income, difficult access to health care facilities, disabilities, and beliefs about tuberculosis. On the other hand, other concomitant pulmonary diseases, sputum smear-negative or extra-pulmonary tuberculosis, inexperience of health care workers, weak health care infrastructure, and absence of respiratory symptoms are all factors associated with healthcare delay [17, 18].

The annual burden of the most relevant source of tuberculosis infections is assessed through an epidemiological indicator: the incidence of sputum smear-positive pulmonary tuberculosis cases. This describes the number of new sputum smear-positive cases of pulmonary tuberculosis occurring within 1 year in a defined population at risk of tuberculosis [5, 6, 8]. The average annual risk of latent tuberculosis infection is an important epidemiological indicator that assesses the probability of becoming infected with *Mycobacterium tuberculosis* in a period of 1 year [5, 8].

Clinical and/or radiological signs of latent tuberculosis infection cannot be detected in the majority of the cases. However, it can only be presumptively diagnosed through a tuberculin skin testing (TST) and/or an interferon-γ release assay (IGRA), which identifies a persistent adaptive immunological reactivity against mycobacterial antigens [14, 19–22]. The tuberculin skin test, performed with an intradermal injection of purified protein derivative (PPD) obtained from *Mycobacterium tuberculosis* culture filtrate, has been considered as the gold standard diagnostic tool for identifying individuals latently infected with *Mycobacterium tuberculosis*. The main limitation of this technique is the low specificity, due to antigenic similarities between PPD, Bacille Calmette-Guerin (BCG) strains, and nontuberculous mycobacterial strains. Furthermore, its application and interpretation significantly depend on the health worker responsible for the test [14, 19, 22].

Interferon Gamma Release Assays (IGRA), available in two different commercial forms, that is, QuantiFERON-TB Gold In Tube (QFT-IT; Cellestis Ltd., Chadstone, Australia) and T-SPOT.TB (Oxford Immunotec, Abingdon, UK), are tools for the in vitro diagnosis of latent tuberculosis infection. They detect cellular immune reactivity toward *Mycobacterium tuberculosis*–specific antigens (ESAT-6, CFP-10, and TB7.7). As such, contrary to PPD, they can discriminate between people previously immunized with Bacille Calmette-Guerin vaccination or those infected by non-tuberculous mycobacteria from tuberculosis patients [14, 19–22].

The average annual risk of tuberculosis infection can be estimated with direct or indirect methods:

- The direct evaluation assesses the proportion of individuals who convert to tuberculin skin test in a defined period of time. The main limitations are related to the sample size, which should be large enough in order to obtain an adequate statistical power, and to the boosting phenomenon (an apparent conversion related to repeated tests).

– The indirect evaluation relies on the relationship between the prevalence of tuberculosis infection in a specific time-point and the average annual risks of infection.

The probability of being infected is estimated to be very low in high-income countries (from 0.1% to 0.01%) especially among the young. On the other hand, the prevalence of infection is high in most developing countries, due to ongoing transmission [5, 8]. Overall, one third of the global population is estimated to be infected by *Mycobacterium tuberculosis* [5, 6, 8].

Disease

Active tuberculosis is the outcome of infection with *Mycobacterium tuberculosis* [1, 5, 8, 23]. As for exposure and infection, the likelihood of clinical tuberculosis is strictly dependent on risk factors, the impact of which is not only related to their effect on the pathogenesis of the disease but also to their prevalence in the general population [5, 8].

After the establishment of tuberculosis infection the probability of developing a disease is higher in the first 12–24 months after the entry of *Mycobacterium tuberculosis* into the human body, and decreases as time elapses from the infection. It is generally estimated that the risk of disease is about 5% in the first 1–2 years after infection, and up to 5% in the subsequent years [1, 5, 8]. The age of primary infection is an important determinant of the cumulative risk, as this is the highest in young individuals (from 35% to 50% in children aged less than 15 years and those who are in close contact with a sputum smear-positive patient) [5, 8].

The key pathogenic factor favoring the development of tuberculosis is the quantitative and/or functional deficiency of the innate and/or adaptive immune system, which is responsible for the control of tuberculosis infection [2, 5, 8, 23]. The host cellular and molecular mechanisms triggered by mycobacterial infections are only partially known. After reaching the lower respiratory tract, the mycobacteria grow, partially controlled for the first few weeks by nonspecific immunity. CD4+ lymphocytes producing interferon-γ are recruited after 2 weeks but their number is not sufficient to activate the pool of alveolar macrophages. Their priming in the draining lymph nodes occurs only after the first 7–10 days because mycobacteria are initially phagocytosed by non-motile phagocytic cells. T-cell priming and accumulation could additionally be delayed through the activation of regulatory T lymphocytes. Continuous activation of T lymphocytes is vital to control mycobacteria in the granuloma; nevertheless, several experimental data demonstrated that once exposed to chronic mycobacterial stimulation, T lymphocytes can experience a functional collapse [1, 23]. Numerous medical conditions can modify one or more components of the immune response against *Mycobacterium tuberculosis*, compromising the control of the infection, and, then, increasing the probability of development of disease [1, 5, 8]. For these reasons and due to its global burden, HIV/AIDS is recognized as the most important risk factor for the development of tuberculosis [1, 2, 5, 8, 24, 25]. Numerous studies demonstrated that the lower the CD4+ lymphocytes counts the higher the risk of developing pulmonary as well as extrapulmonary disease. The qualitative and quantitative role of CD4+ lymphocytes in the adaptive immune response against mycobacteria is crucial, particularly their interaction with macrophages, which need to be activated in order to be effective. The risk of development of tuberculosis disease is 5–15% per year if the mycobacterial infection precedes HIV transmission. On the contrary, such risk could be significantly higher if mycobacterial infection develops in a seriously immunocompromised host. Generally, the probability of developing tuberculosis is 20–37-fold higher in HIV-infected compared to HIV-negative individuals. The relative risk is 20.6 in countries with a generalized HIV epidemic, 26.7 in countries with concentrated epidemics, and 36.7 in low HIV-prevalence countries [1, 2, 5, 8, 26].

Another medical condition that could compromise the immune key players against mycobacterial strains is diabetes mellitus. The estimated prevalence of diabetes is 180 million individuals globally, a figure expected to increase to 360 million by 2030 [5, 27]. The probability of developing tuberculosis is 1.5–8 times higher among individuals with diabetes compared to healthy subjects. No significant differences in terms of risk of disease have been detected between low- and high-income countries [28–32]. Experimental studies showed that hyperglycemic mice have higher mycobacterial loads, impaired production of interferon-γ and interleukin-12 together with an impaired T helper 1 response. Leukocyte bactericidal activity is reduced in diabetic patients compared to healthy individuals. Diabetic patients have an altered chemotaxis and oxidative killing of neutrophils. Furthermore, impaired immunity is frequently associated with other relevant risk factors like chronic renal failure, malnutrition, and pulmonary microangiopathy [28, 33, 34]. Nevertheless, several epidemiological analytical studies assessing the relative risk or the odds ratio of diabetic patients are biased because routine data sources were used, which do not allow controlling for all confounders. The population-attributable risk for diabetes is similar to that of HIV/AIDS; although the relative risk related to HIV/AIDS is highest, ranging from 6.5 to 26 (i.e., 2–9 times higher than the relative risk of diabetes), it is less prevalent than diabetes [28–33, 35].

Chronic renal failure and the hemodialytic treatment increase the risk of developing tuberculosis from 6.9 to 52.5 times compared to the general population [5, 36]. Some epidemiological studies highlight a greater incidence of tuberculosis in the first year of dialysis due to the occurrence of a severe immune depression. It is difficult to understand the role of chronic renal failure because the majority of the studies enrolled principally hemodialytic patients. The crucial pathogenetic feature is the impaired immune response to mycobacterial isolates, in particular type 1 helper T-cell responses involving interleukin-12 and interferon-γ production and co-stimulatory function of antigen-presenting cells. In the final stages of a chronic renal failure, a high rate of anergy to mycobacterial antigens administered intracutaneously has been documented (i.e., from 32% to 40%). Other relevant elements are the continual inflammatory phase of monocytes/macrophages due to the uremia and to the dialysis, malnutrition (mainly related to vitamin D intake), and hyperparathyroidism [5, 36–38].

Silicosis increases the likelihood of pulmonary tuberculosis from 2.8 to 39 times if compared to healthy controls whereas the probability of developing extra-pulmonary tuberculosis is 3.7-fold greater (the pleural form is the most common form, 61% of cases) [1, 5, 24, 39–41]. The average time elapsing between the diagnosis of silicosis and the development of tuberculosis is 6.8 years. The incidence of tuberculosis seems to be proportional to the severity of silicosis and the intensity of exposure to crystalline silica dust. Occupational activities in the mines seem to modify considerably the risk of developing tuberculosis: in particular, drilling has been associated with a greater intensity of exposure. Some studies identified a higher relative risk (i.e., 1.1–4.0) of developing tuberculosis in miners exposed to silica but without silicosis. These epidemiological findings support the results of experimental studies demonstrating that silica modifies the pulmonary immune response, and impairs the metabolism and the function of pulmonary macrophages till their apoptosis after long exposure [5, 24, 39–41]. The incidence of silica-related tuberculosis has been increased by the increased prevalence of HIV infection in resource-limited countries. The incidence of pulmonary tuberculosis in South African gold miners is 3,000 per 100,000 population. For this reason, several authors strongly recommend the treatment for latent tuberculosis infection in individuals with and without silicosis (especially for HIV-infected people) [5, 39–41].

Smoking is another relevant modifier, being associated to a relative risk of developing tuberculosis ranging from 2.3 to 2.7 [1, 5, 24, 42, 43]. Furthermore, it increases the risk of infection (relative risk of 1.7). The relative risk for latent tuberculosis infection and disease is not independent: the increase of the risk of infection directly increases the proportion of individuals at risk of disease. Therefore, the independent relative risk for tuberculosis in an infected population is obtained computing the ratio of the two relative risks (i.e., 1.4–1.6). Smoking reduces adaptive immune responses, decreasing CD4+ cell counts, altering macrophages, and mechanically disrupting the cilia in the upper respiratory airways. A dose–response effect has been documented: the likelihood of tuberculosis is directly linked to the number of cigarettes smoked daily.

Being treated with immunosuppressive drugs could increase the risk of tuberculosis, as recently demonstrated with the anti-TNFα therapies for the treatment of chronic inflammatory diseases (for instance, rheumatoid arthritis, psoriasis and psoriatic arthritis, ankylosing spondylitis, juvenile idiopathic arthritis, and inflammatory bowel disease) [44]. TNF is a protein crucial for granuloma initiation and stability in the lungs. It fosters the phagocytic and killing activities of the alveolar macrophages and favors the gathering of cells at the inflammatory site inducing adhesion molecules. Four monoclonal anti-TNF antibodies are currently available and are used for the treatment of chronic inflammatory diseases: adalimumab, certolizumab pegol, golimumab, and infliximab. The risk of developing tuberculosis is amplified by 1.6–25.1 times and varies depending on the clinical setting and type of anti-TNF therapy prescribed [44–46].

The role of a corticosteroid treatment in the pathogenesis of tuberculosis is debated [5]. Daily dose of 10 mg or higher of corticosteroids, prescribed for a few days, appears to not be associated with an increased risk of developing tuberculosis but experimental studies on rabbits documented their role in increasing the risk of disease.

Numerous studies evaluated the impact of several tumors as risk factors for tuberculosis, but only lymphomas, pulmonary neoplasia, and head/neck cancers seem associated to the development of the disease [5]. Other relevant diseases and medical events that could increase the risk of tuberculosis are malnutrition, substance abuse, gastrectomy, and jejunoileal by-pass [5, 24]. Intravenous drug and alcohol abuse could increase the risk of tuberculosis by compromising the immune system. Malnutrition, meat and fish deficiencies, or a vegetarian diet are epidemiologically associated to increased probability of tuberculosis [5, 24].

Mortality

The wide availability of anti-tuberculosis drugs has decreased mortality from tuberculosis and case fatality in developed countries. In the pre-antibiotic era, 30% and 66% of patients with tuberculosis died after 1 and 5 years from the notification of the disease, respectively [5, 8]. The most important prognostic variables are the site and type of tuberculosis, the early and adequate administration of anti-tuberculosis treatment, and adherence to treatment [3, 5, 8]. The majority of deaths caused by tuberculosis are attributable to sputum smear-positive tuberculosis [5, 8].

Patient's delay and/or health-care system's delay are associated to increased mortality. Patient's delay can be reduced with health education measures, decreasing tuberculosis stigma, and improving the health-care network. The delay of the health-care system could be related to several factors, particularly poor availability of diagnostic means and misdiagnosis; furthermore, negative prognostic health-care–associated factors are poor accessibility to antituberculosis drugs, incorrect prescription of anti-tuberculosis regimens in terms of dosages, time of exposure, and drug susceptibility of the infecting mycobacterial strains.

Tuberculosis/HIV Coinfection

HIV infection has dramatically changed the epidemiology of tuberculosis in high- and low-income countries, mostly in sub-Saharan Africa, Southeast Asia, and Eastern Europe. It was estimated that tuberculosis incidence in 2008, if compared with the rate computed in 1990, increased by 11% due to the HIV pandemic. In sub-Saharan Africa, tuberculosis mortality has tripled in the last two decades and 29% of tuberculosis-related deaths are associated to HIV infection [1, 2, 5, 8, 47, 48]. The increase in tuberculosis notification rates mirrored the increase in the HIV prevalence with a delay of 4–7 years [2, 48–50].

By the end of 2008, 33.2 million individuals were estimated to be living with HIV. Incidence and mortality in 2007 were 2.7 million and 2.1 million, respectively [48, 49]. Tuberculosis is estimated to be the cause of 26% of the deaths in people living with HIV. The majority of TB/HIV patients is located in sub-Saharan Africa but in numerous high HIV prevalence areas only 20% HIV-infected individuals know their HIV status. The infection is principally transmitted through sexual relationships, exchange of contaminated syringes among intravenous drug users, and transfusion of infected blood. The risk of developing tuberculosis ranges from 3% to 15% per year in HIV coinfected patients, while the cumulative lifelong risk in non-HIV-infected individuals is 10%. However, surveys carried out in high HIV prevalence areas estimated an annual probability of developing tuberculosis up to 30% in advanced immunocompromised patients [1, 2, 5, 8, 48–50].

The risk of developing tuberculosis is directly correlated to the degree of immunodeficiency, particularly with the quantitative impairment of CD4+ cell counts, which are considered a relevant surrogate marker of infection together with the HIV-viral load. For this reason tuberculosis is the main opportunistic disease and cause of death in HIV + individuals globally [1, 2, 5, 8, 48, 50].

It is crucial to detect HIV-positive individuals and to start antiretroviral therapy (ART) to decrease the probability of developing tuberculosis and, consequently, the probability of transmitting mycobacterial strains to susceptible individuals. The risk of developing tuberculosis is reduced by 70–90% in patients treated with ART if compared with untreated HIV-positive individuals. In 2008, more than four million individuals were treated with antiretroviral drugs [1, 2, 5, 8, 48–54].

The role of HIV infection on tuberculosis incidence is strictly related to the prevalence of latent tuberculosis infection in the population, particularly in those aged 15–54 years. In industrialized countries, where the prevalence of the mycobacterial infection is low in all age groups, incidence and prevalence of tuberculosis in HIV + individuals are also low. However, in some metropolitan areas an epidemiological scenario comparable to that described in some low-income countries, with high proportions of coinfected individuals has been documented [2, 5, 8, 47, 48].

It is estimated that 30% of HIV + individuals are infected by mycobacterial strains worldwide, with a wide variability from 14% in Europe to 46% in Southeast Asia.

A series of tuberculosis/HIV collaborative activities have been developed to address tuberculosis/HIV coinfection (Table 12.1) [2, 3, 48, 49, 51, 55, 56]:

- Establishment of a tuberculosis/HIV coordinating body to plan and carry out common activities, such as monitoring and evaluation of the epidemiological indicators and clinical outcomes.
- Surveillance of the prevalence of HIV infection in patients with tuberculosis.
- Intensified tuberculosis case finding aimed at identifying tuberculosis symptoms and clinical signs in HIV-positive individuals attending specialized health-care settings (HIV clinics, sexually transmitted diseases, clinics, etc.).

Table 12.1 Tuberculosis/HIV collaborative activities [3]

1. *Establishment of the mechanisms for collaboration*
• Set up a coordinating body for tuberculosis/HIV collaborative activities that could be effective at all levels
• Carry out surveillance of HIV-prevalence among tuberculosis cases
• Carry out joint tuberculosis/HIV planning
• Conduct monitoring and evaluation
2. *To decrease the burden of tuberculosis in people with HIV/AIDS*
• Implement intensified tuberculosis case finding
• Introduce isoniazid preventive therapy
• Ensure the control of latent tuberculosis infection in health-care and congregate settings
3. *To decrease the burden of HIV in patients with tuberculosis*
• Provide HIV testing and counseling
• Introduce HIV-prevention methods
• Introduce cotrimoxazole preventive therapy
• Ensure care and support for people with HIV/AIDS
• Introduce antiretroviral therapy

- Isoniazid preventive therapy (IPT) in HIV-positive individuals latently infected by mycobacterial strains. In these cases, it is strongly recommended to microbiologically exclude tuberculosis before the administration of the anti-tuberculosis drug.
- Infection control measures in health care and congregate settings to avoid the spread of mycobacterial strains.
- HIV counseling together with HIV rapid testing in tuberculosis patients.
- Introduction of HIV-prevention methods.
- Cotrimoxazole preventive therapy.
- Wide availability of antiretroviral (ARV) drugs.

Delivery of ART to HIV-positive patients with tuberculosis is deemed one of the most important measures for a successful DOTS strategy as it can prevent the shift from latent tuberculosis infection to active tuberculosis following the immunological recovery. However, several issues can hamper the efficacy of ART in tuberculosis patients: pharmacokinetic interactions between anti-HIV drugs and anti-tuberculosis drugs and immune-reconstitution syndrome (IRIS) could increase morbidity and mortality, for instance, decreasing the adherence and, then, the clinical efficacy of the specific treatment regimens [2, 48, 50–54].

The combination of three of the above mentioned tuberculosis/HIV activities, that is, Intensified Case Finding, Isoniazid Preventive Therapy and Infection Control measures, known as the "Three I's", can have a significant role in the prevention of mycobacterial infections and tuberculosis, and in the early identification of tuberculosis cases [2, 3, 55, 56].

Intensified Case Finding helps to rapidly identify tuberculosis suspects using questionnaires focused on symptoms and clinical signs of tuberculosis. Individuals with or at high risk of HIV infection or in congregate settings (such as mines, prisons, etc.) are regularly screened and counseled by HIV services and/or by community-based organizations supporting HIV-positive patients.

Isoniazid Preventive Therapy can be administered to HIV-positives without tuberculosis, reducing the risk of developing the disease by 33–67% for up to 4 years. The criteria for treatment include HIV-positive individuals living in areas with a proportion of latent tuberculosis infection above 30% or HIV-positives with documented latent tuberculosis infection or exposure to a contagious tuberculosis index case. The efficacy of its combination with the ART in preventing the development of tuberculosis has been proven.

Tuberculosis Infection Control measures can prevent the spread of mycobacterial strains to vulnerable patients, health-care workers, the community, and those living in congregate settings.

However, slow implementation of the "Three I's" has been documented because of the difficulty faced by some national programs to integrate these components in the framework of HIV care. Some national tuberculosis programs do not support the implementation of some activities (for instance, the use of isoniazid preventive therapy) and gaps in policy and operational guidance remain a major obstacle in some settings. Infection control still represents a serious challenge at national-, regional-, and facility-level for the absence of a coordinating body or for multiple coordinating bodies, lack of technical expertise, trained health care workers, and laboratory biosafety. Furthermore, inadequate diagnostic tools for latent tuberculosis infection and tuberculosis have hampered the implementation of intensified tuberculosis case finding and the delivery of isoniazid preventive therapy predominantly in resource-limited settings [2, 48, 55, 56]. Nevertheless, it was estimated that 1.4 million tuberculosis cases were tested globally for HIV infection in 2008, which is a 64-fold increase if compared with the estimates in 2002. The increase was significant in sub-Saharan Africa (from 4% to 45% in 2004 and 2008, respectively). A significant improvement has been documented in the number of HIV-infected people screened for tuberculosis: from 600,000 in 2007 to 1.4 million in 2008. Improvements were also recently recorded on the provision of preventive and therapeutic drugs, particularly trimethoprim-sulfamethoxazole preventive treatment (200,000 HIV-positives with tuberculosis in 2008), isoniazid preventive therapy (50,000 HIV-positives in 2008), and ART (100,000 HIV-positives with tuberculosis in 2008) [2, 48, 49].

All countries with a high HIV prevalence should be committed to rapidly implement and scale-up all tuberculosis/HIV collaborative activities.

Future investments in the translational research aimed at identifying new diagnostics and therapeutics as well as international and national political commitment could improve the epidemiology of tuberculosis/HIV coinfection [2, 5, 8, 48, 50, 53].

Drug-Resistant Tuberculosis

Development of resistance to anti-tuberculosis drugs has been documented since the early years of introduction of chemotherapy for the treatment of tuberculosis. The large majority of patients treated with streptomycin in the first Medical

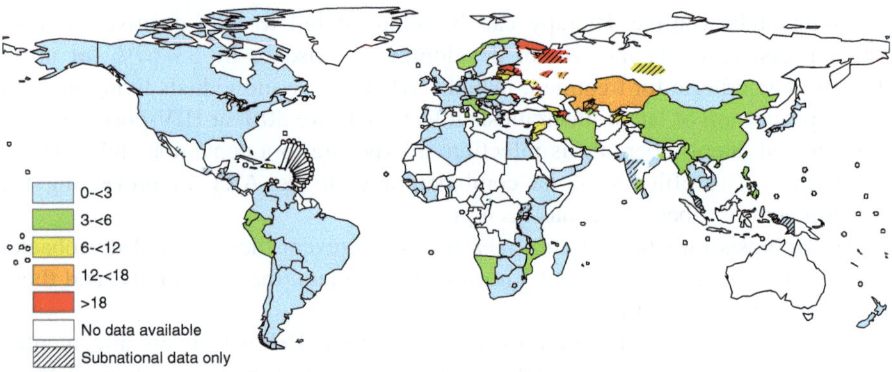

Fig. 12.1 Distribution of MDR-TB among new TB cases, 1994–2010

Research Council randomized clinical trial in the 1940s acquired resistance to that drug [57]. The spread of drug-resistant strains was soon recognized, and a survey of clinics in England in the 1950s found that over 5% of patients with tuberculosis without a history of previous treatment had strains resistant to at least one of the three major drugs in use at that time [58]. If even once the three effective drugs are used in combination, the development of drug resistance is theoretically impossible [59]. However, despite the introduction of combination regimens throughout the world many years ago, drug resistance has been progressively documented [60].

In the early 1990s, several reports were published on the emergence of multidrug resistant TB (MDR-TB), but the populations surveyed were not comparable, and methods used to quantify the problem were not standardized thus making it difficult to estimate the global magnitude of drug-resistant TB [61–64].

In 1994, the Global Project on Anti-tuberculosis Drug Resistance Surveillance was established in order to estimate the global burden of drug-resistant TB worldwide using standardized methodologies so that data could be compared across and within regions [65]. The Project aimed to monitor trends in resistance, evaluate the performance of TB control programs, and advise on drug regimens. The Supranational TB Reference Laboratory Network (SRLN), a network now consisting of 29 laboratories globally, was developed in order to provide quality assurance to drug resistance surveys including panel testing before the start of a survey and rechecking of isolates during the survey [66]. Since 1994 drug resistance data have been systematically collected and analyzed from 119 countries worldwide (62% of all countries of the world). Out of them, 48 countries can rely on continuous surveillance systems based on routine diagnostic drug susceptibility testing of all patients. The remaining 71 countries have relied on special surveys of representative samples of patients.

The distributions of multidrug-resistant TB (MDR-TB), defined as TB caused by strains of *Mycobacterium tuberculosis* that are resistant to at least isoniazid and rifampicin, are given in Figs. 12.1 and 12.2, for new and previously treated TB cases, respectively. Proportions of MDR-TB exceeding 18% among new TB cases

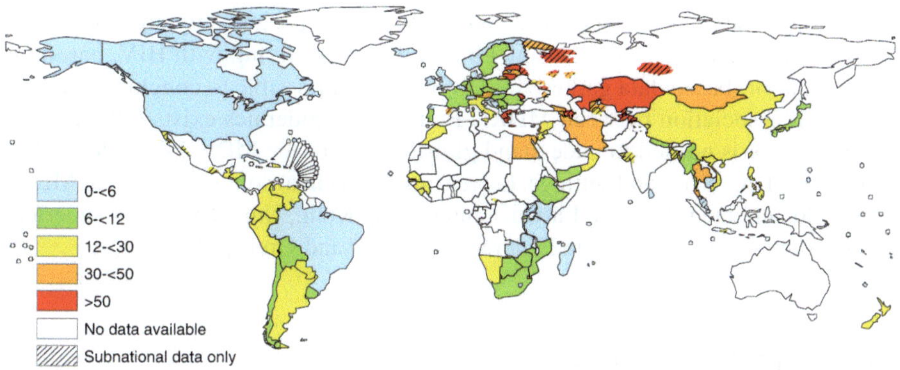

Fig. 12.2 Distribution of MDR-TB among previously treated TB cases, 1994–2010

(in countries reporting more than ten MDR-TB cases) have been documented in Estonia (22.0%), Russian Federation (Arkhangelsk Oblast, 25.7%, Belgorod Oblast, 19.8%, Ivanovo Oblast, 20.3%, Kaliningrad Oblast, 22.3%, Murmansk Oblast, 28.9%, Pskov Oblast, 24.3%, and Vladimir Oblast, 20.9%). Proportions of MDR-TB exceeding 50% among previously treated TB cases (for countries reporting more than ten MDR-TB cases) are found in Estonia (51.6%), Lithuania (51.5%), Russian Federation (Arkhangelsk Oblast, 58.8%, Belgorod Oblast, 51.6%, Ivanovo Oblast, 57.7%, and Tomsk Oblast, 53.8%), and Tajikistan (Dushanbe city and Rudaki district, 61.6%, 95% CI: 52.5–70.2) [67, 68].

Since 2006, WHO has also been collecting and analyzing data on resistance to second-line anti-TB drugs. Extensively drug-resistant TB (XDR-TB), defined as MDR-TB plus resistance to a fluoroquinolone and at least 1 second-line injectable agent – amikacin, kanamycin, and/or capreomycin-, has been documented in 69 countries globally [67].

Overall, there were an estimated 390,000–510,000 cases of MDR-TB (primary and acquired) arising in 2008, with the best estimate at 440,000 cases. Among all incident TB cases globally, 3.6% (95% CI: 3.0–4.4) are estimated to have MDR-TB.

The estimated global number of incident MDR-TB episodes among new and relapse TB cases in 2008 was between 310,000 and 430,000 episodes, with the best estimate at 360,000 episodes. The estimated global number of incident acquired MDR-TB episodes was between 83,000 and 110,000 episodes, with the best estimate at 94,000 episodes. Almost 50% of MDR-TB cases worldwide are estimated to occur in China and India. These estimates refer to cases of MDR-TB that arose in 2008 and do not reflect the number of prevalent cases of MDR-TB. The number of prevalent cases of MDR-TB in many parts of the world is estimated to be much higher than the number arising annually.

An estimated 150,000 deaths caused by MDR-TB occurred globally in 2008, including those with HIV infection (range: 53,000–270,000). The estimated number of MDR-TB deaths excluding those with HIV infection was 97,000 (range: 6,000–220,000) [69].

Although the association of HIV and MDR-TB has been widely documented in hospital outbreaks of drug-resistant TB among people living with HIV, based on the population-based data gathered till now, it is not possible to conclude whether an overall association between MDR-TB and HIV epidemics exists [70–72].

More work is needed to understand global trends of the MDR-TB epidemic but in a group of countries and settings, decreasing trends in absolute numbers of MDR-TB are documented (Estonia, Latvia, China, Hong Kong, United States of America, and two Oblasts in the Russian Federation, Tomsk and Orel) [69].

Control of Tuberculosis

In the second half of the nineteenth century more than one third of deaths globally were due to infectious diseases and 15% were due to tuberculosis. In particular, more than 30% of men and women during their productive years died of tuberculosis.

In industrialized nations, before the availability of the first anti-mycobacterial antibiotics in the second half of 1900, the incidence of tuberculosis notification dramatically decreased because of the reduction or elimination of some risk factors favoring the spread of mycobacterial strains in the community (for instance, improvement of the social and economic conditions of many families) and the implementation of public health interventions, such as the identification and isolation of contagious tuberculosis patients [1, 5, 8, 16, 73].

The first sanatorium was built in Germany in 1857 in order to isolate contagious tuberculosis patients and to provide them with adequate food, rest, sunlight, and fresh air. The dispensary system was introduced in 1897 in Scotland, England, following the basic German principle of the separation of an infectious patient from the community. After the Italian, Forlanini proved the efficacy of artificial pneumothorax to increase the likelihood of cure in tuberculosis cases in 1907, sanatoria increased their surgical activities to create pulmonary collapse (for instance, plombage and thoracoplasty) [1, 5, 8, 16, 74, 75].

The wide distribution of antituberculosis drugs and the scale-up of therapeutic combination, together with the establishment of the first National Tuberculosis Programs (NTPs) and the further improvement of the socio-economic situation, have been significantly associated with a rapid decline of tuberculosis incidence and prevalence in the USA and in numerous European countries [1, 5, 8, 16]. However, in the mid-1980s the erroneous feeling of imminent elimination of tuberculosis, supported by the persistent decline of incidence and prevalence, has led to a reduction of tuberculosis control measures with numerous sanatoria closed or converted into non-tuberculosis specialized facilities. The escalation of the HIV/AIDS epidemic, the increasing number of migrants from high tuberculosis incidence countries, socioeconomic changes due to the collapse of the Soviet Union contributed to the rapid increase of tuberculosis incidence in low-incidence countries [1, 5, 8, 16, 76, 77]. Epidemiological trends in low-income countries are partially known but in the last decades the improvement of surveillance systems has increased the confidence on morbidity and mortality estimates [5, 8].

The World Health Organization has played a crucial role in the fight against tuberculosis worldwide. In 1993, tuberculosis was declared a global public health emergency and reiterated in 2000 by the United Nations Millennium Development Goals (MDGs). The Millennium Development Goal mainly focused on the tuberculosis issue aims at combating HIV, malaria, and other infectious diseases.

In the last 20 years, two public health strategies have changed the natural history of tuberculosis worldwide: DOTS strategy since 1996 and Stop TB Strategy since 2006. Both of them were launched after interactive and proactive debates and discussions between governmental and non-governmental organizations and national representatives of high tuberculosis incidence countries (tuberculosis control program managers and their staffs) [8, 78–82].

DOTS strategy is based on finding and treating contagious cases. It consists of five components: government commitment to tuberculosis control, bacteriological diagnosis through smear microscopy (mainly on individuals complaining of characteristic tuberculosis symptoms), short-course antituberculosis therapy that has to be standardized and supervised (i.e., directly observed, at least during the initial phase), uninterrupted high-quality drug supply, and individual outcome evaluation through a standardized recording and reporting system. Its targets were to detect 70% of sputum smear-positive tuberculosis cases and to successfully treat 85% of them by 2005. DOTS can reduce annual tuberculosis notifications from 6% to 8%, decreasing the incidence, the prevalence, and the mortality.

In 2005, the World Health Organization launched the new Stop TB Strategy (Table 12.2) [82]. The former strategy was revisited to pursue DOTS expansion, adding six other components to be implemented to reach the 2015 Millennium Development Goal related to tuberculosis control, that is, to reduce prevalence of and mortality due to tuberculosis by 50% relative to those of 1990.

The new Stop TB Strategy can be summarized in the following elements [82]: (1) pursuing high-quality DOTS expansion and enhancement; (2) addressing tuberculosis/HIV coinfection, multi-drug resistant tuberculosis, and other challenges; (3) contributing to the strengthening of health-care systems; (4) engaging all care providers; (5) empowering people and communities with tuberculosis; and (6) enabling and promoting research.

This new Strategy addresses several issues that emerged during the implementation and scale-up of the DOTS strategy: spread of tuberculosis/HIV coinfection and increased rates of multi-drug resistant tuberculosis, weak health care systems, exclusion of the private sector from national tuberculosis control strategies (often managing poor and marginalized individuals), passive involvement of tuberculosis patients and communities, few resources in research and development of new diagnostics, drugs, and vaccines.

The first component, that is, "to pursue quality DOTS expansion and enhancement" is aimed at improving the previous strategy: political commitment should be strong in terms of financial resources available for tuberculosis control at national- and regional-level as well as for strengthening of health system. National laboratory

Table 12.2 WHO-recommended Stop TB Strategy [82]

1. *Pursue high-quality DOTS expansion and enhancement*
   - Secure political commitment, with adequate and sustained financing
   - Ensure early case detection, and diagnosis through quality-assured bacteriology
   - Provide standardized treatment with supervision, and patient support
   - Ensure effective drug supply and management
   - Monitor and evaluate performance and impact
2. *Address TB/HIV, MDR-TB, and the needs of poor and vulnerable populations*
   - Scale up collaborative TB/HIV activities
   - Scale up prevention and management of multidrug-resistant TB (MDR-TB)
   - Address the needs of TB contacts, and of poor and vulnerable populations, including women, children, prisoners, refugees, migrants, and ethnic minorities
3. *Contribute to health system strengthening based on primary health care*
   - Help improve health policies, human resource development financing, supplies, service delivery, and information
   - Strengthen infection control in health services, other congregate settings, and households
   - Upgrade laboratory networks, and implement the Practical Approach to Lung Health (PAL)
   - Adapt successful approaches from other fields and sectors, and foster action on the social determinants of health
4. *Engage all care providers*
   - Involve all public, voluntary, corporate, and private providers through Public–Private Mix (PPM) approaches
   - Promote use of the International Standards for TB Care (ISTC)
5. *Empower people with TB, and communities through partnership*
   - Pursue advocacy, communication, and social mobilization
   - Foster community participation in TB care
   - Promote use of the Patients' Charter for TB care
6. *Enable and promote research*
   - Conduct program-based operational research, and introduce new tools into practice
   - Advocate for and participate in research to develop new diagnostics, drugs, and vaccines

networks should be improved with the introduction of culture methods and drug susceptibility testing as well as new effective evidence-based diagnostic technologies. In order to obtain high-treatment rates and to avoid the emergence and spread of drug-resistant strains treatment support should be provided through a patient-centered approach. A national body should manage the continuous supply of quality-assured anti-tuberculosis drugs. Continuous monitoring of the national TB program and evaluation of the tuberculosis control activities should be regularly implemented.

The second component of the strategy describes all the interventions that should be implemented to control and avoid further emergence of the rising issues: tuberculosis/HIV coinfection and multi-drug resistant tuberculosis. Tuberculosis and HIV national programs should implement collaborative activities. Patients with drug-resistant tuberculosis should have access to diagnosis and proper treatment with second-line anti-TB drugs. More attention should be given to groups at highest risk of developing tuberculosis (such as prisoners, immigrants, etc.).

The third component is focused on strengthening of health system, highlighting all the elements necessary to improve a national health system: human and financial resources, information systems, and national policies.

The fourth component underlines the importance of the engagement of all health-care providers, including those outside national health system (public–private mix approach) and those working in prisons, general hospitals, and medical colleges.

The fifth component is aimed at describing the empowerment of people with tuberculosis and of communities, sharing some tuberculosis activities like treatment support. Furthermore, communities could have a role in reducing tuberculosis stigma.

The sixth component highlights the importance of the operational and biomedical research aimed at improving program performance and at developing new diagnostic, therapeutic, and preventive tools.

The Stop TB Strategy addresses all the risk factors associated to the development of mycobacterial infection and tuberculosis, including the reduction of the social and economic consequences of the disease. It supports the Global Plan to Stop TB (2011–2015), developed by the Stop TB Partnership [81]. The plan is aimed at identifying political and public health strategies to be implemented, and financial needs and existing gaps to address in order to scale up the Stop TB Strategy. Finally, International Standards of Tuberculosis Care (ISTC) have been issued to help both public and private health care providers to deliver quality DOTS services [83–85]. The aim is to describe an evidence-based standard of care for health-care workers managing confirmed or suspected tuberculosis individuals. The following principles of care are highlighted: early and accurate bacteriological diagnosis; administration of standardized and effective anti-tuberculosis drug regimens as well as appropriate treatment support and supervision; and monitoring of treatment outcomes.

Global Epidemiological Situation

Incidence

According to the last global estimates the number of new cases of tuberculosis in 2009 was 9.4 million, ranging from 8.9 million to 9.9 million (137 cases per 100,000 inhabitants). During the recent years the incidence trend has slowly decreased but the absolute frequency has increased due to population growth. The regions contributing the most to the global estimated incidence of tuberculosis are the South-east Asian region with 35% of cases (3,300,000), the Western Pacific region with 20% (1,900,000 cases), the African region with 30% (2,800,000 cases) followed by the Eastern Mediterranean region (7%), the European region (4%), and the Region of the Americas (3%) [4, 6].

The World Health Organization identified 22 so-called high burden countries (HBCs), accounting for more than 80% of the incident cases estimated globally. India accounts for 21% of all tuberculosis patients and China for 14%. The highest incidence rates in 2009 were estimated in India (range: 1.6–2.4 million), China (range: 1.1–1.5 million), South Africa (range: 0.40–0.59 million), Nigeria (range: 0.37–0.55 million), and Indonesia (range: 0.35–0.52 million). The incident tuberculosis/HIV coinfection was estimated in 1.1/9.4 million (12%; 1.0–1.2 million, 11–13%) individuals, with the vast majority of patients in the African region. Among the incident cases estimated in 2008, 440,000 (range: 390,000–510,000) patients were infected by multi-drug resistant mycobacterial strains, that is, with resistance in vitro to at least isoniazid and rifampicin (MDR-TB).

Prevalence

The estimated tuberculosis prevalence in 2009 was 200 cases per 100,000 inhabitants as a consequence of 14 million (range: 12–16 million) prevalent patients.

Mortality

In 2009 about 1.7 million individuals (i.e., 26 cases per 100,000 inhabitants) died of tuberculosis, with 400,000 (range: 320,000–450,000) and 1.3 million (range: 1.2–1.5 million) deaths among tuberculosis/HIV coinfected and HIV-negative patients, respectively.

Case Notifications

National tuberculosis programs registered 5.8 million tuberculosis patients in 2009: most of them were new cases (94.8%) and 0.3 million were relapses (i.e., previously treated cases whose most recent anti-tuberculosis treatment outcome was "cure" or "treatment completed" but who were afterward diagnosed with sputum smear and/or culture-positive tuberculosis); 57% of the pulmonary tuberculosis cases was considered highly infectious, that is, sputum smear-positive. Among the 5.5 million new cases notified in 2009, 47.3% (2.6 million) was sputum smear-positive, 36.4% (2.0 million) sputum smear-negative while 16.4% (900,000) was affected by extrapulmonary tuberculosis.

The causes behind the global under-reporting could be linked to the patients (limited knowledge of tuberculosis; economic and/or geographical barriers to access health care; etc.) and/or to the health systems (poor clinical and laboratory capacities; few trained health care workers; poor compliance to national laws on notification after tuberculosis diagnosis).

Treatment Outcomes

The global treatment success ("cured" plus "treatment completed") rates for new sputum smear-positive cases treated during the years 2007 and 2008 were higher than the target set in 1991, that is, 85%.

In 2008 three WHO regions contributed to this result: Eastern Mediterranean Region, the South-east Asian Region, and the Western Pacific Region; the worst treatment success rate was documented in the European Region (66%), followed by the Region of the Americas (77%) and the African Region (80%).

Future Directions

Elimination of tuberculosis, or reducing the incidence of new sputum smear-positive cases below one per one million inhabitants, is the goal for the next decades. The target endorsed by Stop TB Partnership is the elimination of tuberculosis as a public health problem by 2050 [6–88].

The objectives of the tuberculosis elimination strategy are:

- Reduction of the incidence of latent tuberculosis infection.
- Reduction of the prevalence of latent tuberculosis infection.

The aim is to decrease the burden of future new tuberculosis cases.

In low tuberculosis-incidence countries, the lowest prevalence and incidence of latent tuberculosis infection are detected in the autochthonous youngest groups. The epidemiological projections indicate that each generation will be substituted by a generation with a lower risk of infection and, then, characterized by a lower prevalence of latent tuberculosis infections [5, 8, 86, 88].

Early detection and adequate treatment of contagious tuberculosis patients are of paramount importance in order to achieve a positive epidemiological trend. Moreover, the progression to tuberculosis in individuals infected by mycobacterial strains should be prevented, particularly focusing on high-risk groups (migrants, HIV-positives, ethnic minorities, prisoners, elderly people, and close contacts of tuberculosis index cases) [5, 8, 86].

This combined approach is deemed essential to reach the elimination goal, and it should be supported by international and national policies, embracing the following components [2, 5, 8, 86, 87]:

- Political commitment toward control and elimination, particularly providing human resources, facilities, and funds
- Case detection through case-finding among symptomatic individuals presenting at health services and active case finding in special groups through quality-assured laboratories and trained health-care providers
- Access to tuberculosis diagnostic and treatment services, implementing collaborative activities with health bodies or organizations working with migrants, prisoners, homelessness, etc.

- Standard approach to treatment of tuberculosis and latent tuberculosis infection, following international evidence-based recommendations
- Surveillance and treatment outcome monitoring, following standard internationally accepted definitions

However, tuberculosis indicators could potentially worsen in the next decades due to the following public health challenges:

- Drug-resistant mycobacterial strains (mostly multi/extensively drug-resistant mycobacteria)
- HIV/AIDS
- Immunocompromised individuals
- Political unrests and wars, natural disasters, and starvation

Adequate funds, constant international and national political commitment, and an effective supply of anti-tuberculosis and anti-HIV drugs and diagnostics represent relevant components for a successful strategy. In particular, the research and development of new effective anti-tuberculosis drugs with an improved toxicity, tolerability, and efficacy profile and the evaluation of the activity of potential adjunctive treatments, such as vitamin D, as well as of new primary preventive measures (i.e., vaccines) are crucial priorities for the near future. Basic and applied research would need to be followed by operational research, as new diagnostic, therapeutic, and preventive approaches must be tested at the program-level (field trials) after clinical trials [89].

Bibliography

1. Fitzgerald D, Haas DW (2005) *Mycobacterium tuberculosis*. In: Mandell GL et al (eds) Principles and practice of infectious diseases, 6th edn. Churchill Livingstone, Philadelphia, pp 212–2886
2. Nunn P, Williams B, Floyd K, Dye C, Elzinga G, Raviglione M (2005) Tuberculosis control in the era of HIV. Nat Rev Immunol 5(10):819–826
3. Raviglione MC (2006) The global plan to stop TB, 2006–2015. Int J Tuberc Lung Dis 10(3):238–239
4. World Health Organization (2010) Global tuberculosis control. A short update to the 2009 report. WHO, Geneva, 2010
5. Rieder H (1999) Epidemiologic basis of tuberculosis control. International Union Against Tuberculosis and Lung Disease, Paris
6. World Health Organization (2010) Global tuberculosis control 2010. World Health Organization Document 2010, Publication No. WHO/HTM/TB/2010.7
7. Everitt BS, Palmer C (2011) Encyclopedic companion to medical statistics, 2nd edn. Wiley, Chichester
8. Styblo K, Raviglione MC (1997) Tuberculosis, public health aspects. In: Encyclopedia of human biology, vol 8, 2nd edn. Academic, San Diego, pp 537–558
9. Stýblo K, Danková D, Drápela J, Galliová J, Jezek Z, Krivánek J, Kubík A, Langerová M, Radkovský J (1967) Epidemiological and clinical study of tuberculosis in the district of Kolin, Czechoslovakia. Report for the first 4 years of the study (1961–64). Bull World Health Organ 37(6):819–874

10. Krivinka R, Drápela J, Kubík A, Danková D, Krivánek J, Ruzha J, Miková Z, Hejdová E (1974) Epidemiological and clinical study of tuberculosis in the district of Kolín, Czechoslovakia. Second report (1965–1972). Bull World Health Organ 51(1):59–69
11. Borgdorff MW, Nagelkerke NJ, van Soolingen D, Broekmans JF (1999) Transmission of tuberculosis between people of different ages in The Netherlands: an analysis using DNA fingerprinting. Int J Tuberc Lung Dis 3(3):202–206
12. Rieder HL (1999) Socialization patterns are key to the transmission dynamics of tuberculosis. Int J Tuberc Lung Dis 3(3):177–178
13. Stead WW (1998) Tuberculosis among elderly persons, as observed among nursing home residents. Int J Tuberc Lung Dis 2(9 Suppl 1):S64–S70
14. Leung CC, Rieder HL, Lange C, Yew WW (2011) Treatment of latent infection with *Mycobacterium tuberculosis*: update 2010. Eur Respir J 37(3):690–711
15. WHO (2009) WHO policy on infection control in health-care facilities, congregate settings and households. World Health Organization, Geneva. WHO/HTM/TB/2009.419
16. Migliori GB, Sotgiu G, Lange C, Centis R (2010) Extensively drug-resistant tuberculosis: back to the future. Eur Respir J 36(3):475–477
17. Storla DG, Yimer S, Bjune GA (2008) A systematic review of delay in the diagnosis and treatment of tuberculosis. BMC Public Health 8:15
18. Ngadaya ES, Mfinanga GS, Wandwalo ER, Morkve O (2009) Delay in tuberculosis case detection in Pwani region, Tanzania. A cross sectional study. BMC Health Serv Res 9:196
19. European Centers of Disease Prevention and Control (2011) Use of interferon-γ release assays in support of TB diagnosis. ECDC, Stockholm, 2011
20. Cellestis (2006) QuantiFERON-TB Gold (In-Tube Method) Package Insert. Valencia, 2006
21. Oxford_Immunotec (2009) T-Spot. TB Package Insert. 2009
22. Diel R, Goletti D, Ferrara G, Bothamley G, Cirillo D, Kampmann B, Lange C, Losi M, Markova R, Migliori GB, Nienhaus A, Ruhwald M, Wagner D, Zellweger JP, Huitric E, Sandgren A, Manissero D (2011) Interferon-γ release assays for the diagnosis of latent *Mycobacterium tuberculosis* infection: a systematic review and meta-analysis. Eur Respir J 37(1):88–99
23. Torrado E, Robinson RT, Cooper AM (2011) Cellular response to mycobacteria: balancing protection and pathology. Trends Immunol 32(2):66–72
24. Creswell J, Raviglione M, Ottmani S, Migliori GB, Uplekar M, Blanc L, Sotgiu G, Lonnroth K (2010) Tuberculosis and non-communicable diseases: neglected links, missed opportunities. Eur Respir J 37(5):1269–1282
25. Lönnroth K, Jaramillo E, Williams BG, Dye C, Raviglione M (2009) Drivers of tuberculosis epidemics: the role of risk factors and social determinants. Soc Sci Med 68(12):2240–2246
26. Schutz C, Meintjes G, Almajid F, Wilkinson RJ, Pozniak A (2010) Clinical management of tuberculosis and HIV-1 co-infection. Eur Respir J 36(6):1460–1481
27. World Health Organization (2006) Diabetes fact sheet No. 312
28. Jeon CY, Murray MB (2008) Diabetes mellitus increases the risk of active tuberculosis: a systematic review of 13 observational studies. PLoS Med 5(7):e152
29. Stevenson CR, Forouhi NG, Roglic G, Williams BG, Lauer JA, Dye C, Unwin N (2007) Diabetes and tuberculosis: the impact of the diabetes epidemic on tuberculosis incidence. BMC Public Health 7:234
30. Stevenson CR, Critchley JA, Forouhi NG, Roglic G, Williams BG, Dye C, Unwin NC (2007) Diabetes and the risk of tuberculosis: a neglected threat to public health? Chronic Illn 3(3):228–245
31. Harries AD, Billo N, Kapur A (2009) Links between diabetes mellitus and tuberculosis: should we integrate screening and care? Trans R Soc Trop Med Hyg 103(1):1–2
32. Dooley KE, Chaisson RE (2009) Tuberculosis and diabetes mellitus: convergence of two epidemics. Lancet Infect Dis 9(12):737–746
33. Martens GW, Arikan MC, Lee J, Ren F, Greiner D, Kornfeld H (2007) Tuberculosis susceptibility of diabetic mice. Am J Respir Cell Mol Biol 37(5):518–524

34. Stalenhoef JE, Alisjahbana B, Nelwan EJ, van der Ven-Jongekrijg J, Ottenhoff TH, van der Meer JW, Nelwan RH, Netea MG, van Crevel R (2008) The role of interferon-gamma in the increased tuberculosis risk in type 2 diabetes mellitus. Eur J Clin Microbiol Infect Dis 27(2):97–103
35. Lienhardt C, Rodrigues LC (1997) Estimation of the impact of the human immunodeficiency virus infection on tuberculosis: tuberculosis risks re-visited? Int J Tuberc Lung Dis 1(3): 196–204
36. Hussein MM, Mooij JM, Roujouleh H (2003) Tuberculosis and chronic renal disease. Semin Dial 16(1):38–44
37. Woeltje KF, Mathew A, Rothstein M, Seiler S, Fraser VJ (1998) Tuberculosis infection and anergy in hemodialysis patients. Am J Kidney Dis 31(5):848–12
38. Girndt M, Sester U, Sester M, Kaul H, Köhler H (1999) Impaired cellular immune function in patients with end-stage renal failure. Nephrol Dial Transplant 14(12):2807–2810
39. Rees D, Murray J (2007) Silica, silicosis and tuberculosis. Int J Tuberc Lung Dis 11(5):474–484
40. Barboza CE, Winter DH, Seiscento M, Santos Ude P, Terra Filho M (2008) Tuberculosis and silicosis: epidemiology, diagnosis and chemoprophylaxis. J Bras Pneumol 34(11):959–966
41. Hnizdo E, Murray J (1998) Risk of pulmonary tuberculosis relative to silicosis and exposure to silica dust in South African gold miners. Occup Environ Med 55(7):496–502
42. Bates MN, Khalakdina A, Pai M, Chang L, Lessa F, Smith KR (2007) Risk of tuberculosis from exposure to tobacco smoke: a systematic review and meta-analysis. Arch Intern Med 167(4):335–342
43. Arcavi L, Benowitz NL (2004) Cigarette smoking and infection. Arch Intern Med 164(20):2206–2216
44. Solovic I, Sester M, Gomez-Reino JJ, Rieder HL, Ehlers S, Milburn HJ, Kampmann B, Hellmich B, Groves R, Schreiber S, Wallis RS, Sotgiu G, Schölvinck EH, Goletti D, Zellweger JP, Diel R, Carmona L, Bartalesi F, Ravn P, Bossink A, Duarte R, Erkens C, Clark J, Migliori GB, Lange C (2010) The risk of tuberculosis related to tumour necrosis factor antagonist therapies: a TBNET consensus statement. Eur Respir J 36(5):1185–1206
45. Ehlers S (2005) Tumor necrosis factor and its blockade in granulomatous infections: differential modes of action of infliximab and etanercept? Clin Infect Dis 41(Suppl 3):S199–S203
46. Bekker LG, Freeman S, Murray PJ, Ryffel B, Kaplan G (2001) TNF-alpha controls intracellular mycobacterial growth by both inducible nitric oxide synthase-dependent and inducible nitric oxide synthase-independent pathways. J Immunol 166(11):6728–6734
47. Raviglione MC, Harries AD, Msiska R, Wilkinson D, Nunn P (1997) Tuberculosis and HIV: current status in Africa. AIDS 11(Suppl B):S115–S123
48. Getahun H, Gunneberg C, Granich R, Nunn P (2010) HIV infection-associated tuberculosis: the epidemiology and the response. Clin Infect Dis 50(Suppl 3):S201–S207
49. Joint United Nations Programme on HIV/AIDS (2008) Report on the global AIDS epidemic. Joint United Nations Programme on HIV/AIDS, Geneva, 2008
50. Corbett EL, Watt CJ, Walker N, Maher D, Williams BG, Raviglione MC, Dye C (2003) The growing burden of tuberculosis: global trends and interactions with the HIV epidemic. Arch Intern Med 163(9):1009–1021
51. Reid A, Scano F, Getahun H, Williams B, Dye C, Nunn P, De Cock KM, Hankins C, Miller B, Castro KG, Raviglione MC (2006) Towards universal access to HIV prevention, treatment, care, and support: the role of tuberculosis/HIV collaboration. Lancet Infect Dis 6(8):483–495
52. World Health Organisation (2006) Antiretroviral therapy for HIV infection in adults and adolescents in resource limited settings: towards universal access. World Health Organisation, Geneva
53. World Health Organization (2010) Treatment of tuberculosis guidelines, 4th edn. World Health Organization, Geneva, Document WHO/HTM/TB/2009.420
54. Abdool Karim S, Naidoo K, Grobler A, et al. Initiating ART during TB treatment significantly increases survival: results of a randomised controlled clinical trial in TB/HIV co-infected

patients in South Africa. Presented at 16th conference on retroviruses and opportunistic infections. Montreal, 8–11 Feb 2009
55. WHO (2011) Guidelines for intensified tuberculosis case-finding and isoniazid preventive therapy for people living with HIV in resource constrained settings. WHO, Geneva, 2011
56. WHO Three I's Meeting. 2008
57. Medical Research Council (1948) Streptomycin treatment of pulmonary tuberculosis. A Medical Research Council investigation. BMJ 2:769–782
58. Fox W, Wiener A, Mitchison DA, Selkon JB, Sutherland I (1957) The prevalence of drug-resistant tubercle bacilli in untreated patients with pulmonary tuberculosis: a national survey, 1955–1956. Tubercle 38:71
59. Zhang Y, Yew WW (2009) Mechanisms of drug resistance in *Mycobacterium tuberculosis*. Int J Tuberc Lung Dis 13(11):1320–1330
60. Wright A, Zignol M, Van Deun A, Falzon D, Gerdes SR, Feldman K, Hoffner S, Drobniewski F, Barrera L, van Soolingen D, Boulabhal F, Paramasivan C, Kam KM, Mitarai S, Nunn P, Raviglione M (2009) For the Global Project on Anti-Tuberculosis Drug Resistance Surveillance. Epidemiology of antituberculosis drug resistance 2002–07: an updated analysis of the Global Project on Anti-Tuberculosis Drug Resistance Surveillance. Lancet 373(9678):1861–1873
61. Ellner JJ, Hinman AR, Dooley SW et al (1993) Tuberculosis symposium: emerging problems and promise. J Infect Dis 168:537–551
62. Frieden TR, Sterling T, Pablos-Mendez A et al (1993) The emergence of drug-resistant tuberculosis in New York City. N Engl J Med 328:521–526
63. Rastogi N (1993) Emergence of multiple-drug-resistant tuberculosis: fundamental and applied research aspects, global issues and current strategies. Res Microbiol 144:103
64. Sbarbaro JA (1993) TB control in the 21st century. Monaldi Arch Chest Dis 48:197–198, Editorial
65. Cohn DL, Bustreo F, Raviglione MC (1997) Drug-resistant tuberculosis: review of the worldwide situation and the WHO/IUATLD global surveillance project. Clin Infect Dis 24(Suppl 1):S121–S130
66. Van Deun A, Wright A, Zignol M, Weyer K, Rieder HL (2011) Drug susceptibility testing proficiency in the network of supranational tuberculosis reference laboratories. Int J Tuberc Lung Dis 15(1):116–124
67. World Health Organization (2011) Towards universal access to diagnosis and treatment of multidrug-resistant and extensively drug-resistant tuberculosis by 2015 (WHO/HTM/TB/2011.3). Geneva, 2011
68. Ministry of Health and Social Development of the Russian Federation (2010) Tuberculosis in the Russian Federation 2009. An analytical review of the TB statistical indicators used in the Russian Federation. The Russian Federation, Moscow, 2010
69. World Health Organization (2010) Multidrug and extensively drug-resistant TB (M/XDR-TB) - 2010 global report on surveillance and response (WHO/HTM/TB/2010.3). Geneva, 2010
70. Edlin BR, Tokars JI, Grieco MH, Crawford JT, Williams J, Sordillo EM, Ong KR, Kilburn JO, Dooley SW, Castro KG, Jarvis WR, Holmberg SD (1992) An outbreak of multidrug-resistant tuberculosis among hospitalized patients with the acquired immunodeficiency syndrome. N Engl J Med 326:1514–1521
71. Moro ML, Gori A, Errante I, Infuso A, Franzetti F, Sodano L, Iemoli E (1998) An outbreak of multidrug-resistant tuberculosis involving HIV-infected patients of two hospitals in Milan, Italy. Italian Multidrug-Resistant Tuberculosis Outbreak Study Group. AIDS 12:1095–1102
72. Wells CD, Cegielski JP, Nelson LJ, Laserson KF, Holtz TH, Finlay A, Castro KG, Weyer K (2007) HIV infection and multidrug-resistant tuberculosis: the perfect storm. J Infect Dis 196(Suppl 1):S86–S107
73. Raviglione M (2006) XDR-TB: entering the post-antibiotic era? Int J Tuberc Lung Dis 10(11):1185–1187
74. Dawson JJ, Devadatta S, Fox W, Radhakrishna S, Ramakrishnan CV, Somasundaram PR et al (1966) A 5-year study of patients with pulmonary tuberculosis in a concurrent comparison of

home and sanatorium treatment for one year with isoniazid plus PAS. Bull World Health Organ 34(4):533–551
75. Styblo K, Meijer J, Sutherland I (1969) Tuberculosis surveillance research unit report no. 1: the transmission of tubercle bacilli; its trend in a human population. Bull Int Union Tuberc 42: 1–104
76. Dahle UR, Eldholm V, Winje BA, Mannsåker T, Heldal E (2007) Impact of immigration on the molecular epidemiology of *Mycobacterium tuberculosis* in a low-incidence country. Am J Respir Crit Care Med 176(9):930–935
77. McKenna MT, McCray E, Onorato I (1995) The epidemiology of tuberculosis among foreign-born persons in the United States, 1986 to 1993. N Engl J Med 332(16):1071–1076
78. World Health Organisation (2015) The global plan to stop TB 2006–2015
79. World Health Organization (2005) Fifty-eighth world health assembly: resolutions and decisions. WHA58/2005/REC/1. WHO, Geneva, 2005
80. WHO (2004) Report on the meeting of the second ad hoc Committee on the TB Epidemic (WHO/HTM/STB/2004.28). World Health Organization, Geneva, 2004
81. WHO (2006) Stop TB partnership and the World Health Organization. The global plan to stop tuberculosis, 2006–2015. WHO/HTM/STB/2006.35. WHO, Geneva, 2006
82. Raviglione MC, Uplekar MW (2006) WHO's new stop TB strategy. Lancet 367(9514):952–955
83. Migliori GB, Hopewell PC, Blasi F, Spanevello A, Raviglione MC (2006) Improving the TB case management: the international standards for tuberculosis care. Eur Respir J 28(4):687–690
84. Tuberculosis Coalition for Technical Assistance (2006) International standards for tuberculosis care (ISTC). Tuberculosis Coalition for Technical Assistance, The Hague, 2006
85. Tuberculosis Coalition for Technical Assistance (2009) International standards for tuberculosis care (ISTC), 2nd edn. Tuberculosis Coalition for Technical Assistance, The Hague, 2009
86. Clancy L, Rieder HL, Enarson DA, Spinaci S (1991) Tuberculosis elimination in the countries of Europe and other industrialized countries. Eur Respir J 4(10):1288–1295
87. Broekmans JF, Migliori GB, Rieder HL, Lees J, Ruutu P, Loddenkemper R, Raviglione MC (2002) World Health Organization, International Union Against Tuberculosis and Lung Disease, and Royal Netherlands Tuberculosis Association Working Group. European framework for tuberculosis control and elimination in countries with a low incidence. Recommendations of the World Health Organization (WHO), International Union Against Tuberculosis and Lung Disease (IUATLD) and Royal Netherlands Tuberculosis Association (KNCV) Working Group. Eur Respir J 19(4):765–775
88. Migliori GB, Loddenkemper R, Blasi F, Raviglione MC (2007) 125 years after Robert Koch's discovery of the tubercle bacillus: the new XDR-TB threat. Is "science" enough to tackle the epidemic? Eur Respir J 29(3):423–427
89. Migliori GB, D'Arcy Richardson M, Sotgiu G, Lange C (2009) Multidrug-resistant and extensively drug-resistant tuberculosis in the West. Europe and United States: epidemiology, surveillance, and control. Clin Chest Med 30(4):637–665

Chapter 13
Waterborne Diseases of the Ocean, Enteric Viruses

Jacquelina W. Woods

Glossary

Attenuate	To reduce or weaken a pathogen.
Cultivate	The ability of viruses to replicate under ideal conditions.
Enteric virus	Viral particle associated with human feces. Enteric viruses range in size of 20–80 nm and contain RNA or DNA enclosed by a protein capsid.
Environmental stability (microorganism)	The ability of an organism to withstand degradation when exposed to environmental factors, such as temperature and sunlight.
Epidemiology	The study of the causes, distribution, and patterns of health and illness in a population.
Gastroenteritis	Inflammation of the stomach and large and small intestines.
Genogroup	Related viruses within a genus.
Icosahedral	Having a geometric structure that contains 20 identical equilateral triangular faces, 30 edges, and 12 vertices.
Open reading frame (ORF)	A DNA sequence that does not contain a stop codon. ORFs are presumptive genes.

This chapter, which has been modified slightly for the purposes of this volume, was originally published as part of the Encyclopedia of Sustainability Science and Technology edited by Robert A. Meyers. DOI:10.1007/978-1-4419-0851-3

J.W. Woods (✉)
US Food and Drug Administration, Center for Food Safety and Applied Nutrition, Gulf Coast Seafood Lab, Dauphin Island, AL 36528, USA
e-mail: Jacquelina.Woods@fda.hhs.gov

P. Kanki and D.J. Grimes (eds.), *Infectious Diseases: Selected Entries from the Encyclopedia of Sustainability Science and Technology*, DOI 10.1007/978-1-4614-5719-0_13,

Serotype	Distinct antigenic variations within a subgroup or subspecies of bacteria or viruses.
Viremia	The presence of viruses in the blood spread through primary and secondary transmission.

Definition of the Subject

Human *enteric viruses* pose a significant health threat in the aquatic environment since they are transmitted via the fecal-oral route. Human activities such as faulty septic systems, agricultural runoff, urban runoff, sewage outfall, and wastewater discharge from vessels are ways enteric viruses are introduced into the environment. There are approximately 140 enteric viruses found in humans, and approximately one billion per gram of feces where at least 10% of the population can shed these viruses at any given time [1]. Enteric viruses can be transferred throughout the environment by attaching to particulates in groundwater, estuarine water, seawater and rivers, estuaries, shellfish grown in contaminated waters, and by aerosols emitted from sewage treatment plants [2]. The fate of these enteric viruses can take many routes, such as rivers, lakes, sewage, land runoff, estuaries, and groundwater. Humans can be exposed to enteric viruses through various routes: crops grown in land irrigated with wastewater or fertilized with sewage, shellfish grown in contaminated water, sewage-polluted recreational waters, and contaminated drinking water. In a waterborne disease outbreak study between 1946 and 1980, water system deficiencies that contributed to these outbreaks were categorized under five major headings: (1) use of contaminated untreated surface water, (2) use of contaminated untreated groundwater, (3) inadequate or interrupted treatment, (4) distribution network problems, and (5) miscellaneous [3]. Deficiencies in treatment and distribution of water contributed to more than 80% of the outbreaks.

Introduction

The most commonly studied enteric viruses belong to the families of single-stranded RNA viruses (ssRNA) [*Picornaviridae* (enteroviruses, polioviruses, coxsackieviruses, hepatitis A virus, and echoviruses), *Caliciviridae* (noroviruses, caliciviruses, and astroviruses)], double-stranded DNA (dsDNA) [*Adenoviridae* (adenoviruses)], and double-stranded RNA (dsRNA) [*Reoviridae* (reoviruses and rotaviruses)]. These enteric viruses have cellular and molecular structures that make them resistant to current water treatment processes. Emerging enteric viral pathogens like Aichi virus (ssRNA), sapovirus (ssRNA), and picobirnaviridae (bi-segmented dsDNA) have properties similar to currently studied enteric viral pathogens in that they are non-enveloped, resistant to heat inactivation, stable at low pH, resistant to chlorination, or resistant to UV light inactivation [1]. Parvoviruses (the smallest known enteric viruses with ssRNA and high heat resistance) and polyomaviruses (include JC virus, BK virus, and simian virus 40 which are non-enveloped dsDNA viruses) can also be considered emerging viruses but do not cause acute *gastroenteritis* as do the most commonly studied enteric viruses [4–6].

Although enteric virus infections are associated primarily with self-limiting gastroenteritis in humans, they may also cause respiratory infections, conjunctivitis, hepatitis, and diseases that have high mortality rates, such as aseptic meningitis, encephalitis, and paralysis in immunocompromised individuals [7]. In addition, some enteric viruses have been linked to chronic diseases, such as myocarditis and insulin-dependent diabetes [1].

Human enteric viruses can be transmitted by water, food, fomites, and by human contact. They typically have a low infectious dose which makes them an immediate public health concern. In some instances, e.g., norovirus infections, the infectious dose can be as little as one to ten virions with a secondary attack rate of 50% [7]. The risk for infection when consuming viral contaminated water is at least tenfold greater than that for pathogenic bacteria with similar exposures [8]. The World Health Organization (WHO) estimates that the most common waterborne disease, diarrhea, is responsible for 4.6 billion episodes of diarrhea and causes 2.2 million deaths worldwide [9]. Enteric viruses in water are of particular concern because of the potential for contamination from a variety of sources. Because significant advances have been made in the area of environmental virology, enteric viruses have now been recognized as the causative agents in many nonbacterial gastroenteritis cases and outbreaks identified in the past as of unknown etiological origin [2]. Enteric viruses have been detected and linked to many outbreaks from contaminated waters and foodstuff [10–12].

Enteric Viruses Found in the Environment

Norovirus

An outbreak of acute gastroenteritis occurred among students and teachers in a school in Norwalk, Ohio, in 1968 [13]. The initial attack rate had a morbidity of 50%, and a secondary attack produced a higher rate of 82% [14]. Nausea and vomiting occurred in >90% of those affected, while diarrhea occurred in 38% of those affected. The clinical onset of the illness was typically 12–24 h with duration of 12–60 h. Upon subsequent transmission of stool filtrates to human volunteers in 1972, a small round-structured virus (SRSVs) ~27-nm was identified by electron microscopy [15]. Later studies revealed that other SRSVs morphologically similar

Table 13.1 Genogroups, clusters, and strains of human norovirus

GI		GII		GIV	
Cluster	Strain	Cluster	Strain	Cluster	Strain
4	ChibaJPN00	15	J23-USA02	1	Alpha-NLD99
5	MusgroveGBR00	4	Bristol-GB93		
2	SOV-GBR93	12	Wortley-GB00		
6	Hesse-DEU98	1	Hawaii-US94		
1	NY-USA93	16	Tiffin-US03		
3	DSV-USA93	5	Hillingd-GB00		
7	Wnchest-GBR00	2	Msham-GB95		
8	Boxer-USA02	10	Erfurt-DEU01		
		13	Faytvil-US02		
		17	CSE1-US03		
		6	Seacrof-GB00		
		8	Amstdam-NLD99		
		9	VAbeach-US01		
		14	M7-US03		
		7	Leeds-GB00		
		3	Toronto-CAN93		
		11	SW918-JPN01		

to NoV caused gastroenteritis, but Norwalk virus remained the prototype of these fecal viruses [14]. Before the discovery of NoV, most cases of gastroenteritis not attributed to bacteria were thought to be caused by nonbacterial gastroenteritis [16]. In 1981, Greenberg et al. published data proposing that NoV might be a calicivirus. In 1993, Jiang et al. provided molecular evidence that NoV was a calicivirus with a 7.6-kilobase (kb) viral genome consisting of a positive sense, single-stranded, polyadenylated RNA. The genome consists of three *open reading frames* (ORFs), which code for the nonstructural proteins including the RNA polymerase (ORF1), capsid protein (ORF2), and a minor structural protein (ORF3). Noroviruses can be separated into five *genogroups* (GI, GII, GII, GIV, and GV) based on the sequence comparison of the RNA polymerase and capsid regions [17, 18]. Genogroup I infects humans only, genogroup II infects humans and swine, and genogroup IV infects humans and canine. Genogroup III infects bovine animals, and genogroups V infects mice. Of the five genogroups, the classification scheme for the different clusters and strains identifies NoV GI containing 8 clusters, GII containing 17 clusters, and GIV containing 1 cluster (Table 13.1). Noroviruses have an assigned nomenclature where strains are named after the geographic location of the outbreak from which they are first described. The genogroups and genotypes were characterized and classified based on the RNA polymerase region and the complete capsid gene sequences [17, 18]. To date, the most common genogroup implicated in gastrointestinal infections is genogroup II.

Noroviruses are transmitted primarily through the fecal-oral route, consumption of fecally contaminated food or water, or by person-to-person contact. There is a high rate of secondary infection, and this can occur by airborne transmission. Outbreaks commonly occur in schools, nursing homes, hospitals, camps, daycare centers, cruise ships, and any other close community situation. Because norovirus is not listed as a reportable disease to health officials, estimates of the level of infection may not be accurate.

To date, the ability to *cultivate* norovirus has been futile, although several attempts have been made. A Duizer et al. study [19] utilized 27 different cell lines in an unsuccessful attempt to cultivate norovirus. Straub et al. [20] utilized human embryonic intestinal epithelial cells (INT-407) with 3-D tissue culture. The use of INT-407 cells with 3-D tissue culture appeared to be promising as this method provided the closest attempt at mimicking the structure of in vivo cells. The difficulty with cultivation of norovirus may be explained by its specific requirements or receptors needed for attachment to cells in order for replication to occur. Experiments with recombinant norovirus particles and human gastrointestinal biopsies showed preferential binding to epithelial cells of the pyloric region of the stomach and to enterocytes on duodenal villi [19]. Human-specific blood group antigen H1–type expression was shown to be necessary for norovirus attachment to the cells, and most routine cell cultures lack the characteristics of these specialized human intestinal epithelial cells.

The majority of background information on the biological properties of norovirus has been obtained through humans who volunteer for human feeding studies [21, 22]. Infectivity can only be assessed in human dose–response experiments, and the infectious dose had been determined to be around ten virus genomes [22]. This is very critical when considering norovirus survival. Norovirus remains infectious under refrigeration and freezing conditions, it survives well in the environment, and it is resistant to heat and drying conditions. This was demonstrated in an outbreak at a long-term care facility where norovirus survived on fomites and continued to cause infection 2 weeks after the initial peak of illnesses [23]. Norovirus will continue to be a significant health threat worldwide as this virus continues to evolve. The lack of a tissue culture cell line for effective propagation of norovirus will hinder complete understanding of how this virus causes infection in addition to decreased progression on the development of a productive vaccine.

Hepatitis A Virus

Hepatitis A is a non-enveloped RNA virus 27–32 nm in diameter. It has an *icosahedral* symmetry and belongs to the genus *Hepatovirus* of the *Picornaviridae* family. HAV has a positive-polarity single-stranded 7.5-kb genome with a single ORF with three distinct regions (P1, P2, and P3) [24]. Region P1 consists of four capsid proteins VP1–VP4. Region P2 consists of nonstructural proteins 2A–2C, and region P3 consists of nonstructural proteins and virus-specific proteins (VPg) [24]. HAV demonstrates a high degree of antigenic (amino acid) and genetic (nucleotide) conservation throughout the genome [24, 25]. Although this high propensity for conservation exists, there is still enough diversity to define HAV genotypes and sub-genogroups [26]. The genomic regions commonly used to define HAV genogroups include (1) the C terminus of the VP3 region, (2) the N terminus of the VP1 region, (3) the 168-bp junction of the VP1/P2A regions, (4) the 390-bp region of the VP1/P2B regions, and (5) the entire VP1 region [26, 27]. A total of six genotypes have been identified: genotype 1A, 1B, II, III, IV, V, and VI. Genotypes I, II, and III are of human origin, and IV, V, and VI genotypes are of nonhuman primate origin. Genotype I and III are the most prevalent genotypes isolated from humans [28]. Because there is only one *serotype* of HAV, individuals infected by HAV in one part of the world are protected from reinfection by HAV in another part of the world.

Infections with HAV can produce effects that range in severity from asymptomatic to death from fulminant hepatitis. Infections with HAV are typically self-limiting and do not result in chronic liver disease. The virus shed in the feces and peak fecal excretion, hence infectivity, occurs prior to the onset of symptoms [29]. Clinical manifestations can increase with age, and with older children and adults, symptoms are typically present with jaundice occurring in 70% of those infected [29]. An average incubation period is 28 days with a range of 15–50 days. Symptoms include gastroenteritis, fever, malaise, anorexia, nausea, abdominal discomfort, dark urine, and jaundice, all which may last up to 2 months. Chronic liver disease has not been shown to persist, although 15–20% of those infected may have prolonged or relapsing disease lasting up to 6 months [30]. Fulminant hepatitis is a rare complication in HAV infections, only occurring in <1% of those infected, with the highest rates occurring in young children and elderly adults who may have underlying liver conditions [28]. Among those with fulminant disease, reported findings demonstrate nucleotide and/or amino acid substitutions in the 5′ untranslated region, P2 region, and the P3 region of the HAV genome [28, 31].

Cell culture propagation of human and nonhuman HAV has occurred in African green monkey kidney and fetal rhesus monkey kidney cells [32, 33]. Propagation of HAV of human origin is quite different than propagation of other picornaviruses of human origin. HAV of human origin requires an extensive adaptation period or serial passages before it grows in cell culture, and once it has adapted, HAV becomes *attenuated* as demonstrated by not producing disease in experimentally inoculated nonhuman primates [34]. Mutations causing attenuation in viral nucleic acid could play a significant role in the adaptation of HAV in cell culture [24, 33]. The attenuated strain HM-175 was adapted in cell culture, and this strain is currently used as vaccination agent for HAV [24].

HAV is a major cause of acute hepatitis in developed countries, while in developing countries, it can be considered endemic [35]. HAV is transmitted primarily by the fecal-oral route, via person-to-person contact, and by contaminated food or water, while the other common hepatitis viruses (hepatitis B and hepatitis C) are typically transmitted by blood or body fluids. HAV is stable in the environment when associated with organic material, and it is resistant to low pH and heating [36]. The name "hepatitis" is derived from the fact that HAV replicates in the liver and affects the liver. The source of the infectious agent, however, cannot be identified in approximately 50% of reported hepatitis A cases

Table 13.2 Classification of human enteroviruses

Groups of enteroviruses	Species of enteroviruses
Enterovirus-A	Coxsackievirus A2–8, 10, 12, 14, 16
	Enterovirus 71, 76, 89, 90, 91
Enterovirus-B	Coxsackievirus A9
	Coxsackievirus B1–6
	Echovirus 1–7, 9, 11–21, 24–27, 29–33
	Enterovirus 69, 73–75, 77–78, 79–88, 100–101
Enterovirus-C	Coxsackievirus A1, 11, 13, 17–22, 24
	Poliovirus 1–3
Enterovirus-D	Enterovirus 68, 70

in the U.S., partly because of the long incubation before the appearance of symptoms [28]. Because outbreaks of HAV can cause considerable morbidity and even mortality, it is imperative that HAV be isolated and identified as the implicated pathogen.

Enterovirus

Human enteroviruses are members of Picornaviridae family and *Enterovirus* genus. They are icosahedral, non-enveloped, with a diameter of 27–30 nm. The genome is 7.5–8.5 kb and is composed of single-stranded positive polarity RNA. Enteroviruses are further divided into the subgenera/species of poliovirus, coxsackieviruses, echoviruses, and enteroviruses, and they are marked according to their serotypes. The poliovirus group consists of three different serotypes. Types 1 and 3 are recognized as epidemic, while type 2 is endemic [37]. Coxsackieviruses consist of groups A and B, where the A group contains 24 serotypes, and the B group contains six serotypes. In 2003, the International Committee on Taxonomy of Viruses classified *Enteroviruses* into four groups of species based on their molecular properties [38] (Table 13.2).

Infections due to enteroviruses are common, causing a range of disease including pharyngitis and poliomyelitis. In the United States, it is estimated that 30–50 million enterovirus infections occur a year, of which only 5–15 million are symptomatic [37]. The virus is spread by the fecal-oral route and person to person through direct contact with secretions of an infected individual. The incubation period is usually 3–7 days with virus transmission lasting 3–10 days after symptoms appear. Replication occurs in the gastrointestinal track, but can occur in other tissues, such as nerve and muscle [37, 39]. Polioviruses typically infect their host by attacking the central nervous system, causing paralysis (poliomyelitis) in infected individuals. The spread of poliovirus has been limited by the development and use of vaccines. The Sabin trivalent oral live attenuated vaccine (OPV) consists of three live attenuated strains of 1, 2, and 3 serotypes grown in cell culture. The Salk vaccine is a trivalent inactivated polio vaccine (IPV) given by injection. The Sabin vaccine has an advantage over the Salk vaccine in that it elicits secretory IgA antibody production in addition to IgA, IgM, and IgG serum antibody production [40]. Coxsackieviruses have been associated with respiratory infections, gastroenteritis, insulin-dependent diabetes, myocarditis, and pericarditis [1]. Echoviruses are typically less infectious and are usually associated with the common cold and other respiratory diseases. The numbered enteroviruses have not been widely studied, but they are generally associated with bronchiolitis, conjunctivitis, meningitis, and paralysis [7, 41]. Also, enteroviruses are one of few enteric viruses which produce *viremia* in infected patients.

Enteroviruses are resistant to most concentrations of chlorine used in sewage treatment, and they are tolerant to cold and warm temperatures. This makes them ideally suited for survival in the environment. Stability of enteroviruses in the environment is dependent on temperature, humidity, and UV radiation. In order to inactivate 90% of poliovirus in salt water environment, 671 days at 4°C is required; on the other hand, an increase in the temperature to 25°C reduces the inactivation time by 25 days [37]. Symonds et al. [81] collected sewage influent and effluent from several wastewater treatment plants (WWTP) across the United States and revealed that 75% of the sewage influent and 8.3% of the effluent contained enteroviruses. In 2003, the largest European outbreak of enterovirus-related infection occurred in Belarus [42]. Over 1,300 people became ill, and water contaminated with echovirus and coxsackievirus was identified as the source of the infection. Aside from water and sewage samples, enteroviruses have been detected in food samples. In 1914, the first described food-borne outbreak was linked to milk contaminated with poliovirus [43]. After pasteurization of milk was adopted, transmission of enterovirus by contaminated milk decreased dramatically. Despite its demonstrated presence in the environment and sewage, there have been very few food-borne-related outbreaks due to enterovirus.

Adenovirus

Adenoviruses are members of the *Adenoviridae* family and the *Mastadenovirus* genus, which comprises five genera and infects hosts across the extended spectrum of vertebrates [44, 45]. Human adenoviruses are a double-stranded DNA virus containing a non-enveloped icosahedral shell with fingerlike projections forming each of the 12 vertices [46]. Its DNA is linear, about 35 kb in size, and it encodes for more than 30 structural and nonstructural proteins [47]. In 1953, the first adenovirus was isolated from human adenoid tissue [48]. There are 51 serotypes of adenovirus, and they are divided into six species, A–F, based on their hemagglutination properties, their oncogenic potential in rodents, and DNA homology or GC content [44].

Infections with adenoviruses can result in a wide range of clinical symptoms. Subgroups A, D, and F are sites for gastrointestinal infections. Subgroup B is responsible for lung and urinary tract infections. Subgroups C and E are responsible for respiratory tract infections. Serotypes 40 and 41 are the cause of most adenovirus associated–gastroenteritis, and serotypes 4 and 7 are associated with most cases of ARD (acute respiratory disease) in the United States [49]. Most adenovirus infections are self-limiting except cases where the infected individual is immunocompromised. However, in 2007, there were cases of a new emerging strain of adenovirus 14 that caused fatal respiratory disease in healthy individuals [50]. After the primary infection, immunity is conferred to the causative adenovirus serotype.

Human adenoviruses are specific to humans even though adenoviruses infect a range of animals. In 2005, Cox et al. reported no viable human adenovirus detection in feces of cattle, valve, pig, sheep, horse, dog, poultry, wombat, cat, kangaroo, possum, wood duck, rat, wild pig, fox, rabbit, ferry cat, goat, carp, and deer. In domestic sewage worldwide, human adenoviruses have been detected in high concentrations and their detection in sewage seems to have little seasonal variability [2, 44, 51]. As with most enteric viruses, adenovirus survives better in the environment and sewage treatment than the current indicator bacteria. Adenoviruses have increased resistance to UV light, and this increased resistant could be due to the DNA repair mechanism of the host cell. Because of the *environmental stability* of adenovirus, they have been suggested as an indicator of viral pollution [52]. In current literature, adenovirus has been associated with waterborne outbreaks. Foodborne outbreaks have been suspected but not confirmed [49, 53].

Rotavirus

Rotaviruses are a significant cause of infantile viral gastroenteritis, with 114 million annual cases of diarrhea and up to 352,000–592,000 deaths in children younger than five worldwide [54]. Rotavirus' increased activity occurs during peak respiratory virus activity, rising in the winter and peaking in the spring.

Rotaviruses are members of the family *Reoviridae* and are double-stranded RNA with 11 segments and a triple capsid structure. Classification of rotavirus into P and G genotypes/serotypes are based on the outer capsid proteins, VP4 (P protein) and VP7 (G protein). Only 14 G and 15 P genotypes have been identified in humans [55]. There are seven different serogroups of rotavirus (A–G). Although individuals may have multiple rotavirus infections over time, group A rotaviruses cause the greatest numbers of cases of acute gastroenteritis (AGE) in children under 2.

The primary route of transmission of rotavirus is the fecal-oral route; additionally, the persistence of rotaviruses in aquatic environments and their resistance to water treatment and disinfection processes may facilitate their transmission to humans. Rotavirus has been implicated in many waterborne epidemics, and these viruses have been isolated in sewage, river water, groundwater, and drinking water [56–58]. Rotaviruses are resistant to most disinfectants, heat, proteolytic enzymes, and pH with ranges of 3–10. Their ability to survive in various environmental conditions for long periods of time contributes to their high rate of infection.

The burden of rotavirus disease has prompted a need for an effective vaccine. In 1998, FDA-approved RotaShield, a tetravalent human-rhesus genetic reassortant vaccine. RotaShield was withdrawn from the US market after it was shown to be associated with intussusceptions in infants. Vaccine development continued with the license of a monovalent human rotavirus vaccine Rotarix™ and the pentavalent bovine-human reassortant vaccine RotaTeq™. To date, neither of the later derived vaccines have been associated with intussusceptions.

Hepatitis E Virus

Hepatitis E virus infections are a major cause of acute enterically transmitted non-A, non-B (ET-NANB) hepatitis in many developing countries throughout Middle East, Asia, and Africa [59]. This disease is associated with >50% of sporadic acute hepatitis which is a cause for great public health concern [60]. HEV can be both endemic and epidemic with outbreaks generally associated with fecal contamination of drinking water. North America and Europe have generally been considered non-endemic with documented cases from individuals who had previously traveled to endemic areas. It has been implied that HEV disease may be zoonotic in endemic regions and regions where HEV disease is non-endemic [61]. Domestic animals such as cattle, sheep, and pigs all appear to be reservoirs of the virus, which may explain the presence of anti-HEV antibodies in individuals with no obvious or known contact with human strains of the virus.

Hepatitis E is a single-stranded non-enveloped positive-sense RNA virus approximately 27–30 nm in diameter with a genome of about 7.2 kb [62]. HEV is currently classified as the only member of the genus *Hepevirus* in the family *Hepeviridae* [63]. Phylogenetic analysis has identified four significant genotypes of HEV [62]. Genotype 1 (G1) was isolated from Africa and Asia, genotype 2 (G2) from Mexico, and genotype 3 (G3) and genotype 4 (G4) from the United States, China, Japan, Europe, and Vietnam [62]. Genotype 3, which differs from genotypes 1, 2, and 4 by 25.6–26.3%, 25.3–25.5%, and 23.7–24.7%, respectively, is the first animal strain of HEV to be characterized [61]. Additional HEV genotypes may be present as other new strains have not been completely sequenced and therefore cannot be prominently assigned to a genotype at present time.

Recognition of Hepatitis E as a distinctly unique agent of human disease did not occur until the 1980s [64] with the development of serological laboratory tests. Prior to the development of these tests, clinical HEV infection was determined to be the etiological agent by process of elimination. Since the *epidemiology* of hepatitis A had been well characterized, assumptions were plausible that >90% of the adult population in most Asian and Indian countries had previous exposure to HAV and were therefore immune to HAV infection [65]. This statement was proven when serologic tests for the diagnosis of hepatitis A and hepatitis E were applied to stored clinical samples collected during a waterborne epidemic of viral hepatitis in India [65]. Among these stored samples were those of the epidemic that occurred in Delhi, India, in 1955–1956 where in a population of 1.6 million, approximately 29,300 jaundice cases occurred, with an estimated 67,700 non-icteric infections. This was the first reported outbreak of disease where ET-NANB hepatitis was determined to be the etiological agent and virtually 100% of stored serum samples from such epidemics were found to contain IgG anti-HAV but not IgM anti-HAV. This can be considered definitive evidence for the existence of a previously unrecognized hepatitis as the etiological agent in past epidemics [65]. Application of current HEV immunological tests on these epidemic samples can further strengthen the evidence for ET-NANB as the causative agent in these waterborne outbreaks occurring decades ago.

The epidemiology and virology of hepatitis E suggest that HEV is less readily transmitted than HAV [64]. In South Asia, HEV incidence has been characterized by marked seasonality, with outbreaks occurring during the rainy or monsoon seasons [65]. Outbreak investigations often reveal fecally contaminated drinking water as the source of HEV infections. It appears that rain-induced flooding allows sewage to contaminate water supplies which explains the seasonal associations. Communities with inadequate sewage disposal are prone to recurrent HEV outbreaks. Low levels of community sanitation are also strongly associated with both epidemic and endemic HEV. An example of this occurred in Indonesia where the use of river water for cooking and drinking, improper disposal of human waste, and poor personal hygiene were significantly associated with increased anti-HEV IgG seroprevalence. More recent interactions between changing human culture and HEV are seen in epidemics of hepatitis E that occur among refugee camps under substandard living conditions in Ethiopia and Somalia [66]. These findings highlight the risk of HEV to aid workers and military installations stationed in countries where HEV is epidemic. Therefore, prevention and control of HEV in epidemic areas will entail improvements in community sanitation, management of sewage distribution, and treatment of drinking water.

Enterically transmitted Hepatitis E has a secondary attack rate around 2% and a relatively low infectivity compared to hepatitis A, which has a 10–20% secondary attack rate among household contacts. It is not clear why HEV has a lower secondary attack rate than HAV. The lowered environmental stability of HEV may contribute to its lower secondary attack rate [63]. Also, studies of HEV have shown an association between the severity of the disease and inoculum size, where a larger infectious dose of HEV may be required to cause overt disease. There has been no recognizable difference in attack rates or infection rates by sex where there was documented exposure to HEV [65]. However, some studies have shown that adult men in some instances may have a greater risk of developing clinical illness. This could be due to infectious dose since men could have higher exposure rates than women. In addition, societal restrictions placed on women in some cultures may minimize their exposure to HEV [64].

Transmission of hepatitis E from mother to child has been reported with associated morbidity and mortality. A small study was conducted among ten pregnant women who were infected with HEV in their third trimester. Of those who had successful delivery, six transmitted HEV to their child. HEV antigen was found in five of the six infants, and anti-IgM was found in three of six infants [65]. Other studies have also shown a twofold increased risk of clinical HEV disease among pregnant women compared with non-pregnant women of reproductive age

[65]. Both epidemic and clinical studies have shown that pregnant women have an increased likelihood of developing acute hepatitis and even fulminant hepatic failure after HEV infection. Since hepatitis infections involve the liver, it is reasonable to assume that the increased complications of pregnancy due to hepatitis can lead to increased adverse outcomes with risks of neonatal icterus. These reported findings of vertical transmission pose difficulty in accessing the risk of fecal-oral infections in newborns.

Despite our discovery of the hepatitis E, this virus continues to pose a global public health problem. There have been many advances in understanding the clinical, virological, and epidemiological characteristics of HEV, but there are still many unanswered questions. Development of an effective vaccine is currently not possible because a cell culture system that efficiently grows the virus has not been discovered, although there have been different vaccine candidates purposed. As future research of HEV continues, the epidemiologic features of this important virus will be clarified further and a protective vaccine will be developed. Perhaps then we can resolve the reasons for increased morbidity and mortality in pregnant women and determine factors which lead the persistence of protective antibodies. Once these questions have been answered, we will have better control and understanding of hepatitis E.

Detection of Enteric Viruses

In the past, detection of enteric viruses in environmental samples largely depended on whether the agent grew in cell culture. For those types that do, such as enteroviruses and adenoviruses, detection by virus replication in cell culture demonstrated infectivity as well as their presence. The ability to detect viruses by cell culture is a clear advantage when assessing whether environmental samples or foods are microbiologically hazardous. The capacity to do quantitative assays is also a bonus of cell culture. Cytopathic effect or virus-specific killing or lysing of cells is visible by ordinary light microscopy. Cell culture assays were the most widely used protocol for detection of enteric viruses until the 1990s [53, 67, 68]. While cell culture offers quantitative analysis and infectivity, the high cost, long turnaround time, and labor-intensive efforts are drawbacks. There are many cell lines suitable for growing enteric viruses. Buffalo green monkey (BGM) cell line has been shown to give higher plaques forming units per milliliter (PFU/ml) and faster CPE for coxsackieviruses and polioviruses [69].

There are a host of enteric viruses that are capable of growth utilizing cell culture, but two significant enteric viruses, human noroviruses and wild-type HAV, are difficult or unable to propagate in cell culture. Cell culture would apparently not be a proactive monitoring protocol given the length of time required to complete most assays and difficulties of propagation. However, cell culture used in conjunction with detection emerging assays can be used to address the issue of viability while comparing the sensitivity of each representative assay.

Detection of enteric viral pathogens with molecular-based PCR assays has successfully been utilized for a number of years [17, 70–72]. Conventional PCR methodology utilizes a pair of oligonucleotides or primers, each hybridizing to one strand of double-stranded DNA (dsDNA) target. The types of primers used can be (1) random primers – short single-stranded DNA fragments with all possible combinations of bases, (2) polythymine primers – 16-base-long thymine primers that will hybridize with the polyadenine end of the mRNA, and (3) specific primers – only the targeted region specific to the primers will be amplified. The primers act as a substrate for DNA polymerase which creates a complementary strand by the way of sequential addition of deoxynucleotides. The process of PCR can be summarized into three steps: (1) denaturation – dsDNA is separated by an increase in temperature, (2) annealing – the temperature is decreased to allow the primers to anneal to the separated DNA, and (3) extension – the extension of the DNA fragment with the primers attached by addition of deoxynucleotides. For RNA viruses such as enteroviruses, NoV, and HAV, RT-PCR, or the conversion of RNA to cDNA, is necessary. During reverse transcription, a primer is required for the reverse transcriptase (RNA-dependent DNA polymerase) to initiate the synthesis of cDNA from the viral RNA. For DNA viruses, reverse transcription is not a necessary step. The final PCR product is analyzed by electrophoresis in the presence of ethidium bromide in which the correct size of the product can be examined visually by ultraviolet light. Hybridization with digoxigenin-labeled probes or genetic sequencing can be used to further identify PCR products. If the products examined are from environmental samples, cloning of the PCR products may also be required as there can be multiple strains of individual viruses present.

Real-time quantitative PCR or qPCR is used to quantitatively determine the amount of original target present in the sample [73, 74]. During a qPCR assay, the amplicon produced during each cycle can be quantified using SYBR Green (nonspecific attachment to dsDNA), or by using a fluorescent internal probe (specific hybridization) [74]. For SYBR green assays, melt curve analysis is important to ensure that the desired DNA fragment is detected as different DNA fragments have different T_m (melting temperature). For fluorescent internal probes, fluorescence is measured during each cycle, and when the amount of fluorescence exceeds the background level (threshold level), the sample is scored as positive. The number of cycles required to reach the threshold level, commonly referred to as the cycle threshold value (C_t), correlates with the amount of target in the sample prior to amplification [73]. Real-time PCR is an excellent tool for detection of enteric viruses in environmental samples and has been used successfully to determine the concentrations of viral genomes [75, 76].

Multiplex PCR, which utilizes multiple primer sets within a single PCR reaction, can be used to simultaneously detect different groups of viruses. However, this multiple viral detection can be difficult to optimize because of the different annealing temperature requirements of dissimilar primer sets and because of the properties of the viral nucleic acids found between viral groups [41]. Real-time PCR (qPCR) has been quite successful for detection of multiple enteric viruses, because it can analyze each target independently in the same assay by using specific internal probes binding to different fluorochromes that the real-time PCR equipment can analyze independently [77, 78]. Furthermore, the PCR products can be of a similar size, providing better amplification efficiency. For viruses that grow poorly in cell culture, the detection by PCR integrated with cell culture (i.e., ICC-PCR) drastically reduces the time needed for detection [79]. The detection of enteroviruses in water can be between 3 and 14 days using cell culture, 5 days using integrated cell culture, and less than a day using direct real-time PCR. Because viruses are normally present in very low concentrations in environmental samples, the level of sensitivity of most PCR is advantageous for detection of low copy number. While the sensitivity of PCR is beneficial, the presence of inhibitory substances (i.e., humic acid or heavy metals) in concentrated environmental samples is of concern. Internal controls for real-time PCR have been developed to determine the presence of inhibitors in a sample and ensure that reaction conditions are optimal [80].

While there has been tremendous progress in molecular detection assays, complications remain. Even though most molecular-based assays are specific, sensitive, rapid, and cost efficient, there has been no development of a universal method or standardization. Perhaps future development of molecular-based assays that can establish infectivity will combine best of cell culture and PCR when it comes to the detection of enteric viruses in the environment.

Future Directions

Enteric viruses in the aquatic environment have been thoroughly studied over the years. Their persistence and distribution in the environment continues to cause threats to public health. Development of molecular techniques has allowed detection of these viral pathogens, especially those that cannot be readily propagated utilizing cell culture techniques. As we continue to make advances in understanding the pathogenic nature of these viruses, we can be instrumental in preventing disease outbreaks and improve public health.

Bibliography

Primary Literature

1. Griffin DW, Donaldson KA, Paul JH, Rose JB (2003) Pathogenic human viruses in coastal waters. Clin Microbial Rev 16:129–143
2. Bosch A (1998) Human enteric viruses in the water environment: a minireview. Int Microbial 1:191–196
3. Lippy EC, Waltrip SL (1984) Waterborne disease outbreaks 1946–1980: a thirty-five year prospective. J Am Water Works Assoc 76:60–67

4. Bofill-Mas S, Pina S, Girones R (2000) Documenting the epidemiologic patterns of polyomaviruses in human populations by studying their presence in urban sewage. Appl Environ Microbiol 66:238–245
5. Brauniger S, Peters J, Borchers U, Kao M, Borchers U (2000) Further studies of thermal resistance of bovine parvovirus against moist and dry heat. Int J Hyg Environ Health 203:71–75
6. Engelbrecht RS, Weber MJ, Salter BL, Schmidt CA (1980) Comparative inactivation of viruses by chlorine. Appl Environ Microbiol 40:249–256
7. Kocwa-Haluch R (2001) Waterborne enteroviruses as a hazard for human health. Pol J Environ Stud 10:485–487
8. Haas CN, Rose JB, Gerba CP, Regli R (1993) Risk assessment of viruses in drinking water. Risk Anal 13:545–552
9. World Health Organization (2010) Guidelines for drinking-water quality. Recommendations, 3rd edn. World Health Organization, Geneva
10. Beuret C, Kohler D, Baumgartner A, Luthi TM (2002) Norwalk-like virus sequences in mineral waters: one year monitoring of three brands. Appl Environ Microbiol 68:1925–1931
11. Daniels NA, Bergmire-Sweat DA, Schwab KJ, Hendricks KA, Reddy S, Monroe SA, Fankhauser RL, Monroe SS, Atmar RL, Glass RI, Mead P (2000) A foodborne outbreak of gastroenteritis associated with Norwalk-like viruses: first molecular traceback to deli sandwiches contaminated during preparation. J Infect Dis 181:1467–1470
12. Munnoch S, Ashbolt R, Coleman DJ, Walton N, Beers-Debble MY, Taylor R (2004) A multi-jurisdictional outbreak of hepatitis A related to a youth camp-implications for catering operations and mass gatherings. Commun Dis Intell 28:521–527
13. Adler J, Zickl R (1969) Winter vomiting disease. J Infect Dis 119:668–673
14. Atmar R, Estes MK (2001) Diagnosis of noncultivatable gastroenteritis viruses, the human caliciviruses. Clin Microbiol Rev 14:15–37
15. Kapikian AZ, Wyatt RG, Dolin R, Thornhill TS, Kalica AR, Chanock RM (1972) Visualization by immune electron microscopy of a 27-nm particle associated with acute gastroenteritis. J Virol 10:1075–1081
16. Rippey SR (1994) Infectious disease associated with molluscan shellfish consumption. Clin Microbiol Rev 7:419–425
17. Ando T, Monron SS, Gentsch JR, Jin Q, Lewis DC, Glass RI (1995) Detection and differentiation of antigenically distinct small round structured viruses (Norwalk-like viruses) by reverse transcription-PCR and southern hybridization. J Clin Microbiol 33:64–71
18. Zheng D, Ando T, Fankhauser RL, Beard RS, Glass RI, Monroe SS (2006) Norovirus classification and proposed strain nomenclature. Virology 346:312–323
19. Duizer E, Schwab KJ, Neil FH, Atmar RL, Koopmans MPG, Estes MK (2004) Laboratory efforts to cultivate noroviruses. J Gen Virol 85:79–87
20. Straub TM, Honer zu Bentrup K, Orosz-Coghlan P, Dohnalkova A, Mayer BK, Bartholomew A, Valdez CO, Bruckner-Lea CJ, Gerba C, Abbazadegan CM, Nickerson CA (2007) In vitro cell culture infectivity assay for human noroviruses. Emerg Infect Dis 13:396–403
21. Dolin R, Blacklow NR, Dupont H, Formal S, Buscho RF, Kasel JA, Buscho RF, Kasel JA, Chames RP, Hornick R, Channock RM (1971) Transmission of acute infectious nonbacterial gastroenteritis to volunteers by oral administration of stool filtrates. J Infect Dis 123:307–312
22. Teunis PF, Moe CL, Liu P, Miller SE, Lindesmith L, Baric RS, Le Pendu J, Calderone RL (2008) Norwalk virus: how infectious is it? J Med Virol 80:1468–1476
23. Wu HM, Fornek M, Schwab K, Chapin AR, Gibson K, Schwab E, Spencer C, Henning K (2005) A norovirus outbreak of a long-term-care facility: the role of environmental surface contamination. Infect Control Hosp Epidemiol 26:802–810
24. Cohen JL, Rosenblum B, Ticehurst JR, Daemer RJ, Feinstone SM, Purcell RH (1987) Complete nucleotide sequence of an attenuated hepatitis A virus: comparison with wild-type virus. Proc Natl Acad Sci USA 84:2497–2501
25. Lemon SM, Jansen RW, Brown EA (1992) Genetic, antigenic, and biological differences between strains of hepatitis A virus. Vaccine 10(suppl 1):S40–S44

26. Robertson BH, Janson RW, Khanna B, Totsuka A, Nainan OY, Siegl G, Widell A, Margolis HS, Isomura S, Ito K, Ishizu T, Moritsugu Y, Lemon SM (1992) Genetic relatedness of hepatitis A virus strains recovered from different geographical regions. J Gen Virol 73:1365–1377
27. Hutin YJ, Pool E, Cramer EH, Nainan OV, Weth J, Williams LT, Goldstein ST, Gensheimer KF, Bell BP, Shapiro CN, Alter MJ, Margolis HS (1999) A multistate, foodborne outbreak of hepatitis A. National hepatitis A investigation team. N Engl J Med 340:595–602
28. Nainan OV, Armstrong GL, Han X, Williams I, Bell BP, Margolis HS (2005) Hepatitis A molecular epidemiology in the United States, 1996–1997: sources of infection and implications of vaccination policy. J Infect Dis 191:957–963
29. Lednar WM, Lemon SM, Kirkpatrick JW, Redfield RR, Fields ML, Kelly PW (1985) Frequency of illness associated with epidemic hepatitis A virus infections in adults. Am J Epidemiol 122:226–233
30. Glikson M, Galum E, Owen R, Tur-Kaspa R, Shouval D (1992) Relapsing hepatitis A: a review of 14 cases and literature survey. Medicine (Baltimore) 71:14–23
31. Fujiwara K, Yokosuka O, Fukai K, Imazeki F, Saisho H, Omata M (2001) Analysis of full-length hepatitis A virus genome in sera from patients with fulminant and self-limited acute type hepatitis A. J Hepatol 35:112–119
32. Flehmig B (1980) Hepatitis A-virus in cell culture I. Propagation of different hepatitis A-virus isolates in a fetal rhesus monkey kidney cell line (FRhK-4). Med Microbiol Immunol 168:239–248
33. Daemer RJ, Feinstone SM, Gust LL, Purcell RH (1981) Propagation of human hepatitis A virus in African green monkey kidney cell culture: primary isolation and serial passage. Infect Immun 32:388–393
34. Feinstone SM, Kapikian AZ, Purcell RH (1973) Hepatitis A detection by immune electron microscopy of a virus-like antigen associated with acute illness. Science 182:1026–1028
35. Jothikumar N, Cromeans TL, Sobsey MD, Robertson B (2005) Development and evaluation of a broadly reactive Taqman assay for rapid detection of hepatitis A virus. Appl Environ Microbiol 71:3359–3363
36. Hollinger FB, Emerson SU (2001) Hepatitis A virus. In: Knipe DM, Howley PM (eds) Fields virology, 4th edn. Lippincott Williams & Wilkins, New York, pp 799–840
37. Rajtar B, Majek M, Polanski L, Polz-Dacewicz M (2008) Enteroviruses in water environment – a potential threat to public health. Ann Agric Environ Med 15:199–203
38. Khetsuriani N, Lamonte-Fowlkes A, Oberst S, Pallansch MA (2006) Centers for disease control and prevention: enterovirus surveillance-United States, 1970–2005. Morb Mortal Wkly Rep 55:1–20
39. Colbere-Garapin F, Christodoulou C, Crainic R, Pelletier I (1989) Persistent poliovirus infection of human neuroblastoma cells. Proc Natl Acad Sci USA 86:7590–7594
40. Howard RS (2005) Poliomyelitis and the postpolio syndrome. Br Med J 330:1314–1318
41. Fong T, Lipp EK (2005) Enteric viruses of human and animals in aquatic environments: health risks, detection, and potential water quality assessment tools. Microbiol Mol Biol Rev 69: 357–371
42. Amvrosieva TV, Paklonskaya NV, Biazruchka AA, Kazinetz ON, Bohush ZF, Fisenko EG (2006) Enteroviral infection outbreak in the Republic of Belarus: principal characteristics and phylogenetic analysis of etiological agents. Cent Eur J Public Health 14:67–73
43. Jubb G (1915) The third outbreak of epidemic poliomyelitis at West Kirby. Lancet 1:67
44. Jiang X, Wang M, Wang K, Estes MK (1993) Sequence and genomic organization of Norwalk virus. Virology 195:51–61
45. Wigand R, Adrian T (1986) Classification and epidemiology of adenoviruses. In: Doerfler W (ed) Adenovirus DNA. Martinus Nijhoff, Boston
46. Stewart PL, Fuller SD, Burnett RM (1993) Difference imaging of adenovirus – bridging the resolution gap between X-ray crystallography and electron-microscopy. EMBO J 12: 2589–2599
47. Friefeld BR, Lichy J, Field J, Gronostajski RM, Guggenheimer RA, Krevolin MD, Nagata K, Hutwitz J, Horowitz MS (1984) The in-vitro replication of adenovirus DNA in the molecular

biology of adenoviruses. In: Doerfler W (ed) Current topics in microbiology and immunology. Springer, Berlin, pp 221–225
48. Rowe WP, Huebner RJ, Gilmore LK, Parrot RH, Ward TG (1953) Isolation of a cytopathogenic agent from human adenoids undergoing spontaneous degeneration in tissue culture. Proc Soc Exp Biol Med 84:570–573
49. Center for Disease Control and Prevention (2005) Adenoviruses. CDC, Atlanta [Ref Type: Pamphlet]
50. Center for Disease Control and Prevention (2007) Acute respiratory disease associated with adenovirus serotype 14 – four states, 2006–2007. Morb Mortal Wkly Rep 56:1181–1184
51. Carter MJ (2005) Enterically infecting viruses: pathogenicity, transmission and significance for food and waterborne infection. J Appl Microbiol 98:1354–1380
52. Pina S, Puig M, Lucena F, Jofre J, Girones R (1998) Viral pollution in the environment and in shellfish human adenovirus detection by PCR as an index of human viruses. Appl Environ Microbiol 64:3376–3382
53. Goyal SM (2006) Viruses in food. Springer, New York
54. Grimwood K, Buttery JP (2007) Clinical update: rotavirus gastroenteritis and its prevention. Lancet 370:302–304
55. Martinez-Laso J, Roman A, Head J, Cervera I, Rodrigues M, Rodrigues-Avial I, Picazo J (2009) Phylogeny of G9 rotavirus genotype: a possible explanation of its origin and evolution. J Clin Virol 44:52–57
56. Grassi T, Bagordo F, Idolo A, Lugoli F, Gabutti G, De Donno A (2010) Rotavirus detection in environmental water samples by tangential flow ultrafiltration and RT-nested PCR. Environ Monit Assess 164:199–205
57. Abad FX, Pinto RM, Bosch A (1994) Survival of enteric viruses on environmental fomites. Appl Environ Microbiol 60:3704–3710
58. Gratacap-Cavallier B, Genoulza O, Brengel-Pese K, Soule H, Innocenti-Francillard R, Bost M, Gofit L, Zmirou D, Seigneurin M (2000) Detection of human and animal rotavirus sequences in drinking water. Appl Environ Microbiol 66:2690–2692
59. Ooi W, Gawoski JM, Yarbough PO, Pankey GA (1999) Hepatitis E seroconversion in United States travelers abroad. Am J Trop Med Hyg 61:822–824
60. Casares-Clemente P, Pina S, Buti M, Jardi R, Martin M, Bofill-Mass S, Girones R (2003) Hepatitis E virus epidemiology in industrialized countries. Emerg Infect Dis 9:448–454
61. Engle RE, Yu C, Emerson SU, Meng XJ, Purell RH (2002) Hepatitis E virus (HEV) capsid antigens derived from viruses of human and swine origin are equally efficient for detecting anti-HEV by enzyme immunoassay. J Clin Microbiol 40:4576–4580
62. Li T-C, Saito M, Ogura G, Ishibashi O, Miyamura T, Takeda N (2006) Serological evidence for hepatitis E virus infection in mongoose. Am J Trop Med Hyg 74:932–936
63. Emerson SU, Purcell RH (2003) Hepatitis E virus. Rev Med Virol 13:145–154
64. Purcell RH (1994) Hepatitis virus: changing patterns of human disease. Proc Natl Acad Sci USA 91:2401–2406
65. Labrique AB, Thomas DL, Stoszek SK, Nelson KE (1999) Hepatitis E: an emerging infectious disease. Epidemiol Rev 21:162–179
66. Isaacson M, Frean J, He J, Seriwatana J, Innis BL (2000) An outbreak of hepatitis E in northern Namibia, 1983. Am J Trop Med Hyg 62:619–625
67. Farrah SR, Goyal SM, Gerba CP, Wallis C, Melnick JL (1977) Concentration of enteroviruses from estuarine water. Appl Environ Microbiol 33:1192–1196
68. Rao VC, Seidel KM, Goyal SM, Metcalf TG, Melnick JL (1984) Isolation of enteroviruses from water, suspended solids, and sediments from Galveston Bay; survival of poliovirus and rotavirus adsorbed to sediments. Appl Environ Microbiol 48:404–409
69. Chonmaitree T, Ford C, Sanders C, Lucia H (1985) Comparison of cell cultures for rapid isolation of enteroviruses. J Clin Microbiol 26:2576–2580
70. Caro V, Guillot S, Delpeyroux F, Crainic R (2001) Molecular strategy for 'serotyping' of human enteroviruses. J Gen Virol 82:79–91

71. Croci L, Medici DD, Morace G, Fiore A, Scalfaro C, Beneduce F, Toti L (1999) Detection of hepatitis A virus in shellfish by nested reverse transcription-PCR. Int J Food Microbiol 48: 67–71
72. Chapron CD, Ballester NA, Fontaine JH, Frades CN, Margolin AB (2000) Detection of astroviruses, enteroviruses, and adenovirus types 40 and 41 in surface waters collected and evaluated by the information collection rule and an integrated cell culture-nested PCR procedure. Appl Environ Microbiol 66:2520–2525
73. Gibson UE, Heid CA, Williams PM (1996) A novel method for real time quantitative RT-PCR. Genome Res 6:995–1001
74. Mackay IM, Arden KE, Nitsche A (2002) Real-time PCR in virology. Nucleic Acids Res 30:1292–1305
75. Donaldson KA, Griffin DW, Paul JH (2002) Detection, quantitation of enteroviruses from surface waters and sponge tissue from the Florida Keys using real-time RT-PCR. Water Res 36:2505–2514
76. Heim A, Ebnet C, Harste G, Pring-Akerblom R (2003) Rapid and quantitative detection of human adenovirus DNA by real-time PCR. J Med Virol 70:228–239
77. Kageyama T, Kojima S, Shinohara M, Uchida K, Fukushi S, Hoshino FB, Takeda N, Katayama K (2003) Broadly reactive and highly sensitive assay for Norwalk-like viruses based on real-time quantitative reverse transcription PCR. J Clin Microbiol 41:1548–1557
78. Logan C, Leary JJ, O'Sullivan N (2006) Real-time reverse transcription PCR detection of norovirus, sapovirus and astrovirus as causative agents of acute viral gastroenteritis. J Clin Microbiol 44:3189–3195
79. Reynolds KA, Gerba CP, Abbaszadegan M, Pepper LL (2001) ICC/PCR detection of enteroviruses and hepatitis A in environmental samples. Can J Microbiol 47:153–157
80. Burkhardt W, Woods JW, Calci KR (2005) Evaluation of wastewater treatment plants' efficiency to reduce viral loading using real-time RT-PCR. In: 105 general meeting (ASM Abstract)
81. Symonds EM, Griffin DW, Breitbart M (2009) Eukaryotic viruses in wastewater samples from the United States. Appl Environ Microbiol 75:1402–1409

Books and Reviews

Berg G (1983) Viral pollution of the environment. CRC Press Boca Raton, FL, New York
Cox P, Griffith M, Angles M, Derre D, Ferguson C (2005) Concentration of pathogens and indicators in animal feces in the Sydney watershed. Appl Environ Microbiol 71:5929–5934
Gerba CP, Smith JE Jr (2005) Sources of pathogenic microorganisms and their fate during land application of wastes. J Environ Qual 34:42–49
Greenberg HB, Valdesuso JR, Kalica AR, Wright RG, McAuliffe VJ, Kapikian AZ, Chanock RM (1981) Proteins of Norwalk virus. J Virol 37:994–999
Hurst CJ (2000) Viral ecology. Academic, California
Jiang SC (2006) Human adenoviruses in water: occurrence and health implications: a critical review. Environ Sci Technol 40:7132–7140
Koopmans M (2005) Outbreaks of viral gastroenteritis: what's new in 2004? Curr Opin Infect Dis 18:295–299
Koopmans M, Duizer E (2004) Foodborne viruses: an emerging problem. Int J Food Microbiol 90:23–41
Leclerc H, Schwartzbrod L, Dei-Cas E (2002) Microbial agents associated with waterborne diseases. Crit Rev Microbiol 28:371–409
Lund E (1982) Waterborne virus disease. Ecol Dis 1:27–35
Rheinheimer G (1992) Aquatic microbiology, 4th edn. Wiley, Chichester
Varnam AH, Evans MG (2000) Environmental microbiology. ASM Press, Washington, DC

Chapter 14
Waterborne Infectious Diseases, Approaches to Control

Alan Fenwick, Albis Francesco Gabrielli, Michael French, and Lorenzo Savioli

Glossary

Neglected tropical diseases	A group of poverty-promoting and disfiguring diseases disproportionately affecting the poorest populations in developing countries. They are termed neglected because they lack the visibility, research support, and funding of other, more high profile, infections, such as malaria, HIV/AIDS, and tuberculosis. Several can be treated with inexpensive, effective treatment, while for others new drugs are needed.
Waterborne infectious diseases	Those diseases that rely on, or are heavily associated with, water for at least one stage of their life cycle or transmission, or affect populations closely related to areas of water.

This chapter, which has been modified slightly for the purposes of this volume, was originally published as part of the Encyclopedia of Sustainability Science and Technology edited by Robert A. Meyers. DOI:10.1007/978-1-4419-0851-3

A. Fenwick (✉) • M. French
Department of Infectious Disease Epidemiology, Schistosomiasis Control Initiative, Imperial College London, St. Mary's Campus, Norfolk Place, London W2 1PG, UK
e-mail: a.fenwick@imperial.ac.uk

A.F. Gabrielli • L. Savioli
Department of Control of Neglected Tropical Diseases, World Health Organization, Geneva, Switzerland
e-mail: saviolil@who.int

P. Kanki and D.J. Grimes (eds.), *Infectious Diseases: Selected Entries from the Encyclopedia of Sustainability Science and Technology*, DOI 10.1007/978-1-4614-5719-0_14, © Springer Science+Business Media New York 2013

Definition of the Subject

Waterborne diseases represent a significant and substantial burden on human health that disproportionately affects those living in the developing world. Human populations are attracted to live near surface water, particularly in areas such as sub-Saharan Africa (SSA) where it may represent a scarce resource. The development of water resources to exploit their potential has undoubtedly been of great benefit to some human populations via, for example, an increase in the amount of land suitable for agriculture provided by irrigation schemes, or the provision of hydroelectric power following the construction of dams and reservoirs. However, as an unintentional consequence such development may have unwittingly increased the extent of human waterborne infections by increasing the areas of suitable habitat for disease vectors and/or intermediate hosts, as well as attracting human populations to congregate, thereby aiding the life cycles of many infectious organisms [1–3].

The tools to treat, control and, in some cases, eliminate and eradicate some of these diseases already exist and there are examples of considerable, but not complete, success. Redoubling of efforts against such infections is advocated. In this entry, the major waterborne diseases that impact on human health, their extent, and the current position with regard to control are examined.

Introduction

Water- and Sanitation-Related Diseases Considered (in alphabetical order)

- Cholera
- Dengue and dengue hemorrhagic fever
- Guinea worm disease (Dracunculiasis)
- Lymphatic filariasis (Elephantiasis)
- Malaria
- Onchocerciasis (River blindness)
- Schistosomiasis (Bilharzia)
- Soil-transmitted helminthiasis (STH – ascariasis, trichuriasis, and hookworm)
- Trachoma
- Typhoid and paratyphoid enteric fevers

These diseases represent a diverse taxonomic group of infections stretching from viruses to multicellular macro-parasitic helminth infections. A brief description of the main infectious organisms is presented in Table 14.1.

While malaria has (rightly) long had strong visibility and powerful political advocacy behind its control, many of the other diseases have not traditionally

Table 14.1 Classification of major waterborne diseases affecting human health. MDA: suitable for control via mass drug administration. *STH* soil-transmitted helminthiasis, *M* million, *DALYs* disability adjusted life years

Family	Disease	Type and species	DALYs	Infected globally
Protozoa	Malaria	Cerebral (*Plasmodium falciparum*) Others (*P. vivax; P. ovale; P. malariae*)	46.5M [12]	515M clinical cases of *P. falciparum* malaria [13]
Helminth (MDA)	Schistosomiasis (Bilharzia)	Intestinal (*Schistosoma mansoni, S. japonicum*) Urogenital (*S. haematobium*)	13–15M [14]	207M [2]
	STH (Worms)	Roundworm (*Ascaris lumbricoides*)	10.5M [15]	807M [15]
		Whipworm (*Trichuris trichiura*)	6.4M [15]	604M [15]
		Hookworm (*Necator americanus*; *Ancylostoma duodenale*)	22.1M [15]	576M [15]
	Onchocerciasis (River blindness)	*Onchocerca volvulus*	0.5M [16]	37M [16]
	Lymphatic filariasis (Elephantiasis)	*Wuchereria bancrofti*; *Brugia malayi*	5.8M [17]	120M [17]
Helminth (non-MDA but eradicable)	Guinea worm	*Dracunculus medinensis*	<0.1M [18]	<1,800 [18]
Bacteria (MDA)	Trachoma	*Chlamydia trachomatis*	2.3M [19]	84M [19]
Bacteria (non-MDA)	Cholera	*Vibrio cholera*		3–5 M cases annually [20, 21]
	Typhoid and paratyphoid enteric fever	*Salmonella enterica*		27.1M [22]
Virus	Dengue and dengue hemorrhagic fever	*Flavivirus*	528,000 [23]	50M [23]

benefitted from such a profile. However, in the last decade, a number of individuals advocating on behalf of each of these diseases have changed the perception of them internationally. The World Health Organization has endorsed these increases in recognition and by establishing a "Department of Neglected Tropical Diseases" within the organization, it has encouraged a growing realization that each of the hitherto under-recognized diseases requires specific expertise, targets, and interventions in order to effectively control them. However, it has also been recognized that where overlaps in control approaches exist between infections, not least with regard to the targeted populations, and the frequency and mode of treatment, then an integrated approach to control can be employed which should be more cost-effective than the separate, individual, vertical control programs which have previously been employed. Indeed, there is a growing momentum behind such an approach [4–10]. The progress made in the control of each of these waterborne diseases to date is outlined in Table 14.2; alongside the aim of control (i.e., whether it be morbidity control, local elimination, global eradication [see [11] for definitions of these terms]) and the control strategies employed.

Malaria

Malaria is without doubt the world's most feared and important parasitic disease as measured by the extent of its distribution, the annual number of cases, and the death toll it exerts, particularly in infants [35–36]. It is estimated that 300–500 million cases of malaria occur each year, leading to over one million deaths [35–36], of which 75% occur in children under the age of 5 [37] (Fig. 14.1). The symptoms of malaria vary by parasite species and by individual human host. Fever, chills, headache, muscle aches, tiredness, nausea, and vomiting are common symptoms, but infection can also result in anemia and jaundice [25]. In the absence of effective treatment, convulsions, coma, severe anemia, and kidney failure can also occur. In the case of *Plasmodium falciparum,* the most common species of malaria in Africa, the severe cerebral form can develop which will lead to death without treatment. As a result, cerebral malaria is one of the leading causes of death in under-5-year-old children in Africa. Fortunately, many people living in areas with regular malaria transmission survive malaria and develop some protective immunity after repeated infections. Recently the field of malaria control has been challenged by Bill and Melinda Gates to begin aiming for global elimination of the infection, a target that has not been seriously attempted for decades [24, 38].

Input Photo of Malaria Effects

Four species of the protozoan genus *Plasmodium* infect man (*Plasmodium falciparum, P. vivax, P. ovale,* and *P. malariae*). All four are transmitted by the bite from an infected mosquito of the genus *Anopheles* and all four undergo similar life cycles. When a mosquito bites an infected person, it ingests blood, and takes up any malaria parasites in that blood meal. The parasite then develops in the mosquito passing through a number of stages until after between 9 and 30 days, the parasite is

Table 14.2 A summary of the current status of the major waterborne infectious diseases, aims and approaches to control. *ACT* artemisinin combination therapy, *DEC* diethylcarbamazine, *APOC* African Program for Onchocerciasis Control, *MDA* mass drug administration

Disease	Current status (2011)	Control aim	Control strategy
(1) Malaria	300–500 million cases per year and up to one million deaths [13]. *Plasmodium falciparum* in Africa causes the most fatalities	Transmission control and, recently, eradication [24]	Preventing mosquito bites using long-lasting insecticide-impregnated bed nets. Presumptive treatment, speedy home-based treatment, ACTs, vector control [25]
(2a) Schistosomiasis (Bilharzia)	Over 200 million people infected; over 85% in Africa [2]. Irrigation projects and dams exacerbate the problem – treatment coverage needs to be expanded from current level of 10–20%	Morbidity control and, recently, local elimination [26]	Annual or biannual treatment with praziquantel protects children from future consequences [27]. Improved socioeconomic status, water supplies, and sanitation are necessary for elimination
(2b) Soil-transmitted helminthiasis (STH)	Over one billion of the poorest children and adults infected globally. Annual treatment coverage is now approaching 50%	Morbidity control	Six-monthly or annual deworming treatment of preschool and school-age children with albendazole or mebendazole [28]
(2c) Onchocerciasis (River Blindness)	Great progress in recent decades. 19 countries treated by APOC. Still 50 million infections in Africa because hypo-endemic areas are not treated. Elimination in some foci in Africa thought possible. Elimination being approached in Central and South America [29, 30]	Interruption of transmission (Americas)/Morbidity control (elsewhere)	Multiple annual MDA rounds with ivermectin (Mectizan®)
(2d) Lymphatic filariasis (Elephantiasis)	120 million infected globally. A serious economic problem in Africa and the Indian subcontinent, but progress is being made toward elimination. 500 million people treated annually [31]	Global elimination as a public health problem	Multiple annual MDA rounds with albendazole and ivermectin (Mectizan®) (in Africa) or albendazole and DEC (elsewhere)
(3) Guinea worm	Rapidly approaching eradication. Fewer than 400 cases reported during the first quarter of 2011. Sudan, Ethiopia, Mali, and Chad are	Global eradication	Individual case finding and case containment, clean water provision and filtration, vector control (abate). Regular surveillance of endemic villages

(continued)

Table 14.2 (continued)

Disease	Current status (2011)	Control aim	Control strategy
	the countries with most residual infections (www.cartercenter.org)		
(4) Trachoma (preventable blindness)	80 million infected, 8 million visually impaired [19]. Treatment expansion satisfactory: 70 million due to be treated in 2011 [32]. Has been eliminated from Morocco	Global elimination as a public health problem	Annual treatment (for 3–5 years) with azithromycin (Zithromax®), as part of a "SAFE" strategy (*S*urgery, *A*ntibiotics, *F*ace washing, and *E*nvironmental improvement) [33]
(5) Cholera	3–5 million cases and 120,000 deaths annually. A disease which follows disruption of services, clean water and sanitation after natural disasters [20]	Minimization and control of outbreaks	Up to 80% of cases can be successfully treated with oral rehydration salts. Effective control measures rely on prevention via provision of safe water and sanitation. Oral cholera vaccines can be used for control, but should not replace conventional control measures
(6) Typhoid and paratyphoid enteric fever	Conservatively estimated at 27 million infection and 215,000 deaths annually, mostly in SSA and Southeast Asia	Minimization and control of outbreaks	Improving sanitation, identification and treatment of carriers, and use of typhoid vaccines to reduce susceptibility to infection [34]
(7) Dengue and dengue hemorrhagic fever	250 million at risk and 50 million cases annually in over 100 countries, and increasing. Can be fatal [23]	Transmission and morbidity control	Effective clinical management. Replacement of fluids and sometimes blood transfusions. Vector control also required

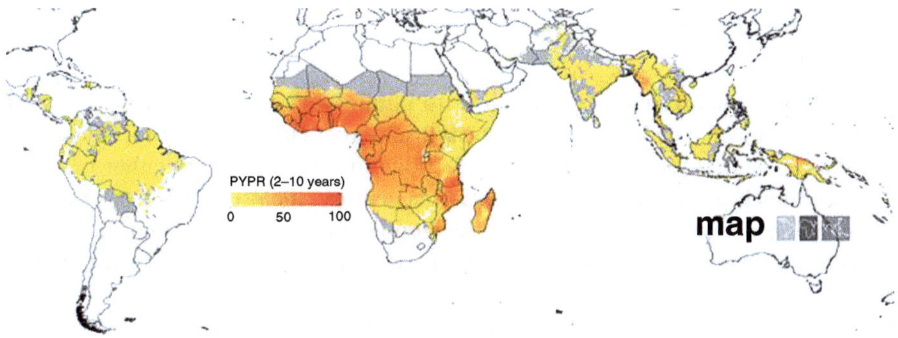

Fig. 14.1 The spatial distribution of *P. falciparum* malaria endemicity. Reproduced from work carried out by the Malaria Atlas Project [40]. *Pf*PR (2–10) corresponds to the *P. falciparum* parasite rate in 2–10-year olds

ready to pass back to another human. The form infective to humans (the sporozoite) is to be found in the mosquito's salivary glands from where it will be injected into the next human host when the mosquito takes future blood meals.

Once back in the human host, the sporozoites migrate to the liver where they multiply and spread into the bloodstream. The liver phase can last between 8 days and several months, depending on the malaria species. Further growth and multiplication then takes place inside red blood cells, and it is the regular destruction of these red blood cells which causes the characteristic intermittent fevers. The released parasites then invade new blood cells, and the cycle of multiplication and fever continues, with most people suffering their first fever 10 days to 4 weeks following infection. Malaria kills when the *P. falciparum* parasite invades the brain and causes cerebral malaria.

Malaria occurs mostly in tropical and subtropical countries, particularly in sub-Saharan Africa (90% of morbidity), Southeast Asia (India, Sri Lanka, Vietnam, and the Solomon Islands are endemic), and the forest fringe zones in South America (Brazil and Colombia are endemic), but its distribution is heavily dependent on the availability of suitable water bodies which support the mosquito larvae. The mosquito species vary considerably in their water and ecological requirements, in regard to levels of sunlight, aquatic vegetation, water flow, brackishness, and turbidity in a particular microhabitat. Several human factors have changed the distribution of the mosquito, for example (relatively) clean and dry urban cities may no longer support mosquitoes and therefore reduce transmission, while climate change may allow malaria to thrive in higher altitude areas than previously. The construction of irrigation systems and reservoirs, particularly in Africa, and changes in rice cultivation can lead to dramatic increases in the extent of areas suitable for malaria transmission. Indeed, malaria, once thought to have been eliminated in some countries, has returned as a result of the poor maintenance of water facilities and irrigation systems.

The Control of Malaria

In the 1960s, malaria was thought close to elimination after the widespread and effective use of DDT (dichlorodiphenyltrichloroethane) to control mosquito populations and chloroquine to treat infections in humans [39]. Neither of these interventions is as effective in the twenty-first century; the use of DDT has stopped due to environmental concerns and the development of resistance [39]. Equally, the parasite has developed resistance to chloroquine, and to a succession of antimalarial drugs which followed it.

In 2011, the WHO strategy for malaria comprises four main interventions, and is implemented through the Global Malaria Program (GMP) and the Roll Back Malaria (RBM) partnership:

- Reducing mortality, particularly among children, by early case-detection and prompt treatment with effective antimalarial drugs
- Promoting the use of insecticide-treated bed nets, especially by children and pregnant women
- Prevention of malaria in pregnancy by applying intermittent preventive therapy
- Ensuring early detection and control of malaria epidemics, especially in emergency situations

In addition, vector control can be an effective control tool; mosquito breeding sites can be removed by filling in and/or draining suitable water bodies and through other environmental management schemes, while the vectors can be controlled using chemical (IRS - Indoor Residual Spraying) and biological control in certain circumstances.

Current Drugs: Artemisinin-Based Combination Therapies (ACTs)

Widespread parasite resistance has arisen to the majority of antimalarials and the management of resistance presents a key battle in the fight for malaria control and eventual eradication. The newest class of drugs, artemisinins, has a very different mode of action to conventional antimalarials, which makes them particularly useful in the treatment of resistant infections [41]. However, in order to control the development of resistance to this drug, its use is only recommended in combination with another nonartemisinin-based therapy. Artemisinin produces a very rapid amelioration of clinical symptoms and is known to cause a reduction in the transmission of gametocytes, which may help decrease the potential for the spread of resistant alleles. To date there have been very few reported side effects to drug usage, however, one great worry is that resistance to Artemisinin may develop and indeed some resistant strains may be emerging [42].

All the treatment combinations outlined below have been recommended by WHO for treatment of *P. falciparum* [43].

Artesunate and amodiaquine (*Coarsucam* and ASAQ). This combination has been tested and proved to be efficacious in many areas where amodiaquine retains some efficacy.

Artesunate and mefloquine (*Artequin* and *ASMQ*) has been used as an efficacious first-line treatment regimen in areas of Thailand for many years. Mefloquine adverse reactions seem to be reduced when the drug is combined with artesunate possibly due to a delayed onset of action of mefloquine. This is not likely to be an option for long-term use in Africa due to the long half-life of mefloquine, which potentially could exert a high selection pressure on parasites.

Artemether and lumefantrine (*Coartem Riamet*, *Faverid*, *Amatem* and *Lonart*). This combination has been extensively tested and is effective in children under 5. It has been shown to be better tolerated than artesunate plus mefloquine combinations. There are no serious side effects documented but the drug is not recommended in pregnant or lactating women due to limited safety testing in these groups. This is the most viable option for widespread use and is available in fixed-dose formulae, which should increase compliance and adherence.

Artesunate and sulfadoxine/pyrimethamine (*Ariplus* and *Amalar plus*). This is a well-tolerated combination but the overall level of efficacy still depends on the level of resistance to sulfadoxine and pyrimethamine thus limiting its usage.

Dihydroartemisinin-piperaquine (*Duo-Cotecxin*, *Artekin*). Has been studied mainly in China, Vietnam, and other countries in Southeast Asia. The drug has been shown to be highly efficacious (greater than 90% clinical response rate).

Pyronaridine and artesunate (*Pyramax*). Manufactured by Shin Poong Pharmaceutical. Has been tested and demonstrated a clinical response rate of 100% in one trial in Hainan (an area with high levels of *P. falciparum* resistance to Pyronaridine). A multicenter phase III trial conducted in Africa found a 99.5% response rate [44].

Current Prophylactics

For travelers to malaria endemic areas who have never been exposed to malaria before, a course of prophylactics is recommended as a protection. Several options exist, as detailed in Table 14.3.

Table 14.3 Current prophylactic malaria tablets available for exposure-naïve individuals

Proganil	100 mg tablets are supplied as **Paludrine**
Chloroquine	150 mg tablets are supplied as **Nivaquine** or **Avloclor**
Mefloquine	250 mg tablets are supplied as **Larium**
Malarone	A combination of **Atovaquone** 250 mg and **Proguanil** 100 mg
Doxycycline	100 mg capsules or tablets

Bed Nets

Insecticide-treated nets (ITNs) were developed in the 1980s to combat malaria by preventing the mosquito from biting humans while they sleep. More recently, longer lasting insecticide nets (LLINs) have been developed which are impregnated with insecticides such as permethrin or deltamethrin. The latest production technologies (e.g., Olyset or DawaPlus) allow for a new generation of LLINs, which release insecticide for approximately 5 years. Such ITNs have been shown to be an extremely effective and cost-effective method of malaria prevention [45], aiding the drive toward Millennium Development Goal 6, which specifically calls for reduced malaria infection. These nets, which are typically available for $2.50–$3.50 (€2–€3) are generally purchased by donor groups like the Bill and Melinda Gates Foundation (BMGF), the President's Malaria Initiative (PMI) and the Global Fund and distributed through in-country distribution networks. One debate that has been ongoing for several years is whether or not international organizations should distribute ITNs and LLINs to people for free in order to maximize coverage and reduce price barriers. Some argue that cost-sharing between the international organization and recipients would lead to greater usage of the net because people will value them if they pay for them. Irrespective of the specifics of the approach, the main target has to be in assisting people, who most need the bed nets, to receive them and use them correctly.

Current Insecticides

Insecticides against malaria are used in two main ways: First, to kill the larvae breeding in water bodies close to human habitation; second, to kill the adult mosquitoes to prevent them biting humans inside a house. Fogging machines can be used to safely distribute insecticide in community-wide spraying in towns and tourist resorts. Unfortunately, sprayed insecticides are often expensive and the effects are only transitory, allowing mosquitoes to reinvade. There are two currently used insecticides: (1) **Malathion** – an organophosphate often used to treat crops against a wide array of insects. It can be sprayed directly onto vegetation, such as the bushes where mosquitoes like to rest, or used in a 5% solution to fog open areas. In the small amounts used for mosquito control, it poses no threat to humans or wildlife. (2) **Permethrin** is a synthetic form of a natural insecticide (pyrethroid) which can be mixed with oil or water and applied as a mist. Both malathion and permethrin are also available in sprays for use for Indoor Residual Spraying (IRS). IRS kills adult mosquitoes that land on sprayed surfaces. Its two primary effects are to reduce the life span of vector mosquitoes so that they can no longer transmit malaria, and to reduce the density of the vector mosquitoes [46]. IRS is currently being funded by the PMI and the Global Fund. Historical and programmatic documentation has clearly established the impact of IRS [47] and expansion of coverage is likely to continue.

DDT has long been the cheapest insecticide and the one with the longest residual efficacy against malaria vectors (6–12 months depending on dosage and substrate), although environmental and resistance problems have largely curtailed its use. Other insecticides have relatively shorter residual effect (pyrethroids: 4–6 months; organophosphates and carbamates: 2–6 months).

Schistosomiasis

Three major species of schistosome worms infect humans [48]:

- *Schistosoma japonicum* (intestinal schistosomiasis) is found only in China and the Far East and infects humans and domestic animals. The intermediate host are amphibian snails of the genus *Onchomelania.*
- *S. mansoni* (intestinal schistosomiasis) originated in Africa, but was carried by humans to South America and the Caribbean where a suitable snail host was able to support transmission. Primates may also become infected. The intermediate hosts are snail species of the genus *Biomphalaria.*
- *S. haematobium* (urogenital schistosomiasis) is found only in Africa and the Middle East and there is no animal reservoir. The snail intermediate hosts are certain species of the genus *Bulinus.*

The worms of each species live in the blood vessels of humans and the female worm lays eggs which either escape through the urine (*S. haematobium*) or feces (*S. mansoni*/*S. japoncium*) to continue the life cycle or remain in the body trapped in the intestine wall, bladder wall, or liver tissues. In female hosts, the worms and eggs can invade the genital system and cause damage [49]. Once the eggs have exited the body they will hatch only when they reach freshwater. The emerging free swimming larvae (miracidia) need to find and invade the correct snail intermediate host species in order to survive. Once there, they undergo asexual reproduction in their host snail over a period of a month developing through the "sporocyst" and "daughter sporocyst" intermediate stages. Eventually the next free living larval stage (cercariae) emerges from the snail into water as swimming organisms with a body and a tail. They have a short life during which time they are attracted to any human skin immersed in the water. They will attach to and penetrate unbroken skin, shedding the tail in the process. The body (now a schistosomulum) migrates to the lungs and passes through the diaphragm to reach the liver. About 6 weeks after initially penetrating the host, the worm in the liver will reach maturity – and about one centimeter in length. The male and female adult worms pair and the male carries the female to their selected blood vessels where they can live and produce eggs, typically for a period between 2 and 10 years [50]. *S. haematobium* worms reside around the bladder, and hence the eggs pass out in the urine. *S. japonicum* and *S. mansoni* worms reside in the blood

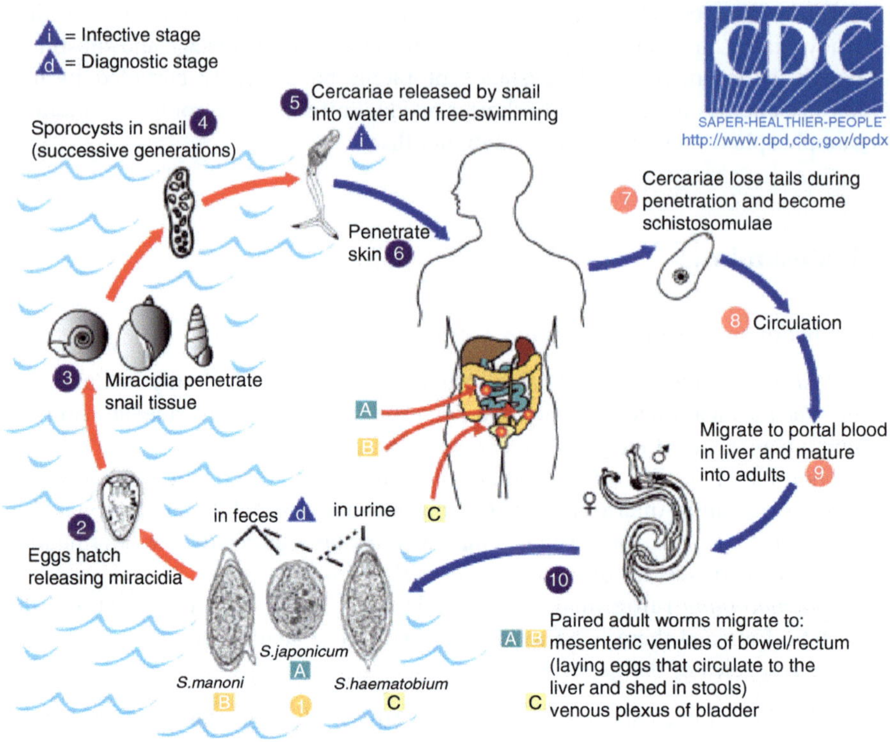

Fig. 14.2 Life cycle of the three major schistosome species that infect humans (Taken from http://www.dpd.cdc.gov/dpdx/html/schistosomiasis.htm, Centers for Disease Control)

vessels in the mesentery and hence their eggs breaking into the intestine and being voided with the feces [51–53] (Figs. 14.2 and 14.3).

An estimated 207 million people are infected globally with schistosomiasis [2] with almost 90% of those infected found in Africa [54]. The first symptom of infection is irritation at the site of penetration and is rarely noticed. A classic symptom of heavy infection is caused by the passage of schistosomula through the lungs 5–10 days after infection, but this is only seen when many worms are passing through concurrently, as in the case of a single heavy exposure. Subsequent symptoms occur as the worms mature and start laying eggs. This can lead to blood in the urine in *S. haematobium* cases (a finding that is also used as a major diagnostic feature). *S. japonicum* and *S. mansoni* eggs breaking into the intestine can cause diarrhea and blood in the stool but this is less obvious and therefore, not a definitive diagnostic feature [48, 51–52].

Schistosomiasis is a multifaceted disease. The worms can live for up to 30 years [55, 56] and the eggs that collect in the body cause increasing organ damage over time. Heavy infections lead to severe symptoms and death from

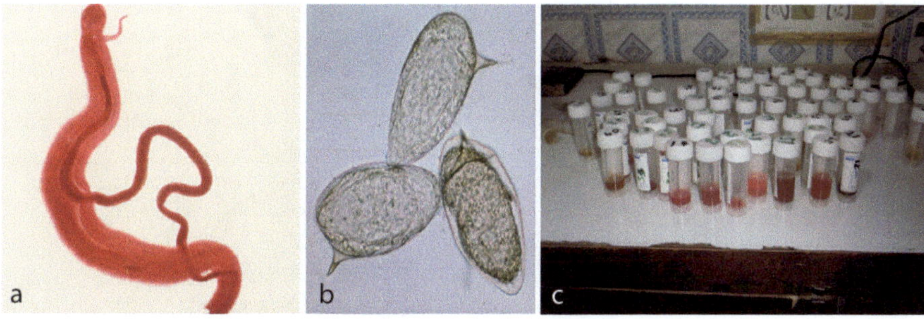

Fig. 14.3 (**a**) A pair of schistosome adult worms, showing the female (thin worm) present in the male's gonotrophic canal, (**b**) Schistosome eggs, and (**c**) the characteristic sign of urogenital impact, blood in the urine

bladder cancer in *S. haematobium* infections and from portal hypertension in *S. japonicum* and *S. mansoni* infections. Uninfected people exposed to an initial heavy infection by swimming in a water body containing many infected snails can suffer extreme early symptoms, while light infections may never be detected. The norm for people residing in endemic areas is for chronic infection lasting years to occur before serious consequences are recognized. This happens because the chronic infection stage can cause significant but often undiagnosed subtle morbidity, including anemia, stunting, urinary and intestinal problems, lassitude and reduced growth, poorer school attendance and performance. Meanwhile, overt and severe organ damage can gradually occur over the course of this worsening process [48, 51, 52, 57–58] (Fig. 14.4).

Treatment

Currently there seems to be little chance of elimination of schistosomiasis in its transmission heartland of rural Africa because to date, elimination has only been possible in countries that have undergone significant socioeconomic development, such as Japan and Puerto Rico. Substantial improvements in water and sanitation conditions in rural African areas are required in order to permanently interrupt transmission. On a wider scale, sustained economic development is needed to eventually rid the world of this disease. Until then, the currently followed strategy is to focus on morbidity control via the large-scale administration of the drug praziquantel to at-risk groups (although recently this has been shown to also have an effect on transmission [59–60]) [61]. An annual round of treatment reaching school-age children would have a substantial effect on improving the quality of life of treated children, and protecting them from the serious consequences of their infections in later life [7].

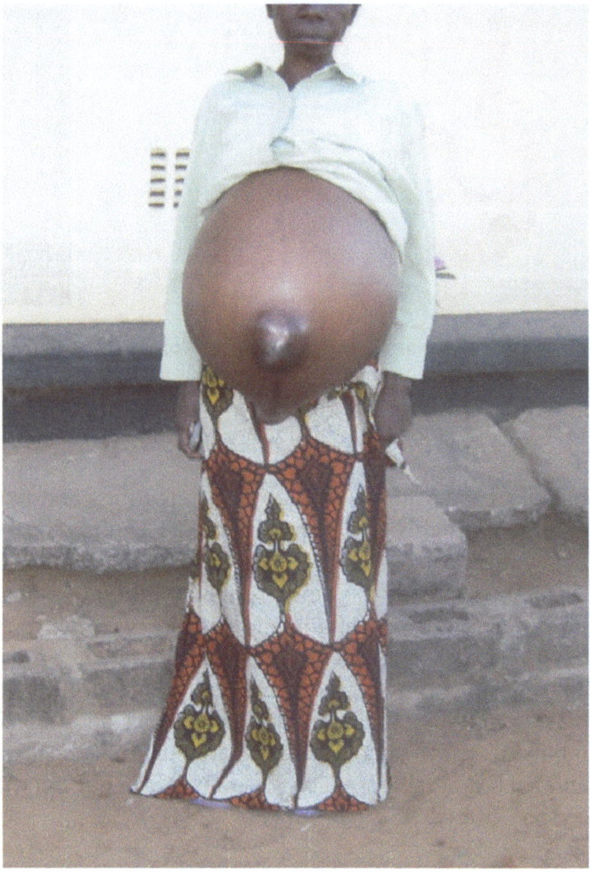

Fig. 14.4 Late stages of schistosomiasis infection, showing severe morbidity

Schistosomiasis can be treated with the generic drug praziquantel [62] (administered at 40 mg/kg body weight) which kills the adult worms. Praziquantel is currently available off-patent at a cost of about 8 USD cents (5 GBP pence) per tablet from a number of manufacturers, most of which are in India, South Korea, and China. On average 2.5 tablets are needed to treat a child and 4 tablets per adult, equating to 20 cents for the tablets needed to treat a child and 32 cents for an adult. The pharmaceutical company Merck Serono has recently committed to a drug donation program that pledges 200 million tablets over 10 years, enough to treat eight million children per year in high-prevalence areas in Cameroon, Malawi, Mozambique, Nigeria, and Senegal. It is hoped that further donations may be forthcoming in future; however, the situation is complicated as the active ingredient for the tablets is the constraining factor in the supply chain due to uncertainty in market demand. WHO is working with the drug companies to attempt to stabilize the market over the medium- to long-term, because Africa is desperately short of praziquantel ([63] and website www.schisto.org).

Improved Funding

During the 1980s several schistosomiasis control programs were started in Africa using donated drugs (by the German developmental organization GTZ) but once the drug donations ended, the programs proved to be unsustainable and ceased. In the 1990s, three countries (Egypt, Brazil, and China) applied to the World Bank for funding and as a result they all reduced their schistosomiasis burden to below what constituted a public health problem, using a combination of drug treatment and molluscicides. The cost of molluscicides, however, proved a barrier to any expansion of their use.

At the start of this century, the Bill and Melinda Gates Foundation (BMGF) made significant funds available for treatment (from 2003 to 2006) through the Schistosomiasis Control Initiative (SCI; www.schisto.org). Six countries (Burkina Faso, Mali, Niger, Tanzania, Uganda, and Zambia) benefited from this donation and led the way in Africa for the development and implementation of national schistosomiasis programs. Thanks to the UK's Department for International Development (DFID) (2008–2013) and the United States Agency for International Development (USAID) (2006–2013), more resources have been made available (primarily through the Schistosomiasis Control Initiative [SCI] and Research Triangle International [RTI]) to expand coverage of schistosomiasis treatment throughout Africa. By working together several countries will receive the support they need to reach national coverage. Other countries plan to start implementation programs in 2011 (Côte d'Ivoire, Liberia, Malawi, and Mozambique). However, even given this expansion, many countries in need of national control programs have yet to start.

Although elimination of infection is not a target in mainland Africa, Zanzibar is aiming for elimination in 5 years from 2011; thanks to a project supported by the Gates Foundation, DFID and WHO which aims to identify the approaches (chemotherapy, water and sanitation improvement, health behaviors) needed to achieve this end point [26]. The chances of success are thought to be high because of many years of control interventions and the insular island biogeography that contributes to preventing importation of cases. Implementation will be organized by the Zanzibar government with assistance from the Ivo de Carneri Foundation on Pemba Island, The WHO, The University of Athens-Georgia, the London Natural History Museum, SCI Imperial College, and the Swiss Tropical Institute.

STH (*Ascaris, Trichuris,* and Hookworm Infections)

There are three separate worms which together constitute the soil-transmitted helminths (STH), so named because their life cycles do not involve an intermediate host. They are the roundworm (*Ascaris lumbricoides*), the whipworm (*Trichuris trichiura*), and the hookworm (*Necator* and *Ancylostoma* spp.). These worms inhabit the human intestine, with their eggs passed out in the feces [15–51, 52, 64].

Roundworm and whipworm cause infections associated with poor hygiene: good sanitation would prevent eggs from contaminating the environment, and regular hand washing when preparing or eating food would prevent the eggs from being ingested. Unfortunately in poor rural areas devoid of toilets, feces tend to be deposited anywhere and the roundworm and whipworm eggs are able to develop on the soil and be ready to reinfect a human host by ingestion, usually on food. Heavy infections lead to malnutrition and growth stunting in children, reduced educational achievement, and loss of future earning power [15, 65–68].

The life cycle of hookworm differs because the eggs that are passed in the feces hatch and the larvae need to pass through some developmental stages before they are ready to reinfect a human host. Infection is acquired when the free-living larvae attach themselves to the feet and ankles of passers-by. They then penetrate the skin, migrate around the body, and end up in the intestine where they attach and gorge on blood. Those individual hosts who have very heavy infections will almost certainly develop anemia as the major result, which has a serious effect on young children but also on women of childbearing age because anemia is the major cause of poor birth outcomes, and infant and maternal mortality [52, 69–71].

The nationwide elimination of these worm infections goes hand in glove with socioeconomic development, for example, these three worms used to be common in Europe and the USA, but with widespread access to safe water and sanitation, they have been eliminated from the majority of developed countries. In South Korea and Japan, infection rates were over 80% following the Second World War but during a period of 15 years this rate was reduced to less than 0.5% [72].

It is estimated that globally over a billion people in 2011 do not have the water and sanitation levels needed to eliminate these STH and are infected with one or more worms in developing countries.

Treatment

The intestinal worms can be expelled from the body with a single 400 mg tablet of albendazole (or a 500 mg tablet of mebendazole) [62, 73]. In endemic areas where prevalence of infection is above 20%, an annual dose (or two doses when prevalence is $\geq$50%) of deworming tablets throughout a child's life (from the age of 1 to 14) will have an amazingly positive effect on their growth and nutritional status, their school attendance, and cognitive ability [61].

While for many years STH deworming for preschool-age children has been included into immunization interventions throughout the world [73], school-age children are usually targeted through school-based treatment interventions. A boost to STH deworming coverage has been "unintentionally" provided by the lymphatic filariasis (LF) elimination program (see LF below) because one of the drugs which has been delivered in ever increasing numbers since 2000 to eliminate LF is albendazole [74]. Approximately two billion albendazole tablets have been delivered globally

Table 14.4 The contribution of the pharmaceutical industry to the control of waterborne infectious diseases

Company	Commitment
MERCK	Merck has reconfirmed their commitment (now already over 25 years) to donate ivermectin (Mectizan®) for as long as needed against both onchocerciasis and lymphatic filariasis in Africa
gsk GlaxoSmithKline	GSK has already donated almost two billion tablets of albendazole against lymphatic filariasis and will continue until elimination is achieved. In October 2010, GSK committed an additional 400 million tablets a year for 5 years to deworm school-age children worldwide
Johnson & Johnson	Johnson & Johnson has, for several years, donated 50 million tablets of mebendazole per year for intestinal worms – from 2012 this will be increased to 200 million tablets per year
Pfizer	Pfizer committed to provide doses of azithromycin (Zithromax®) for trachoma In 2009 alone they donated 50 million tablets; in 2011 that number will reach 70 million
Eisai	In October 2010, EISAI committed to provide two billion tablets of DEC against lymphatic filariasis which is used with albendazole outside of Africa
MEDPHARM	Medpharm (a generic manufacturer) has contributed praziquantel against schistosomiasis in the years 2004–2006
MERCK	Merck Serono has agreed to donate 200 million tablets of praziquantel over 10 years

since the program began [46], which means that the ancillary effect on these worms can be assumed to be significant.

Two pharmaceutical companies have pledged drug donations specifically to reduce the prevalence of STH (for a summary of these, and other drug donations, see Table 14.4). Johnson & Johnson has been donating 50 million tablets a year through the nongovernmental development organization (NGDO) "Children Without Worms" since 2005 and in 2010 they pledged to increase their donation of mebendazole to 200 million tablets a year, which will be mainly used in Asia and the Americas. "Feed the Children International" also donate several million

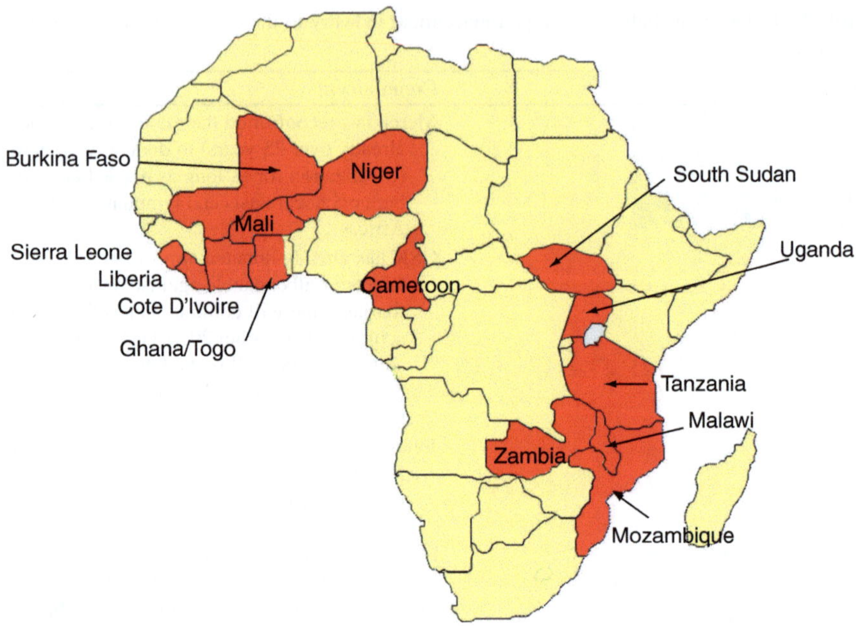

Fig. 14.5 Countries in sub-Saharan Africa currently operating national or subnational schistosomiasis control programs

mebendazole tablets through another organization, "Deworm the World," which directs the drug to where it is needed mostly through Ministries of Education.

For Africa, GSK announced in October 2010 that from 2012 they will donate an additional 400 million albendazole tablets a year to target school-age children throughout the world. Over the next 5 years this will make a huge difference to the health and quality of life of the world's children.

Meanwhile, both DFID and USAID support these donations by providing essential logistical and financial support for the distribution of the tablets. DFID currently funds delivery in eight African countries (Côte d'Ivoire, Liberia, Malawi, Mozambique, Niger, Tanzania [including Zanzibar], Uganda, and Zambia), and USAID in ten countries (see Fig. 14.5) and coverage is likely to expand during 2012–2015.

Onchocerciasis (River Blindness)

Onchocerciasis is a "filarial" worm that is endemic mainly in Africa and the Americas near fast-flowing rivers or streams where the host fly breeds. The adult worm lives in nodules in the human body and gives birth to larvae (microfilariae) which migrate around the skin waiting to be picked up when a black fly (*Simulium*

Fig. 14.6 Statue at the World Health Organization headquarters in Geneva, of a child leading an adult blinded by onchocerciasis through the village (Taken from http://www.cdc.gov/globalhealth/ntd/diseases/oncho_burden.html, Photo Credit: Carter Center)

species) bites. The larvae develop in the black fly and are then transmitted back to another human host when the fly next bites [51, 52].

Prior to the 1970s this parasite was devastating to African populations. The larvae were responsible for great discomfort due to the itching they caused. However, much worse was the widespread blindness caused by the infection near the great rivers in Africa because the larvae produced by the adult worms in the human body migrate across the eye. As recently as 1970, up to 50% of adult populations living on the banks of fast moving rivers suffered impaired vision due to the persistent biting of the black fly [75–78].

The control of river blindness represents a major success in disease control, perhaps, matched only by smallpox eradication in 1979–1980 and latterly the progress toward the elimination of LF (see below). Three major programs (OCP, APOC, and OEPA), an insecticide (DDT), a pharmaceutical company (Merck), and a drug (ivermectin, Mectizan®) contributed to this success which is commemorated by a statue in front of the WHO headquarters in Geneva [79] (Fig. 14.6).

The OCP (Onchocerciasis Control Program) was initiated in the 1970s in West Africa to control onchocerciasis and provided DDT to be sprayed into rivers to kill the black fly larvae which breed in these fast-moving waters and therefore, to eliminate disease transmission. Nine countries implemented control as part of the OCP and as a result their endemic onchocerciasis was controlled and over a 20-year-period infection rates and blindness were reduced to very low levels [80–81].

Sadly the widespread use of DDT was shown to be environmentally unacceptable, but a major breakthrough saved the program when the drug ivermectin

(Mectizan®) was found to have the effect of destroying microfilariae for a year after a single annual dose [82]. The drug manufacturer, Merck and Co., Inc., recognized that poor people in Africa would never be able to afford to purchase this drug, and so agreed to donate the drug to all those who live in endemic areas "for as long as needed" [29]. Today, some 25 years later, this donation continues, and in October 2010 the Merck commitment was renewed. Since 1986 Mectizan® has been distributed widely in OCP countries (and subsequently in 19 African Program for Onchocerciasis Control [APOC] countries) reaching over 50 million people annually, and thereby saving millions of people from losing their sight [83].

The OCP closed down when its objectives had been met by the year 1996 but the APOC was then launched to oversee the distribution of Mectizan® in 19 additional African countries with the help of a number of NGOs (The Carter Center, Sightsavers, CBM, Helen Keller International, World Vision) [84]. The APOC mandate has recently been renewed to continue treatment until 2020, and during that time will look at the possibility of a changed strategy (doubling the dosage) to bring about elimination in some foci instead of merely control.

APOC and its partners have not managed complete coverage in Africa due to a complication to what should be a simple control program. Another parasitic filarial worm, *Loa loa,* which also produces live larvae that circulate in the body, coexists with onchocerciasis in some areas in Central Africa. Some patients with *Loa loa* when treated with Mectizan® can suffer disastrous consequences, such as encephalopathy [85]. Thus in parts of Central Africa where *Loa loa* is endemic, it is not safe to control onchocerciasis (nor LF) using MDA because of the possible dangerous complications. This is a matter yet to be resolved and further discussion is advocated [86].

The Onchocerciasis Elimination Program for the Americas (OEPA) is a regional initiative in the Americas with the goal of eliminating morbidity of river blindness and interrupting transmission in six endemic countries: Brazil, Colombia, Ecuador, Guatemala, Mexico, and Venezuela. The OEPA strategy is to encourage the endemic countries to provide sustained Mectizan® mass treatment every 6 months with the aim of reaching at least 85% of 500,000 persons estimated to be at risk of the disease [87]. All six endemic countries have established effective national programs and there are no new cases of blindness attributable to onchocerciasis in the American region. As of 2011, transmission is continuing in only two countries, Brazil and Venezuela, with surveillance in place in the remaining four [88].

In Africa, the plan for the future is continuation with the successful control program and annual Mectizan® treatments, but for the first time WHO are considering with selected governments the idea of intensifying the strategy with the aim of interrupting transmission permanently. It is thought that in many countries that have been offering annual Mectizan® for many years, the worms are beginning to die out and so by increasing frequency of treatment for a short period elimination may be achievable [89].

Lymphatic Filariasis

Two filarial worms *Brugia malayi* and *Wuchereria bancrofti* infect 120 million people globally with lymphatic filariasis (LF), with a further 600 million at risk of infection. The intermediate host of these worms are mosquitoes of several genera, and the global distribution is very extensive – Africa, the Indian subcontinent, many Pacific Islands, the Caribbean, and tropical America [90]. The deformity and misery caused by this disease are horrific. The adult worms live in and block the lymphatic system, which prevents the drainage of lymph fluid. This causes swelling of lower limbs, secondary infections, and, in the case of some men, the scrotum becomes grotesquely swollen [91]. Meanwhile, the millions of larvae which the females produce circulate in the bloodstream to be picked up by a mosquito for the transmission cycle to be completed.

The Global Alliance for the Elimination of Lymphatic Filariasis (GAELF) is supported by many partners (see http://www.filariasis.org/who_we_are/partners.html) and is housed at the Centre for Neglected Tropical Diseases (CNTD) in Liverpool. This alliance is probably the most successful disease control alliance ever formed as evidenced by the expansion of treatment coverage against this terrible affliction (see [31, 90] and website www.filariasis.org). The principle is that an annual combined dose of albendazole with Mectizan® in Africa and albendazole with DEC outside of Africa given to the whole community will effectively prevent transmission for a year. This is because the concentration of microfilariae in the blood will be so reduced that the biting mosquitoes will not pick them up [74]. Since the filarial adult worms live for an average of just 6 years elimination can be a reality if transmission can be interrupted for six consecutive years [92–94], resulting in huge health and economic benefits [95–97].

The elimination program is being underpinned by the donation pledges of pharmaceutical companies, dating back to the late 1990s. GSK made a commitment to donate albendazole, and Merck & Co. Inc. extended their onchocerciasis donation of Mectizan® to include LF in Africa [98]. Elsewhere DEC was purchased until 2010 when EISAI made their donation for the future. Currently, approximately 600 million people receive treatment annually against LF [46]. Delivery is being funded by individual governments with assistance from the BMGF, USAID and DFID. This emphasizes the capacity of major pharmaceutical companies to work together on a major global health problem.

One unfortunate note is that many people in Africa who need treatment for LF are not yet receiving it as some governments are yet to embrace the value that the LF program provides. Approximately 80 million people are currently being treated annually with free drugs in Africa but because of the *Loa loa* complication [99] and because of the civil unrest in many parts of Africa poor coverage in some areas has not yet been resolved.

Nevertheless, the LF elimination program has achieved some fantastic results: Egypt and Zanzibar [100], and most recently Costa Rica, Suriname and Trinidad and Tobago have stopped MDA because they have successfully achieved their

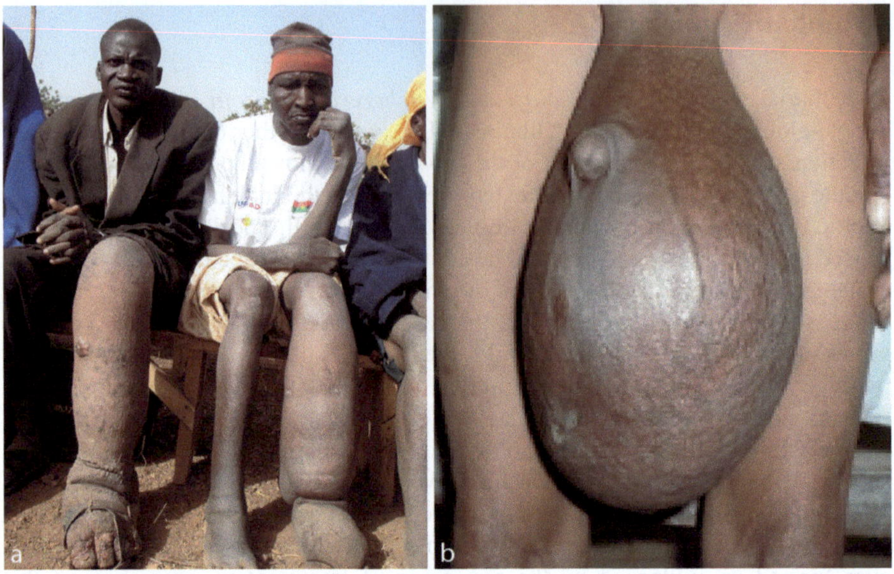

Fig. 14.7 Morbidity caused by long-term lymphatic filariasis to the (**a**) legs, lymphedema and (**b**) scrotum, hydrocele

elimination targets [31, 90, 101]. The new donation from EISAI in Japan, amounting to two billion tablets a year, will be sufficient to treat at least the Indian subcontinent (and possibly the four countries in the American continent that are still endemic), and raising hopes of elimination of LF by 2020.

However, the drug treatments do not reverse the disfiguring morbidity that has already developed. For these cases, an individual care approach is needed based on hygiene and surgery, where appropriate (Fig. 14.7).

Case management of such patients has expanded and improved, and leg washing has increased the care of elephantiasis patients and has led to reductions in leg swellings by preventing secondary infections. Hydrocele surgery is slowly becoming more widespread as "surgery camps" are being established in different countries [102]. For example, in Niger, SCI supported over 600 hydrocele operations in 2010. These have been extremely successful and hundreds of men have come forward requesting the free-of-charge 20-min surgical intervention, performed under local anesthetic. The cost of each intervention at one of SCI's camps averages less than $200, which includes pre- and postoperative treatment (www.schisto.org).

Trachoma

Blindness due to trachoma is caused by the after effects of conjunctivitis due to *Chlamydia trachomatis* infections carried by flies. Trachoma is simply a disease of

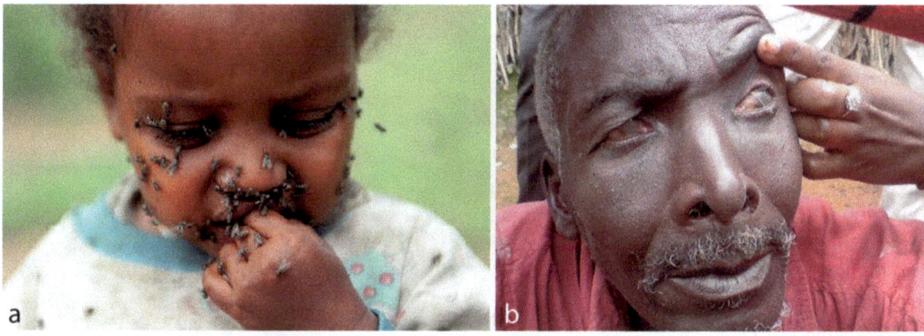

Fig. 14.8 (**a**) Trachoma is caused by the bacteria *Chlamydia trachomatis* which is often spread by flies in areas of poor sanitation; (**b**) Trachoma's major effects include corneal damage and irreversible blindness

poor hygiene, combined with dry and dusty conditions, and poverty. After severe conjunctivitis the eyelids become affected by scarring, and eyelashes are turned inward toward the eye and damage, and eventually destroy the cornea over time, leading to irreversible blindness. Improved water and sanitation would soon lead to elimination of trachoma. Meanwhile, treatment with Zithromax® and regular hand and face washing is all that is needed for prevention of new cases [19, 103] (Fig. 14.8).

For existing trichiasis cases simple surgery can correct the problem. Face washing and improved water and sanitation are essential components in a trachoma elimination strategy. The World Health Organization promotes the global elimination of blinding trachoma by 2020 (GET 2020) through an integrated approach named the SAFE strategy – S for surgery, A for Antibiotics, F for face washing, and E for environmental improvements to reduce the infestation of flies. The International Trachoma Initiative (ITI), established in 1999, and a number of other NGDOs are very active in promoting eye care and control of trachoma, possibly combined with that of onchocerciasis where relevant. Excellent progress has been made in some countries, and ITI's flagship program has contributed to eliminating trachoma from Morocco [103].

Early treatment is effective: azithromycin – Zithromax®, an antibiotic for active infection, which is currently donated by Pfizer needs to be administered annually. In 2011, it is estimated that approximately 70 million doses of Zithromax® will be distributed. A new website www.trachoma.org has been developed to provide information about trachoma and the treatments provided by ITI. The target for reaching elimination of blinding trachoma globally remains as "2020." As of this year:

- About 41 million people have active trachoma needing treatment.
- About 8.2 million people are estimated to have an advanced stage infection and therefore are blind or at risk of blindness.
- Fifty-seven countries are endemic, with up to 1.2 billion people living in at-risk areas.
- Since 1999, Pfizer has donated 225 million doses of Zithromax®.

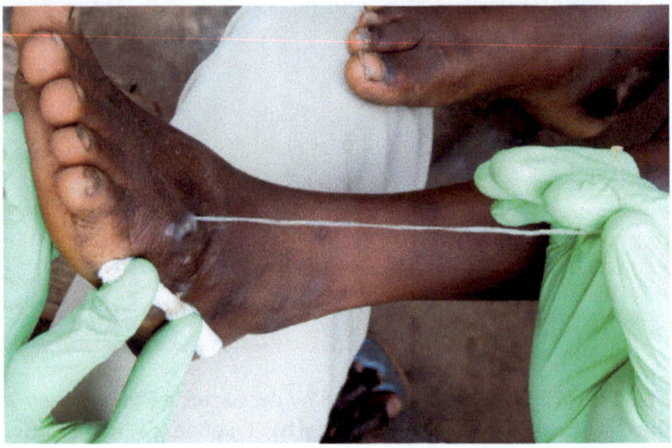

Fig. 14.9 A photograph of the adult Guinea worm (dracunculiasis) slowly being extracted from its emergence point, here on the top of the foot (Taken from http://www.maximsnews.com/news 20081209guineawor meradication 10812090803.htm, Photo Credit: Photo © The Carter Center/Louise Gubb, Ghana, 2007)

Guinea Worm: A Worm on Its Own

Imagine having a worm a meter in length crawling around your body just under your skin. When the female adult worm is ready to lay her eggs, she emerges through the skin around the ankle. This subsequent burning sensation causes the victim to bathe the leg in water to cool it down. This is exactly what not to do because the worm emerges and deposits thousands of larvae into the water. These larvae infect water fleas where they develop and subsequently reinfect humans when the fleas are ingested when water is directly used for drinking. This is Guinea worm (Fig. 14.9).

The life-cycle of Guinea worm involves no animal host, thereby reducing the number of possible control targets. The control strategy is therefore based on identification of infected individuals, reporting and containment of cases, on behavioral change so those infected individuals do not contaminate the water, and on improved water supplies. Such combined efforts could and should lead to the eradication of Guinea worm disease. People can only get infected by swallowing infected water fleas. This happens only when people are forced to drink water from open ponds which contain these fleas, and this only happens when the water they drink has not been filtered.

Thus eradication of this unique worm was recognized to be possible some 20 years ago, and WHO and the Carter Center have been working toward this end ever since. However, there is no drug treatment for Guinea worm; the only course open to an infected person is the slow extraction of the very long worm from the skin by gently winding the worm around a stick and pulling it out over several days.

The Guinea Worm program is now moving toward eradication. The number of annual cases has been dramatically reduced from over one million in 1988 to fewer than 2,000 in 2010; and the disease remains endemic in just four countries – Chad, Ethiopia, Mali, and Sudan (www.cartercenter.org; [104]). A number of countries have been certified as free of transmission, and there are several more countries which have not reported cases during the previous year (Burkina Faso, Cote d'Ivoire, Ghana, Kenya, Niger, Nigeria, Togo) and are considered to be in precertification phase. Once a country has been certified, there is a continued need for surveillance – in order to prevent reestablishment of transmission until global eradication is achieved.

Cholera and Typhoid

The bacterial infections cholera (caused by *Vibio cholerae*) and typhoid (caused by *Salmonella typhi*) are endemic in countries where poor water and sanitation prevail. However, they cause their greatest damage when they reach epidemic outbreaks during times of crisis such as after natural disasters: flood, earthquakes, and tsunamis. During these periods many displaced persons congregate in temporary camps with poor sanitation, and where drinking water sources are contaminated. These really are diseases that thrive when basic services have broken down, and can quickly overwhelm public health services in such locations [105]. Cholera bacteria are most often associated with waterborne transmission, but can also be transmitted by contaminated food [105]. The classic symptoms of cholera are profuse, watery diarrhea, and vomiting, which can rapidly lead to severe dehydration and death in hours if left untreated. In addition to these severe sequelae of infection, there will likely be many, unreported cases that are asymptomatic or result in nonspecific mild symptoms [106]. Treatment of cholera via oral and intravenous rehydration therapy has resulted in reductions in fatality rates, but it still remains a dreaded disease [20]. The consequences of infection are vomiting and diarrhea, and treatment is patient care and rehydration. In epidemics about 5% of those infected will die, but if rehydration can be provided, recovery is usually achieved. Typhoid fever is caused by the bacterium *Salmonella enterica* serotype Typhi (*S. typhi*). Improvements in water and sanitation virtually eliminated typhoid from the majority of the developed world, yet it still exerts a significant public health burden in the developing world [107]. The most common mode of transport is through contaminated water, or food that has come into contact with contaminated water, and thus is heavily associated with poor water and sanitation. The burden of disease is thought to be highest in children under the age of 5 who suffer the most serious sequelae of infection [108]. The emergence of drug resistance to a succession of antibiotics, most recently to quinolones, has put great pressure on public health systems in resource-poor settings [107].

Dengue and Dengue Hemorrhagic Fever (DHF)

These are acute febrile diseases, found in the tropics, and caused by four closely related virus serotypes, transmitted by the *Aedes* mosquito which bites during the day. The disease is now endemic in more than 100 countries in Africa, the Americas, the Eastern Mediterranean, Southeast Asia, and the Western Pacific [109]. Southeast Asia and the Western Pacific are the most seriously affected. Before 1970 only nine countries had experienced DHF epidemics, a number that had increased more than fourfold by 1995. Not only is the number of cases increasing as the disease is spreading to new areas, but explosive outbreaks are also occurring. In 2007, Venezuela reported over 80,000 cases, including more than 6,000 cases of DHF [110].

- During epidemics of dengue, infection rates among those who have not been previously exposed to the virus are often 40–50%, but can reach 80–90%.
- An estimated 500,000 people with DHF require hospitalization each year, a very large proportion of whom are children. About 2.5% of those affected die.
- Without proper treatment, DHF fatality rates can exceed 20%. Wider access to medical care from health providers with knowledge about DHF – physicians and nurses who recognize its symptoms and know how to treat its effects – can reduce death rates to less than 1%.

Early treatment with therapy to tackle shock due to hemoconcentration reductions and bleeding is crucial. Increased oral fluid intake is recommended to prevent dehydration. In extreme cases, internal gastrointestinal bleeding may occur requiring a transfusion.

Unfortunately, progress on control of this infection has been poor. Dengue fever is becoming more prevalent in some areas, which is cause for great concern. The spread of dengue is attributed to expanding geographic distribution of the four dengue viruses and their mosquito vectors, the most important of which is the predominantly urban species *Aedes aegypti*. A rapid rise in urban mosquito populations is bringing ever greater numbers of people into contact with this vector, especially in areas that are favorable for mosquito breeding, for example, where household water storage is common and where solid waste disposal services are inadequate.

Future Directions

Great progress has been made in some areas of the battle against waterborne diseases. Control of malaria has long captured the scientific community's attention and with the recent Bill and Melinda Gates call to aim for eradication hopefully significant further progress will be accomplished. Control and elimination of some of the other waterborne diseases, such as some of the NTDs (schistosomiasis, STH, onchocerciasis, LF, trachoma), can be achieved through inexpensive (or even donated), effective, and easily administered drugs. Even more promisingly, Guinea worm is on the verge of becoming the world's second major human infectious disease to be eradicated from the globe (following smallpox [111]). This has been achieved with no available drug; but purely through health education and behavior change. For other diseases, like dengue, cholera, and typhoid, the picture is bleaker; increases in the number of cases and their distribution demand that further tools be developed and greater effort is expended in their control.

Clearly, for all these waterborne diseases discussed here, permanent change will be aided hugely by improvements in water and sanitation, particularly in the transmission heartlands of sub-Saharan Africa and Southeast Asia. Naturally, people will continue to be drawn to areas providing sources of water, and the exploitation of such resources will play a crucial role in socioeconomic development of less-developed countries. A multisectorial approach to disease control is advocated, for example, when carrying out the development of water resources (drinking water, sanitation, dams, hydroelectric power), a greater integration between engineering and health sectors is required to effectively combat these waterborne diseases permanently. In the mean time, it is advocated to use whatever tools are currently available to reduce infection and transmission and ease the burden of these diseases.

Bibliography

1. Fenwick A (2006) Waterborne infectious diseases – could they be consigned to history? Science 313:1077–1081
2. Steinmann P et al (2006) Schistosomiasis and water resources development: systematic review, meta-analysis, and estimates of people at risk. Lancet Infect Dis 6:411–425
3. Keiser J, Singer BH, Utzinger J (2005) Reducing the burden of malaria in different eco-epidemiological settings with environmental management: a systematic review. Lancet Infect Dis 5:695–708
4. Fenwick A et al (2009) The Schistosomiasis Control Initiative (SCI): rationale, development and implementation from 2002–2008. Parasitology 136:1719–1730
5. Hotez PJ et al (2009) Rescuing the bottom billion through control of neglected tropical diseases. Lancet 373:1570–1575
6. Hotez PJ et al (2006) Incorporating a rapid-impact package for neglected tropical diseases with programs for HIV/AIDS, tuberculosis, and malaria. PLoS Med 3:e102
7. Lammie PJ, Fenwick A, Utzinger J (2006) A blueprint for success: integration of neglected tropical disease control programmes. Trends Parasitol 22:313–321
8. Zhang Y et al (2010) Control of neglected tropical diseases needs a long-term commitment. BMC Med 8:67–75
9. Kolaczinski JH et al (2007) Neglected tropical diseases in Uganda: the prospect and challenge of integrated control. Trends Parasitol 23:485–493
10. Linehan M et al (2011) Integrated implementation of programs targeting neglected tropical diseases through preventive chemotherapy: proving the feasibility at national scale. Am J Trop Med Hyg 84:5–14
11. Molyneux D, Hopkins DR, Zagaria N (2004) Disease eradication, elimination and control: the need for accurate and consistent usage. Trends Parasitol 20:347–351

12. WHO (2004) World health report 2004: changing history. World Health Organization, Geneva. http://www.who.int/whr/2004/en
13. Snow RW et al (2005) The global distribution of clinical episodesk of *Plasmodium falciparum* malaria. Nature 10:214–217
14. King C (2010) Parasites and poverty: the case of schistosomiasis. Acta Trop 113:95–104
15. de Silva NR et al (2003) Soil-transmitted helminth infections: updating the global picture. Trends Parasitol 19:14–551
16. Dadzie Y, Neira M, Hopkins D (2003) Final report of the conference on the eradicability of onchocerciasis. Filaria J 2:2
17. Zagaria N, Savioli L (2002) Elimination of lymphatic filariasis: a public health challenge. Ann Trop Med Parasitol 96:S3–S13
18. Centre C (2011) Dracunculiasis eradication: the final inch. http://www.cartercenter.org/news/documents/doc2229.html
19. Kumaresan JA, Mecaskey JW (2003) The global elimination of blinding trachoma: progress and promise. Am J Trop Med Hyg 69:S24–S28
20. WHO (2006) Cholera, 2005. Wkly Epidemiol Rec 81:297–308
21. WHO (2003) Cholera unveiled. Global Task Force on Cholera Control, Editor, Geneva. http://whqlibdoc.who.int/hq/2003/WHO_CDS_CPE_ZFK_2003.3.pdf
22. Crump JA, Luby SP, Mintz ED (2004) The global burden of typhoid fever. Bull World Health Organ 82:346–353
23. Suaya JA, Shepard DS, Beatty ME (2006) Dengue: burden of disease and cost of illness. Working paper for the scientific working group on dengue research, convened by the special programme for research and training in tropical diseases, Geneva, 1–5 October 2006
24. Roberts L, Enserink M (2007) Did they really say. . .eradication? Science 318:1544–1555
25. WHO (2008) Global malaria control and elimination: report of a technical review. World Health Organization, Geneva
26. Knopp S et al (in press) From morbidity control to transmission control: time to change tactics against helminths on Unguja Island, Zanzibar. Acta Trop. doi: 10.1016/j.actatropica.2011.04.010
27. Richter J (2003) The impact of chemotherapy on morbidity due to schistosomiasis. Acta Trop 86:161–183
28. WHO (2002) Prevention and control of schistosomiasis and soil-transmitted helminthiasis. In: World health organization technical report. World Health Organization, Geneva, pp I–v1
29. Thylefors B (2008) The Mectizan donation program (MDP). Ann Trop Med Parasitol 102(S1):S39–S44
30. Basáñez MG et al (2006) River blindness: a success story under threat? PLoS Med 3:e371
31. GAELF (2010) Half-time in LF elimination: teaming up with NTDs. In: Sixth meeting of the global alliance to eliminate lymphatic filariasis, Seoul, 1–3 June 2010
32. Mariotti SP, Pascolini D, Rose-Nussbaumer J (2009) Trachoma: global magnitude of a preventable cause of blindness. Br J Ophthalmol 93:563–568
33. West S (2003) Blinding trachoma: prevention with the SAFE strategy. Am J Trop Med Hyg 69:S18–S23
34. Crump JA, Mintz ED (2010) Global trends in typhoid and paratyphoid fever. Clin Infect Dis 50:241–246
35. Breman JG (2001) The ears of the hippopotamus: manifestations, determinants, and estimates of the malaria burden. Am J Trop Med Hyg 64S:1–11
36. Snow RW et al (1999) Estimating mortality, morbidity, and disability due to malaria among Africa's non-pregnant population. Bull World Health Organ 77:627–640
37. Brooker S et al (2008) Malaria in African schoolchildren: options for control. Trans R Soc Trop Med Hyg 102:304–305
38. Greenwood B (2009) Can malaria be eliminated? Trans R Soc Trop Med Hyg 103(S1):S2–S5
39. Kapp C (2000) WHO wins reprieve for DDT against malaria. Lancet 356:2076

40. Hay SI et al (2009) A world malaria map: *Plasmodium falciparum* endemicity in 2007. PLoS Med 6(3):e1000048
41. Chawira AN et al (1987) The effect of combinations of qinghaosu (artemisinin) with standard antimalarial drugs in the suppressive treatment of malaria in mice. Trans R Soc Trop Med Hyg 8:554–558
42. Lim P et al (2009) Pfmdrl copy number and artemisinin derivatives combination therapy failure in falciparum malaria in Cambodia. Malar J 8:11
43. WHO (2006) Guidelines for the treatment of malaria. World Health Organization, Geneva
44. Tshefu AK et al (2010) Efficacy and safety of a fixed-dose oral combination of pyronaridine-artesunate compared with artemether-lumefantrine in children and adults with uncomplicated *Plasmodium falciparum* malaria: a randomised non-inferiority trial. Lancet 375:1457–1467
45. Maxwell CA et al (2002) Effect of community-wide use of insecticide-treated nets for 3–4 years on malarial morbidity in Tanzania. Trop Med Int Health 7:1003–1008
46. WHO (2011) Malaria: Indoor residual spraying. http://www.who.int/malaria/vector_control/irs/en/index.html
47. Pluess B et al (2010) Indoor residual spraying for preventing malaria (review). The Cochrane Libr 4:CD006657
48. Gryseels B et al (2006) Human schistosomiasis. Lancet 368:1106–1118
49. Poggensee G, Feldmeier H (2001) Female genital schistosomiasis: facts and hypotheses. Acta Trop 79:193–210
50. Fulford AJ et al (1995) A statistical approach to schistosome population dynamics and estimation of the life-span of *Schistosoma mansoni* in man. Parasitology 110:307–316
51. Crompton DW, Savioli L (2007) Handbook of helminthiasis for public health. Taylor and Francis, Boca Raton
52. Cook G, Zumla A (eds) (2008) Manson's tropical diseases, 22nd edn. W.B. Saunders, London
53. Mahmoud AAF (2001) Schistosomiasis. Imperial College Press, London
54. Fenwick A (2006) New initiatives against Africa's worms. Trans R Soc Trop Med Hyg 100:200–207
55. Warren KS et al (1974) *Schistosomiasis mansoni* in Yemeni in California: duration of infection, presence of disease, therapeutic management. Am J Trop Med Hyg 23:902–909
56. Harris ARC, Russell RJ, Charters AD (1984) A review of schistosomiasis in immigrants in western Australia, demonstrating the unusual longevity of *Schistosoma mansoni*. Trans R Soc Trop Med Hyg 78:385–388
57. de Clerq D et al (1998) The relationship between *Schistosoma haematobium* infection and school performance and attendance in Bamako, Mali. Ann Trop Med Parasitol 92:851–858
58. Vennerveld BJ, Dunne DW (2004) Morbidity in schsistosomiasis: an update. Curr Opin Infect Dis 17:439–447
59. Miguel E, Kremer M (2004) Worms: identifying impacts on education and health in the presence of treatment externalities. Econometrica 72:159–217
60. French MD et al (2010) Observed reductions in *Schistosoma mansoni* transmission from large scale administration of praziquantel in Uganda: a mathematical modelling study. PLoS Negl Trop Dis 4:e897
61. WHO (2006) Preventive chemotherapy in human helminthiasis. Coordinated use of anthelminthic drugs to control interventions: a manual for health professionals and programme managers. World Health Organization, Geneva
62. WHO (2009) WHO model formulary. Based on the 15th model list of essential medicines 2007. World Health Organization, Geneva
63. Hotez P et al (2010) Africa is desperate for praziquantel. Lancet 376:496–498
64. Bethony J et al (2006) Soil-transmitted helminth infections: ascariasis, trichuriasis, and hookworm. Lancet 367:1521–1532
65. Nokes C et al (1991) Geohelminth infections and academic assessment in Jamaican children. Trans R Soc Trop Med Hyg 85:272–273

66. Nokes C et al (1992) Parasitic helminth infection and cognitive function in school children. Proc R Soc Lond B 247:77–81
67. Simeon D et al (1994) School performance, nutritional status and trichuriasis in Jamaican schoolchildren. Acta Paediatr 83:1188–1193
68. Hutchinson SE et al (1997) Nutrition, anaemia, geohelminth infection and school achievement in rural Jamaican primary school children. Eur J Clin Nutr 51:729–735
69. Albonico M et al (2006) Intervention for the control of soil-transmitted helminthiasis in the community. Adv Parasitol 61:311–348
70. Brooker S, Hotez PJ, Bundy DAP (2008) Hookworm-related anaemia among pregnant women: a systematic review. PLoS Negl Trop Dis 2:9
71. Smith JL, Brooker S (2010) Impact of hookworm infection and deworming on anaemia in non-pregnant populations: a systematic review. Trop Med Int Health 15:776–7959
72. Yokogawa M (1985) JOICFP's experience in the control of ascariasis within an integrated programme. In: Crompton DW, Nesheim MC, Pawlowski ZS (eds) Ascaris and its public health significance. Taylor and Francis, London, pp 265–278.
73. Albonico M et al (2008) Controlling soil-transmitted helminthiasis in pre-school-age children through preventive chemotherapy. PLoS Negl Trop Dis 2:e126
74. Ottesen EA (2006) Lymphatic filariasis: treatment, control and elimination. Adv Parasitol 61:395–441
75. Kirkwood B et al (1983) Relationships between mortality, visual acuity and microfilarial load in the area of the onchocerciasis control programme. Trans R Soc Trop Med Hyg 77:862–868
76. Prost A (1986) The burden of blindness in adult males in the Savanna villages of West Africa exposed to onchocerciasis. Trans R Soc Trop Med Hyg 80:525–527
77. Murdoch ME et al (2002) Onchocerciasis: the clinical and epidemiological burden of skin disease in Africa. Ann Trop Med Parasitol 96:283–296
78. Little MP et al (2004) Incidence of blindness during the onchocerciasis control programme in western Africa, 1971–2002. J Infect Dis 189:1932–1941
79. Boatin BA, Richards FO (2006) Control of onchocerciasis. Adv Parasitol 61:349–394
80. Hougard J-M et al (1997) Twenty-two years of blackfly control in the onchocerciasis control programme in West Africa. Parasitol Today 13:425–431
81. Boatin B (2008) The onchocerciasis control programme in West Africa (OCP). Ann Trop Med Parasitol 102(S1):S13–S17
82. Dadzie KY et al (1991) Onchocerciasis control by large-scale ivermectin treatment. Lancet 337:1358–1359
83. Tielsch JM, Beeche A (2004) Impact of ivermectin on illness and disability associated with onchocerciasis. Trop Med Int Health 9:A45–A56
84. Amazigo U (2008) The African programme for onchocerciasis control (APOC). Ann Trop Med Parasitol 102(S1):19–22
85. Gardon J et al (1997) Serious reactions after mass treatment of onchocerciasis with ivermectin in an area endemic for *Loa loa* infection. Lancet 350:18–22
86. Padgett JJ, Jacobsen KH (2008) Loiasis: African eye worm. Trans R Soc Trop Med Hyg 102:983–989
87. Sauerbrey M (2008) The onchocerciasis elimination program for the Americas. Ann Trop Med Parasitol 102(S1):S25–S29
88. Onchocerciasis Elimination Program for the Americas (OEPA) (2011) Distribución geográfica de la oncocercosis y situación de la transmisión en las Américas. http://www.oepa.net/epidemiologia.html
89. Cupp EW, Sauerbrey M, Richards FO Elimination of human onchocerciasis: history of progress and current feasibility using ivermectin (Mectizan®) monotherapy. Acta Trop 120 S1:S100–S108
90. GAELF (2011) Global programme to eliminate lymphatic filariasis: progress report on mass drug administration, 2010. Wkly Epidemiol Rec 86:377–388

91. Nienga SM et al (2007) Chronic clinical manifestations related to *Wuchereria bancrofti* infection in a highly endemic area in Kenya. Trans R Soc Trop Med Hyg 101:439–444
92. Gyapong JO et al (2005) Treatment strategies underpinning the global programme to eliminate lymphatic filariasis. Expert Opin Pharmacother 6:179–200
93. Gambhir M et al (2010) Geographic and ecologic heterogeneity in elimination thresholds for the major vector-borne helminthic disease, lymphatic filariasis. BMC Biol 8:22
94. Michael E et al (2006) Mathematical models and lymphatic filariasis control: endpoints and optimal interventions. Trends Parasitol 22:226–233
95. Ottesen EA et al (2008) The global programme to eliminate lymphatic filariasis: health impact after 8 years. PLoS Negl Trop Dis 2:e317
96. Hooper PJ et al (2009) The global programme to eliminate lymphatic filariasis: health impact during its first 8 years (2000–2007). Ann Trop Med Parasitol 103(S1):S17–S21
97. Chu BK et al (2010) The economic benefits resulting from the first 8 years of the global programme to eliminate lymphatic filariasis. PLoS Negl Trop Dis 4:e708
98. Hopkins AD, Molyneux DH (2009) A decade of ivermectin-albendazole donation for lymphatic filariasis. Ann Trop Med Parasitol 103(S1):S3
99. Molyneux DH (2009) Filaria control and elimination: diagnostic, monitoring and surveillance needs. Trans R Soc Trop Med Hyg 103:338–341
100. Mohammed KA et al (2006) Progress towards elimination lymphatic filariasis in Zanzibar: a model programme. Trends Parasitol 22:340–344
101. Rawlins SC et al (2004) Evidence for the interruption of transmission of lymphatic filariasis among schoolchildren in Trinidad and Tobago. Trans R Soc Trop Med Hyg 98:473–477
102. Brantus P (2009) Ten years of managing the clinical manifestations and disabilities of lymphatic filariasis. Ann Trop Med Parasitol 103(S1):S5–S10
103. Knirsch C (2007) Trachoma: ancient scourge, disease elimination, and future research. Curr Infect Dis Rep 9:21–28
104. Barry M (2007) The tail end of guinea worm – global eradication without a drug or a vaccine. N Engl J Med 356:2561–2564
105. Sack DA et al (2004) Cholera. Lancet 363:223–233
106. Glass RI et al (1982) Endemic cholera in rural Bangladesh. Am J Epidemiol 116:959–970
107. Bhutta Z (2006) Endemic cholera in rural Bangladesh, 1966–1980. Br Med J 333:78–82
108. Sinha A et al (1999) Typhoid fever in children aged less than 5 years. Lancet 354:734–737
109. WHO (1997) Dengue haemorrhagic fever: diagnosis, treatment prevention and control. World Health Organization, Geneva
110. San Martin JL et al (2010) The epidemiology of dengue in the Americas over the last three decades: a worrisome reality. Am J Trop Med Hyg 82:128–135
111. Henderson DA (1976) The eradication of smallpox. Sci Am 235:25–33

Chapter 15
Waterborne Parasitic Diseases in Ocean

Robin M. Overstreet

Glossary

Autoinfection	Reinfection of a host by the progeny of a parasite already present in the same host individual without exposure to the external environment.
Cyst/Encapsulation	A cyst is a structure to aid dissemination or protection of a parasitic stage derived from the parasite and in some cases additionally from the host. A capsule differs by being derived from the host, often as a cellular response.
Erythema	Abnormal redness of the skin due to local congestion as in inflammation, caused by irritation or injury to the tissue.
Hermaphrodite parasite	An individual containing both male and female gametes that can be functional concurrently or with one following the other; some able to form zygotes.
Hypersensitivity reaction	A damaging and sometimes fatal reaction produced by the normal immune system, which requires a pre-sensitized, immune state of the host.
Life cycle of parasite	The orderly sequence of distinct stages through which the agent progresses in the course of development to maturity or sexual stage.

This chapter, which has been modified slightly for the purposes of this volume, was originally published as part of the Encyclopedia of Sustainability Science and Technology edited by Robert A. Meyers. DOI:10.1007/978-1-4419-0851-3

R.M. Overstreet (✉)
Department of Coastal Sciences, Gulf Coast Research Laboratory, University of Southern Mississippi, 703 East Beach Drive, Ocean Springs, MS 39564, USA
e-mail: robin.overstreet@usm.edu

P. Kanki and D.J. Grimes (eds.), *Infectious Diseases: Selected Entries from the Encyclopedia of Sustainability Science and Technology*,
DOI 10.1007/978-1-4614-5719-0_15, © Springer Science+Business Media New York 2013

Life history of parasite	The life cycle of an agent including facultative paratenic hosts, feedback strategies, interaction with the environment, and other ecological influences.
Paratenic host	A specific host, also termed "transport host," which acquires a stage of an agent that does not develop to the subsequent stage and can be either acquired by another paratenic host or by a final host; that host can be critical in completing a life cycle.
Pathogenesis	The production or development of a disease, specifically the cellular reactions and other pathologic mechanisms occurring in the progression of the disease.
Reservoir host	A definitive host that serves as an alternant relative to the host of interest for an agent that disseminates infective stages.
Salinity	An expression of concentration of salts dissolved in water, including primarily chloride, sodium, magnesium, sulfur, calcium, and potassium but with other elements in low concentrations. It is usually expressed as parts per thousand (ppt) or parts per million (ppm) and usually based on the electrical conductivity ratio of the sample to "Copenhagen water," an artificial seawater manufactured to serve as a world "standard" or more recently in practical salinity units (psu) as the conductivity ratio of a seawater sample to a standard KCl solution. Full strength seawater is considered 35 ppt, or 35 g of salt per liter of solution; salinity in estuaries can fluctuate from 0 to over 35 ppt daily, seasonally, or yearly, depending on winds, rain, currents, temperature, and geography.
Urticaria	An allergic reaction comprising pale, pink, focal swellings, or wheals, on the skin that itch, burn, or sting; also referred to as "hives."
Zoonosis	An animal disease transmissible to humans under natural conditions or a human disease transmissible to animals.

Definition of the Subject and Its Importance

Several parasites that infect marine and coastal animals or contaminate the marine and coastal waters can infect humans and present a significant public health risk. Historically, parasites from the marine environment were ignored as representing a risk. More recently, more researchers have investigated actual or potential human marine infections; easily observable worms and other metazoans have been experimentally studied to determine if they can infect model mammals; some humans are now more likely to come into contact with infective parasites; immunocompromised people are more susceptible to parasitic infections; molecular methods have been used to differentiate known or potential infective parasitic species; improved molecular and serological methods have been developed to detect known protozoan and metazoan parasites; and infectious parasites have been investigated in detail. As discussed in this chapter, some parasites involved with marine waters and seafood products can kill large numbers of people, other fatal ones are rare, many cause human illness, and infections or the possibility of infections keeps people from eating seafood or entering marine waters. Moreover, the economic loss of seafood products from actual or perceived problems involving parasites in seafood caused by media exposure can be significant [1, 2] just as it is when involving perceived contamination of seafood by oil or other toxicants (gulfseagrant.tamu.edu/oilspill/pdfs/latourismPerception_BPOilWave1.pdf).

Introduction

Textbooks once considered parasites of marine origin or those in the marine environment as posing no public health risk. Over the last few decades, more and more parasites in different groups have been determined to be zoonotic agents that either are known to infect humans or can serve as a potential public health risk based on experimental infections in nonhuman mammalian hosts or are closely related to known human-infecting agents. These parasites include metazoans and "protozoans" that occur in water or in seafood products. Protozoa occurs in quotation marks because its historic higher taxonomic groups are not necessarily closely related [3] and some include stages with more than one cell. For example, a single myxosporidian spore may develop from a stage consisting of 13 cells, and molecular data demonstrate no relationship between Myxosporea and any group consisting of members with single cells and recognized as a protozoan. At least one metazoan taxonomic group, ascaridoid nematodes, has members that produce excretions and secretions (ES) that are potentially harmful to susceptible people.

Numerous factors are involved with risk of infection. For example, the likelihood of a visitor or traveler getting infected by eating raw seafood in well-known urban Japanese or other Asian restaurants is relatively low, but dining in rural local restaurants or street shops has a higher risk of infection with a variety of parasites [4]. The highest worldwide risk for a helminth infection is from preparation of raw or inadequately cooked seafood at home. This is usually because restaurants typically use products historically known not to be infected in the region from which they were collected or the products were frozen and then thawed. Public health risk also occurs for personal contact with parasites for people using coastal and marine water for either occupation or recreation.

An apparent increase in human infections has occurred because of increased opportunities to get infected. (1) A global spread and increased popularity of "exotic cuisines" at least partially relates to increased human infections because of the increase in various media such as television and magazines. This has led to an increase in homemade raw seafood products, which are much more likely to contain infective parasites than products served in restaurants. (2) The increased popularity of raw products also relates to an increase and translocation of people throughout the world. People that move or visit abroad take their cuisines with them. They may eat infected products in their new surroundings that are similar to the uninfected products they ate in their native country. They can also spread their diseases as they travel, allowing new intermediate hosts and reservoirs to become established. (3) Much of the world's population is becoming more affluent, and with that more people are able to try new cuisines. Also, there is more opportunity to partake of established cuisines. For example, in Vietnam the improved economy has increased the traditional social activities that are conducive to eating more raw seafood and, consequently, acquiring more parasites. (4) With the resulting increase in people and demand for healthy fishery products, there has been increased utilization of different fishing grounds, species, and stocks. Even though some stocks are overfished, methods are now available for ships to go farther to sea or deeper to catch replacement products, possibly with improved transportation and storage systems. When a specific fish is assumed to be free from parasites harmful to humans, when the same species is caught in a different location, when it is maintained on a vessel longer, or when it is produced on fish farms, that resulting fish product may have or it may acquire harmful parasites. Also, fishing a previously underutilized species may result in obtaining an infected stock. (5) Regulations such as those protecting specific animals may result in an increased public health risk. For example, protecting seal pups and reducing seal fashions increases the final host population for *Pseudoterranova* spp., which results in a heavily infected cod intermediate host population. Also, the need to discharge fisheries wastes from commercial and subsistence vessels results in increased infections in fishes and marine mammals that feed on the wastes. Alternatively, regulations in some countries are lacking in regard to fishery products or for imported fishery products but not necessarily all meats, leaving consumers with the false impression that the products are safe. (6) Climatic and environmental health conditions are always changing, and, with those changes, infection dynamics of parasites quickly change. For example, during the El Niño of 1997–1998, the junction of the warmwater Kuroshio Current from the south with the cool Oyashio Current from the north moved 3,000 km northward from Kyushu to Hokkaido, Japan. This migration of the current's junction and its temperature elevation of 3.4°C caused an increase in abundance of krill, the intermediate host of *Pseudoterranova azarasi* and members of the *Anisakis simplex* complex. The sea lion final host of *P. azarasi* occurred in Hokkaido but not Kyushu, and, in addition, protected cetaceans that are the final hosts of *A. simplex* sensu lato migrated north so as to feed heavily on krill or on fish that ate the krill, acquiring the juvenile nematodes from them. Consequently, the Japanese used to eating nearly parasite-free seafood from Hokkaido began acquiring infections. In contrast, pollution from Southeast Asia kept some fish intermediate hosts of *A. simplex* sensu lato from migrating to Japan, especially to southern Kyushu, reducing the number of anisakiasis cases there [5]. Each few years, climate in some regions worldwide changes enough to influence the seafood production and the parasitic infections. (7) Reduced resistance in some people because of disease, organ transplants, or immunosuppressive drug therapy relate to an increased susceptibility to various parasites, especially parasites that can replicate in or on humans with being challenged by an additional dose of parasites.

Increased documentation of human infections results from increased technical abilities. (1) Improved media coverage allows people to know when an epidemic occurs. It also stimulates potentially infected people to get examined, sometimes unnecessarily and to the detriment of the seafood industry. (2) Diagnostic tests are becoming better and more available both because of research plus technical advancements and because various funds are made available to survey regions or satisfy needs. (3) The same can be said for accurate identifications. Molecular means make it easy to detect parasites that cannot be seen grossly or a juvenile that cannot be morphologically associated with an adult, upon which the species has been named and described. (4) With good identifications and detection methods, more critical examinations can be obtained and compiled, allowing for useful epidemiological assessments.

The organization involves five sections. The first treats parasites that people acquire from the marine environment by eating seafood. The second treats those where people get infected with parasites by contact with them, either directly by active penetration by the parasites or indirectly by passive association with the agents. In the third, people get disease from allergens from parasites or parasite products. This is followed by a section on management, control, and treatment of the infectious parasites. The final section briefly examines what is required in the future to reduce, eliminate, or keep infections under control. In the first section on acquisition of parasites by eating inadequately prepared food, examples are divided into those that mature in humans and those that do not. The reason for this is that worms that mature in humans are voided or produce eggs that can be detected in fecal examinations. Those parasites that do not mature in the patient or produce cyst dispersal stages are more difficult to detect without immunological or invasive techniques. Moreover, the consumer not able to provide proof of a parasitic infection is less likely to find medical help, and the medical community is less likely to specifically diagnose, report the condition, or show a follow-up interest in the case. Several of these agents that have not yet been reported are treated as potential public health risks. The first two sections are each divided by taxonomic groups. In some cases the same parasites or members of the same parasitic group fit into more than one section, and the possible confusion is discussed in the different cases.

Parasites Acquired When People Eat Infected Seafood Products

Zoonotic parasitic diseases acquired by eating inadequately prepared infected seafood products include most of those treated by disease researchers and reported in recent literature. Some of the responsible agents are seen grossly, such as 5–10 m of senile cestode strobilae protruding out of the anus of alarmed seafood consumers or a 4-cm-long ascaridoid nematode being vomited or passing out through the nose of someone recently eating uncooked cod. Most zoonotic diseases are recognized by the signs of a disease or fortuitous findings of diagnostic features. Not all infections are diseases. A few hundred small marine heterophyids probably can line the human intestine without apparent symptoms, but more may cause pathological alterations and result in disease. A lot of factors involving the patient and the history of the infection can influence the alterations. Perhaps easier to envision would be a non-marine example of the hookworm *Necator americanus*. A barefoot boy can pick up a few dozen juveniles penetrating his feet from the warm, moist soil. There may be a slight itching and rash where the juveniles enter, but as they are carried to the lungs, pass up the respiratory tract to the mouth, are swallowed, and ultimately attach to the intestine as 7- to 11-cm-long adults, they produce no symptom. On the other hand, increase the number, say above 50, and pathological changes can occur during all the phases. Most important, once attached, the blood loss becomes enough to produce anemia and protein deficiency, and the resulting loss of iron and protein may retard growth and mental development. This will be associated with abdominal pain, diarrhea, loss of appetite and weight, and "disease." Numbers of worms alone can shift an "infection" without disease to a "disease."

Parasites That Mature in Humans

Helminths (Worms in General)

Most agents that mature in humans are helminths. Helminth diseases consist largely of those acquired by eating inadequately prepared seafood products usually infected with platyhelminths and nematodes, but occasionally with other agents. Some agents are freshwater parasites that, because of their complicated life cycles, have one or more stages that enter the coastal or marine environment. Presently, several of these are treated as freshwater cases because of the epidemiological aspects of the case. As more cases are studied, more human infections that involve strictly marine parasites will be demonstrated.

Many more cases of marine helminth infections probably exist than are reported or recognized by physicians. Surveys involving fecal examinations exhibit adult worms or their eggs, but, for a variety of reasons, the samples are difficult to acquire and the corresponding infections are difficult to correctly diagnose or get reported. Most medical attention is not likely to culminate in confirmed reports of identified agents because physicians or technicians not familiar with proper techniques, means to correctly identify the agents, or people to consult are interested in treatments and not the specifics of a parasitic infection. Even when treated, the patient never knows what was causing the signs of disease because the agents are killed and digested or degenerate. On the other hand, in some countries where considerable raw or insufficiently prepared seafood is consumed, attention by researchers and the medical community results in an abundance of case reports. The Japanese and Koreans provide good examples of documented cases of infections by marine parasites.

Platyhelminths consist mostly of hermaphroditic soft-bodied worms. Those hermaphrodites of medical marine significance are cestodes (tapeworms) and trematodes (flukes), and most of those that mature in humans were consumed as larval stages. For many, the taxonomic identification is confused and the life cycle is partially or fully unknown. Moreover, infections are often undiagnosed because fecal examinations are not always conducted or are conducted by those without parasitological training. Moreover, the eggs of many species are similar in appearance.

Cestodes

Cestodes infective to humans from marine hosts are members of the family Diphyllobothridiidae in the Diphyllobothriidea, previously classified in the artificial assemblage Pseudophyllidea, now represented as comprising two unrelated clades, "Bothriocephalidea" and "Diphyllobothriidea" [6]. *Diphyllobothrium latum* is the best known member of the group, and it is considered a freshwater member with a cyclopoid-freshwater fish-bear/human/vertebrate cycle. Unpublished studies attempting to infect Mississippi marine fishes with the agent were unsuccessful; most copepod species developed a strong cellular response to the larva, and a mixture of "infected" copepods did not produce an infection in the Atlantic croaker. However, five human infections reported from eating the estuarine mullet *Liza haematocheila* in Korea were identified as *D. latum* on the basis of morphological characteristics [7]. Molecular tools should be able to confirm that identification as those tools have confirmed or corrected identifications of other species mentioned later.

Many human case reports have been identified as *Diphyllobothrium pacificum*, which typically matures in the pinnipeds. A recent taxonomic assessment of *D. pacificum* [8] involved numerous available specimens of variously misidentified species, including *D. latum* and *D. pacificum*. As it turned out, the classic publication [9] reporting human infections by *D. pacificum* in Peru was actually dealing with *Diphyllobothrium arctocephalinum*. Both *D. pacificum* and *D. arctocephalinum* occur along the South American Pacific coast in fur seals and secondarily in humans, with *D. pacificum* ranging north into Alaska and Japan and *D. arctocephalinum* restricted to the Southern Hemisphere, ranging south to South Australia and South Africa [8]. In one review, there are ten reported species of *Diphyllobothrium* from humans [10], including 6 from Japan [11], and in another there are 15 species [12]. Of the latter 15, not including *D. arctocephalinum*, there are eight nominal species in marine final hosts that also have been reported from humans: *Diphyllobothrium cameroni* (Hawaiian monk seal), *Diphyllobothrium cordatum* (Arctic seals, walruses), *Diphyllobothrium hians* (Arctic seals), *Diphyllobothrium orcini* (killer whale), *D. pacificum* (sea lions, eared seals),

Diphyllobothrium scoticum (leopard seal, southern sea lion), and *Diphyllobothrium stemmacephalum* (harbor porpoise, bottlenose dolphin).

The species commonly infecting humans in Japan and previously referred to there as *D. latum* has been confirmed as *Diphyllobothrium nihonkaiense* ([8, 12], and others) (Fig. 15.1). The brown bear (*Ursus arctos*) is its natural definitive host, but it commonly infects humans, with some preserved worms measured as long as 8.8 m. Since infections are acquired from Pacific salmons (mainly cherry, pink, and chum salmons plus the Japanese huchen), either from the sea or just leaving the sea, it is considered a marine infection. Molecular studies have shown that human infections in Far East Russia up to the Kamchatka Peninsula reported as *Diphyllobothrium klebanovskii* are conspecific, identical to *D. nihonkaiense*. Molecular sequences also suggest the existence of two genotypes, biologically homogeneous populations not defined by locality [13]. Infections also involve consumers in Canada [14]. Moreover, salmon are exported to many countries, resulting in genetically confirmed infections in consumers in France and Switzerland [15] as well as New Zealand [16]. Another "freshwater" species, *Diphyllobothrium ursi*, also infects salmons and both brown and black bears as well as humans and dogs that feed on salmon or eat salmon-liver paste [17]. Cestodes and other parasites maturing in terrestrial definitive hosts that infect anadromous fishes such as Pacific salmonids or their freshwater fish or invertebrate hosts can be eaten by marine fishes, which in turn, serve as paratenic hosts. On the other hand, freshwater fishes can become paratenic hosts of helminths of marine origin. They can feed on infected marine fishes entering brackish river mouths or lower reaches of the rivers, or they can feed on anadromous fishes migrating upstream. A confirmed example is the plerocercoid of *Diphyllobothrium alascense* of dogs and rarely humans from the anadromous boreal smelt (*Osmerus mordax*) being transferred to the paratenic freshwater burbot (*Lota lota*) [18].

Members of *Diplogonoporus* have been reported from over 200 people in the twentieth century. Most cases have been identified as *D. balaenopterae*, or the names are probably synonyms of that species, a species that infects consumers in Japan but also Korea [12, 19]. A case from Spain probably represents infection from fish imported from the Far East [20]. Genetic analyses of 18S rDNA, ITS1, and cox1 nucleotide sequences of isolates were obtained from five Japanese cases [21]. The phylogenetic analysis of that material revealed little divergence and a close relationship to *Diphyllobothrium stemmacephalum*, a species maturing in the harbor porpoise and bottlenose dolphin rather than in whales as in *D. balaenopterae*. Data suggest that not only is *Diplogonoporus* paraphyletic, but those genetic data show that so are *Spirometra folium* and *Spirometra decipiens* [22].

Coastal Peruvians of ancient times had infections of *Diphyllobothrium* sp. as determined from coprolites [23], but infections of *D. pacificum* recently seem to be less prevalent in the northern Pacific Ocean, resulting from decreasing numbers of marine mammals and commercially important fishes like the pollack [8].

Not only have diphyllobothriid species been misidentified in the past, but human infections have been and will probably continue to be missed in many cases. They are most easily detected when a human expels lengthy portions of senile strobilae.

Fig. 15.1 Author in Meguro Parasitological Museum, Tokyo, Japan, standing next to an 8.8-m-long specimen (> 10 m before fixation) of the tapeworm *Diphyllobothrium nihonkaiense* and the curators, the late Shunya Kamegai on *right* and Jun Araki on *left*

Signs are variable but usually not severe [11, 24]. Most cases are probably not reported. The author did not experience any symptoms from an infection of four long specimens of *D. latum* other than mild discomfort several hours prior to periodic release of senile strobilae. The treatment of choice for the adult tapeworm infections in humans is a single dose of praziquantel (10–20 mg/kg) plus 30 g of magnesium sulfate as a purgative, if specimens are to be adequately studied.

Trematoda (Digenea, Flukes)

Trematodes have a molluscan first intermediate host from which free-living cercariae are shed. The cercaria of most groups encysts or becomes encapsulated as a metacercaria in a second intermediate host, and this metacercaria matures as a hermaphroditic individual in a vertebrate final host when the second intermediate host is eaten. Because there are so many trematode taxonomic groups and species, there are numerous exceptions to the basic life cycle, and some of these involve species that can infect humans. For example, the well-known freshwater sheep liver fluke, *Fasciola hepatica*, encysts as small white spherules (metacercariae) on vegetation. Most people that get infected do so by eating raw watercress on which the metacercariae have encysted. Schistosomes do not encyst but develop through the juvenile stage (schistosomula) in the final host where they ultimately mature in a blood vessel as separate male and female worms.

Depending on the taxon, the larval stage in the egg (miracidium) either hatches and penetrates its molluscan host or hatches after being eaten by a specific mollusk. The germ cells develop through a series of two or more asexual phases before shedding a continual production of hundreds to millions of infective cercariae. Some species have broad specificity in the molluscan, second intermediate host, or final host. Second intermediate hosts differ by family of trematode, but those discussed below include fishes, crustaceans, insects, mollusks, and none. Most of those marine/estuarine species reported from humans have a broad specificity of mammal or avian hosts. Most are rather small intestinal worms that necessitate a microscope to detect them unless hundreds are present, and each of these produces a continuous flow of a large number of small (25–30 μm long) eggs in the human feces for weeks or years. Most species are difficult to obtain in a healthy condition after medicinal treatment, are not transferred from the medical community to taxonomic parasitologists, are difficult to distinguish from one another, and produce minimal symptoms in light infections. Consequently, the number of human-infecting species and the prevalence and intensity of infections with marine trematodes are most certainly underestimated. Most of what is known has come from parasitologists within the medical community of Japan and Korea, where consumers eat an abundance of raw and "inadequately" prepared seafood products.

According to recent literature [25], an estimated 18 million people are infected with fish-borne intestinal trematodes. This was considered an underestimate, and it included some freshwater species. In the Republic of Korea, 17 species represent five families (Heterophyidae, Echinostomatidae, Plagiorchiidae, Neodiplostomidae, and Gymnophallidae) [26]. Of those, ten are heterophyids and, of those, seven are prevalent among residents who consume raw flesh from estuarine fishes in south and west coastal areas [25–27]. Those are *Heterophyes nocens*, *Heterophyopsis continua*, *Pygidiopsis summa*, *Stellantchasmus falcatus*, *Stictodora fuscata*, *Stictodora lari*, and *Acanthotrema felis*. Documentation of all food-borne trematodes in humans includes 70 species in 14 families [25], but several are not from marine hosts and presently infected people are estimated to surpass 50 million.

Heterophyidae (Opisthorchioidea). As with zoonotic infections from eating fishes in Japan, Korea, China, and Thailand, such infections from Vietnam are now being recognized as a food safety risk in a country whose people have a strong tradition of eating raw fish [28]. The heterophyids use a snail first intermediate host, a fish second intermediate host, and either a bird or mammal definitive host. A prevalence of heterophyid infections with *Haplorchis pumilio* (100%), *Haplorchis taichui* (70%), *Haplorchis yokogawai* (6%), and *S. falcatus* (3%) was determined from expelled (25 mg/kg praziquantel, followed by a saturated solution of magnesium sulfate) adult worms in a survey of residents from coastal Nam Dinh Province south of Hanoi in 2005. The recent finding of those parasites in Vietnam suggested to the surveyors [28] that some may be introduced rather than endemic, increased in magnitude with intensification of aquaculture, or increased because of increased consumption and importation of raw fish. Difficulty presently exists in determining from which fishes the infections are acquired in people throughout Vietnam as well as the rest of the world where people eat raw fishes. Human-infecting species are beginning to be recognized in food fishes as those fishes are examined for parasites. For example, examination of wild and cultured grouper (*Epinephelus bleekeri*, duskytail grouper, and *Epinephelus coioides*, orange-spotted grouper, both marine species) have revealed *Heterophyopsis continua* and *Procerovum varium* (recognized as

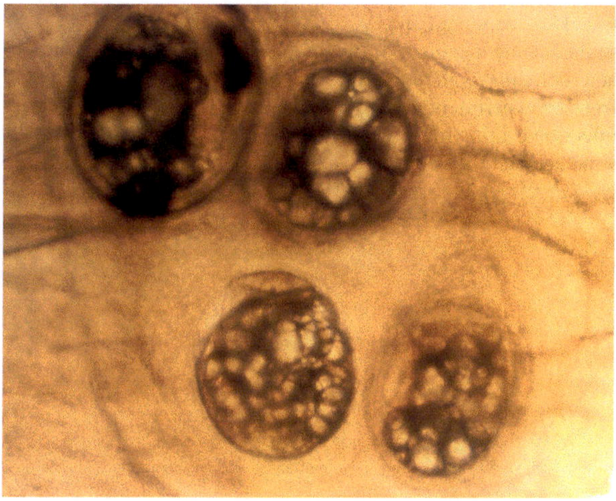

Fig. 15.2 Living encysted metacercariae of the heterophyid *Phagicola nana* in fillet of largemouth bass, *Micropterus salmoides*, from Mississippi estuary. The cysts average 0.30 mm across

a freshwater parasite), and examination of mullet (*Mugil cephalus*) have revealed *Pygidiopsis summa* and *H. continua* from brackish water in Khanh Hoa Province in central coastal Vietnam [29]. The long tradition of eating raw fish has intensified with increasing affluence; a higher prevalence between men (69%) and women (23%) can be explained because the social gatherings where consumption of raw or pickled fish occurs have historically been male-oriented [28]. Concern exists for the exportation of infected fresh fishes, some freshwater species, because aquaculture in rural Vietnam is increasing in economic importance [30].

Human infections have been known from Japan for decades even though most are freshwater species [25]. All three species of *Metagonimus* are freshwater species in regard to the snail host, but, in Korea, one, *Metagonimus yokogawai*, infects a marine fish, which consumers eat raw. Actually, an amphidromous fish, the sweetfish, *Plecoglossus altivelis*, spawns in rivers near the sea. Then, the fish fry feed on plankton at sea until 5–7 cm TL at which time they enter into the river mouths. Infections can be acquired from the fish in estuarine habitats. Sequence data readily separate the three species [31].

In the Western Hemisphere, there are several heterophyids that present a human health risk. For example, in the USA, species of *Phagicola* represent one of several potential infectious genera (not accepting *Phagicola* as a junior synonym of *Ascocotyle*). Knowledge about human infections from the Southeast US has not progressed much beyond anecdotal mention [1], but *Phagicola nana* (Fig. 15.2) from several fishes [32], *Phagicola longa* [33] from mullets (Fig. 15.3), and several other species (Fig. 15.4) probably occur in many consumers who eat inadequately prepared fish. *Heterophyes heterophyes*, a heterophyid from the Middle East has been reported from fish-eating patients in Florida never having left the USA [34, 35]; these infections were most likely species of *Phagicola*. Actually, even capillarid eggs get misidentified as those of *H. heterophyes* in Egypt, where the trematode is common [36]. Unlike many heterophyids in the USA that exhibit

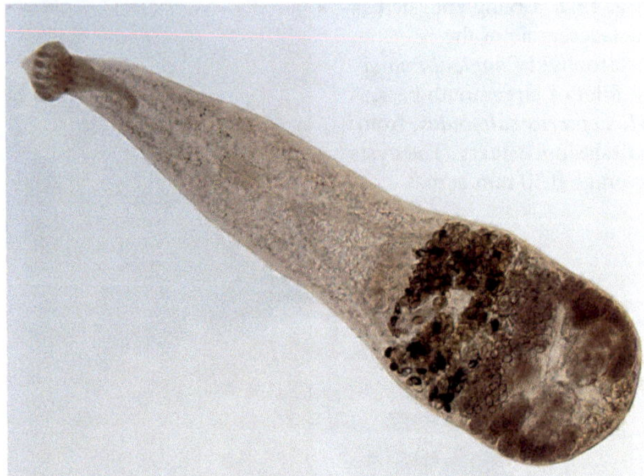

Fig. 15.3 Mounted and stained adult specimen of the heterophyid *Phagicola longa* from an experimental infection originating from a metacercaria in *Mugil cephalus* from Mississippi

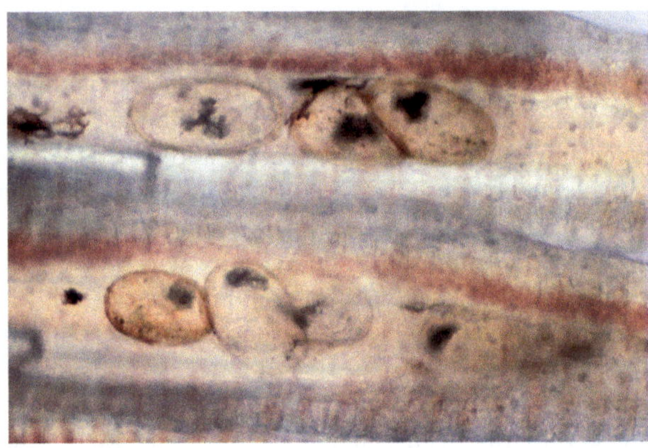

Fig. 15.4 Living encysted metacercariae of the heterophyid *Phagicola diminuta* in gills of Gulf killifish, *Fundulus grandis*, a bite-sized fish from Mississippi estuary

a high "site specificity" in the second intermediate fish hosts, meaning found in one or few specific organs or tissues, *P. nana* occurs in many tissues such as the muscular fillets in addition to the viscera, gills, and other sites [32]. Moreover, *P. nana* infects a few different fishes such as the largemouth bass (*Micropterus salmoides*) and several sunfishes (Centrarchidae), popular recreational fishes in estuaries of the Southern US that are consumed in a variety of cuisines. Most heterophyid species infect specific organs in specific small fishes that are eaten raw in their entirety by specific groups of people, if eaten at all. As with several different parasites, some heterophyids that constitute public health risks constitute species complexes that are difficult to distinguish by morphological features. For example, what is commonly reported as *Centrocestus formosanus* is a mixture of morphologically similar species. Members of this complex are common in the Far and Middle East, but some apparently have been introduced into Hawaii, USA mainland, and Mexico through introduced infected snails (*Melanoides tuberculatus*) in the aquarium trade on vegetation and with fishes [37]. Members of this basically freshwater heterophyid complex infect a large number of fish hosts, including some estuarine species.

The reason for stressing heterophyids is because, even though most species measure 1 mm long or less, they often occur in large numbers, influence public health, and can result in severe pathological changes in humans. A series of studies in the 1930s based on human infections in the Philippines was summarized in 1940 [38]. Five heterophyids were identified from 33 of 297 autopsy cases, not all randomly chosen: *H. yokogawai* (16 cases, 13 alone), *H. taichui* [5], *Haplorchis calderoni* [2], *Haplorchis vanissima* [1], and *Diorchitrema pseudocirrata* [6]. These occurred either singly or with other species. The other species consisted of the microphallid *Microphallus brevicaecum*, which was originally identified as a heterophyid (*Heterophyes brevicaeca* but later recognized as *Spelotrema brevicaeca*) (in 11 cases), an echinostomid identified as *Euparyphium ilocanum* [2], and an unidentified plagiorchiid (*Plagiorchis* sp.) [1], plus the common heterophyids already listed. All the trematodes occur normally in the intestine, where lesions were not found to be severe. More important, eggs of the heterophyids *H. yokogawai* and *H. taichui* and the microphallid *M. brevicaecum* were also associated with acute and chronic lesions in the heart and other visceral organs such as liver, spleen, kidneys, brain, spinal cord, and lungs. *Diorchitrema pseudocirrata* also occurred in some of those lesions in mixed infections but never by itself. Worm eggs, especially from degenerated adults, enter into the blood and lymphatic streams and are carried to different visceral tissues where they provoke the cellular reactions. Adhesions and hemorrhaging were common complications. Of 34 positive cases (1 did not exhibit an intestinal infection), 15 (44%) had visceral complications (1 in brain with fatal hemorrhaging, 1 in spinal cord, and 13 in heart). Of 89 enlarged hearts in these charity patients in the Philippines, the 13 (15%) with myocardial or valvular lesions associated with worm eggs were assumed to have caused the heart dysfunction, a value attributed to heterophyid-caused fatal heart disease [39]. Adult worms (Fig. 15.5) identified as *Heterophyes heterophyes* also encapsulate in the human brain [40].

An enormous number of infective metacercariae can occur in some habitats. To appreciate the life cycle, each adult out of the thousands that can be present in the intestine of a human or in that of a domestic dog or cat produces thousands of eggs, with several hundred laid daily for several weeks or months. One dog was reported to host 13,000 specimens of *H. heterophyes*. Each egg produces a miracidia that can infect a snail. In the lakes along the Nile Delta and in the Bardawil Lagoon, over 90% of the host snail population can be infected (Figs. 15.6, 15.7). Just short of 500 cercariae can be released from each of the abundant snails daily for a year, but with decreasing cercarial production over time. Each cercaria survives 1–2 days, and mullets are common hosts. Eleven emaciated specimens of *Liza ramada* contained 1,730–6,000 metacercariae per gram for an estimated 582,000 metacercarial cysts in one 255-mm-long fish. A specimen of *Mugil cephalus* from the same location had 1,136 metacercariae per gram [41]. Fishermen, seafood dealers, children of those people, and those who taste fish being prepared commonly become infected because the rules established for protection of the consumers buying the products

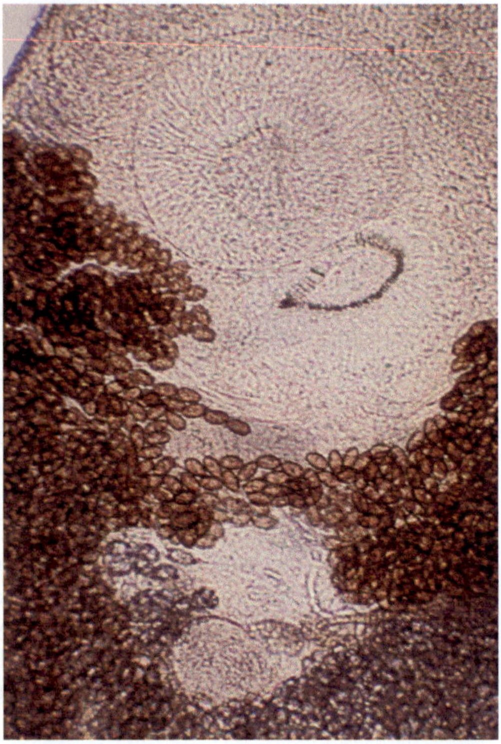

Fig. 15.5 Close-up of living 0.8-mm-long adult specimen of the trematode *Heterophyes* sp. from Bardawil Lagoon, Northern Sinai, Egypt, showing diagnostic sclerites in the modified genital sinus, often misstated as a "gonotyl," below the ventral sucker

Fig. 15.6 Street vendors selling fresh fish in El Arish, Northern Sinai, Egypt

are not always followed by the handlers. Moreover, fishermen do not always completely cook their fish.

Microphallidae (Microphalloidea). Members of small intestinal trematode complexes other than Heterophyidae also have been misidentified. Microphallidae constitutes one of those families. One or more microphallid members also have

Fig. 15.7 Close-up of activities in Fig. 15.6, showing vendor displaying salted mullet. A regulation at the time was to keep fish salted for 8 days before selling them to the public. Eight days was necessary for the heavy infections of heterophyid metacercariae to lose their infectivity

a relatively broad specificity of the final host and can also cause severe potentially fatal pathologic lesions [38, 42]. As mentioned above, *Microphallus brevicaecum* from shrimp has been determined to be the cause of human mortalities in the Philippines. Microphallids are superficially similar to heterophyids and also infect birds and mammals and occasionally fishes, reptiles, and amphibians, but the microphallids have entirely different cercariae, differ phylogenetically, and use crustaceans such as crabs and palaemonid and peneid shrimps as second intermediate hosts rather than fishes.

Troglotrematidae (Gorgoderoidea). *Nanophyetus salmincola* is another small, about 1-mm-long, trematode that infects the intestine of numerous mammals, including humans. In fact it is probably unknowingly the most common helminth in people of the North American Pacific Northwest. It produces gastrointestinal distress, fatigue, loss of weight, and peripheral eosinophilia when numerous specimens occur in humans, and can be diagnosed easily from its relatively large, 0.090–0.100-mm-long, diagnostic eggs that appear in stools about a week after eating infected fish. No symptom occurs in most light infections. Basically a freshwater parasite, cercariae form metacercariae in several salmons and steelhead trout (*Oncorhynchus mykiss*, also known as sea-run rainbow trout) as well as freshwater fish and amphibians that do not make oceanic migrations. Infections in the Pacific Northwest are attributed to eating raw, incompletely cooked, cold-smoked salmon and steelhead trout and their eggs [43]. A second species, *Nanophyetus schikhobalowi*, needs to be studied molecularly to determine if it is conspecific with *N. salmincola*. It had been demonstrated to infect as many as 98% of the people in some eastern Siberian villages.

The trematode in the Pacific Northwest serves as host for the rickettsial agent, *Neorickettsia helminthoeca*, which in turn causes "salmon poisoning disease" of canines. Dogs that are fed infected salmon may die, but humans and other non-canines are not known to be affected. Nevertheless, the loss of dogs in the Arctic can be a serious problem.

Gymnophallidae (Gymnophalloidea). Gymnophallids mature in shore birds, gulls, and diving ducks. However, members of Gymnophalloides also infect mammals. Human infections with *Gymnophalloides seoi* are known only from Korea, but most are tiny, <1 mm, and it and other species all use a variety of mollusks as first and second intermediate hosts and probably infect people throughout the world where mollusks are consumed raw. *Gymnophalloides seoi* is acquired primarily from eating the raw oyster, *Crassostrea gigas*, known as the Japanese oyster, Miyagi oyster, giant oyster, immigrant oyster, and giant Pacific oyster, and infects the intestine and possibly pancreatic duct of human consumers [25, 27]. Nevertheless, like *G. seoi*, which infects the Palearctic oystercatcher, *Haematopus ostralegus*, most members of this genus can be experimentally administered into a variety of birds and mammals, suggesting the human health risk. Studies with mice demonstrate that *G. seoi* produces focal responses for 2–3 weeks, at which time the mucosal integrity is restored. If provided with immunosuppressive treatment, the mice underwent minimal goblet cell hyperplasia, extended retention of worms, and invasion of the submucosa. Even though genetics of hosts also play a role in this sequence, immunosuppressed humans are at risk for transference of eggs and worms to remote organs as with heterophyid infections. With a relatively low production of eggs relative to heterophyids, a total of 100 specimens would be necessary to produce 8,400 eggs per day, considerably fewer than most heterophyids [25].

Echinostomatidae (Echinostomatoidea). Members of some other trematode families also infect the small or large intestine but grow slowly and reach larger size (roughly 3–7 mm) compared with those mentioned above (0.5–2.0 mm); a few marine members are known to infect humans. A total of 21 members of Echinostomatidae are listed as causing human infections, some resulting in gastroenteritis [25]. Of those, all were from freshwater mollusks except two. One, *Acanthoparyphium tyosenense*, can be acquired in Korea from eating at least the raw bivalves *Mactra veneriformis* and *Solen grandis* and the gastropod *Neverita bicolor* available at a coastal village seafood market [44]. The gastropod first intermediate hosts *Lunatia fortuni* and *Glassaulax didyma* shed cercariae that infect the gills of the popularly eaten bivalves. Ducks serve as the natural definitive host, but experimental infections in chick and seagull have confirmed the adult infection [25]. Related species, and there are several, such as what was identified as *Acanthoparyphium spinulosum* in the oyster *Crassostrea virginica* from Texas, USA, are probably also commonly infective to humans, but such infections are limited only by the infrequency with which the molluscan second intermediate hosts are eaten raw [45].

Plagiorchiidae (Plagiorchioidea). Human infections with *Plagiorchis* spp. are uncommon, but 12 cases have been reported. *Plagiorchis vespertilionis* occurred in a man from coastal Republic of Korea who often ate raw mullet and gobies and was not known to eat caddisfly larvae, mayfly larvae, or dragonfly nymphs. The insects were suggested as possible intermediate hosts and are known hosts for many other species of *Plagiorchis*. The patient also contained *Heterophyes nocens*, acquired from fish [46]. Nevertheless, fish were suggested intermediate hosts, and perhaps the infection was transmitted by eating raw estuarine fish that had recently fed on infected dragonflies or some infected insect larva, joining the freshwater parasite with the coastal habitat. Members of *Plagiorchis* and related genera mature in birds and mammals plus accidentally in amphibians and reptiles. They are relatively large (3–5 mm) and usually occur in low numbers in the final hosts.

Didymozoidae (Hemiuroidea). Humans do not get infected by didymozoids! These trematodes of fishes are unusual in that they occur usually in pairs of either hermaphroditic individuals or as separate sexes like mammalian schistosomes and occur encysted in the flesh, encysted in other tissues, or even free in body spaces other than the lumen of the alimentary tract like most trematodes. The group is mentioned because people eat infected fish products and diagnosticians find the small reniform didymozoid eggs (roughly 0.015–0.030 mm long but in some species up to 0.040 mm) of the worms in human stool samples, leading those diagnosticians to assume either an active trematode or protozoan infection. Such misidentifications have been reported from people eating marine fishes in Japan, Taiwan, and the Philippines, but cases could occur throughout the world where infected fish are abundant, especially in pelagic tropical and subtropical fishes [47].

Nematoda (Roundworms)

Nematodes, or roundworms, comprise both free-living and parasitic groups. Parasitic groups that have marine members exhibiting a public health risk also fit into different groups: those that mature in humans, those that have larvae that can infect humans eating inadequately prepared seafood, those that produce excretions or secretions that result in an immunological response in humans, and those that can be acquired by being in the water with free-living stages or handling infected products. When consumers think of parasites in seafood, they typically think of anisakiasis involving grossly apparent nematode juveniles. Of the many nematode groups, each may have a different strategy in completing a life cycle. An example cycle will be mentioned for each major group of worms.

Trichinellidae (Dorylaimia, Trichinellida). Trichinellosis has received considerable attention over the last 150 years of control efforts, but in the past 20–30 years, there has been a dramatic reemergence as an emerging zoonosis [48]. The reasons are severalfold involving changing agricultural, marketing, environmental, and economic practices and conditions, but basically prior attentions involved *Trichinella spiralis* in domestic pork. In the 1940s and 1950s, a large number of marine and terrestrial mammals in Alaska were examined for what was then known as *T. spiralis*, and most had infections [49]. A total of 320 native individuals from three Alaskan areas given intradermal tests showed that 28% from Wainwright, Alaska, tested positive. There are now at least ten recognized species or genotypes, and the involvement of the sylvatic cycle of some of those is responsible for new human cases. For purposes of this review, *Trichinella nativa* and to a lesser degree *Trichinella britovi* play a role. *Trichinella nativa* infects the carnivores that inhabit the marine environs: polar bear (*Ursus maritimus*), walrus (*Odobenus rosmarus*), Arctic fox (*Alopex lagopus*), and wolf (*Canis lupus*), but it is even known from China in domestic pigs. This species differs from the others in that it can survive the freezing process, the process that keeps domestic pork safe to eat. Uncertainty exists in how long the frozen juveniles remain infective, but, when juveniles from walrus meat were stored at −20°C for up to 20 months, some remained infective to experimental guinea pigs [50]. There was a gradual degradation process that rendered some noninfective. The freezing resistance appears to be related to the parasite-nurse cell-host complex, and the nurse cells yield a cryoprotectant when occurring in carnivore muscle tissue but not when in rodents [51]. Outbreaks are common in the Canadian Arctic where uncooked walrus meat is eaten raw, frozen, and aged by Native Americans as "igunaq" [52], but some that get infected also eat the products cooked [53]. This product is produced by placing meat and fat tissues into skin bags, which are sewn shut and aged about a meter under a cache of rocks or gravel along a beach. When meat from experimentally infected seal was prepared as igunaq and allowed to ferment, as air-dried meat (nikku), and as both raw and partially cooked sausage (core temperature less than 50°C), it remained infective for at least 5 months under laboratory conditions. The food product was fed to cats and the extracted juveniles were orally inoculated into mice [54]. Whale is usually eaten as muktuk, a preparation of fermented skin and blubber containing no muscle [53].

Trichinella britovi also occurs in temperate areas of the Palearctic region where people obtain infections primarily from eating wildlife. Apparently, it cannot resist freezing for a long period; the juveniles of *T. spiralis*, which have been reported from a variety of marine mammals, die after 4 days at −10°C. The status of infections from marine mammals requires additional research. The cycles of all the species are similar to each other but different from other nematodes in this chapter. The definitive host, including human, eats uncooked infected meat from another definitive host, and the juveniles are released when the cyst is digested in the stomach. The juveniles invade the mucosa of the upper intestine and develop into mature adults within 2 days. After 4 days, the mated female starts depositing motile juveniles, a process that continues for 4–16 weeks and produces up to 1,500 juveniles in the nonimmune host. These juveniles pass within the lymphatic system or mesenteric venules to muscles, usually striated ones, where it encysts. Striated muscle tissue stimulates development of nurse cells and capsule formation within 2–3 weeks (Fig. 15.8). The coiled juveniles can survive in human muscle for years, even though calcification may occur within 6–9 months. Infections can be fatal with as few as five juveniles per gram of muscle, but infections can be much higher, especially if the initial dose in an immunologically naïve consumer contains hundreds of cysts in striated muscle. A low initial dose establishes a protective immunity to challenge doses.

Paracapillaria philippinensis (until recently [55], a trichinellid that was and still is referred to by some as *Capillaria philippinensis*) seems to be spreading

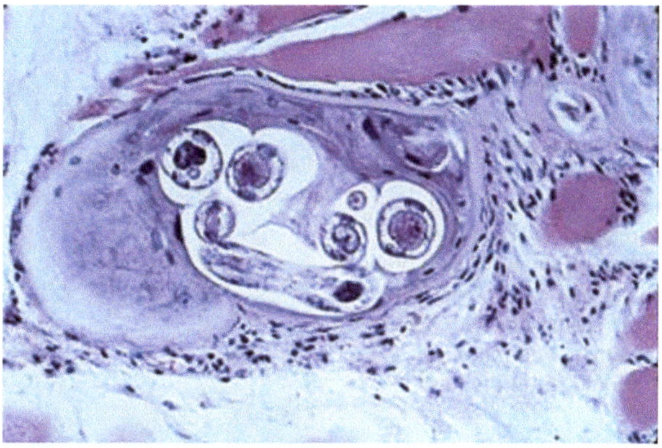

Fig. 15.8 Section of coiled specimen of juvenile *Trichinella spiralis* encapsulated in striated muscle tissue

geographically. This small intestinal nematode produces mortality in humans. More than 2,000 human cases have been reported from the Philippines and Thailand, with several of those from more recent cases in Korea, Japan, Taiwan, India, Iran, United Arab Emirates, Egypt, Italy, Spain, and the United Kingdom [56], but with some of those acquired from eating freshwater fishes or imported fishes. The number of human cases is being reduced in Thailand, but infections probably contribute significantly to mortality there [57]. The life cycle probably includes a bird definitive host as shown by experimental infections in a fish-eating rail and heron, even though many mammals can host transient infections, and humans and the Mongolian gerbil can die from infections in less than 2 months [58]. Eggs passed by the definitive host hatch in a variety of freshwater and estuarine fish species. At least when the gerbil or human eat the infected fish, the juveniles develop into males and females that release a first-stage juvenile from a thin-shelled egg, which in turn can autoinfect the host and produce adults, with a female that produces the typical capillarid thick-walled eggs that are released in the feces. Even though most second generation females are oviparous, a few females still produce juveniles that can progress to autoinfection and hyperinfection, a condition in which the produced juveniles penetrate the intestinal mucosa and proceed with migration. At autopsy, 1 l of bowel fluid contained an estimated 200,000 specimens in all stages [58].

Signs of infection in humans are varied: they can include abdominal pain, diarrhea, vomiting, heart irregularities, edema, and weight loss, with the blood exhibiting low protein, calcium, potassium, and sodium, and with high immunoglobulin E (IgE) with diminished levels of IgG, IgM, and IgA for several months after treatment with albendazole. Albendazole seems more effective than thiabendazole and mebendazole because there is no relapse, presumably because only albendazole affects the larval stages [58].

Epidemiological studies are lacking in most reports, but where infections are common, elderly aboriginal groups with infections in Taiwan prefer Chou-Bao, which includes soaking raw, small entire fish in fermented millet [56], and infected Filipinos consume Kinilaw, which consists primarily of raw pilchards,

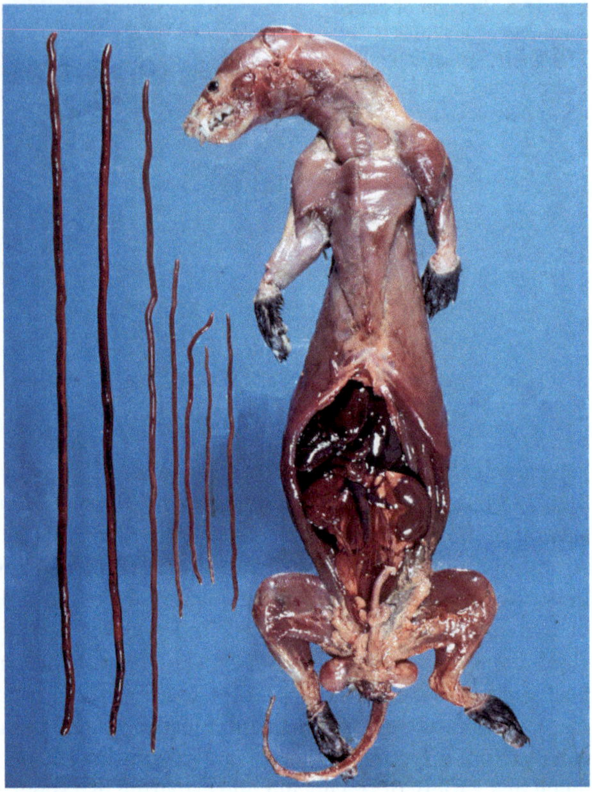

Fig. 15.9 Specimens of the giant kidney worm, *Dioctophyme renale*, removed from a mink's abdominal cavity in Louisiana. This species usually devours all the parenchyma in the right kidney of the mink and other fish-eating carnivores, but these specimens ended up in the cavity and matured without killing the host

or other small fishes from coastal lagoons, marinated in vinegar, garlic, onions, ginger, tomato, and various peppers. The small fishes throughout SE Asia also are eaten whole as raw products, and some native groups (Ilocanos of Northern Luzon, Philippines) use animal organs and intestinal juices to season rice and other foodstuffs [58]. Most human cases have been traced to consumers that eat small, raw entire fish, but exact causes are not known. When diagnosis is delayed, the disease can be severe, with up to 30% fatality in some localities in the Philippines.

Dioctophymatidae (Dorylaimia, Dioctophymatida).The giant kidney worm, *Dioctophyme renale*, rarely infects humans, but the potential because of aquaculture and new cuisines is increasing. Even though usually considered a freshwater parasite, it also is common in the estuarine environment [1]. Its life cycle involves an oligochaete and a fish, amphibian, or reptilian paratenic host. Its counterpart *Eustrongylides* will be discussed later because it does not mature in humans. *Dioctophyme renale* matures in and completely devours the kidneys of its definitive host. The site is usually restricted to the right kidney, probably because of migratory behavior of the acquired juvenile. If it does not encounter the right kidney, it can mature in the peritoneal cavity, where it also causes harm (Fig. 15.9); it can also become encapsulated and form a subcutaneous nodule [59]. Human infections are usually diagnosed by the worm eggs in urine, followed up by radiological exams to detect enlarged or calcified kidneys. The worm has a cosmopolitan occurrence, but most human infections are from Iran and from freshwater habitats. In fact, human coprolites dated about 3,375 YBC from an archeological site in Switzerland suggest that human infections were more prevalent in the Neolithic than now [60]. In the USA, sylvatic infections are common in the mink (*Mustela vison*) in brackish marshes of southern Louisiana [1].

Acanthocephalans (Spiny-Headed Worms)

Acanthocephalans from marine products are not known to mature in humans. However, a few from terrestrial arthropods such as cockroaches do produce eggs in humans. In fact, eggs were found in human coprolites from a cave in Utah dated about 1869 BC ± 160 years and 20 AD ± 240 years [61]. On the other hand, juveniles can attach in the lumen of the intestine without maturing, and, because they are in the lumen and can be voided in the feces, they are treated here. This seems to be the case for *Corynosoma strumosum* in a person from Alaska [62] and *Bolbosoma* sp. from a Japanese fisherman [63]. The first case was noticed when the infected male was treated with Atabrine. The worm matures in seals but is known as a juvenile in dogs, otters, birds, and other fish-eating birds. When fresh marine fish were fed to mink, some mink died [64]. Even though not routinely detected in people, *C. strumosum*, common as adults in pinnipeds and as juveniles from birds and other mammals in the Holarctic, was considered probably to be common in humans but for a short duration [18, 62]. The second case was detected because the worm had penetrated the intestine and was surgically removed because the disease was initially diagnosed as appendicitis.

Acanthocephalans are like nematodes because they both have separate males and females. However, all species are parasitic, and molecular data place them with rotifers. The life cycle includes an arthropod or crustacean and a vertebrate. In the case of the two palaeacanthocephalans mentioned above, the crustacean is unknown, but it could be small or large. The life history is typically completed by seal, whale, or sea otter by feeding on a paratenic fish host containing an encysted or encapsulated larva or juvenile. Most acanthocephalans have a rather specific crustacean and final host and a larger but still rather specific range of paratenic hosts. Even though the definitive host is quite specific, the worm may attach in the intestine of non-closely-related hosts. For example, two palaeacanthocephalan species that mature in a few teleost fishes can be acquired by local dasyatid stingrays as juveniles from amphipods and attach to the spiral valve without developing further or without migrating to the viscera or muscles as paratenic parasites [65]. Stingrays are not typical final hosts for any acanthocephalan.

Information on additional juvenile acanthocephalans that penetrate through the alimentary tract and cause a host response, sometimes severe, appears later in the next section of this chapter.

Parasites with Larvae That Can Infect Humans Eating Inadequately Prepared Seafood

The best-studied parasitic infections in humans eating raw marine seafood are caused by members of the ascaridoid genus *Anisakis* and first recognized in the 1960s in the Netherlands, Japan, and the USA, even though Native Americans in Alaska passed specimens of both *Anisakis* sp. and *Pseudoterranova decipiens* sensu lato in feces [66]. There is now confusion in the names given the human disease because ascaridoids other than species of *Anisakis* also induce similar infections. These agents plus infections and potential infections by non-ascaridoid members of a variety of classes, based on the ability of such species to cause disease in nonhuman mammals, are also discussed in this chapter. In 1988, a group interested in standardizing nomenclature of ascaridoid diseases recommended calling disease involving a member of *Anisakis* as "anisakiasis" and disease caused by any member of the family Anisakidae as "anisakidosis," even though some authors use a name referring to the responsible generic agent or the host response, such as "eosinophilic granuloma." DNA-based techniques have allowed accurate diagnosis, species differentiation, and phylogenetic relationships. Diseases caused by members of other higher taxa also create some confusion with terminology of the names.

Anisakidae (Spirurina, Ascaridomorpha). Species of *Anisakis* are the most widespread cause of notable and severe human disease. Until recently, human infections were divided into two types of juveniles, *Anisakis* type I and type II [67] based on the presence of a caudal spine and the length of the ventriculus, a glandular structure separating the muscular esophagus and the intestine. A group of 21 species based on adults were divided into three primary species [68], including *Anisakis simplex* with ten junior synonyms, *Anisakis typica* with one synonym, and *Anisakis physeteris* with three synonyms. With improved molecular tools, some of those junior synonyms have been accepted, and additional species have been recognized. *Anisakis* type I juveniles comprise the *Anisakis simplex* complex: *Anisakis simplex* sensu stricto, *Anisakis pegreffii*, *Anisakis simplex* C [69], *Anisakis ziphidarum*, and *A. typica*. *Anisakis* type II juveniles comprise *A. physeteris, Anisakis brevispiculata*, and *Anisakis paggiae*. All of these can be differentiated and are considered human pathogens. The differences in pathogenesis and location in the human probably reflects the different species, at least in part. Recent polymerase chain reaction amplification (PCR) methods show that *A. pegreffii* comprises the source of most infections in consumers in Italy but that *A. simplex* sensu stricto is the most common species in people in Japan (Figs. 15.10 and 15.11) even though fish in the Sea of Japan are known to be infected with *A. pegreffii* [70]. Also, *A. pegreffii* seems to cause gastric distress in humans and *A. simplex* seems to mostly affect the intestine.

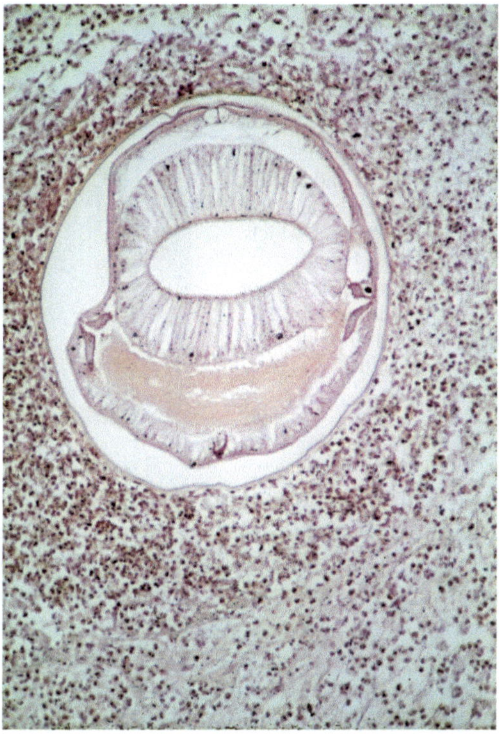

Fig. 15.10 Section of *Anisakis simplex* in ileum of human patient in Japan. Note the characteristic eosinophilic response, referred to as "eosinophilic granuloma," surrounding the diagnostic section of the worm

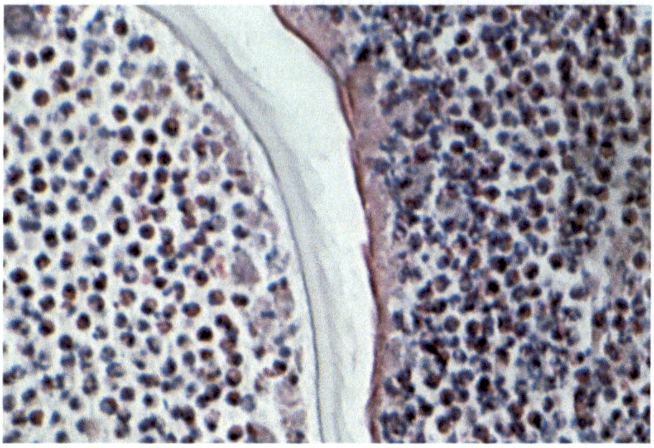

Fig. 15.11 Section through degenerating specimen of *Anisakis* sp. in stomach of experimentally infected miniature pig showing juvenile with infiltrated eosinophils and surrounding infiltrate consisting almost exclusively of eosinophils

Exceptions involve the abdominal cavity, mesentery, and oral cavity [71]. Juveniles typically measure 1–3 cm in length.

The different species can be differentiated by cetacean host and geographical locality. *Anisakis schupakovi* infects the Caspian seal, *Pusa caspica*, and belongs in the genus as do a couple other species from delphiniids in South America rivers and other isolated habitats. Tabularized reviews list which species of *Anisakis* have

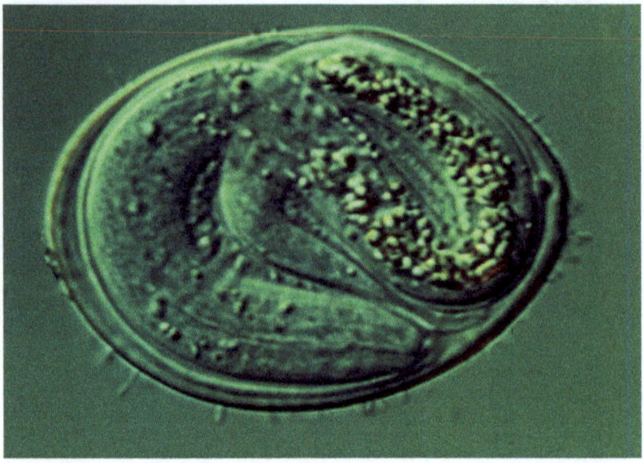

Fig. 15.12 Egg of *Hysterothylacium reliquens* showing infective third-stage juvenile representative of ascaridoids infective to humans. In this species, the juvenile hatches and the free juvenile is eaten by an appropriate copepod, which in turn is eaten by an appropriate fish. That fish or the copepod is eaten by a paratenic host or by the definitive fish host

been confirmed molecularly from different cetaceans, cephalopods, and fishes [70]. The life cycle of *Anisakis* spp. includes a cetacean final host, from which worm eggs (Fig. 15.12) are deposited in feces and embryonate in seawater; the active, hatched third-stage juvenile (the same as for *Pseudoterranova decipiens*), also referred to as a larva, is eaten by a crustacean [72]. The juvenile hatching from most nematode groups is a second-stage juvenile. Euphausids are known to host the infective third-stage juvenile. A variety of paratenic hosts, including specific fishes, squids, and crustaceans can acquire and maintain the infective juveniles by feeding on the euphausid or some undetermined crustacean hosts. The appropriate cetacean feeds on any appropriate animal with an infective juvenile. In some cases, there can be a large cluster of these worms, involving third-, fourth-, and fifth-stage juveniles and adults attached within a stomach ulcer. As with several ascaridoids in a variety of genera and host groups, individuals not embedded in the ulcerated tissue will leave the ulcer and forage among the digesting prey in the stomach only to return at a later time.

Cultured fish usually do not exhibit ascaridoid infections present in their wild counterparts because they are fed a diet that does not contain infective juvenile worms. Such is not always the case. Infected prey can enter a sea pen or be fed to the cultured fish in fish farms or cages. When juvenile specimens of the greater amberjack, *Seriola dumerilli*, and three-line grunt, *Parapristipoma trilineatum*, were imported into Japan from China for grow-out in fish farms, they were determined to be heavily infected with *Anisakis pegreffii* [73]. The Japanese Ministry suggested the fish were fed wild infected fish before being exported. In any event, it required that the products, originally grown for sashimi (raw fish), be frozen prior to taking them to market, considerably reducing their value.

Humans become infected with juveniles of *Anisakis* when they eat uncooked or inadequately prepared seafood containing an infective third- or fourth-stage juvenile. Symptoms and signs of infection probably vary by species of *Anisakis*, condition of worm, and condition and genetics of patient. Some patient complaints include sudden epigastric pain, nausea, diarrhea, and urticaria occurring within an hour but usually at about hour 6. Because infections are difficult to diagnose, the acute infection caused by a migrating or encapsulated juvenile develops into a chronic case with intermittent abdominal pain lasting from weeks to several years. There may be a slight fever and moderate leukocytosis up to 15,000 leukocytes per cubic milliliter. Peripheral eosinophilia can vary from 4% to 41% and will be discussed later in the chapter. The condition presenting other vague symptoms cannot be diagnosed with various immunologic assays because the sera cross-reacts with related nematodes or even with human sera [74], but more specific serological tests are being perfected and various molecular tools are being developed [75, 76]. Most prior cases have been identified morphologically after misdiagnoses of appendicitis, ileitis, cholecystitis, Crohn's disease, and tuberculosis peritonitis, and a variety of neoplasms resulted in resection of the corresponding lesions. In most cases, the worms penetrate or embed in the ileum or duodenum but occasionally penetrate the stomach as for *A. pegreffii*. Anisakiasis occurs in the Netherlands and other European countries, Japan, the USA, and Chile, and cases are remaining high. When the estimated prevalence of *A. simplex (A. pegreffii?)*-specific IgE assay in the population of Madrid, Spain, was tested by enzyme-linked immunosorbent assay (ELISA) as well as for cross-reactivity, 12% of the healthy population tested positive compared with 22% in southern Spain determined with a somewhat different technique [76] and without any indication of cross-reactivity. Those testing positive in Madrid were more likely to admit to being habitual consumers of fresh (Fig. 15.13) or undercooked fish as opposed to eating frozen and cooked products like reported for the negative population.

As with "*Anisakis simplex*," "*Pseudoterranova decipiens*" also was recognized as a single species. It infected fishes and humans and could be identified by a morphological diagnostic juvenile and a variable adult, but the adult matured in pinnipeds rather than cetaceans. Recently, however, this agent of human infections that was supposed to occur worldwide has been shown to include several species, most in specific pinnipeds and consequently in specific geographic regions. Molecular and biochemical tools, backed with morphological features, have shown this species to represent a complex of at least six species: *Pseudoterranova decipiens* sensu stricto, *Pseudoterranova krabbei*, *Pseudoterranova bulbosa*, *Pseudoterranova azarasi*, *Pseudoterranova cattani*, and *Pseudoterranova decipiens* E. There are also exceptions to the strict host specificity within this genus: *Pseudoterranova ceticola* and *Pseudoterranova kogiae* both infect the cetaceans *Kogia simus* and *Kogia breviceps* (dwarf and pigmy sperm whales, respectively) [70].

When consumers eat infective juveniles of *Pseudoterranova* spp., the condition is usually not serious. Specimens are typically longer than those of *Anisakis* spp., roughly 4 cm versus 2 cm but dependent on age and development of worm, species, and temperature, and they usually have a more reddish or darker coloration. Whether they occur in the flesh or viscera of the fish depends on the species of worm and species of fish for both *Anisakis* spp. and *Pseudoterranova* spp. Specimens are often noninvasive and void in the feces or vomited; they tickle the throats of those eating inadequately prepared seafood when they migrate from

Fig. 15.13 Italian fisherman aboard fishing boat with catch including sardines infected with *Anisakis pegreffii*. Fresh fish from boats in the Port of Civitavécchia, Italy, are boxed, iced, and sold to restaurants, seafood dealers, and others

the stomach back up the esophagus into the oropharynx, or they are coughed up into the mouth. *Pseudoterranova* spp. that cause disease usually involve the mucosal or submucosal tissues of the stomach. The common name of *P. decipiens* from the harbor seal had been "codworm" because the species was so abundant in cod, the second intermediate/paratenic host, especially during strict protection of harbor seal pups, which contained heavy infections. Actually, the "sealworm" consists of *P. decipiens* sensu stricto and *P. krabbei*, which can occur in mixed infections in *Phoca vitulina* and *Halichoerus grypus*, and both infect the cod, *Gadus morhua* [70]. Infections of large, 2–4-mm-long, unidentified fourth-stage juveniles of species of *Pseudoterranova* have become more prevalent and better recognized in Chile during recent years. Of 15 patients, all but one appeared to lose the infection through the mouth after irritation; one was voided through the anus with intense diarrhea. The patients all enjoyed seviche or fried fish [77]. Differing clinical signs between human infections in the USA and Japan may be different because the cause is from separate species. These may be *P. bulbosa* and *P. azarasi* in Japan and *P. decipiens* sensu stricto and *P. bulbosa* in the USA (Fig. 15.14) [70].

The life history of *Pseudoterranova decipiens* sensu lato is more studied than that of known species of *Anisakis* sp. [78]. The life cycle differs slightly from that of known species of *Anisakis* sp. The first intermediate host is a small crustacean such

Fig. 15.14 Commercial fillet with conspicuous specimens of *Pseudoterranova decipiens* sensu lato

as several mycidacean species that spend part of their life in the benthos [79]. The hatched third-stage juveniles are not active. The agent in the crustacean appears to require a fish or large invertebrate paratenic host to complete the history in the pinniped.

Members of *Contracaecum* have been reported from both pinnipeds and fish-eating birds. Even though morphologically they appear similar, those from the pinnipeds and birds are molecularly separated, perhaps with those species from seals representing *Phocascaris*, which is paraphyletic within *Contracaecum* [70]. Like for *Pseudoterranova decipiens, Contracaecum osculatum* from phocid seals consists of a complex of at least five genetically recognized, but not all named, species. There are 11 mammalian species that have been sequenced. Because they are from temperate and polar areas, there is less known about human infections, but they are discussed based on known genetic information [70]. Two human cases of *C. osculatum* sensu lato have been reported from Hokkaido, Japan [80].

Species of *Contracaecum* from birds appear to be phylogenetically separate from the seal species and confused taxonomically. Each of the well-known species actually comprises a separate complex [81]. Members of the genus were not originally thought to infect humans, but at least two human cases of unidentified species from a bird or seal have been reported [82], and at least fourth-stage juveniles of a member of the *Contracaecum multipapillatum* complex (Fig. 15.15) can infect and mature in experimentally infected kittens but not rats, ducks, and chickens [83]. The worms, some attached, were in the intestine, causing hemorrhaging and associated with small ulcers.

Because members of *Hysterothylacium*, many of which were originally placed into the genus *Contracaecum*, mature in fishes, no one expected them to infect humans. Juveniles of *Hysterothylacium aduncum* have been suspected as the cause for eosinophilic granuloma in humans [84, 85], and *Hysterothylacium* type MB larva can produce pathological alterations, including eosinophilic granuloma, consistently in rodents (mice and rats) and rhesus monkey (*Macaca mulatta*) (Figs. 15.16 and 15.17) and probably humans [86]. The juvenile is small (1.4–3.1 mm long) but occurs in large numbers in some of its several secondary intermediate

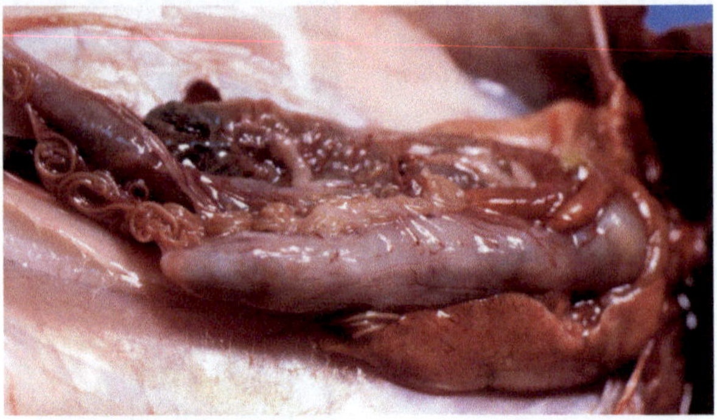

Fig. 15.15 Red drum, *Sciaenops ocellatus*, exhibiting infective juveniles of *Contracaecum multipapillatum* sensu lato in mesentery. The same species is also common in fillets or viscera of other fish species in coastal Mississippi

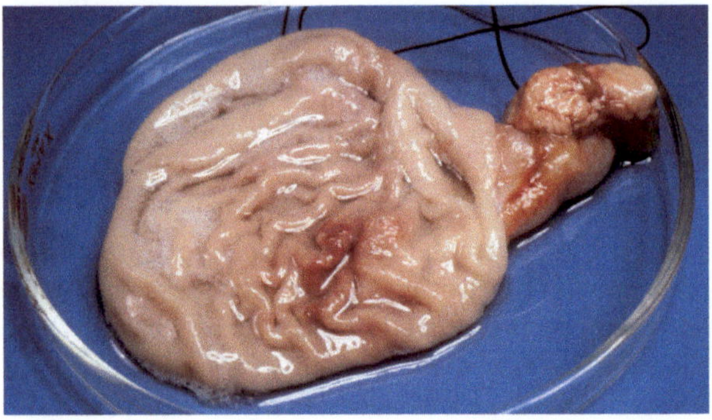

Fig. 15.16 Stomach of rhesus monkey, *Macaca mulatta*, 5 h post-feeding with juveniles of *Hysterothylacium* type MB showing hemorrhagic lesions where juveniles penetrated

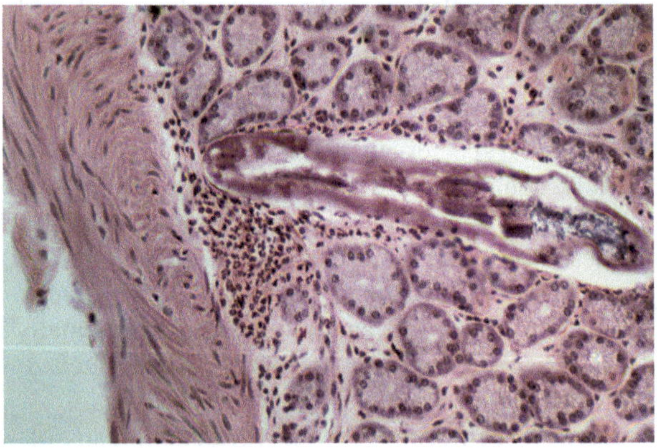

Fig. 15.17 Section showing juvenile of *Hysterothylacium* type MB penetrating near muscularis mucosae of rhesus monkey stomach surrounded by eosinophils at 1.5 h post-feeding

or paratenic fish and invertebrate hosts [87]. Worms penetrated the stomach or duodenum within an hour or so, with the attraction of an abundance of eosinophils accompanied by an apparent decrease of those leukocytes in the peripheral blood. Not all species of *Hysterothylacium* will penetrate the alimentary tract or cause lesions in mammals [87, 88], and juveniles of some species approach 3 cm long. A Japanese male passed a living young adult female specimen of *H. aduncum* after suffering chronic abdominal pain and diarrhea for a month. Whether the worm was acquired as a juvenile or an adult was not clear, but he had eaten raw fish, *Gadus macrocephalus* fillet pressed against kelp (Tara no kobujime) a month before onset of symptoms and *Oncorhynchus nerka* sliced and quick frozen (Ruibe), a week before onset [89]; both the cod and salmon host *H. aduncum*. The symptoms cleared following excretion of the worm. Whether *Terranova* type HA juveniles from Hawaiian bony fishes mature in sharks or in a marine mammal is not known, but, when gavaged into the laboratory rat, it migrates into the submucosa and forms a granuloma but does not pass through muscularis mucosae [90].

Other Nematodes

Gnathostomatidae (Spirurina, Gnathostomatomorpha). Gnathostome nematodes are easily recognized by their spiny anterior swelling in both juveniles and adults. They are best known as public health risks of freshwater species but are mentioned here because the fishes that host the infective juveniles move into estuarine and marine habitats where they infect mammals and humans in coastal areas. *Gnathostoma spinigerum* occurs throughout Asia with infected humans primarily in Thailand and Japan, but *Gnathostoma hispidum, Gnathostoma doloresi*, and *Gnathostoma nipponicum* have all been identified from human cases in Japan [4, 91]. *Gnathostoma binucleatum* infects humans eating seviche made with estuarine fish in Sinaloa, Mexico [92], as do other species. With infections of *G. spinigerum* and *G. binucleatum*, the juveniles usually produce a migrating erythema (larval migrans) on the peripheral portions of the body for a few days to several years. Juveniles of the three other species usually migrate into surface skin, forming a serpiginous eruption on the trunk before disappearing spontaneously within 3 months. The juvenile can migrate into the central nervous system (CNS) and other vital organs such as lungs, intestine, genital organs, ear, and nose, occasionally producing a fatal disease, especially when involving the CNS [93, 94].

The typical life cycle starts with adults within a tumor-like mass in the stomach wall and serosa (Fig. 15.18) of the mammal host. Eggs extrude into the stomach and pass out with the feces. First-stage juveniles hatch within 12 days and develop when eaten by *Cyclops* sp. or presumably some other cyclopoid copepods. Development progresses in the copepod hemocoel to the third-stage juvenile and are infective to a variety of fishes, amphibians, reptiles, birds, and mammals, including humans. Perhaps in some species, a specific fish serves as a second intermediate host, but fish serve as a primary paratenic host (Fig. 15.19), allowing the juvenile to migrate or encapsulate in any of the previously indicted paratenic hosts necessary to maintain the population until eaten by the proper definitive host. Human infections can come from any of the paratenic hosts and probably from drinking freshwater containing infective copepods. The importance of marine or estuarine hosts depends on the area, since infections can occur in freshwater or estuaries and freshwater hosts can migrate into estuarine or marine habitats.

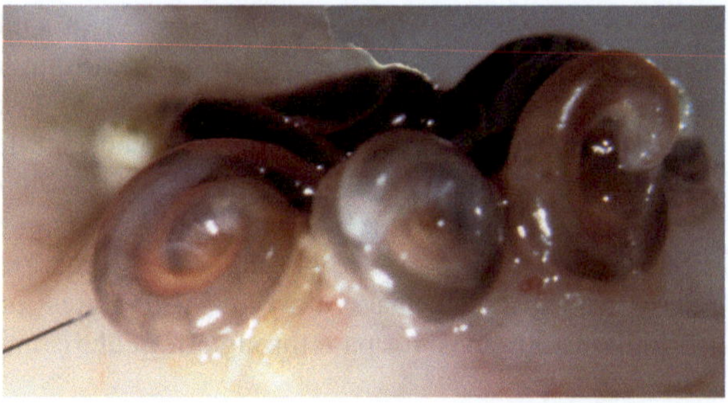

Fig. 15.18 Tumor-like mass containing a few adult individuals of *Gnathostoma* sp. in the serosa of the stomach of the raccoon in Mississippi

Fig. 15.19 Infective third-stage juvenile of *Gnathostoma* sp. in mesentery of the Gulf killifish, *Fundulus grandis*, in open marine lagoon of offshore barrier island, Horn Island, Mississippi

The true marine counterpart of *Gnathostoma* is *Echinocephalus*, and at least *Echinocephalus sinensis* can probably infect humans. It matures in rays and sharks in the South Pacific and is transmitted by at least the oyster *Crassostrea gigas* and scallops. It was observed throughout the year in Hong Kong, but juveniles administered to kittens, monkeys, and puppies were infective to those mammals from August through October only [95]. When a dose of 600 worms was given to a kitten, most were apparently vomited, but the host died at 16 h. A total of 134 worms were recovered in various visceral organs. All but one of nine other kittens administered with fewer juvenile worms, 130–350, survived, but exhibited visceral larval migrans during the first 40 h. Kittens examined at days 3, 4, and 9 revealed no juvenile worm. The two different monkey species given 80 or 200 juveniles and dogs given larger doses each showed similar results of worms located throughout various visceral organs for the first 40 h post-feeding and no worm after 3 days. Substantial infections were also achieved at 18 h in kittens when fed additional worms acclimated in oysters at 28°C and 33°C [96]. Humans are potentially at risk, especially if eating heavily infected mollusks and any worms enter the CNS. There are several other species of *Echinocephalus*, and some may also have the ability to

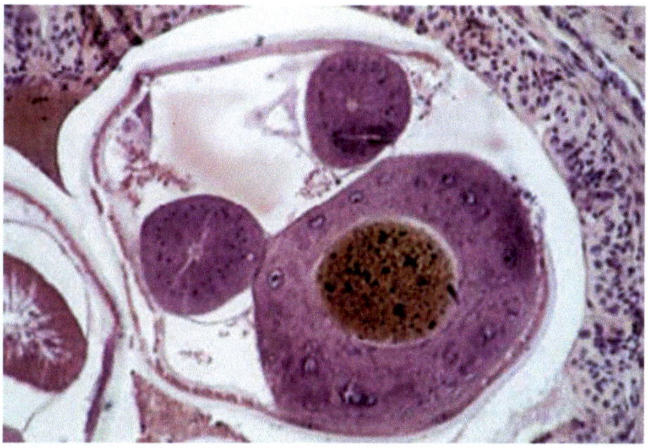

Fig. 15.20 Section of adult female of *Angiostrongylus cantonensis* associated with other males and females in pulmonary artery of rat definitive host. The paired creamy-white uterine branches spiral around the hematin-filled intestine giving a diagnostic barber-pole appearance to this 20–30-mm-long worm when alive

penetrate the alimentary tract of mammals and undergo a visceral or cutaneous migration. Gnathostomes cause both cutaneous and visceral migration, and human cases are probably much more common than diagnosed because specimens of the fast-moving worms are difficult to collect or biopsy; a refined serological method should be able to determine the presence of gnathostomes, and specific diagnoses should allow for a better understanding of marine associations.

A related unidentified juvenile spiruroid nematode known as suborder Spirurina type X has been known to cause creeping eruption in at least 28 Japanese [97]. The juvenile is acquired by eating raw tiny squid (*Watasenia scintillans*) as sashimi in Japan; the serpiginous erythematous eruption along the abdomen took 4 weeks to become apparent. All infections were confirmed with an indirect immunofluorescence test.

Metastrongyloidea (Rhabditina, Strongyloidea). *Angiostrongylus cantonensis* is usually considered a freshwater or terrestrial disease, even though it was originally thought to be acquired from improperly prepared marine fishes. There are several ways that this rat metastrongyle nematode can infect humans that involve estuarine or fully marine situations, and they have been reviewed [98]. The species matures in the pulmonary arteries of rats (Fig. 15.20) and has a rather complicated life cycle. Eggs lodge in the arteries and capillaries of the rat lung where they hatch (Fig. 15.21); the first-stage juvenile migrates through the alveoli and up the trachea before being swallowed and ultimately passed out in the feces. The juvenile develops to a third stage in a molluscan intermediate host. When the juvenile in the mollusk, its slime trail, or a paratenic host feeding on the host or the free juvenile is eaten by the rat, it migrates to the subarachnoid space adjacent to the brain (Fig. 15.22). After a couple of weeks, it forces its way into the venous system to be carried to the pulmonary arteries. Humans are accidental hosts, and the infective juvenile still goes into the brain cavity, but it remains there and often enters the brain tissue (Fig. 15.23) and causes behavior problems and even death in the patient. Because of abnormal behavior in infected humans, many never receive medical attention. Even though the infective juveniles do not necessarily survive as long in marine waters or in marine invertebrate or vertebrate hosts as freshwater

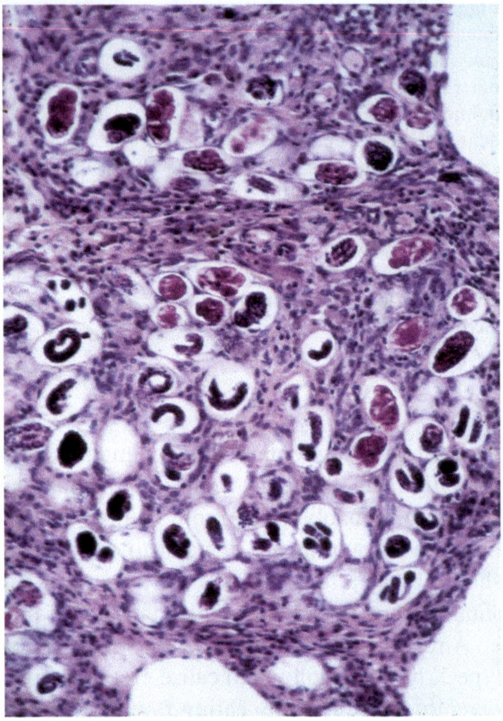

Fig. 15.21 Section of rat lung showing deposited eggs and hatching juveniles of *Angiostrongylus cantonensis* infective to mollusks

counterparts, they survive long enough to transmit the infections. The eastern oyster, *Crassostrea virginica*, and northern quahog, *Mercenaria mercenaria*, can serve as intermediate hosts in 15 ppt, and large amounts of the oyster are eaten raw. There are many possible ways for humans to acquire the infection, but it is difficult to discern individual cases. In Tahiti, special dishes involve a sauce prepared from uncooked shrimp hepatopancreas (liver/digestive gland) often containing infective juveniles mixed with grated coconut and "coconut milk." Infections originated in the Indo-Pacific, but have spread by rats aboard ships to the Northern Hemisphere. Infections have become recognized in Hawaii, Cuba, Puerto Rico, and the USA through Louisiana.

Dioctophymatidae (Dorylaimia, Dioctophymatida). Juveniles of *Eustrongylides* spp., like those of *Dioctophyme renale*, are relatively large and red, making ingestion seem unlikely. However, when fish intermediate or paratenic hosts are eaten alive whole, a severe peritoneal infection can occur. In the USA, eating live fish is usually a behavior to gain attention or a hazing ritual. Consequently, fishermen eating their bait, tavern clientele taking a wager, and students being initiated as well as unaware consumers eating homemade sushi are likely victims [99]. Unlike *D. renale*, which matures in mammals, all species of *Eustrongylides* mature in birds. Three well-known species are recognized, *Eustrongylides ignotus* from North and South America plus New Zealand, usually infecting herons and egrets; *Eustrongylides tubifex* from North and South America, Europe, and into Asia in ducks, loons, and cormorants; and

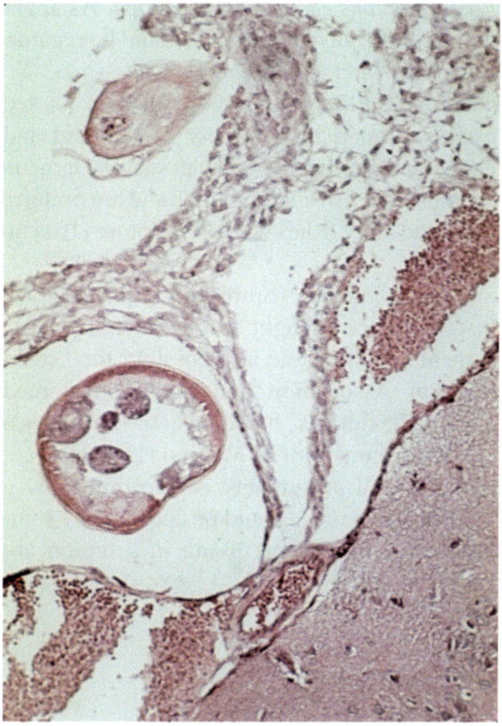

Fig. 15.22 Section of fourth-stage juvenile of *Angiostrongylus cantonensis* developing normally in meninges of brain of rat after 29 days without any apparent inflammatory response

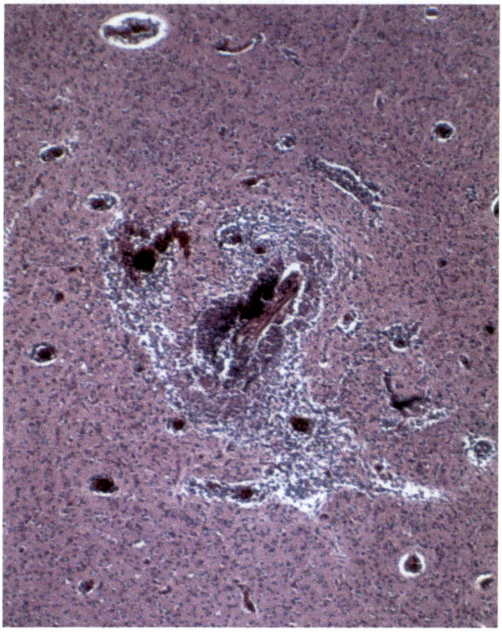

Fig. 15.23 Section of juvenile of *Angiostrongylus cantonensis* after abnormal migration into brain of a Thai, showing atypical inflammatory response in human brain

Eustrongylides excisus from Europe, Asia, and Australia in ducks, cormorants, and herons [100]. Juveniles of all should be regarded as a health risk, and all occur in both freshwater and estuaries. In Mississippi, *E. ignotus* occurs mostly in estuarine conditions and commonly spreads to its second intermediate mosquito fish and *Fundulus* hosts by a variety of tubificid and other oligochaete first intermediate hosts. Because paratenic hosts such as largemouth bass and striped bass commonly eat mosquito fish and fundulids and are preferred seafood products, humans dining on these and other fishes wild or in culture [101] in an uncooked state are at a high risk for infection.

Dracunculidae (Spirurina, Spiruromorpha, Dracunculoidea). *Dracunculus insignis*, once thought to be a synonym of *Dracunculus medinensis*, the "fiery serpent" of the Bible and possibly the "snakes" in the caduceus of the American Medical Association and many other medically related societies, constitutes a public health risk. It occurs in the hind legs of the raccoon in the brackish marshes of the northern Gulf of Mexico [1] as well as freshwater habitats throughout North America and presumably is transmitted by paratenic fish and frog hosts. Consequently, *D. insignis* could be acquired by someone eating raw fish, not presently the habit of most persons living in a region abundant in seafood typically eaten in a cooked state.

Other Parasites

Acanthocephala (Spiny-Headed Worms) Acanthocephalans, especially those that mature in the alimentary tract of marine mammals, pose a threat to those who eat raw fishes. Two palaecanthocephalans were discussed above as species that occur in the intestine of humans, even though they did not mature in the patients. They also did not penetrate the human digestive tract. Another case of what was tentatively identified as *Bolbosoma* sp., presumably the same marine mammal parasite species mentioned earlier [63], penetrated through the intestine and became encapsulated in a granulomatous mass on the serosa of the ileum in the peritoneal cavity of a 16-year-old boy in Japan [102]. Another species, one that matures in a fish, *Acanthocephalus rauschi*, was discovered in the peritoneum of a native in Alaska. The frequency of acanthocephalan infections in humans, especially native groups from the Arctic where fishes containing infective juveniles are eaten raw, is difficult to discern but is probably substantial [18, 62].

The spiny-headed worm with hooks on the everting proboscis can produce severe lesions if it penetrates through the intestine or occurs in large numbers. An experimental study feeding rats with *Centrorhynchus* sp. from a paratenic host resulted in the worms penetrating the intestine on day 1, entering the muscles by day 5, and producing granulomas on day 20 [103]. When captive primates feed on intermediate hosts or are fed paratenic hosts, they can die when the worms invade beyond the serosa but otherwise produce a focal inflammatory response without harming the host [104]; when large die-offs occur, the invading worms are usually associated with secondary bacterial involvement.

Fig. 15.24 Nymphs of *Sebekia mississippiensis* partially pulled from body cavity of intermediate host, the western mosquito fish, *Gambusia affinis*, which is infective to paratenic fish, turtle, and mammalian hosts, including humans, or to the alligator definitive host

Members of the Archiacanthocephala, *Moniliformis moniliformis* from rats and *Macracanthorhynchus hirudinaceus* from pig have been reported from man throughout much of the world, sometimes with serious complications. They are acquired from terrestrial arthropods such as roaches or paratenic hosts. Nine specimens of the relatively large related *Macracanthorhynchus ingens*, known from the raccoon, *Procyon lotor*, and a few other carnivores, was reported from the intestine of a 1-year-old child in Texas [105]. The worm is common in coastal areas, so it could be transmitted to humans in estuarine, freshwater, or terrestrial paratenic hosts as well as from wood roaches and millipedes, if eaten raw.

Sebekidae (Crustacea, Pentastomida, Porocephalida). Nymphs of pentastomes, a crustacean group perhaps derived from and of equal status as Branchiura (argulid parasites of fish and amphibians) also can migrate in and become encapsulated in human visceral organs. A marine threat would be *Sebekia mississippiensis*, a parasite from the lungs of the American alligator. Nymphs are abundant in the western mosquito fish (*Gambusia affinis*) (Figs. 15.24 and 15.25) and present in the Atlantic croaker and gulf killifish, which may or may not have acquired the infection from the mosquito fish from estuarine and marine habitats, like its alligator final host. When any of the intermediate or paratenic fish, snake, or turtle hosts and possibly eggs is eaten by a mammal, the nymphs migrate and usually become encapsulated by the mammal. The opossum and river otter were infected in Mississippi, suggesting a human potential host [106], since humans are known to host nymphs of other pentastomes. This and related species infect other crocodilians. At least one case report for a related species of

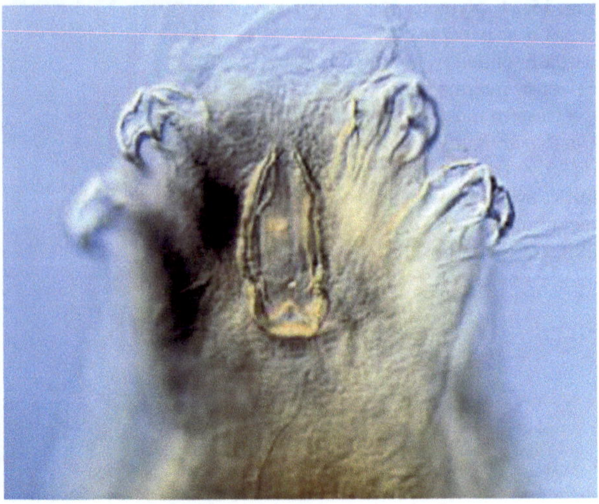

Fig. 15.25 Anterior end of *Sebekia mississippiensis*, a pentastome from the lung of the American alligator in Mississippi

Sebekia involved skin dermatitis in a woman in Costa Rica [107]. An autopsy of a woman from Georgia, USA, with presumptive acquired immunodeficiency syndrome (AIDS) revealed the cuticle of an unidentifiable pentastome in her pericardial sac and epicardium associated with tuberculous pericarditis [108]. Non-marine-inhabiting pentastome species of several genera present an equal risk to infect people.

Protozoans Sarcocystidae (Alveolata, Apicomplexa). The coccidian *Toxoplasma gondii* has final hosts that contain a sexual, coccidian-like phase, restricted to cats and other felines, all terrestrial mammals. Human infection results from shed oocysts in feline feces, by ingesting tissue cysts in warm-blooded mammals and birds, by eating passive vector hosts, and by congenital transmission. Infected humans, predominantly immunocompromised patients or transplant recipients on immunosuppressive therapy, can develop encephalitis, mental retardation, and blindness. If infection occurs for the first time during pregnancy, the parasite can cross the placenta into the brain or eye, and possibly cause miscarriage. The reason for the large number of human infections, perhaps a significant portion in some geographic regions, in the marine environment, may be the low number of oocysts necessary to establish a human infection. As an example, a pet cat sheds oocysts for only a few days during its lifetime, and it produces about 25 g of feces per day. However, in each fecal deposit, tens of millions of oocysts may be produced, and as few as one oocyst can infect a mouse or pig [109]. The small floating oocyst can sporulate into the infectious agent at 15 or 32 ppt and remain infectious for long periods, up to 6 months at 15 ppt [110]. These oocysts, resistant to most disinfectants, however, have not been reported from coastal waters even though present in marine and coastal animals [111, 112]. When experimentally exposed to the eastern oyster, the agent in oyster tissue was infectious to mice for 6 days [113], but, even though experimentally exposed mussels (*Mytilus galloprovincialis*) tested positive by PCR assay, mice fed with various PCR-positive tissues as old as 21 days demonstrated infections for at least 3 days only when fed digestive gland tissue [114]. At least 13 marine mammals, including dolphins, have shown a positive reaction by agglutinations and indirect fluorescent antibody tests [111, 115]. The California sea otter (*Enhydra lutris nereis*) has been studied most extensively because *T. gondii* has been associated with meningoencephalitis and extensive mortalities in the California population. There are three dominant genotypes, two of which have been identified from human infections; the otter had two genotype isolates in the brain and heart, and one of those was one that was known from humans and the other a rare one [111]. The otter feeds on mollusks, and evidence indicates land-based runoff as the source of the otter infections, suggesting "potent implications for human health" [116]. Perhaps crustaceans also serve as vectors. Additional epidemiological evidence deals with Inuit women of childbearing age in Northern Quebec, Canada, where seals are infected [117]. About half were seropositive to *T. gondii*, and seropositive women were at least four times more likely to have eaten dried seal meat or seal liver than seronegative counterparts. Seronegative women during pregnancy are now instructed to avoid all uncooked or dried meat, particularly seal and caribou, and to refrain from skinning animals.

Microspora. Microspora is a large phylum with obligate intracellular members that infect and obtain their energy from primarily invertebrates and lower vertebrates, cold-blooded vertebrates. The life cycles of microsporidans include several patterns including a direct cycle, a cycle that needs a true intermediate host, a transovarian cycle, and a cycle that passes through a vector without undergoing any development. Members of few genera infect mammals, with *Encephalitozoon cuniculi* being the best known. A few species have been reported from humans, and most of those represent immunologically incompetent individuals [118]. These include *Encephalitozoon intestinalis* and *Encephalitozoon hellem*. Others, such as *Pleistophora ronneafiei*, *Trachipleistophora hominis*, and *Anncaliia vesicularium*, involve species with fish or other cold-blooded hosts. These species are named for material taken from human AIDS or otherwise immunologically incompetent patients, and they may have already been named earlier from a cold-blooded host, even a marine one. *Pleistophora ronneafiei* was suspected to have been injected into a prisoner in Florida with a syringe containing infective material from a fish [1, 118]. A study conducted on other microsporidans to determine human risk from spores reaching coastal and marine waters tested the viability of the three indicated species of *Encephalitozoon* [119]. As expected for spores of species that can infect humans, the viability decreased as temperature increased from 10°C to 20°C. More proliferation of the parasites tested at 0, 10, 20, and 30 ppt occurred at the lower salinity concentrations, with *E. hellem* and *E. intestinalis* surviving better than *E. cuniculi*, in which survival lasted 1–2 weeks at 10 ppt. Those former two still proliferated at 4 weeks, with some proliferation of *E. intestinalis* at 10 ppt at 12 weeks. Consequently, all three species could infect humans and marine mammals or contaminate shellfishes. At least *E. hellem* and *E. intestinalis* can contaminate water with spores passed by aquatic birds [120]. *Enterocytozoon bieneusi* may be the most common microsporan of man. A total of 81 different genotypes with 111 different genotype names have been identified, with 26 exclusively in humans, but with none from marine animals [121]. The species without a characterized genotype was identified with PCR from an ill bottlenose dolphin (*Tursiops truncatus*) [122].

Myxozoa. Myxozoans, actually multicellular parasites that get placed in most textbooks as protozoans, seem to be bilaterian metazoans, possibly in a sister group to the Nematoda [123], or a trachylinan Cnidaria; members are not thought to infect humans. Spores occur primarily in fishes and amphibians, but atypical species have been reported from a variety of other hosts. Sexual development takes place in oligochaetes and a variety of other invertebrates. Because people often eat fish infected by myxosporidians, they can exhibit spores from the product passed in their feces. There is no indication that any myxosporan has actually infected healthy humans and undergone vegetative reproduction. Two patients were reported to exhibit spores and diarrhea, but at least one of those was an immunosuppressed, HIV-positive patient infected with the human coccidian *Isospora belli* and the other had human helminths *Strongyloides stercoralis, Hymenolepis nana*, and *Ascaris lumbricoides* [124]. Other case reports, such as some in Australia involving eating fish that had been frozen, exist but probably do not represent human infections [125]. Spores of *Henneguya salminicola* also have been reported from human diarrheic feces [126, 127]. Patients had been eating salmon, and, in one case, the spores were misdiagnosed as human sperm, resulting in scandalous consequences [126].

Parasites Acquired When People Have Direct or Indirect Contact with Them

Parasites acquired by contact consist of both helminths and protozoans, but helminths usually actively enter humans by direct contact. In some cases, difficulty exists distinguishing between active and passive contact. Also diseases acquired from agents in marine water fit into different groups. Some originate from estuaries and marine habitats, but most seem to have a freshwater origin but can tolerate marine or estuarine conditions.

Direct (Active) Contact with Disease-Causing Agents

Trematoda

Schistosomatidae. Members of the Schistosomatidae exhibit atypical features among trematodes such as having separate male and female individuals rather than being hermaphroditic and living in blood. They also do not have a distinctive second intermediate host, but there occurs a juvenile stage, schistosomula, that develops from the cercaria after it penetrates the definitive host. Those species that mature in humans use as their first intermediate host freshwater snails only, but some members in marine snails occurring throughout much of the world that infect birds and mammals produce cercariae that accidentally penetrate humans and establish a hypersensitivity dermatitis.

Dermatitis, known as "swimmer's itch," "clam digger's itch," and other terms, results when humans are challenged by epithelial penetration of the cercariae of some specific schistosome species occurring in the mesenteric blood vessels of the avian and mammalian marine and freshwater hosts. Swimmer's itch is best known for freshwater species; however, several marine species result in similar human responses. Some freshwater cercariae can survive well in low-salinity conditions, but salt usually kills the snail hosts. For example, the cercariae of *Bolbophorus damnificus* can tolerate 2.5 ppt sodium chloride, a concentration toxic for the freshwater snail host *Planorbella trivolvis* that sheds it [128].

For cases of most dermatitis caused by schistosome cercariae, the cercaria hangs in or near the surface of the water where it comes into contact with the appropriate bird or mammalian (such as raccoon) host. Consequently, a pruritic maculopapular rash occurs on the human body in contact with the surface, often around the knees or ankles for those occurring in shallow water. Typically, when the cercaria penetrates human skin that it had previously penetrated with no or limited discernment by the patient, it produces a more apparent prickling sensation for several minutes. Depending on the number of cercariae and the host immune system, this condition may be associated with "macular eruption" (color change in 5–10 mm focal area without elevation or depression) with diffuse erythema or urticaria. Later in the day an itchy "macropapular eruption" (elevated colored area, 5–10 mm) develops and the papules "progress" to fluid-filled vesicles that may become secondarily infected because of the tendency to scratch them. Over the next week to month, the lesions become pigmented and apparent. The "hypersensitivity" reaction is not a true allergic one, and, when the same cercarial species or one with cross-reactivity penetrates the skin in the future, it produces a more severe dermatitis. Occasionally, systemic signs such as fever lymphadenopathy and edema develop [129].

A large number of marine schistosome species probably produce human dermatitis. The number of reported species, however, has remained low because of the inability until now with molecular tools to identify the species, the paucity of studies associating specific species with dermatitis, and the lower likelihood of human contact with cercariae in the marine environment when compared with infective species in freshwater. Swimmers in several freshwater lakes are more prone to continual visits to cercarial/snail-laden portions of the lake. Moreover, the fact that a variety of agents in marine habitats can produce a rash, apparently including non-schistosome cercariae, makes diagnosis of the cause of the rash difficult.

Austrobilharzia variglandis is probably the best known marine cause of the dermatitis, but what is identified as *A. variglandis* probably consists of a complex of several species from California [130], Hawaii [131], Australia [129], and elsewhere as well as the clam beds of Rhode Island [132] and Connecticut [133] where it is transmitted by *Ilyanassa obsoleta* to a variety of gulls plus the double-crested cormorant and Canada goose. At least *Austrobilharzia terrigalensis* in the silver gull *Larus novaehollandiae* and *Batillaria australis* (as *Pyrazus australis*) in South Australia and Western Australia [134] has been distinguished from misidentified Australian *A. variglandis*. The snail *I. obsoleta* ranges along the Atlantic coast to Florida. *Austrobilharzia variglandis* was reported from the migratory American white pelican in Mississippi [135]. A second species from the northern Gulf of Mexico, *Austrobilharzia penneri*, occurs in *Cerithidea scalariformis*. Gulls are important definitive hosts of marine schistosomes, especially species of *Austrobilharzia*, and since gulls and shorebirds commonly occur in close association with humans, there is a tendency for infections in Hawaii, Australia, and China [129, 136]. There is need to accurately differentiate species worldwide because most or all produce dermatitis.

Members of other blood fluke genera also produce dermatitis. *Gigantobilharzia huttoni* in both the brown pelican and American white pelican and the Antilles glassy-bubble, *Haminoea antillarum* (Cephalaspidea), occurs in Florida [135, 137, 138]; *Gigantobilharzia acotylea* caused dermatitis in the Venice Canal, Italy [139]; *Ornithobilharzia* cf. *canalicula* infects terns, and *Batillaria minima* from the Gulf of Mexico in Florida, USA [140, 141], although the dermatitis attributed to this cercariae was rightfully questioned [138]. Unidentified species producing dermatitis were reported from southern California in *Littorina keehae* (as *L. planaxis*) [142], from eastern Australia in the limpet *Siphonaria denticulate* on rocky shores [143], and from other regions such as Japan [144]. One from California occurs in the Japanese bubble snail (*Haminoea japonica*), which was introduced into San Francisco Bay in about 1999 [136]. Comparative sequence data did not allow information on whether the blood fluke was native to California, introduced from Asia in the snail, or a species from Asia acquired from migrating birds.

As indicated above, most schistosome species responsible for dermatitis are shed from freshwater snails in freshwater. Ten genera contain these species and are cited [145] compared with just three genera from salt or brackish waters. Studies on species of *Trichobilharzia* have provided some information that may relate to human infections. When bird species are experimentally exposed to a mouse, rather than mammalian skin trapping and eliminating the cercariae, it allows some to produce schistosomula. In the case of *Trichobilharzia regenti*, the agent typically penetrates the skin of the bird host and migrates to and becomes established in nasal tissues after passage through the peripheral and central nervous system [146]. When in a mouse, even a small number of the agent in the CNS can produce severe responses [147]. Members of *Trichobilharzia* as well as of *Gigantobilharzia* and *Ornithobilharzia*, especially on the initial exposure to experimental mammals, can progress to the lungs and CNS where they can produce severe pathological alterations [145].

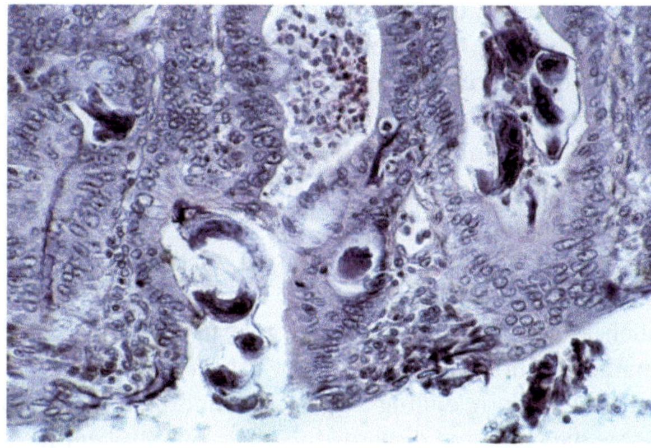

Fig. 15.26 Juveniles of *Strongyloides stercoralis* passing from human intestinal crypts and representing zoonotic species, some released as eggs and others as hatched juveniles

Nematoda

Strongyloididae (Rhabditina, Rhabditoidea). Some members of the nematode genus *Strongyloides* cause a creeping eruption (larval migrans) or rash in people exposed to juveniles and then challenged at a later date. Two species proven to cause the dermatitis in man are *Strongyloides myopotami* from the nutria (*Myocastor coypus*) and *Strongyloides procyonis* from the raccoon (*Procyon lotor*). These are probably the principal agents of what trappers, hunters, fishermen, and oil workers in the northern Gulf of Mexico call "marsh itch." A human volunteer was exposed in his arm to a relatively small number of third-stage juveniles of *S. myopotami* on seven occasions over 11 months before a minor rash (pruritus with small papules) erupted and lasted about 5 days [148]. Following that time, the volunteer was sensitized and juveniles of both species produced a reaction. Those from the raccoon allowed to penetrate the opposite arm produced more extensive reaction at the site of penetration for a few days, but, after disappearing for several days, it reappeared 5 cm from the site. Future exposures responded similarly but the recurrence resulted in a sinuous creeping eruption that spread from the arm to the chest. The juveniles migrating near the surface differs from the reaction by the larvae of blood flukes mentioned earlier. The species from the raccoon is morphologically similar to the human species, *Strongyloides stercoralis* (Fig. 15.26), which can undergo autoinfection. When a challenge dose of 185 juveniles of the raccoon species were exposed, at least one was able to mature and produce juveniles for a few days. The normal life cycle of species of *Stronglyloides* differs from that of most nematodes. The threadlike parasitic female in the mucosa of the intestine deposits eggs passed in the feces as either eggs or first-stage juveniles. In the external environment, the juvenile develops into a feeding rhabditiform juvenile, which in turn develops into an infective third-stage filariform juvenile. The non-feeding filariform can invade the definitive host or develop further into a single free-living generation of both males and females. These mate and produce infective juveniles, even though under optimal conditions, the offspring may develop into a second free-living generation. Infective juveniles penetrate the skin and pass by the cutaneous blood vessels to the lungs, where they break out of the alveoli and migrate by the respiratory tree to the pharynx and intestine, maturing in mucosal epithelium in about 2 weeks. The parasitic female produces juveniles without being fertilized and is parthenogenic. Most species of *Stronglyloides* infect one or few related amphibian, reptile, mammal, and bird hosts. Probably few species would be capable of producing creeping eruption in humans, but at least two from mammals in the Southern US do. Unpublished research in Mississippi showed that the free-living infective filariform juveniles of *S. myopotami* and *S. procyonis* can tolerate seawater well.

Miscellaneous agents. Unusual circumstances can result in involvement by a parasite not expected to affect humans. For example, a diving accident resulted in such a case with an adult philometrid (Dracunculoidea) nematode, a parasite that matures in non-intestinal sites in fishes. A scuba diver in Hawaii accidently stabbed himself when reloading a spear gun and then later cleaned a fish to eat. He noticed some parasites and discarded the fish, but 3 h later he sensed a 1-cm portion of a red worm protruding from the stab wound. He tried to kill and remove this firmly still-attached worm with vinegar, rubbing alcohol, and forceps. Because of the pain, he had the worm surgically removed by a physician [149]. Other parasites or free-living marine invertebrates can also take advantage of wounds or create their own. Isopods (Crustacea, Peracarida) in the Aegidae, Cymothoidae, Cirolanidae, and other flabelliferan families can produce lesions and urticaria, and are especially annoying. Even an amphipod has produced acute urticaria in a diver in Hawaii [1].

Protozoans

Acanthamoebidae (Amoebozoa). A few different free-living amoebae can cause severe disease if they can enter a human. Best known is *Naegleria fowleri* (Excavata, Heterolobosea, Vahlkampfiidae), which causes fatal "primary amoebic meningoencephalitis (PAM)," but it is restricted to freshwater habitats and moist soil throughout the world. Several species of *Acanthamoeba*, however, occur in marine waters or along marine beaches as well as in freshwater habitats. Species are known to cause chronic granulomatous amoebic encephalitis (GAE), granulomatous skin and lung lesions, and amoebic keratitis. The life cycle of *N. fowleri* consists of an amoeboid, feeding, and replicating trophozoite. During periods of low nutrients or desiccation, the motile amoeboid stage forms either a biflagellated, non-feeding, non-replicating trophozoite or a cyst until conditions improve. Human tissues contain only the amoeboid stage. The trophozoite apparently enters through the nasal cavity and migrates into the brain and central nervous system (CNS) along the olfactory nerve. The PAM typically progresses rapidly from a sudden headache and fever to coma and death, without much possibility of diagnosis before death. Species of *Acanthamoeba* do not have a flagellated trophozoite, the amoeboid trophozoite has spike-like pseudopodia and is more sluggish, and three-layered cysts can occur in human tissues. The portal of entry may differ for different species. Invasion appears to be from a primary focus in the skin, lower respiratory tract, or nasopharynx, with dissemination of the trophozoite reaching the CNS through the circulatory system and possibly by crawling along the outside of blood vessels [150]. Signs of GAE progress much slower and initially are less severe than for PAM; they take weeks or months to result in death, and infections are difficult to diagnose. In contrast with PAM, which affects healthy people, GAE is usually associated with chronically ill, immunocompromised, or otherwise debilitated patients. Predisposing factors for infections include broad-spectrum antibiotics, steroid or antineoplastic therapy, radiation therapy, alcoholism, pregnancy, and bone marrow or renal transplantation. Many of the earlier diagnosed infections, especially in non-compromised patients, were associated with *Balamuthis mandrillaris*, a morphologically similar free-living but genetically different species that has been considered a soil-inhabiting species.

Which species of *Acanthamoeba* infect humans in the marine environment and how they infect the people have not been established. Several isolates of the *Acanthamoeba* complex have been cultured from marine waters or beaches [151–153], and eight strains are pathogenic to mice and potentially to humans. Of 17 species of free-living amoeba cultured from salt water in Northwest Spain, only *Acanthamoeba polyphaga* was pathogenic to mice [154]. Seven species have been associated with human disease, and of these, *Acanthamoeba castellanii, Acanthamoeba culbertsoni, Acanthamoeba hatchetti, Acanthamoeba rhysodes, A. polyphaga*, and *Acanthamoeba griffini* have been identified from amoebic keratitis, a hard-to-treat disease of the cornea [155]. This disease causes severe ocular pain, photophobia, recurrent epithelial breakdown, and edema, and confirmed by trophozoites and cysts in a corneal scrape or biopsy. It was recognized in 1973 and not common until 1985 when contact lenses became popular. Lenses are maintained in a saline or tap water solution in a reusable storage case that can get contaminated.

Members of *Acanthamoeba* also have the ability to harbor viruses and bacteria, perhaps serving as vectors or reservoir hosts for a variety of agents. Isolates of *Acanthamoeba* from water sources in the Canary Islands revealed four different serotypes of adenoviruses, one related to ocular disease and the others unknown, probably respiratory, gastroenteritis, or neurologic [156]. A large DNA mimivirus was identified from *A. polyphaga*. Experimental studies [157] have shown that coxsackie B3 viruses can be adsorbed on the surface and accumulated within *A. castellanii*. The same marine amoeba can also host the bacteria *Chlamydia pneumoniae* and *Legionella pneumophila*. It did not prey on *Vibrio parahaemolyticus*, but it secreted a factor that promoted survival of the bacterium in coculture, suggesting the amoeba provides a survival to the extracellular pathogen in the environment [158]. These and other facultative parasites probably contribute greatly to transmission of microbial disease agents as can obligate parasites [159].

Nonparasitic: But Confused with Parasites

In the marine environment, there are, in addition to schistosome cercariae, a variety of nonparasitic agents that cause similar appearing rashes. The stinging nematocysts of several members of the phylum Cnidaria (previously called

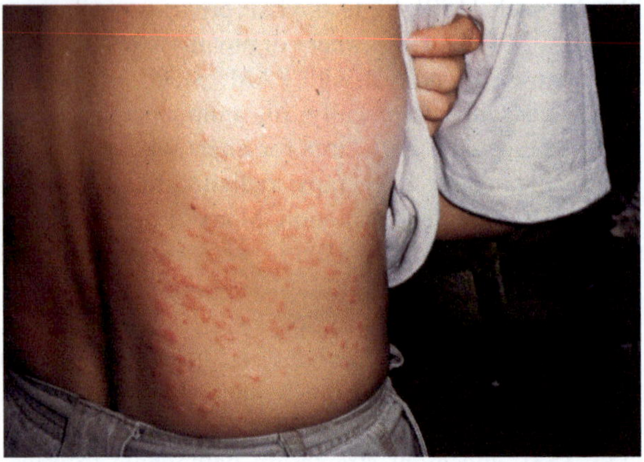

Fig. 15.27 Visiting, teaching marine biologist at San Salvador, The Bahamas, exhibiting back with hypersensitive rash developed after and presumably caused by exposure to challenge dose of planula larvae of the thimble jelly, *Linuche unguiculata*

Coelenterata) have been held responsible for "seabather's rash" or "seabather's eruption." Members usually go through an alteration of generations with asexual polyploidy and sexual medusoid generations, but there are variations of this life cycle. Divers can brush against the nematocysts of fire corals (Hydrozoa, Anthomedusae, *Millepora* spp.) with their encrusting calcareous coral-like skeletons, while bathers can become entwined in expandable fishing tentacles lined by batteries of nematocysts of the "man-of-war" (Hydrozoa, Siphonophora, *Physalia* spp.) hanging from the gastrozooid, maintained on the surface by a gas-filled float. This floating colony appears like a medusa but is actually a highly modified polyp. When the long, up to 13 m, tentacle of the Atlantic *Physalia physalis* washes onto sandy beaches, it gets ground up in the surf, leaving minute pieces with undischarged batteries of nematocysts to get into bathing suits or work gear. Medusa of several jellyfishes (Scyphozoa) have been associated with stings and rashes. Moreover, some sea wasps and box jellyfishes (Cubozoa), exhibiting square cross sections of their medusa, have been shown to cause human fatalities. The planula larvae of true jellyfishes, such as the thimble jelly, *Linuche unguiculata*, get under swimsuits, and challenge exposures produce a hypersensitive rash reaction (Fig. 15.27). Cases are common along the Southeastern Atlantic US, Mexico, and the Caribbean Sea, but exact identifications and prevalence of these and other cases worldwide require study. Systemic symptoms from the planulae are considered more common than with schistosome dermatitis [129]. The sting from the planula larva of the burrowing anemone (e.g., Anthozoa, *Edwardsia lineata* as *E. leidyi*), appearing as a "tiny pink egg" [160] attached to the inner lining of bathing suits, has been associated with seabather's eruption in Long Island, New York, from mid-August through early September. The planula also parasitizes ctenophores, which can get caught in swimming apparel and transmit the condition. Not all rashes attributed to blood flukes or cnidarians come from those agents. Many cases also result not from ejected toxins but from rubbing between the bathing suits and human skin of molluscan opisthobranch gastropod shells, sponge spicules, and spines of zoea and megalops larvae of crabs and shrimps, swarms of which can occur seasonally in heavy concentrations. Rashes attributed to "sea lice" are misleading; the term "sea lice" is a common name usually attributed to a variety of parasitic copepods but primarily *Lepeophtheirus salmonis* infesting salmon. None of these copepods produces human rashes, but some marine dermatitis caused by cnidarians has been referred to as caused by sea lice. Some polychaetes (Annelida) produce an irritating rash as does contact with the excrement of some sea cucumbers (Holothuroidea) or eating some inadequately cooked species. Contact with Cyanobacteria (previously referred to as blue-green algae), such as *Lyngbya majuscula* growing on sea grasses in Hawaiian waters, has been identified as causing "swimmer's itch" or "seaweed dermatitis." The agent is called fireweed in Australia. Rashes also result from some "harmful algal blooms" (red tide) throughout the world. The Internet includes some incorrect or misleading information, but researchers are tackling some of the cases, especially since they have a large impact on recreational and occupational activities in the marine and coastal environments.

Indirect (Passive) Contact with Disease-Causing Agents

Eating products contaminated with an agent or drinking contaminated water are herein considered differently than eating infected products because the food or water contaminated with the organism serves as a mechanical or passive vector of the agent rather than as a host that the organism requires for its survival. Also, the agent does not actively penetrate the skin as achieved by an agent of swimmer's itch or marsh itch.

Trematodes

Humans occasionally get infected with trematodes in a passive process. A person does not knowingly eat a product, and the parasite does not actively penetrate a host. For example, a seafood handler who came in contact with an abundance of fresh killed coho salmon (*Oncorhynchus kisutch*) known to be heavily infected with the trematode *Nanophyetus salmincola* developed an infection exhibiting acute clinical signs. He denied eating raw or cold-smoked fish and rarely ate any seafood products. Alternatively, he could have transferred the minute metacercariae on his hand to his mouth while smoking [1, 43].

Protozoans

Giardiinae (Excavata, Diplomonadida). Species of the flagellate *Giardia* have long been associated with severe diarrheal disease in humans. They are typically considered as freshwater or terrestrial species acquired when people drink contaminated water, but they have recently been found to be abundant in marine waters. The agents were thought to be confined to those discharged in raw sewage disposal and from recreational swimmers throughout the world, both raising public health risks. Consequently, infectious cysts or the trophozoites of the flagellates have been detected along some marine beaches near sewage outfall and canals influenced by runoff, where bathing is common in Hawaii [161], as well as in Panama and Hong Kong [111]. The trophozoite is a microscopic pear-shaped organism with four pairs of flagella. Two of these occur within the resistant cyst, and when the vertebrate host acquires the cyst, it excysts, releasing the trophozoites which reproduce asexually by binary fission and remain in the lumen or attach to the mucosa of the small intestine. Under the appropriate conditions, encystment occurs and both the cysts and trophozoites pass out with the feces. The taxonomic classification of members in the genus is not well established, but there are five recognized species (e.g., [162]), and the genetic characterization of *Giardia duodenalis*, also referred to as *Giardia intestinalis* or *Giardia lamblia* and the species with the widest host-specificity, groups at least seven genetic assemblages (genotypes) including Assemblage A, the one most commonly found in humans [163]. *Giardia* sp. occurs widespread in the river otter (*Lontra canadensis*) from the marine waters of Puget Sound Georgia Basin of Washington [164], but species also occur abundantly in the feces of a variety of temperate and Arctic marine mammals such as the right whale (*Eubalaena glacialis*, 71% prevalence), bowhead whale (*Balaena mysticetus*), ringed seal (*Phoca hispida*, 65%), California sea lion (*Zalophus californianus*), and harbor seal (*Phoca vitulina*) but not the beluga whale (*Delphinapterus leucas*) or bearded seal. Assemblage A occurred in the harp seal (*Phoca groenlandica*), gray seal (*Halichoerus grypus*), and harbor seal (*Phoca vitulina*) but not the beluga whale [165]. Moreover, despite the common occurrence of infections in adult seals and cyst contamination on ice floes, cysts were not encountered in seal pups less than 1 year of age. However, the epidemiology and potential for zoonotic transmission in marine systems remains virtually unknown [164], but the genotypes of *G. duodenalis* in the harbor seal in Puget Sound, Washington, were the same as from a canine, from other sites, and were mostly novel genotypes [166]. Cysts remain infective for about 56 days in relatively deep freshwater cool lake water, but at least those of *Giardia muris* become inactivated by the combination of salinity and sunlight in warm Hawaiian marine beach water within 3–6 h [167]. In the Netherlands where the salinity remains stable at 28–31 ppt, *Giardia* cysts were detected in what was considered to be a low estimate of 12% of oysters, where viability was presumed to be preserved, representing some public health risk to those consumers who ate the oysters raw [168]. Cysts have also been filtered out of the water and concentrated by the blue mussel, *Mytilus edulis*, and other bivalves, and this tool has been used to monitor *G. lamblia* in Ireland [169] and for that and other species or genotypes elsewhere [112, 170]. Depending on the size and species of bivalve, it can filter as much as 20–100 l of water per day. Much information on human health risks from beaches and marine waters, animal health, source of agents, dispersal of agents, and primary vectors can be established by a variety of surveillance programs. For example, a survey of live and dead, stranded and bycatch, birds, dolphins, seals, and fishes was conducted in the Northwest Atlantic using PCR techniques [171]. With the exception of infections in seals, those of *Giardia* spp. were more frequent than those of *Cryptosporidium* spp., to be discussed next.

Coccidiasina (Chromalveolata, [Alveolata, Apicomplexan]). Coccidian species in the genus *Cryptosporidium* pose a stronger human risk for infections than those in *Giardia* [112], especially species in estuaries compared with freshwater [169]. Typically, marine mammals, invertebrates, and marine water are examined for the agents concurrently with *Giardia* because the agents are transmitted by the fecal-oral route from humans and other mammals. Thick-walled oocysts are passed in the feces as a resting cyst, each with four sporozoites infective to the vertebrate host, where development occurs. Additionally, there is a thin-walled cyst that cannot survive long in a harsh environment, but it may lead to endogenous autoinfection. This seems to be especially true for persistent infections in immunocompromised patients. A total of 14–15 species are accepted [172, 173], and several are known to infect humans: *Cryptosporidium hominis* (previously *Cryptosporidium parvum* genotype 1), *C. parvum*, *Cryptosporidium meleagridis, Cryptosporidium felis, Cryptosporidium canis*, and *Cryptosporidium muris*, with most human infections from *C. hominis* and *C. parvum*. Human agents have been detected in more marine recreational areas than *Giardia*. They occurred in Hawaii, Panama, Puerto Rico, Australia (tidal area of Georges River), and Hong Kong [111]. In fresh deionized water, oocysts of *C. parvum* can survive for at least 6 months at 0–20°C, 3 months at 25–30°C, 1 week at 35°C [172]. As many as 10^9 to 10^{10} can be excreted from bovine calves or non-compromised humans per week, and as few as 30 oocysts have been known to infect a human volunteer [174]. Bioassays in mice show that oocysts of *C. parvum* can survive 12 weeks in 30 ppt at 10°C and at 20°C for 12 at 10 ppt, 8 weeks at 20 ppt, and 4 weeks at 30 ppt, which was long enough for the oocysts to be removed by filter feeding bivalves such as the eastern oyster (*Crassostrea virginica*) [170]. Literature reports consist of a variety of bivalve mollusks from seven countries that contained *Cryptosporidium* as detected by immunofluorescence microscopy, bioassay in mice, or molecular methods [111]. In North America, oocysts have been detected from New Brunswick to Texas. Gill washings from the oyster near a large cattle farm and septic tanks were positive for the oocysts and infective to mice [170]. The agents occur in hematocytes as well as on gill surfaces, and they were more abundant after rainfall events, with two of the four identified species (*C. hominis* and *C. parvum*) infective to humans [111]. The same two species plus *C. meleagridis* also occurred in North American commercial products [111]. Species have also accumulated in hard clams, bent mussels, and zebra mussels [175]. To date, no human cryptosporidiosis cases have been linked to eating raw shellfish [111], even though some human genotypes have been concentrated in various bivalves. There is a higher prevalence of infection of both *Cryptosporidium* spp. and *Giardia* spp. in the ringed seal and right whale than in reported terrestrial mammals, and those infections might result from ingestion of contaminated water or from prey that had concentrated oocysts [176]. Other hosts of *Cryptosporidium* spp. include the California sea lion, bowhead whale, dugong (*Dugong dugon* in Australia), and marine foraging river otter but not the bearded seal and beluga whale [164, 176]. Isolates from the ringed seal from Nunavik, Quebec, were genetically characterized using two gene fragments and actin loci as *C. murus* plus two novel genotypes that were distantly related [177]. The risk for recreational bathers to come into contact with *C. parvum* was significantly greater on weekends than on weekdays, when the number of bathers was lower. The same was true in that Chesapeake Bay marine beach for *G. duodenalis* and *Enterocytozoon bieneusi*, a microsporidian to be discussed below [178]. The proportion of water containing the agents also correlated significantly with enterococci counts. The study suggested using those counts to indicate the presence of the parasites and recommended preventing diapered children from entering the water, restricting the number of bathers in the recreational areas, advising those with gastroenteritis to avoid bathing, and using showers prior to and after bathing.

Birds contract cryptosporidiosis with at least five avian species and numerous avian genotypes, and some of these have been investigated in regard to infecting mammals. They can also serve as vectors of *Cryptosporidium* spp., *Giardia* spp., and microsporidians infective to humans. In fact, aquatic birds seem to play a substantial role in contaminating water around the world, and managing water resources could benefit by incorporating protection measures for pathogens linked to these birds [120, 179]. In contrast, little research has been conducted to evaluate the public health threat of species of *Cryptosporidium* and related genera described from marine and freshwater fishes and from amphibians and reptiles. The various taxa and isolates have been reviewed [180].

Entamoebidae (Amoebozoa). Some strains or species of amoeba *Entamoeba histolytica* complex also are zoonotic threats. Poorly documented records of infections in the bottlenose dolphin (*Tursiops truncatus*) from Cuba, Mexico, and other locations suggest a relationship among the marine mammals, sewage, recreational beaches, and human infections. Researchers at the Gulf Coast Research Laboratory and the American Type Culture Collection have seen cysts but were unable to culture material from the bottlenose dolphin (*Tursiops truncatus*) from Mexico. Because there were slight morphological differences between it and *E. histolytica* sensu stricto, this subject deserves a critical investigation to determine if the dolphin form is restricted to marine mammals and whether it can cause fulminating dysentery, bloody diarrhea, fatigue, and abdominal pain in humans. Possibly, the agent was acquired by the dolphin from humans during associated recreational activities or from human sewage.

Parasites That Produce Excretions or Secretions That Result in an Immunological Response in People

Ascaridoid juvenile nematodes produce excretions and secretions (ES products) that elicit immunological responses. These responses include both the allergic response by the host to living worms migrating into or through tissues and true anaphylactic reactions triggered by dead or living worms by food-borne, airborne, and skin contact routes [75]. Ascaridoids are unusual among nematodes to produce such reactions. Most concerning is that some sensitized individuals respond to infected cooked seafood products and even chicken that has been fed fish meal made from infected fish products [181].

The tissue migration by ascaridoids is made possible by potent proteolytic enzymes released from the esophageal gland and excretory cell through separate anterior openings. These contain hyaluronidase, serine proteases, anticoagulants, and numerous other identified and unidentified substances, but only specific ones probably produce the well-defined, erosive, hemorrhagic lesions in the gastric mucosa and these and others produce other effects in the human accidental host. Metabolic products released by the migrating juvenile also produce humeral and cellular responses, especially involving the acute lesions. These products, also including surface and somatic components of the juvenile, form insoluble immune responses with antibody. Some also cause direct IgE-independent degranulation of mast cells, at least in sensitized mice, and chemotaxis of eosinophils associated with thermolabile factors from the parasite, and produce the characteristic local response in the digestive tract but not systemically like in most other helminth infections [75]. When either living or dead individuals of *A. simplex* are ingested, cholinergic hyperactivity and adrenergic blockade can be rapidly induced, resulting in focal reactions associated with living worms but widespread along the entire bowel if the worm becomes ruptured or disintegrated. These activities may explain why 70% of the symptoms in *A. simplex*-induced anaphylactic reactions involve the digestive tract.

Anaphylaxis appears to be complex. Rodent studies demonstrate a mixed Th1/Th2 pattern of cytokines when the host is sensitized and challenged intravenously with certain parasites, but not when challenged orally. When the antigen is an extract of *A. simplex* proteins or a live worm, a strong Th2 response occurs, resulting in scratching, irritability, diarrhea, and puffiness around the eyes within 1 h. The multifaceted immune hypersensitivity reactions after induction by members of the *Anisakis simplex* complex have been investigated in terms of a wide range of parasite products and host cells; a concise helpful review treats all the known facets [75]. Hypersensitivity is usually diagnosed by skin prick tests and in vitro confirmation (specific IgE, histamine release, and basophil activation test). The skin prick test for *A. simplex* was first used in 1995 and now is used by physicians as an important test for cases of urticaria and anaphylaxis [75, 182, 183].

Recognition of allergic signs has been emphasized because of cases in Spain and South Africa. In Spain where fish is consumed abundantly, what was referred to as *A. simplex* is the most important hidden food allergen in the adult population suffering acute urticaria and anaphylaxis. It comprises as much as 10% of the anaphylaxis previously diagnosed as idiopathic [75], if the specific immunoglobulin E (IgE) detection by ImmunoCAP assay did not overestimate the number of sensitized subjects. Anisakiasis in Spain is caused primarily by eating pickled anchovy (*Engraulis encrasicholus*). That fish plus European hake (*Merluccius merluccius*) and cod (*Gadus morhua*, just in Atlantic Ocean rather than in both the

Mediterranean Sea and Atlantic Ocean) are eaten by allergic patients. Half the 64 allergic patients in one report required emergency treatment [184]. Half of the patients presenting signs of infection said they ingested raw fish, but the remainder ate cooked fish or, in rare cases, canned fish. Allergic cases requiring hospitalization demonstrated respiratory arrest, severe shock, and persistent angioedema.

In Spain, the matter of allergic responses in "gastroallergic anisakiasis" remains complicated and seems to be reported primarily from some regions in Spain. Patients entering the hospital emergency room with a severe acute epigastric pain and positive *Anisakis* prick tests plus total and specific IgE assays were divided into two groups. One group was diagnosed by the presence of at least one worm in the stomach, and the other group which had no worm seen in the stomach. Most patients had just eaten raw or pickled anchovies. Both groups averaged about 5 h, but up to 26 h, between intake of raw fish and onset of hypersensitivity symptoms, and no significant difference occurred between the two in their allergic symptoms of urticaria, angioedema, erythema, bronchospasm, and anaphylaxis, suggesting a borderline condition between a food allergy and parasitic disease supporting quick removal of the worm from the stomach. Moreover, only 26 of 40 patients required drug therapy to manage the allergic reaction, with only three requiring it for more than 1 day, suggesting an acute, self-limiting disease. In contrast, anaphylaxis can be life threatening, but abdominal symptoms disappeared within a few hours after removal of the worm [185]. When a portion of the *Anisakis*-sensitized groups with gastric worms, without gastric worms, and with no gastroscopy conducted (when symptoms did not persist for more than 8 h) were challenged with frozen worms from the blue whiting (*Micromesistius poutassou*, from either Atlantic or Mediterranean sources) and later from undocumented hosts on multiple occasions, there was no patient who suffered a reaction. This lack of a response to the thermostable proteins suggested that live worms had to be present for the allergic symptoms to occur [186]. Because the specimens from the Mediterranean Sea were from the anchovy, they were *A. pegreffii* [70] and not *A. simplex* as reported, even though if some material came from the Atlantic Ocean, both species could have been present. However, human infections with *A. pegreffii* are typically gastric in nature, and those of *A. simplex* are usually intestinal, based on findings from Japan, where 10% of healthy adults were seropositive against the *Anisakis* antigen [187]. The fact that some individuals exhibit allergic episodes when no viable nematode is present may express a difference between the effects of *A. pegreffii* and some other species of *Anisakis* or may be variation in responses by individual patients.

The cases in Western Cape Province, South Africa, involved workers in fish-processing plants. An epidemiological study of 578 workers from two large-scale plants reported 30 workers (5%), who said they had allergic symptoms to seafood. A total of 87% said symptoms occurred after eating fish, 40% after handling them, and 17% after smelling them. A prevalence of sensitization was tested as 6% with a higher value of 8% sensitive to *Anisakis* [188, 189]. *Anisakis*-specific IgE reactivity in the workers was determined to be associated with bronchial hyperreactivity and dermatitis. In corresponding mice studies, the juvenile worm induced a striking Th2/type 2 response. The investigation suggested that consumers acquiring an infection most likely can acquire sensitization to anaphylaxis, but that exposure to *Anisakis* proteins alone may be enough to elicit allergic reactions in sensitized individuals. In other words, occupational exposure to infected fish or fish meal constitutes a risk factor for developing sensitization to *Anisakis* or other ascaridoid-related allergic disease.

Historically, eating fish fillets infected with the plerocercoids of trypanorhynchean cestodes has been considered harmless with rare exception. All members of the group mature in elasmobranchs. In fact, some infected intermediate host products are preferred over noninfected counterparts; one example is the flesh of the Atlantic pomfret, *Brama brama*, also known by a variety of other common names, infected with plerocercoids of *Gymnorhynchus gigas* and preferred by some Portuguese [99]. Recently, the rat and mouse were orally inoculated with plerocercoids of *G. gigas* to test for anti-worm IgG, M, and A (H + L) levels in intestinal fluids and serum as well as specific serum IgE levels by enzyme-linked immunosorbent assay (ELISA) [190]. Levels of all increased in the challenged mouse, producing distress. The rat had an increased expression of heat shock proteins in the intestine and spleen. Repeated exposure to the worm in the rodents produced clinical signs appearing progressively more rapid and lasting longer, suggesting that feeding on infected fish triggered production of anaphylactic-type antibodies in rat, mouse, and, by implication, human.

Management, Control, and Treatment of Parasites Affecting People

Management

Management starts with education about parasites. Parasites are or may be involved with a recognized or unrecognized problem but are not necessarily harmful or bad, if certain practices are recognized or followed by regulating agencies, funding agencies, the seafood industry, consumers, and those using marine waters for occupation or recreation. Harm can be done if partial data deludes people not to eat seafood or not to enter the water.

Being alerted to public health risks allows people to make choices about what product to eat, how it should be prepared, when to avoid a product, when not to enter specific bodies of water, and when to visit a physician. If the industry provides a safe and sanitary product to its customers, the number of customers will increase.

Control

The best way for processors and restaurants to control infections with most helminths and most protozoans is to freeze or heat the product. Some restaurants flash-freeze products sold as fresh to avoid infective products, and apparently few customers can detect the difference from the never frozen products. If products with few or no parasites can be provided to these groups by fishermen or fish farmers, that product will be in higher demand, able to meet regulations, and more valuable. Controlling infections by interrupting a parasite life history link and harvesting from a noninfected region are easy ways to avoid future problems.

Heating and Freezing

Historically, the US Food and Drug Administration (FDA) [191] has recommended food-service industries to cook fish products to an internal temperature of at least 63°C by conventional methods or to an internal temperature of 74°C by a microwave process rotated midway through the process and allowed to stand for 2 min afterward. Any process sufficient to kill bacterial pathogens also will kill parasites.

However, the FDA, under the seafood Hazard Analysis and Critical Control Point (HACCP) regulations, will allow for a facility to submit to it an alternative proposal with justification to modify the recommendations. When ten specimens of *Anisakis simplex* sensu lato were embedded into an approximate 1.8-cm-thick portion of a fillet of the arrowtooth flounder (*Atheresthes stomias*) and microwaved, the viability of the worms was assessed. Survival was determined as 31% at 60°C, 11% at 65°C, 3% at 74°C, and 0% at 77°C [192]. Thick salmon fillets with *Diphyllobothrium* spp. and *Anisakis* spp. require additional cooking. Microwaving pork chops with *Trichinella spiralis* still had active juveniles at 82°C, and microwaving can produce temperatures differing by 20°C when measured just 1 cm apart. The purpose of HACCP Regulation, as implemented in 1997, is for processors of fish and fisheries products to develop and implement reasonable plans for safe and sanitary procedures. Recommendations are established at each point that specific parasites can be detected, eliminated, or treated, and FDA will guide and work with those people and facilities in the industry to provide their specific appropriate plans and systems (www.fda.gov/Food/FoodSafety).

Freezing infected fisheries products is also not clear because the effectiveness depends on the temperature of the process, the length of time undergoing freezing, the length of time held frozen, the fat content of the product, and the species and stage of the parasite. For example, most cestode plerocercoids are more susceptible to freezing than nematode juveniles, which in turn are more susceptible than encysted trematode metacercariae. FDA recommends the following: freezing and storing at −20°C or below for 7 days, freezing at −31°C or below until solid and then storing at that temperature for 15 h, or at −4°C for 24 h. The fishing industry has used a process called "blast freezing," which rapidly reduces the temperature to −40°C, and it has shown to effectively kill parasites and not have much influence on flavor or texture of the product. Also, most US sushi bars freeze their fish [1].

Other communities have different recommendations or regulations. The European Food Safety Authority presented a scientific opinion on food safety related to parasites in fishery products [193]. The presence of *Anisakis* was of major concern for products intended for marinating, salting, and eating fresh. Treatments should provide an equivalent level of protection as freezing at −20°C for ≥24 h, at −35°C for ≥15 h, or −15°C for ≥96 h at the product's core; heat treatment should be >60°C for at least 1 min. These are less rigorous than those in the USA but more than some other regulations. Emphasis focused on the insufficiency of many traditional marinating and cold-smoking methods to kill the juvenile worms. Some metacercariae can tolerate more-harsh temperatures and treatments than other parasites.

Detection of Parasites and Culling

Large parasites and those that contrast well with the infected seafood product are those most easy to detect. Many of these are not harmful to consumers, but they will have an influence on whether consumers will purchase the products or eat them in a "raw" state. Some, such as *Anisakis* spp. and *Pseudoterranova* spp., constitute a public health risk, and these have attracted research in detection methodology. The efficiency of successively more accurate methods for detecting these nematodes range from (1) gross visual inspection, (2) candling on a light table, (3) candling of belly flaps (ventral portion of fish), (4) ultraviolet illumination (UV) of frozen fillets, (5) pepsin/hydrochloric acid degradation, and (6) UV illumination of frozen remains [194]. Candling or passing a product over a light table, either sandwiched between two glass or plastic plates, sliced fillet, or whole fillet, has been a popular method for several decades and usually the recommended method during industrial processing of marine fish intended for human consumption. European Union regulation 91/493/EEC and Norwegian fish quality regulations require removal of any visible parasites seen in gross inspection of commercial products by spot checking [194]. The more efficient methods indicated above are occasionally used for scientific purposes, and in one article [194] examining Norwegian spawning herring, mackerel, and blue whiting from the Northeast Atlantic Ocean for nematode juveniles in fillets, the detection efficiency for candling was 10% or less. After testing with the other listed methods, the highest values were obtained with UV illumination of remains after enzymatic digestion. Size and thickness of fillet as well as texture and color of flesh influence the efficiency. Pressing 2–3 mm layers of fillet with a hydraulic and manual press between 12-mm-thick acryl glass sheets, deep-freezing the removed cake in a plastic bag, and examining the sample in the bag with UV light at 366 nm more than doubles the recovery rate of candling at 1,500 lx; this method is rapid and good for surveys [195]. Imaging spectroscopy provides a promising method to detect spectral and spatial information from nematodes in cod as deep as 0.8 cm below the surface of the fillet. That is deeper than can be determined by manual inspection, but the instrumentation has not been perfected enough for commercial use [196].

Fish Farming Practices

In Europe, apart from farmed Atlantic salmon, sufficient monitoring data on marine products are not available [193]. A wealth of information occurs in the literature and on Internet sites about how to detect and treat parasites and control infections, usually with focus on health of the cultured product rather than on a human consuming or coming into contact with the product. This chapter does not recommend or discuss the numerous specific methods for use in fish farms and other aquaculture operations, but parasites are best managed in those facilities by breaking the link in the life cycle that can disrupt or eliminate the infections. Fish in culture can have more or fewer parasites than wild products, and the degree relates to several factors [1].

Treatment

This chapter mentions a few chemotherapeutic treatments for infected patients but purposely does not attempt to provide compounds for all taxa. People are becoming resistant to some compounds, some compounds have serious side effects, the Internet contains both good and poor or misleading advice, and new compounds are continually being tested or made available. Because a reader should use or recommend the most reliable treatment for the agent or strain, the reader should seek professional advice. Recent information has been compiled such as that found in the most recent edition by Garcia listed in the "Books and Reviews" or some other recent source. The Centers for Disease Control and Prevention (CDC) has a website with much helpful information and a willingness to respond to individuals that contact it. On the other hand, methods to avoid infections by properly freezing or cooking a product occur above. Of course, because of well-entrenched customs, the wonderful flavor of raw or specially prepared seafood, and the lack of knowledge about most parasites, inadequately prepared and uncooked products will always pose a risk of infection.

Future Directions

As seen already in this chapter, a wealth of information has been acquired in the last few decades regarding parasites that pose a public health risk. Many questions remain unanswered and lifetimes of research will be required to answer these and additional questions regarding the risks and what to do about them. Some of the suggested paths for future research occur below in five overlapping categories:

1. Harmful parasites presently unknown or unrecognized still need to be discovered. Because of new cuisines, opportunities to get infected, methods to characterize agents, and methods to detect agents as well as variability in human responses to parasites, those parasites need to be determined that are capable of infecting humans. As seafood products are overfished or the consumers increase in number, seafood products need to come from new areas or new products. Consequently, the risk for acquiring an infection from a product historically recognized as safe increases without the consumer's knowledge. Protozoans cannot be seen without a microscope, and infections by marine species and freshwater species flushed into the marine environment are being recognized as important agents affecting human health, and the marine environment probably contains many more of these than recognized.
2. An improvement in taxonomy of many parasitic groups will allow the scientific community and public to know what species or genotypes are involved with diseases in different areas resulting from different parasite life histories. Morphologically similar species can elicit entirely different responses, and the knowledge of what specific agents are where and what signs of disease result from which of those agents comprises a fundamental cornerstone of biology and research. Taxonomic research will involve morphological treatment of quality specimens from both humans and natural hosts backed up with corresponding molecular characteristics. Molecular tools (PCR, quantitative PCR, sequence analysis, and other techniques) allow better detection and diagnosis of some of these parasites and consequently improve our understanding of the epidemiology of human infections, the geographic range of the agents, the hosts of the agents, and the longevity of the parasites. Taxonomic studies matching genetic sequences with corresponding morphological data help identify or establish sister and cryptic (closely related) species or strains. This is true for helminths as well as protozoan groups as indicated in the discussion of specific taxa. In addition to needing supplemental molecular research on taxonomy, identifications, and diagnoses, molecular tools will help determine/confirm life cycles, life histories, mechanisms of pathogenesis of infections, and approaches to manage, control, and treat the specific agents.
3. Once specific agents are characterized, methods to detect and diagnose them need improvement, especially by noninvasive means. This is especially important because many parasites do not mature in humans, thereby not allowing the detection of diagnostic eggs, cysts, and spores in the feces or blood. Even when parasites do mature in patients, they cannot always be specifically identified. For example, most heterophyids have very similar eggs, and biological variation occurs for each species. Cross-reactivity occurs in many of the present serological and molecular assays. In the case of *Paracapillaria philippinensis*, there is a need to identify juvenile stages as well as serologic or other diagnostic methods to assess infections when eggs and adult specimens are not apparent in fecal examinations. Improved molecular assays could also detect specific infected fishes in specific locations that are responsible for most of the resulting human infections. More sensitive methods are necessary to detect general and specific

isolates of *Cryptosporidium*, *Giardia*, and *T. gondii* in bivalves, where some unknown and nonremovable components inhibit the presently used PCR reactions [168]. As indicated above, many more protozoans probably infect humans, and new methods will allow their detection.

4. Surveillance methods of known agents require improvement so that infections can be reduced, and causes can be detected. In Europe, the EFSA recommends coordinated studies to improve and implement surveillance and diagnostic awareness of allergic reactions to parasites in fishery products and encourages epidemiological studies throughout Europe to assess the impact of *Anisakis* on human-associated disease, including all allergic forms [193]. All surveillance programs in the marine environment are in an early stage of research, and many could be implemented in conjunction with microbiological studies involved with maintaining healthy waters.
5. There is a need to create ways to avoid parasitic infections without having to reduce seafood consumption or reduce time spent in the water for recreation or occupation. The spread of information in a subtle manner should definitely reduce parasite infections in most countries today. On the other hand, this can be difficult when an improved economy in areas such as Vietnam have increased the social activities that are conducive to eating more raw seafood because it is "healthier when eaten raw" and acquiring more parasites. Also, global climate changes can affect infections as exemplified earlier for the peoples of Hokkaido Island, Japan, who understood parasitic infections and depended on their seafood products.

A few causes for human infection by zoonotic parasites that seem correctable by means other than spreading knowledge about specific parasites include unsanitary defecation habits, use of human excreta as fertilizer, inadequate sewerage systems, and inadequate aquaculture practices. New ideas associated with acquired knowledge about infective parasites and their life histories can contribute to modifications in sources of products and methods of complex food consumption and cooking habits. The latter include economic and sociocultural factors such as beliefs and tradition. Examples of traditional dishes in addition to those already mentioned include raw or partially cooked aquatic products such as raw crab soaked in soy sauce (ke-jang, Korea), raw drunken crabs and raw grass carp in China, and raw fish in Thailand (Koi-pla [raw fish with chopped garlic, lemon juice, chili, rice, and vegetables] and pla som). In contrast, in some industrialized countries like Japan, where eating raw fish is widespread, infections can be coupled with foreign travel and with eating imported foods or exotic delicacies.

Acknowledgments I thank Janet Wright for help with organization of the literature and Kim Overstreet and Jean Jovonovich for reading the draft. Some of the figured histological sections are from the author's old photographs taken of slides in the collection of the late Paul C. Beaver at Tulane Medical School. The review is based on work supported by USDC, NOAA, award no. NA08NOS4730322 and subaward no. NA17FU2841, the NSF under grant no. 0529684, and US Wildlife, Fisheries, and Parks, CIAP, award M10AF2015, MS.R. 798.

Bibliography

Primary Literature

1. Deardorff TL, Overstreet RM (1991) Seafood-transmitted zoonoses in the United States: the fishes, the dishes, and the worms. In: Ward DR, Hackney CR (eds) Microbiology of marine food products. Van Nostrand Reinhold, New York, pp 211–265
2. Fischler C (2002) Food selection and risk perception. In: Anderson GH, Blundell J, Chiva M (eds) Proceedings from the symposium food selection, from genes to culture. Danone Inst., Paris, France, pp 135–151
3. Adl SM, Simpson AGB, Farmer MA, Andersen RA, Anderson OR, Barta JR, Taylor FJR (2005) The new higher level classification of Eukaryotes with emphasis on the taxonomy of protists. J Eukaryot Microbiol 52:399–451
4. Nawa Y, Hatz C, Blum J (2005) Sushi delights and parasites: the risk of fishborne and foodborne parasitic zoonoses in Asia. Clin Infect Dis 41:1297–1303
5. Ishikura H, Takahashi S, Yagi K, Nakamura K, Kon S, Matsuura A, Sato N, Kikuchi K (1998) Epidemiology: global aspects of anisakidosis. In: Tada I, Kojima S, Tsuje M (eds), ICOPA IX 9th International Congress of Parasitology, Bologna, Italy, pp 379–382
6. Kuchta R, Scholz T, Brabec J, Bray RA (2008) Suppression of the tapeworm Order Pseudophyllidea (Platyhelminthes: Eucestoda) and the proposal of two new orders, Bothriocephalidea and Diphyllobothriidea. Int J Parasitol 38:49–55
7. Chung PR, Wm S, Jung Y, Pai SH, Nam MS (1997) Five human cases of *Diphyllobothrium latum* infection through eating raw flesh of redlip mullet, *Liza haematocheila*. Korean J Parasitol 35:283–289, Article in Korean
8. Rausch RL, Adams AM, Margolis L (2010) Identity of *Diphyllobothrium* spp. (Cestoda: Diphyllobothriidae) from sea lions and people along the Pacific coast of South America. J Parasitol 96:359–365
9. Baer KG (1969) *Diphyllobothrium pacificum*, a tapeworm from sea lions endemic in man along the coastal area of Peru. J Fish Res Board Can 126:717–723
10. Adams AM, Rausch RL (1997) Diphyllobothriasis. In: Connor DH, Chandler FW, Schwartz DA, Manz HF (eds) Pathology of infectious diseases, vol II, Appleton and Lange. Stamford, Connecticut, pp 1377–1389
11. Yamane Y, Shiwaku K (2003) *Diphyllobothrium nihonkaiense* and other marine-origin cestodes. Prog Med Parasitol Jpn 8:245–259, Meguro Parasitological Museum (Tokyo)
12. Scholz T, Garcia HH, Kuchta R, Wicht B (2009) Update on the human broad tapeworm (Genus *Diphyllobothrium*), including clinical relevance. Clin Microbiol Rev 22:146–160
13. Arizono N, Yamada M, Fukumoto S, Nakamura-Uchiyama F, Ohnishi K (2009) Diphyllobothriasis associated with eating raw pacific salmon. Emerg Infect Dis 15:866–870
14. Wicht B, Scholz T, Peduzzi R, Kuchta R (2008) First record of human infection with the tapeworm *Diphyllobothrium nihonkaiense* in North America. Am J Trop Med Hyg 78:235–238
15. Wicht B, de Marval F, Peduzzi R (2007) *Diphyllobothrium nihonkaiense* (Yamane et al., 1986) in Switzerland: first molecular evidence and case reports. Parasitol Int 56:195–199
16. Yamasaki H, Kuramochi T (2009) A case of *Diphyllobothrium nihonkaiense* infection possibly linked to salmon consumption in New Zealand. Parasitol Res 105:583–586
17. Margolis L, Rausch RL, Robertson E (1973) *Diphyllobothrium ursi* from man in British Columbia – first report of this tapeworm in Canada. Canadian J Public Health 64:588–589
18. Rausch RL, Adams AM (2000) Natural transfer of helminths of marine origin to freshwater fishes, with observations on the development of *Diphyllobothrium alascense*. J Parasitol 86:319–327
19. Chung DI, Kong HH, Moon CH, Choi DW, Kim TH, Lee DW, Park JJ (1995) The first human case of *Diplogonoporus balaenopterae* (Cestoda: Diphyllobothriidae) infection in Korea. Korean J Parasitol 33:225–230

20. Clavel A, Bargues MD, Castillo JF, Rubio MD, Mas-Coma S (1997) Diplogonoporiasis presumably introduced into Spain: first confirmed case of human infection acquired outside the Far East. Am J Trop Med Hyg 57:317–320
21. Arizono N, Fukumoto S, Tademoto S, Yamada M, Uchikawa R, Tegoshi T, Kuramochi T (2008) Diplogonoporiasis in Japan: genetic analyses of five clinical isolates. Parasitol Int 57:212–216
22. Skeríková A, Brabec J, Kuchta R, Jiménez J, García HH, Scholz T (2006) Is the human-infecting *Diphyllobothrium pacificum* a valid species or just a South American population of the holarctic fish broad tapeworm, *D. latum*? Am J Trop Med Hyg 75:307–310
23. Holiday DM, Guillin S, Richardson DJ (2003) Diphyllobothriasis of the Chiribaya culture (700–1476 AD) of southern Peru. Comp Parasitol 70:167–171
24. Kamo H (1981) Present situation of human diphyllobothriasis in Japan. Yonago Acta Med 25:144–155
25. Chai J-Y (2007) Intestinal flukes. In: Murrell KD, Fried B (eds) Food-borne parasitic zoonoses: fish and plant-borne parasites. Springer, New York, pp 53–115
26. Chai JY, Lee SH (2002) Food-borne intestinal trematode infections in the Republic of Korea. Parasitol Int 51:129–154
27. Cho S-H, Cho P-Y, Lee D-M, Kim T-S, Kim I-S, Hwang E-J, Na B-K, Sohn W-M (2010) Epidemiological survey on the infection of intestinal flukes in residents of Muan-gun, Jeollanam-do, the Republic of Korea. Korean J Parasitol 48:133–138
28. Dung DT, De NV, Waikagul J, Dalsgaard A, Chai JY, Sohn WM, Murrell KD (2007) Fishborne intestinal zoonotic trematodiasis, Vietnam. Emerg Infect Dis 13:1828–1833
29. Vo DT, Murrell D, Dalsgaard A, Bristow G, Nguyen DH, Bui TN, Vo DT (2008) Prevalence of zoonotic metacercariae in two species of grouper, *Epinephelus coioides* and *Epinephelus bleekeri*, and flathead mullet, *Mugil cephalus*, in Vietnam. Korean J Parasitol 46:77–82
30. Perry B, Sones K (2007) Poverty reduction through animal health. Science 315:333–334
31. Yang J-J, Guk S-M, Han E-T, Chai J-Y (2000) Molecular differentiation of three species of *Metagonimus* by simple sequence repeat anchored polymerase chain reaction (SSR-PCR) amplification. J Parasitol 86:1170–1172
32. Font WF, Overstreet RM, Heard RW (1984) Taxonomy and biology of *Phagicola nana* (Digenea: Heterophyidae). Trans Am Microscop Soc 103:408–422
33. Overstreet RM (1978) Marine maladies? Worms, germs, and other symbionts from the northern Gulf of Mexico. Mississippi-Alabama sea grant consortium, MASGP-78-021, Blossman Printing, Ocean Springs, MS, 140p
34. Welberry AE, Pacetti W (1954) Intestinal fluke infestation in a native Negro child. Bull Dade County Med Assoc 24(34):45
35. Adams KO, Jungkind JL, Bergquist EJ, Wirts CW (1986) Intestinal fluke infection as a result of eating sushi. Am J Clin Pathol 86:688–689
36. Youssef FG, Mikhail EM, Mansour NS (1989) Intestinal capillariasis in Egypt: a case report. Am J Trop Med Hyg 40:195–196
37. Mitchell AJ, Overstreet RM, Goodwin AE, Brandt TM (2005) Spread of an exotic fish-gill trematode: a far-reaching and complex problem. Fisheries 30(8):11–15
38. Africa CM, de Leon W, Garcia EY (1940) Visceral complications of intestinal heterophyidiasis of man. Monographic series no. 1, University of the Philippines, Manila
39. Kean BH, Breslau RC (1964) Cardiac heterophyidiasis. In: Parasites of the Human Heart. Grune and Stratton, New York, pp 95–103
40. Deschiens R, Collomb H, Demarchi J (1958) Distomastose cerebrale a *Heterophyes heterophyes*. In: Abstracts of the sixth international congress of tropical medicine and malaria. Lisbon
41. Paperna I, Overstreet RM (1981) Parasites and diseases of mullets (Mugilidae). In: Oren OH (ed) Aquaculture of grey mullets. Cambridge University Press, Cambridge, pp 411–493

42. Heard RW, Overstreet RM (1983) Taxonomy and life histories of two North American species of "*Carneophallus*" (= *Microphallus*) (Digenea: Microphallidae). Proc Helminthol Soc Wash 50:170–174
43. Eastburn RL, Fritsche TR, Terhune CA Jr (1987) Human intestinal infection with *Nanophyetus salmincola* from salmonid fishes. Am J Trop Med Hyg 36:586–591
44. Chai J-Y, Han E-T, Park Y-K, Guk S-M, Lee S-H (2001) *Acanthoparyphium tyosenense*: the discovery of human infection and identification of its source. J Parasitol 87:794–800
45. Little JW, Hopkins SH, Schlicht FG (1966) *Acanthoparyphium spinulosum* (Trematoda Echinostomatidae) in oysters at Port Isabel. Texas J Parasitol 52:663
46. Guk S-M, Kim J-L, Park J-H, Chai J-Y (2007) A human case of *Plagiorchis vespertilionis* (Digenea: Plagiorchiidae) infection in the Republic of Korea. J Parasitol 93:1225–1227
47. Justo MCN, Tortelly R, Menezes RC, Kohn A (2008) First record in South America of *Didymosulcus palati* and *Didymosulcus philbranchiarca* (Digenea, Didymozoidae) with new hosts records and pathological alterations. Mem Inst Oswaldo Cruz 103:207–210
48. Murrell KD, Pozio E (2000) Trichinellosis: the zoonosis that won't go quietly. Int J Parasitol 30:1339–1349
49. Rausch R, Babero BB, Rausch RV, Schiller EL (1956) Studies on the helminth fauna of Alaska. XXVII. The occurrence of larvae of *Trichinella spiralis* in Alaskan mammals. J Parasitol 42:259–271
50. Leclair D, Forbes LB, Suppa S, Proulx J-F, Gajadhar AA (2004) A preliminary investigation on the infectivity of *Trichinella* larvae in traditional preparations of walrus meat. Parasitol Res 93:507–509
51. Pozio E, La Rosa G, Rossi P, Fico R (1989) Survival of *Trichinella* muscle larvae in frozen wolf tissue in Italy. J Parasitol 75:472–473
52. Proulx J-F, MacLean JD, Gyorkos TW, Leclair D, Richter AK, Serhir B, Forbes L, Gajadhar AA (2002) Novel prevention program for trichinellosis in Inuit communities. Clin Infect Dis 34:1508–1514
53. Margolis HS, Middaugh JP, Burgess RD (1979) Arctic trichinosis: two Alaskan outbreaks from walrus meat. J Infect Dis 139:102–105
54. Forbes LB, Measures L, Gajadhar A, Kapel C (2003) Infectivity of *Trichinella nativa* in traditional northern (country) foods prepared with meat from experimentally infected seals. J Food Prot 66:1857–1863
55. Moravec F (2001) Redescription and systematic status of *Capillaria philippinensis*, an intestinal parasite of human beings. J Parasitol 87:161–164
56. Lu L-H, Lin M-R, Choi W-M, Hwang K-P, Hsu Y-S, Bair M-J, Liu J-D, Want T-E, Liu T-P, Chung W-C (2006) Human intestinal capillariasis (*Capillaria philippinensis*) in Taiwan. Am J Trop Med Hyg 74:810–813
57. Saichua P, Nithikathkul C, Kaewpitoon N (2008) Human intestinal capillariasis in Thailand. World J Gastroenterol 14:506–510
58. Cross JH (1992) Intestinal capillariasis. Clin Microbiol Rev 5:120–129
59. Beaver PC, Theis JH (1979) Dioctophymatid larval nematode in a subcutaneous module from man in California. Am J Trop Med Hyg 28:206–212
60. Le Bailly M, Leuzinger U, Bouchet F (2003) Dioctophymidae eggs in coprolites from Neolithic site of Arbon-Bleiche 3 (Switzerland). J Parasitol 89:1073–1076
61. Moore JG, Fry GF, Englert E Jr (1969) Thorny-headed worm infection in North American prehistoric man. Science 163:1324–1325
62. Schmidt GD (1971) Acanthocephalan infections of man, with two new records. J Parasitol 57:582–584
63. Tada I, Otsuji Y, Kamiya H, Mimori T, Sakaguchi Y, Makizumi S (1983) The first case of a human infected with an acanthocephalan parasite, *Bolbosoma* sp. J Parasitol 69: 205–208
64. Nuorteva P (1966) *Corynosoma strumosusm* (Rudolphi) and *C. semerne* (Forssell) (Acanthocephala) as pathogenic parasites of farmed minks in Finland. J Helminthol 40:77–80

65. Buckner RL, Overstreet RM, Heard RW (1978) Intermediate hosts for *Tegorhynchus furcatus* and *Dollfusentis chandleri* (Acanthocephala). Proc Helminthol Soc Wash 45:195–201
66. Hitchcock DJ (1950) Parasitological study on the Eskimos in the Bethel area of Alaska. J Parasitol 36:232–234
67. Berland B (1961) Nematodes from some Norwegian marine fishes. Sarsia 2:1–50
68. Davey JT (1971) A revision of the genus *Anisakis* Dujardin, 1845 (Nematodes: Ascaridata). J Helminthol 45:51–72
69. Mattiucci S, Nascetti G, Cianchi R, Paggi L, Arduino P, Margolis L, Brattey J, Webb SC, D'Amelio S, Orecchia P, Bullini L (1997) Genetic and ecological data on the *Anisakis simplex* complex with evidence for a new species (Nematoda, Ascaridoidea, Anisakidae). J Parasitol 83:401–416
70. Mattiucci S, Nascetti G (2008) Advances and trends in the molecular systematics of anisakid nematodes, with implications for their evolutionary ecology and host-parasite co-evolutionary processes. Adv Parasitol 66:47–148
71. Yoshimura H, Akao N, Kondo K, Ohnishi Y (1979) Clinicopathological studies on larval anisakiasis, with special reference to the report of extra-gastrointestinal anisakiasis. Jap J Parasitol 28:347–354
72. Køie M, Berland B, Burt MDB (1995) Development to third-stage larvae occurs in the eggs of *Anisakis simplex* and *Pseudoterranova decipiens* (Nematoda, Ascaridoidea, Anisakidae). Can J Fish Aquat Sci 52:134–139
73. Yoshinaga T, Kinami R, Hall KA, Ogawa K (2006) A preliminary study on the infection of anisakid larvae in juvenile greater amberjack *Seriola dumerili* imported from China to Japan as Mariculture seedlings. Fish Pathol 41:123–126
74. Sakanari JA, McKerrow JH (1989) Anisakiasis. Clin Microbiol Rev 2:278–284
75. Audicana MT, Kennedy MW (2008) *Anisakis simplex*: from obscure infectious worm to inducer of immune hypersensitivity. Clin Microbiol Rev 21:360–379
76. Puente P, Anadón AM, Rodero M, Romarís F, Ubeira FM, Cuéllar C (2008) *Anisakis simplex*: the high prevalence in Madrid (Spain) and its relation with fish consumption. Exp Parasitol 118:271–174
77. Torres P, Jercic MI, Weitz JC, Dobrew EK, Mercado RA (2007) Human pseudoterranovosis, an emerging infection in Chile. J Parasitol 93:440–443
78. Bowen WD (1990) Population biology of sealworm (*Pseudoterranova decipiens*) in relation to its intermediate and seal hosts. Canadian Bulletin of Fisheries and Aquatic Sciences 222. Canadian Government Publishing, Canada
79. Jackson CJ, Marcogliese DJ, Burt MDB (1997) Role of hyperbenthic crustaceans in the transmission of marine helminth parasites. Can J Fish Aquat Sci 54:815–820
80. Ishikura H, Takahashi S, Sato N, Matsuura A, Nitto H, Tsunokawa M, Kikuchi K (1996) Epidemiology of anisakidiosis and related human diseases and studies on parasites infecting marine mammals, fishes and squids. Bull Mar Biomed Inst Sapporo Med Univ 3:23–37
81. Fagerholm H-P, Overstreet RM (2008) Ascaridoid nematodes: *Contracaecum, Porrocaecum*, and *Baylisascaris*. In: Atkinson CT, Thomas JN, Hunter DB (eds) Parasitic diseases of wild birds. Wiley-Blackwell, Ames, pp 413–433
82. Im KI, Shin HJ, Yoag TS (1989) Twenty cases of gastric anisakiasis. Korean J Parasitol 27:323 (Abstract, in Korean)
83. Vidal-Martinez VM, Osorio-Sarabia D, Overstreet RM (1994) Experimental infection of *Contracaecum multipapillatum* (Nematoda: Anisakinae) from Mexico in the domestic cat. J Parasitol 80:576–579
84. Petter AJ (1969) Enquête sur les nématodes des sardines pêchées dans la région nantaise. Rapport possible avec les granulomes éosinophiles observés chez l'homme dans la région. Ann Parasitol Hum Comp 44:25–35
85. Petter AJ (1969) Enquête sur les nematodes des sardines pêchées dans la région nantaise. Identification des larves d'ascarides parasitant les sardines (en rapport avec les

granulomes éosinophiles observes chez l'homme dans la région). Ann Parasitol Hum Comp 44:559–579

86. Overstreet RM, Meyer GW (1981) Hemorrhagic lesions in stomach of rhesus monkey caused by a piscine ascaridoid nematode. J Parasitol 67:226–235
87. Deardorff TL, Overstreet RM (1981) Larval *Hysterothylacium* (=*Thynnascaris*) (Nematoda: Anisakidae) from fishes and invertebrates in the Gulf of Mexico. Proc Helminthol Soc Wash 48:113–126
88. Gonzales L (1998) Experimental infection of mice with *Hysterothylacium aduncum* (Nematoda: Anisakidae) larvae from marine-farmed trout in Chile. Arch Med Vet 30:139–142
89. Yagi K, Nagasawa K, Ishikura H, Nakagawa A, Sato N, Kikuchi K, Ishikura H (1996) Female worm *Hysterothylacium aduncum* excreted from human: a case report. Jpn J Parasitol 45:12–23
90. Deardorff TL, Kliks MM, Desowitz RS (1983) Histopathology induced by larval *Terranova* (Type HA) (Nematoda: Anisakinae) in experimentally infected rats. J Parasitol 69:191–195
91. Nawa Y (1991) Historical review and current status of gnathostomiasis in Asia. Southeast Asian J Trop Med Public Health 22(Suppl):217–219
92. Diaz Camacho SP, Zazueta-Ramos M, Ponce-Torrecillas E, Osuna Ramirez I, Castro Velazquez R, Flores Gaxiola A, Baquera Heredia J, Willms K, Akahane H, Ogata K, Nawa Y (1998) Clinical manifestations and immunodiagnosis of gnathostomiasis in Culiacan, Mexico. Am J Trop Med Hyg 59:908–915
93. Schmutzhard E, Boongird P, Vejjajiva A (1988) Eosinophilic meningitis and radiculomyelitis in Thailand, cause by CNS invasion of *Gnathostoma spinigerum* and *Angiostrongylus cantonensis*. J Neurol Neurosurg Psychiatry 51:80–87
94. Elzi L, Decker M, Battegay M, Rutishauser J, Blum J (2004) Chest pain after travel to the tropics. Lancet 363:1198
95. Ko R (1976) Experimental infection of mammals with larval *Echinocephalus sinensis* (Nematoda: Gnathostomatidae) from oysters (*Crassostrea gigas*). Can J Zool 54:597–609
96. Ko R (1977) Effects of temperature acclimation on infection of *Echinocephalus sinensis* (Nematoda: Gnathostomatidae) from oysters to kittens. Can J Zool 55:1129–1132
97. Goto Y, Tamura A, Ishikawa O, Miyachi Y, Ishii T, Akao N (1998) Creeping eruption caused by a larva of the suborder Spirurina type x. Br J Dermatol 139:315–318
98. Overstreet RM (2005) Medical importance: Infection by the rat lungworm, *Angiostrongylus cantonensis*. In: Rohde K (ed) Marine parasitology. CSIRO Publishing, Collingwood, Victoria, Australia, pp 442–446, 556–557
99. Overstreet R (2003) Flavor buds and other delights. J Parasitol 89:1093–1107
100. Measures LN (1988) Revision of the genus *Eustrongylides* Jägerskiöld, 1909 (Nematoda: Dioctophymatoidea) of piscivorous birds. Can J Zool 66:885–895
101. Mitchell AJ, Overstreet RM, Goodwin AE (2009) *Eustrongylides ignotus* infecting commercial bass, *Morone chrysops* female X *Morone saxatilis* male, and other fish in the southeastern USA. J Fish Dis 32:795–799
102. Beaver PC, Otsuji T, Otsuji A, Yoshimura H, Uchikawa R, Sato A (1983) Acanthocephalan, probably *Bolbosoma*, from the peritoneal cavity of man in Japan. Am J Trop Med Hyg 32:1016–1018
103. Choi C-J, Lee H-J, Go J-H, Park Y-K, Chia J-Y, Seo M (2010) Extraintestinal migration of *Centrorhynchus* sp. (Acanthocephala: Centrorhynchidae) in experimentally infected rats. Korean J Parasitol 48:139–143
104. Schmidt GD (1972) Acanthocephala of captive primates. In: Fiennes TW, Karker S (eds) Pathology of simian primates part II. Karger, Basel, pp 144–156
105. Dingley D, Beaver PC (1985) *Macracanthorhynchus ingens* from a child in Texas. Am J Trop Med Hyg 34:918–920
106. Overstreet RM, Self JT, Vliet KA (1985) The pentastomid *Sebekia mississippiensis* sp. n. in the American alligator and other hosts. Proc Helminthol Soc Wash 52:266–277
107. Mairena H, Solano M, Venegas W (1989) Human dermatitis caused by a nymph of *Sebekia*. Am J Trop Med Hyg 41:352–564

108. Abadi MA, Stephney G, Factor SM (1996) Cardiac pentastomiasis and tuberculosis: the worm-eaten heart. Cardiovasc Pathol 5:169–174
109. Dubey JP, Lunney JK, Shen SK, Kwok OCH, Ashford DA, Thulliez P (1996) Infectivity of low numbers of *Toxoplasma gondii* oocysts to pigs. J Parasitol 82:438–443
110. Lindsay DS, Collins MV, Mitchell SM, Cole RA, Flick GJ, Wetch CN, Lindquist A, Dubey JP (2003) Sporulation and survival of *Toxoplasma gondii* oocysts in sea water. J Eukaryot Microbiol 50:S687–S688
111. Fayer R, Dubey J, Lindsay D (2004) Zoonotic protozoa: from land to sea. Trends Parasitol 20:531–536
112. Robertson LJ (2007) The potential for marine bivalve shellfish to act as transmission vehicles for outbreaks of protozoan infections in humans: a review. Int J Food Microbiol 120:201–216
113. Lindsay DS, Phelps KK, Smith SA, Flick G, Summer SS, Dubey JP (2001) Removal of *Toxoplasma gondii* oocysts from sea water by eastern oysters (*Crassostrea virginica*). J Eukaryot Microbiol 48:S197–S198
114. Arkush KD, Miller MA, Leutenegger CM, Gardner IA, Packham AE, Heckeroth AR, Tenter AM, Barr BC, Contad PA (2003) Molecular and bioassay-based detection of *Toxoplasma gondii* oocyst uptake by mussels (*Mytilus galloprovincialis*). Int J Parasitol 33:1087–1097
115. Dubey JP, Zarnke R, Thomas JN, Wong SK, Van Bonn W, Briggs M, Davis JW, Ewing R, Mense M, Kwok OCH, Romand S, Thulliez P (2003) *Toxoplasma gondii*, *Neospora caninum*, *Sarcocystis neurona*, and *Sarcocystis canis*-like infection in marine mammals. Vet Parasitol 116:275–296
116. Miller M, Conrad P, Gardner I, Kreuder C, Mazet J, Jessup D, Dodd E, Harris M, Ames J, Worcester K, Paradies D, Grigg M (2004) An update on *Toxoplasma gondii* infections in California sea otters. Vet Parasitol 125:133–134
117. McDonald JC, Gyorkos TW, Alberton B, Maclean JD, Richer G, Juranek D (1990) An outbreak of toxoplasmosis in pregnant women in northern Quebec. J Infect Dis 161:769–774
118. Cali A, Takvorian PM (2003) Ultrastructure and development of *Pleistophora ronneafiei* n. sp., a Microsporidium (Protista) in the skeletal muscle of an immune-compromised individual. J Eukaryot Microbiol 50:77–85
119. Fayer R (2004) Infectivity of microsporidia spores stored in seawater at environmental temperatures. J Parasitol 90:654–657
120. Slodkowicz-Kowalska A, Graczyk TK, Tamang L, Jedrzejewski S, Nowosad A, Zduniak P, Solarczyk P, Girouard AS, Majewska AC (2006) Microsporidian species known to infect humans are present in aquatic birds: implications for transmission via water? Appl Environ Microbiol 72:4540–4544
121. Santín M, Fayer R (2009) *Enterocytozoon bieneusi* genotype nomenclature based on the internal transcribed spacer sequence: a consensus. J Eukaryot Microbiol 56:34–38
122. Rhinehart HL, Townsend FI, Overstreet RM, Visvesvara GS, da Silva A, Pieniazek NJ (1996) First report of microsporidiosis in the bottlenose dolphin, *Tursiops truncatus*. Annual Conference of International Association for Aquatic Animal Medicine 27:13–14
123. Canning EU, Okamura B (2004) Biodiversity and evolution of the Myxozoa. Adv Parasitol 56:43–131
124. Moncada L, López M, Murcia M, Nicholls S, León F, Guío O, Corredor A (2001) *Myxobolus* sp., another opportunistic parasite in immunosuppressed patients? J Clin Microbiol 39:1938–1940
125. Boreham R, Hendrick S, O'Donoghue P, Stenzel D (1998) Incidental finding of *Myxobolus* spores (Protozoa: Myxozoa) in stool samples from patients with gastrointestinal symptoms. J Clin Microbiol 36:3728–3730
126. McClelland R, Murphy D, Cone D (1997) Report of spores of *Henneguya salminicola* (Myxozoa) in human stool specimens: possible source of confusion with human spermatozoa. J Clin Microbiol 35:2815–2818
127. Lebbad M, Wilcox M (1998) Spores of *Henneguya salminicola* in human stool specimens. J Clin Microbiol 36:1820

128. Venable DL, Gaudé AP III, Klerks PL (2000) Control of the trematode *Bolbophorus confusus* in channel catfish *Ictalurus punctatus* ponds using salinity manipulation and polyculture with black carp *Mylopharyngodon piceus*. J World Aquacult Soc 31:158–166
129. Walker J (2005) Medical importance: marine schistosome dermatitis. In: Rhode K (ed) Marine parasitology. CABI, Wallingford, pp 439–442, 555–556
130. Grodhaus G, Keh B (1958) The Marine dermatitis-producing cercaria of *Austrobilharzia variglandis* in California (Trematoda: Schistosomatidae). J Parasitol 44:633–638
131. Chu GWTC, Cutress CE (1954) *Austrobilharzia variglandis* (Miller and Northup, 1926) Penner, 1953, (Trematoda: Schistosomatidae) in Hawaii with notes on its biology. J Parasitol 40:515–552
132. Stunkard HW, Hinchliffe MC (1952) The morphology and life-history of *Microbilharzia variglandis* (Miller and Northup, 1926) Stunkard and Hinchliffe, 1951, avian blood-flukes whose larvae cause "swimmer's itch" of ocean beaches. J Parasitol 38:248–265
133. Barber KE, Caira JN (1995) Investigation of the life cycle and adult morphology of the avian blood fluke *Austrobilharzia variglandis* (Trematoda: Schistosomatidae) form Connecticut. J Parasitol 81:584–592
134. Bearup AJ (1955) A schistosome larva from the marine snail *Pyrazus australis* as a cause of cercarial dermatitis in man. Med J Aust 1:955–958
135. Overstreet RM, Curran SS (2005) Parasites of the American white pelican. Gulf Caribbean Res 17:31–48
136. Brant SV, Cohen AN, James D, Hui L, Hom A, Loker ES (2010) Cercarial dermatitis transmitted by exotic marine snail. Emerg Infect Dis 16:1357–1365
137. Leigh WH (1953) *Cercaria huttoni*, sp. nov., a dermatitis-producing schistosome larva from the marine snail, *Haminoea antillarium guadalupensis* Sowerby. J Parasitol 39:625–629
138. Leigh WH (1955) The Morphology of *Gigantobilharzia huttoni* (Leigh, 1953) an avian schistosome with marine dermatitis-producing larvae. J Parasitol 41:262–269
139. Nobile L, Fioravanti ML, Pampiglione S, Calderan M, Marchese G (1996) Report of *Gigantobilharzia acotylea* (Digenea: Schistosomatidae) in silver gulls (*Larus argentatus*) of the Venice Lagoon: considerations on its possible etiological role in the human dermatitis observed in the same area. Parassitologia 38:267
140. Penner L (1953) The biology of a marine dermatitis producing schistosome cercaria from *Batillaria minima*. J Parasitol 39:19–20
141. Malek EA, Chang TC (1974) Medical and economic malacology. Academic, New York
142. Penner L (1950) *Cercaria littorinalinae* sp. nov., a dermatitis-producing schistosome larva from the marine snail, *Littorina planaxis* Philippi. J Parasitol 36:466–472
143. Ewers WH (1961) A new intermediate host of schistosome trematodes from New South Wales. Nature 190:283–284
144. Komiya Y, Ito J (1952) The morphology of *Cercaria sturniae* Tanabe, 1948 (cercaria of *Gigantobilharzia sturniae* Tanabe, 1951), a cause of cercaria dermatitis in Japan. Jpn J Med Sci Biol 5:215–220
145. Kolářová L (2007) Schistosomes causing cercarial dermatitis: a mini-review of current trends in systematics and of host specificity and pathogenicity. Folia Parasitol 54:81–87
146. Hrádková K, Horák P (2002) Neurotopic behaviour of *Trichobilharzia regenti* in ducks and mice. J Helminthol 76:137–141
147. Kolářová L, Horák P, Cada F (2001) Histopathology of the CNS and nasal infections caused by *Trichobilharzia regenti* in vertebrates. Parasitol Res 87:644–650
148. Little MD (1965) Dermatitis in a human volunteer infected with *Strongyloides* of nutria and raccoon. Am J Trop Med Hyg 14:1007–1009
149. Deardorff TL, Overstreet RM, Okihiro M, Tam R (1986) Piscine adult nematode invading an open lesion in a human hand. Am J Trop Med Hyg 35:827–830
150. Recavarren-Arce S, Velarde C, Gotuzzo E, Cabrera J (1999) Amoeba angeitic lesions of the central nervous system in *Balamuthia mandrilaris* amoebiasis. Hum Pathol 30:269–273

151. Sawyer TK, Visvesvara GS, Harke BA (1977) Pathogenic amoebas from brackish water and ocean sediments with a description of *Acanthamoeba hatchetti*, n. sp. Science 196:1324–1325
152. Daggett P-M (1982) Protozoa from polluted waters; potential human pathogens. In: Colwell RR (ed) Microbial hazards of diving in polluted waters: a proceedings. Maryland Sea Grant Publication UM-SG-TS-82-01. University of Maryland, College Park, pp 39–42
153. Lorenzo-Morales J, Monteverde-Miranda CA, Jiménez C, Tejedor ML, Valladares B, Ortega-Rivas A (2005) Evaluation of *Acanthamoeba* isolates from environmental sources in Tenerife, Canary Islands, Spain. Ann Agric Environ Med 12:233–236
154. Fernandez M-CA, Crespo EP, Mallen MM, Ares MPMP, Casas MC (1989) Marine amoebae from waters of northwest Spain, with comments on a potentially pathogenic euryhaline species. J Eukaryot Microbiol 36:239–241
155. Ledee DR, Hay J, Byers TJ, Seal DV, Kirkness CM (1996) *Acanthamoeba griffini*: molecular characterization of a new corneal pathogen. Invest Ophthalmol Vis Sci 37:544–550
156. Lorenzo-Morales J, Coronado-Álvarez N, Martínez-Carretero E, Maciver SK, Valladares B (2007) Detection of four adenovirus serotypes within water-isolated strains of *Acanthamoeba* in the Canary Islands, Spain. Am J Trop Hyg 77:753–756
157. Mattana A, Serra C, Mariotti E, Delogu G, Fiori PL, Cappuccinelli P (2006) *Acanthamoeba castellanii* promotion of in vitro survival and transmission of coxsackie B3 viruses. Eukaryot Cell 5:665–671
158. Laskowske-Arce MA, Orth K (2008) *Acanthamoeba castellanii* promotes the survival of *Vibrio parahaemolyticus*. Appl Environ Microbiol 74:7183–7188
159. Overstreet RM, Jovonovich J, Ma H (2009) Parasitic crustaceans as vectors of viruses, with an emphasis on three penaeid viruses. Integr Comp Biol 49:127–141
160. Nuzzi R, Zaki MH (1982) Unusual health effects associated with surface waters. New York State J Med 82:1347–1349
161. Johnson D, Reynolds K, Gerba C, Pepper I, Rose J (1995) Detection of *Giardia* and *Cryptosporidium* in marine waters. Water Sci Technol 31:439–442
162. Thompson R (2004) The zoonotic significance and molecular epidemiology of *Giardia* and giardiasis. Vet Parasitol 126:15–35
163. Monis P, Andrews R, Mayrhofer G, Ey P (2003) Genetic diversity within the morphological species *Giardia intestinalis* and its relationship to host origin. Infect Genet Evol 3:29–38
164. Gaydos J, Miller W, Gilardi K, Melli A, Schwantje H, Engelstoft C, Fritz H, Conrad P (2007) *Cryptosporidium* and *Giardia* in marine-foraging river otters (*Lontra canadensis*) from the Puget Sound Georgia Basin ecosystem. J Parasitol 93:198–202
165. Olson M, Appelbee A, Measures L (2004) *Giardia duodenalis* and *Cryptosporidium parvum* infections in pinnipeds. Vet Parasitol 125:131–132
166. Gaydos J, Miller W, Jognson C, Zornetzer H, Melli A, Packham A, Jeffries S, Lance M, Conrad P (2008) Novel and canine genotypes of *Giardia duodenalis* in Harbor Seals (*Phoca vitulina richardsi*). J Parasitol 94:1264–1268
167. Johnson D, Enriquez C, Pepper I, Davis T, Gerba C, Rose J (1997) Survival of *Giardia*, *Cryptosporidium*, Poliovirus and *Salmonella* in marine waters. Water Sci Technol 35:261–268
168. Schets F, Van den Berg H, Engels G, Lodder W, de Roda HA (2007) *Cryptosporidium* and *Giardia* in commercial and non-commercial oysters (*Crassostrea gigas*) and water from the Oosterschelde, the Netherlands. Int J Food Microbiol 113:189–194
169. Lucy FE, Graczyk TK, Tamang L, Miraflor A, Minchin D (2008) Biomonitoring of surface and coastal water for *Cryptosporidium*, *Giardia*, and human-virulent microsporidia using molluscan shellfish. Parasitol Res 103:1369–1375
170. Fayer R, Graczyk T, Lewis E, Trout J, Farley C (1998) Survival of infectious *Cryptosporidium parvum* oocysts in seawater and eastern oysters (*Crassostrea virginica*) in the Chesapeake Bay. Appl Environ Microbiol 64:1070–1074

171. Bogomolni AL, Gast RJ, Ellis JC, Dennett M, Pugliares KR, Lentell BJ, Moore MJ (2008) Victims or vectors: a survey of marine vertebrate zoonoses from coastal waters of the Northwest Atlantic. Dis Aquat Org 81:13–38
172. Fayer R (2004) *Cryptosporidium*: a water-borne zoonotic parasite. Vet Parasitol 126:37–56
173. Appelbee A, Thompson R, Olson M (2005) *Giardia* and *Cryptosporidium* in mammalian wildlife – current status and future needs. Trends Parasitol 21:370–376
174. Dupont HL, Chappell CL, Sterling CR, Okhuysen PC, Rose JB, Jakubowski W (1995) The infectivity of *Cryptosporidium parvum* in healthy volunteers. N Engl J Med 332:855–859
175. Graczyk T, Marcogliese D, de LaFontaine Y, Da Sliva A, Mhangami-Ruwende B, Pieniazek N (2001) *Cryptosporidium parvum* oocysts in zebra mussels (*Dreissena polymorpha*): evidence from the St Lawrence River. Parasitol Res 87:231–234
176. Hughes-Hanks J, Rickard L, Panuska C, Saucier J, O'Hara T, Dehn L, Rolland R (2005) Prevalence of *Cryptosporidium* spp. and *Giardia* spp. in five marine mammal species. J Parasitol 91:1225–1228
177. Santín M, Dixon B, Fayer R (2005) Genetic characterization of *Cryptosporidium* isolates from ringed seals (*Phoca hispida*) in Northern Quebec, Canada. J Parasitol 91:712–716
178. Graczyk TK, Sunderland D, Awantang GN, Mashinski Y, Lucy FE, Graczyk Z, Chomicz L, Breysse PN (2010) Relationships among bather density, levels of human waterborne pathogens, and fecal coliform counts in marine recreational beach water. Parasitol Res 106:1103–1108
179. Graczyk TK, Majewska AC, Schwab KJ (2008) The role of birds in dissemination of human waterborne enteropathogens. Trends Parasitol 24:55–59
180. Ryan U (2010) *Cryptosporidium* in birds, fish and amphibians. Exp Parasitol 124:113–120
181. Armentia A, Martín-Gil FJ, Pascual C, Martín-Esteban M, Callejo A, Martínez C (2006) *Anisakis simplex* allergy after eating chicken meat. J Investig Allergol Clin Immunol 16:258–263
182. Del Pozo MD, Moneo I, Fernández de Corres L, Audicana MT, Munoz D, Fernandez E, Navarro JA, Garcia M (1996) Laboratory determinations in *Anisakis simplex* allergy. J Allergy Clin Immunol 97:977–984
183. Del Pozo MD, Audicana MT, Diez J, Muñoz I, Ansotegui IJ, Fernández E, García M, Etxenaguisa M, Monio I, Fernández de Corres L (1997) *Anisakis simplex*, a relevant etiologic factor in acute urticaria. Allergy 52:576–576
184. Audicana MT, Ansotegui IJ, Fernández de Corres L, Kennedy MW (2002) *Anisakis simplex*: dangerous – dead and alive? Trends Parasitol 18:20–25
185. Daschner A, Alonso-Gómez A, Cabañas R, Suarez-d-Parga J-M, López-Serrano M-C (2000) Gastroallergic anisakiasis: borderline between food allergy and parasitic disease – Clinical and allergologic evaluation of 20 patients with confirmed acute parasitism by *Anisakis simplex*. J Allergy Clin Immunol 105:176–181
186. Alonso-Gómez A, Moreno-Ancillo A, López-Serrano MC, Suarez-de-Parga JM, Daschner A, Caballero MT, Barranco P, Cabañas R (2004) *Anisakis simplex* only provokes allergic symptoms when the worm parasitizes the gastrointestinal tract. Parasitol Res 93:378–384
187. Takahashi S, Ishikura H, Sato M, Iwasa K (1993) Seroimmunodiagnostics of anisakiosis. Ann Rev Immunol Cyugai-Medical, Tokyo, 211–217 (in Japanese)
188. Nieuwenhuizen N, Lopata AL, Jeebhay MF, Herbert DR, Robins TG, Brombacher F (2006) Exposure to the fish parasite *Anisakis* causes allergic airway hyperreactivity and dermatitis. J Allergy Clin Immunol 117:1098–1105
189. Jeebhay MF, Robins TG, Miller ME, Bateman E, Smuts M, Baatjies R, Lopata AL (2008) Occupational allergy and asthma among salt water fish processing workers. Am J Ind Med 51:899–910
190. Vázquez-López C, de Armas-Serra C, Bernardina W, Rodríguez-Caabeiro F (2001) Oral inoculation with *Gymnorhynchus gigas* induces anti-parasite anaphylactic [sic] antibody production in both mice and rats and adverse reactions in challenge mice. Int J Food Microbiol 64:307–315

191. U.S. Food and Drug Administration (1997) Food code, sections 3–401.11 (A1) and 3–401.12. In: 1997 Recommendations of the United States Public Health Administration. U.S. Food and Drug Administration, Washington, D.C.
192. Adams AM, Miller KS, Wekell MM, Dong FM (1999) Survival of *Anisakis simplex* in microwave-processed arrowtooth flounder (*Atheresthes stomias*). J Food Prot 62:403–409
193. EFSA Panel on Biological Hazards (BIOHAZ) (2010) Scientific opinion on risk assessment of parasites in fishery products. EFSA Journal 8:1543 (91 pp)
194. Levsen A, Lunestad BT, Berland B (2005) Low detection efficiency of candling as a commonly recommended inspection method for nematode larvae in the flesh of pelagic fish. J Food Protect 68:828–832
195. Karl H, Leinemann M (1993) A fast and quantitative detection method for nematodes in fish fillets and fishery products. Arch Lebensmittelhyg 44:124–125
196. Heia K, Sivertsen AH, Stormo SK, Wold JP, Nilsen H (2007) Detection of nematodes in cod (*Gadus morhua*) fillets by imaging spectroscopy. J Food Sci 72:E11–E15

Books and Reviews

Anderson RC (2000) Nematode parasites of vertebrates: their development and transmission, 2nd edn. CABI, New York

Beaver PC, Jung RC, Cupp EW (1984) Clinical parasitology, 9th edn. Lea and Febiger, Philadelphia

Brusca RC, Brusca GJ (2002) Invertebrates, 2nd edn. Sinauer, Sunderland, MA

Coombs I, Crompton DWT (1991) A guide to human helminths. Taylor & Francis, London

Crompton DWT, Saviolo L (2007) Handbook of helminthiasis for public health. CRC Press, Boca Raton, FL

Eiras JC, Segner H, Wahli T, Kapoor BG (eds) (2008) Fish diseases, volumes 1 and 2. Science Publishers, Enfield, NH

Fayer R (2010) Taxonomy and species delimitation in *Cryptosporidium*. Exp Parasitol 124:90–97

Garcia LS (2007) Diagnostic medical parasitology, 5th edn. ASM Press, Washington, DC

Grabda J (1991) Marine fish parasitology: an outline. VCH, New York

Horák P, Kolárová L, Adema CM (2002) Biology of the schistosome genus *Trichobilharzia*. Adv Parasitol 52:155–233

Krauss H, Weber A, Appel M, Enders B, Isenbert HD, Schiefer HG, Slenczka W, von Graevenitz A, Zahner H (2003) Zoonoses: infections diseases transmissible from animals to humans, 3rd edn. ASM Press, Washington, DC

Murrell KD, Fried B (2007) World class parasites: volume 11, Food-borne parasitic zoonoses, fish and plant-borne parasites. Springer, New York

Nawa Y, Nakamura-Uchiyama F (2004) An overview of gnathostomiasis in the world. Southeast Asian J Trop Med Public Health 35(Suppl 1):87–91

Orihel TC, Ash LR (1995) Parasites in human tissues. American Society of Clinical Pathologists, Chicago

Rhode K (ed) (2005) Marine parasitology. CABI, New York

Roberts LS, Janovy J (2008) Foundations of parasitology, 8th edn. McGraw-Hill, New York

Williams H, Jones A (1994) Parasitic worms of fish. Taylor & Francis, London

Woo PTK (ed) (2006) Fish diseases and disorders vol 1: Protozoan and metazoan infections, 2nd edn. CABI, Cambridge, MA

Zhou X-N, Lv S, Yang G-J, Kristensen TK, Bergquist NR, Utzinger J, Maline JB (2009) Spatial epidemiology in zoonotic parasitic diseases: insights gained at the 1st International Symposium on Geospatial Health in Lijiang, China, 2007. Parasit Vectors 2:10–25

Index

P. Kanki and D.J. Grimes (eds.), *Infectious DiseasesInfectious Diseases: Selected Entries from the Encyclopedia of Sustainability Science and Technology*, DOI 10.1007/978-1-4614-5719-0,

MIX
Papier aus verantwortungsvollen Quellen
Paper from responsible sources
FSC® C105338

If you have any concerns about our products,
you can contact us on
ProductSafety@springernature.com

In case Publisher is established outside the EU,
the EU authorized representative is:
Springer Nature Customer Service Center GmbH
Europaplatz 3, 69115 Heidelberg, Germany

Printed by Libri Plureos GmbH
in Hamburg, Germany